Lecture Notes in Physics

Editorial Board

R. Beig, Vienna, Austria
J. Ehlers, Potsdam, Germany
U. Frisch, Nice, France
K. Hepp, Zürich, Switzerland
R. L. Jaffe, Cambridge, MA, USA
R. Kippenhahn, Göttingen, Germany
I. Ojima, Kyoto, Japan
H. A. Weidenmüller, Heidelberg, Germany
J. Wess, München, Germany
J. Zittartz, Köln, Germany

AF181933

Managing Editor

W. Beiglböck
Assisted by Ms. M. Eisenächer
c/o Springer-Verlag, Physics Editorial Department II
Tiergartenstrasse 17, D-69121 Heidelberg, Germany

Springer-Verlag Berlin Heidelberg GmbH

The Editorial Policy for Proceedings

The series Lecture Notes in Physics reports new developments in physical research and teaching – quickly, informally, and at a high level. The proceedings to be considered for publication in this series should be limited to only a few areas of research, and these should be closely related to each other. The contributions should be of a high standard and should avoid lengthy redraftings of papers already published or about to be published elsewhere. As a whole, the proceedings should aim for a balanced presentation of the theme of the conference including a description of the techniques used and enough motivation for a broad readership. It should not be assumed that the published proceedings must reflect the conference in its entirety. (A listing or abstracts of papers presented at the meeting but not included in the proceedings could be added as an appendix.)
When applying for publication in the series Lecture Notes in Physics the volume's editor(s) should submit sufficient material to enable the series editors and their referees to make a fairly accurate evaluation (e.g. a complete list of speakers and titles of papers to be presented and abstracts). If, based on this information, the proceedings are (tentatively) accepted, the volume's editor(s), whose name(s) will appear on the title pages, should select the papers suitable for publication and have them refereed (as for a journal) when appropriate. As a rule discussions will not be accepted. The series editors and Springer-Verlag will normally not interfere with the detailed editing except in fairly obvious cases or on technical matters.
Final acceptance is expressed by the series editor in charge, in consultation with Springer-Verlag only after receiving the complete manuscript. It might help to send a copy of the authors' manuscripts in advance to the editor in charge to discuss possible revisions with him. As a general rule, the series editor will confirm his tentative acceptance if the final manuscript corresponds to the original concept discussed, if the quality of the contribution meets the requirements of the series, and if the final size of the manuscript does not greatly exceed the number of pages originally agreed upon. The manuscript should be forwarded to Springer-Verlag shortly after the meeting. In cases of extreme delay (more than six months after the conference) the series editors will check once more the timeliness of the papers. Therefore, the volume's editor(s) should establish strict deadlines, or collect the articles during the conference and have them revised on the spot. If a delay is unavoidable, one should encourage the authors to update their contributions if appropriate. The editors of proceedings are strongly advised to inform contributors about these points at an early stage.
The final manuscript should contain a table of contents and an informative introduction accessible also to readers not particularly familiar with the topic of the conference. The contributions should be in English. The volume's editor(s) should check the contributions for the correct use of language. At Springer-Verlag only the prefaces will be checked by a copy-editor for language and style. Grave linguistic or technical shortcomings may lead to the rejection of contributions by the series editors. A conference report should not exceed a total of 500 pages. Keeping the size within this bound should be achieved by a stricter selection of articles and not by imposing an upper limit to the length of the individual papers. Editors receive jointly 30 complimentary copies of their book. They are entitled to purchase further copies of their book at a reduced rate. As a rule no reprints of individual contributions can be supplied. No royalty is paid on Lecture Notes in Physics volumes. Commitment to publish is made by letter of interest rather than by signing a formal contract. Springer-Verlag secures the copyright for each volume.

The Production Process

The books are hardbound, and the publisher will select quality paper appropriate to the needs of the author(s). Publication time is about ten weeks. More than twenty years of experience guarantee authors the best possible service. To reach the goal of rapid publication at a low price the technique of photographic reproduction from a camera-ready manuscript was chosen. This process shifts the main responsibility for the technical quality considerably from the publisher to the authors. We therefore urge all authors and editors of proceedings to observe very carefully the essentials for the preparation of camera-ready manuscripts, which we will supply on request. This applies especially to the quality of figures and halftones submitted for publication. In addition, it might be useful to look at some of the volumes already published. As a special service, we offer free of charge LaTeX and TeX macro packages to format the text according to Springer-Verlag's quality requirements. We strongly recommend that you make use of this offer, since the result will be a book of considerably improved technical quality. To avoid mistakes and time-consuming correspondence during the production period the conference editors should request special instructions from the publisher well before the beginning of the conference. Manuscripts not meeting the technical standard of the series will have to be returned for improvement.

For further information please contact Springer-Verlag, Physics Editorial Department II, Tiergartenstrasse 17, D-69121 Heidelberg, Germany

P. G. L. Leach S. E. Bouquet
J.-L. Rouet E. Fijalkow (Eds.)

Dynamical Systems, Plasmas and Gravitation

Selected Papers from a Conference
Held in Orléans la Source, France, 22-24 June 1997

Springer

Editors

P. G. L. Leach
Department of Mathematics and Applied Mathematics
University of Natal
4041 Durban, South Africa

S. E. Bouquet
CEA/B3, DIF/DPTA/PPE
BP 12
F-91680 Bruyères-le-Châtel, France

J.-L. Rouet
E. Fijalkow
MAPMO-UMR6628, UFR Sciences
Université d'Orléans
BP 6759
F-45067 Orléans Cedex 2, France

Library of Congress Cataloging-in-Publication Data.

Die Deutsche Bibliothek - CIP-Einheitsaufnahme

Dynamical systems, plasmas and gravitation : selected papers of a conference held in Orléans la Source, France, 22 - 24 June 1997 / P. G. L. Leach ... (ed.). - Berlin ; Heidelberg ; New York ; Barcelona ; Hong Kong ; London ; Milan ; Paris ; Singapore ; Tokyo : Springer, 1999
(Lecture notes in physics ; Vol. 518)

ISSN 0075-8450
ISBN 978-3-662-14246-2 ISBN 978-3-540-49251-1 (eBook)
DOI 10.1007/978-3-540-49251-1

This work is subject to copyright. All rights are reserved, whether the whole or part of the material is concerned, specifically the rights of translation, reprinting, reuse of illustrations, recitation, broadcasting, reproduction on microfilm or in any other way, and storage in data banks. Duplication of this publication or parts thereof is permitted only under the provisions of the German Copyright Law of September 9, 1965, in its current version, and permission for use must always be obtained from Springer-Verlag Berlin Heidelberg GmbH. Violations are liable for prosecution under the German Copyright Law.

© Springer-Verlag Berlin Heidelberg 1999
Originally published by Springer-Verlag Berlin Heidelberg New York in 1999
Softcover reprint of the hardcover 1st edition 1999

The use of general descriptive names, registered names, trademarks, etc. in this publication does not imply, even in the absence of a specific statement, that such names are exempt from the relevant protective laws and regulations and therefore free for general use.

Typesetting: Camera-ready by the authors/editors
Cover design: *design & production*, Heidelberg

SPIN: 10644270 55/3144 - 5 4 3 2 1 0 – Printed on acid-free paper

Preface

In coming together to honour our respected and beloved colleague, Marc Feix, in this colloquium we, his scientific associates, have not been alone in the preparations for the meeting. We have been very fortunate to be assisted by very competent and efficient support staff. In particular we thank Virginie Foucault, the secretary of MAPMO, for acting as conference secretary and cheerfully performing a multitude of other tasks as the occasion required it. We also thank Isabelle Taquin of the École Polytechnique, Palaiseau, for her assistance in the preparations for the colloquium and Michel Billard-Sirakawa, Université d'Orléans, for the reproduction of figures for the conference poster. Their skilled contributions helped make the roles of the scientific personnel much easier.
A conference which attracts attendees from eight countries from North America through Europe and Africa to Asia requires considerable support. We thank the President of the Université d'Orléans for the use of the facilities of the university and the Délégué Régional of the CNRS for the use of its facilities in Orléans. For financial support of the conference we thank the Laboratoire de Mathématique, Applications et Physique Mathématique of the Université d'Orléans, the Commissariat à l'Energie Atomique, the Région Centre, the Centre National de la Recherche Scientifique and the Université d'Orléans. Without this generous support the conference could not have been the success that it was.

Finally we thank Springer-Verlag and its series editor, Professor W Beiglböck, for their guidance and technical advice in the preparation of this celebratory volume honouring the work of our distinguished friend and colleague, Marc Feix.

Orléans la Source
November, 1998

P.G.L. Leach
S.É. Bouquet
J.-L. Rouet
E. Fijalkow
Honorary Editors

Scientific Committee

Prof. C.K. Birdsall (Berkeley)
Dr. J.P. Chièze (CEA)
Academician R. Dautray (Academy of Sciences, France)
Prof. H. Doucet (CEA)
Dr. J.M. Frey (CEA)
Prof. J.P. Gazeau (Paris VII)
Prof. J.P. Hansen (Cambridge)
Prof. J. Hiéblot (CNRS)
Dr. J. Lachkar (CEA)
Prof. P. Lascaux (CEA)
Prof. G. Laval (École Polytechnique)
Prof. R. Pellat (CEA)
Prof. P.A. Raviart (École Polytechnique)
Prof. J. Salmon (CNAM)
Prof. Y. Vandenboomgaerde (CEA)
Academician H. Wilhelmsson (Academy of Sciences, Sweden)

Organising Committee

Pierre Bertrand (Nancy)
Serge Bouquet (CEA)
Jean-Robert Burgan (CEA)
Jean-Max Buzzi (CNRS)
Eric Fijalkow (CNRS)
Peter Leach (Durban)
Ralph Lewis (Dartmouth)
Jean-Louis Rouet (Orléans)
Magdi Shoucri (CCFM)

Conference Participants

1. E. Bonomi
2. K. Vo Khac
3. A. Muriel
4. P. Leach
5. F. Bouchut
6. C.K. Birdsall
7. G. Bonnaud
8. J. Chave
9. J.-C. Andrieux
10. J.-L. Lutton
11. A. Munier
12. S. Bouquet
13. G. Manfredi
14. L. Cairo
15. E. Fijalkow
16. J.-R. Burgan
17. V. Skarka
18. M.R. Feix
19. P. Bertrand
20. B. Izrar
21. J. Llibre
22. M.-L. Bégué
23. M. Lavaud
24. S. Mas-Gallic
25. O. Vallée
26. A. Bonami
27. V. Krasnosselskikh
28. F. James
29. H.R. Lewis
30. J. Goedert
31. B.N. Miller
32. A. Ghizzo
33. E. Minardi
34. H. Wilhelmsson
35. J.-L. Rouet
36. J. Clérouin
37. P. Mineau
38. J. Hiéblot
39. J. Salmon
40. J.-M. Frey
41. M. Sabatier
42. B. Perthame
43. F. Nier
44. C. Géronimi
45. H. Figua
46. H. Giacomini
47. M. Soares
48. E. Gutierrez
49. B. Lefebvre
50. Y. Benhadid

Contents

Part IV Closure

General Introduction

The idea of publishing a book entitled " Dynamical Systems, Plasmas and Gravitation " was born, for two reasons, at the international meeting which took place in Orléans (France) on June 22nd to 24th, 1997.

First, 1997 corresponded to the year of the celebration of the fiftieth anniversary of the research career of Professor Marc Roy Feix (University of Orléans). Second, among the approximately fifty PhD students he has supervised, four of them (J.-R. Burgan, J.-M. Buzzi, J.-L. Rouet and I) and some of his very close collaborators (L. Cairo, E. Fijalkow, P.G.L. Leach and M. Shoucri) decided to organize a meeting to honour his numerous and famous contributions to science (theoretical physics, applied mathematics and numerical simulations).

A special effort was made to ensure a large attendance by researchers from a number of countries. There were sixty-one participants from nine countries including USA, Spain, Italy, Brasil and France. An interesting point is that three generations of scientists comprised the audience: I mean that students of Marc's own students were present. I suspect that maybe even the fourth generation (!) was also represented since there were some very young attendees at the meeting.

The topics presented at this meeting celebrating the studies by Marc Feix as well as the title of this book and its content (this volume includes in written form all the talks given at the conference) were representative of both his work in several fields and his particular views about science: rigour in his approach, extremely relevant intuition but also precise knowledge in science, profound understanding and interpretation of computer simulations (see below for more details), globality in his ideas. He is the author or coauthor of about two hundred publications in plasma physics, stellar dynamics and astrophysics, statistical mechanics, quantum mechanics, information theory, integrability, mathematical methods for physics, numerical methods, etc. Marc is very fond of using analogies between apparently different topics (plasmas and gravitation, information theory and statistical physics, phase space description and quantum mechanics, etc...) and practical approaches. He always keeps in mind the goal of reaching a concrete result or an easily applicable method. Moreover, he is strongly motivated to try to find analytical solutions to problems but always keeps in mind the way a computer could solve them step by step ... and, in addition to all these scientific capabilities, he is a man of tremendous humanity.

Marc is a very fascinating person and I dare to adopt him as my spiritual father. On the other hand, even after having worked for fifty years in physics, he marvels at every new study or result he learns or discovers, forgetting the heritage of fifty years work and knowledge which could stop conceivably the evolution of new ways of tackling a problem not already visited. He is able to continually renew his approaches and to extract the exact relevant core of a (very) complex study! Moreover, the relationships he can develop with his students are very specific to him: I remember that, when I joined his group as a graduate student, he considered me as a full member of the laboratory and he persuaded me to believe I was a pertinent colleague capable of helping him in his research! This was partly due to my naïvety and my ignorance of the incredible richness and complexity of physics, but largely due to his decision to involve me (as well as all of his students) directly in his own work. I had the feeling of having solved an important problem, although before asking me the question I am sure that he already knew the solution. What an optimal way to start to learn research! Now, fifteen years later, I admit that I try to reproduce this attitude and to transmit his "Way of Research" to the students I am working with. I believe that this "legacy" is worthy of being passed on to the forthcoming generations of physicists belonging to the world-wide family of Scientists.

The topics presented in this book are related to Plasmas and Gravitation (Part II) and to Dynamical Systems (Part III). The Parts I and IV correspond respectively to the "Opening" and to the "Closure". The papers range from various fields of physics through numerical computation and applied mathematics.

The "Opening" has been written by Professor P. Bertrand, University of Nancy (France). Pierre Bertrand has the privilege to have been the first graduate student of Marc Feix just after he returned from the USA in 1967. He underlines the very pioneering work developed by Marc until this year and, then, relates much of his activities and his major contributions to mathematical physics (and other fields) from 1967 up to now. While O. Buneman should be considered as the grandfather of plasma simulation, this "child" has undoubtedly three fathers: C.K. Birdsall, J.M. Dawson and M.R. Feix. Obviously, this is the domain for which Marc is most famous. With computer simulations, he looked at nonlinear behaviour and studied processes where many competing phenomena take place. In addition to the notion of the numerical Lagrangian model (particle simulation), he developed the Eulerian approach to numerically solve the nonlinear Vlasov equation. Another exciting idea was the introduction of the so-called "water-bag" model which illustrates that numerical simulations must not go too far from analytical calculations. Several papers related to plasma physics and based on the continuation and the extension of these ideas are presented in Part II. As mentioned above, Marc has also made many other pioneering contributions in various fields. Among them is the invention in mathematical physics of the "rescaling transformation" (initially called "quasi-invariance"). This is a technique initially based on Lie group theory, but which applies very easily to differential (and partial differential) equations and leads to practical results in physics (for instance, derivation of the relevant parameters for a given problem) as well as in numerical simulations. The major part of the time-evolution of a problem is analytically absorbed and the code just has to describe small deviations about this evolution. In applied mathematics, the rescaling transformation can be used to study integrability, to obtain particular solutions, etc. This shows the "universality" of this transformation which is so widely applicable (analytically, numerically and in a many fields of physics). Several examples are given in Parts II and III. An introduction to each of Parts II and III has been written by my friend P.G.L. Leach and each is printed at the beginning of the corresponding Part. I shall, therefore, give a short summary.

Among the most fundamental research areas in physics, plasma sciences take a very important place — and fundamental applications, such as energy production, are also expected from these studies. Several papers dealing with the most advanced work in this area are presented in Part II. The very difficult points of the modelling of a dense plasma, the interaction of a plasma with an intense laser beam, the dynamics of fusion plasmas etc... are studied. Moreover, various problems (plasma accelerators, relativistic behaviour ...) are examined via the use of different kinds of numerical schemes (Lagrangian, Eulerian and semi-Lagrangian simulations, etc.) but also in an analytical manner. On the other hand, since for both plamas and gravitation, the forces decrease as the inverse of the square of the distance, this allows one to develop and to use similar tools for these two topics. Although cross-fertilisation is very efficient in these domains, gravitational physics remains very difficult. Even the formation of a star (which is also a gravitational plasma confinment device!), which could appear elementary, is not completely understood up to now. However, with the construction in the next decade of high energy lasers (National Ignition Facility — NIF — in USA, Laser MégaJoule — LMJ — in France, etc.) the time is right to invest efforts in developing "Laboratory Astrophysics". The physics of plasmas (created by these big lasers) and astrophysics will, therefore, become closely connected with the aim of simulating in the laboratory (for the first time) astrophysical phenomena (supernovae explosions, hydrodynamical instabilities, plasma collisions, etc.) with the plasmas created by these lasers. This will be a very attractive challenge for the beginning of the third millennium.

Part III is devoted to current research on dynamical systems. The topics include integrable, near integrable and nonintegrable evolution equations and dynamical systems. The contributions deal, on one hand, with the concepts of Lie symmetries, first integrals, limit cycles, Liouville integrability, local integrability, etc. and, on the other hand, with the problems of Lotka Volterra systems, Hamiltonian systems, lattice models for forests, telecommunication networks, etc. The leading idea in all these approaches is the use of theoretical techniques for the derivation of major information and results that can be useful and applicable for applied mathematicians and/or physicists. In addition, the tremendous significance of such equations and the associated problems in physics and mathematics leads to the need to understand the notion of integrability at a more precise level and to comprehend the role of integrable systems in mathematical physics. An attempt to make a step in this direction is presented in this Part. Now, how can we determine whether a given system is integrable and in what sense? This is a currently relevant question which is also addressed in this Part. Nevertheless, the important point of the connection between the concept of integrability and the use of the computer is not tackled. In my opinion, the "marriage" of analytical properties and adequate numerical schemes (and tools) will be a very fruitful approach in the near future.

The "Closure" (Part IV) corresponds to a paper written by Professor C. K. (Ned) Birdsall (University of Berkeley). First of all, I would like to express my gratitude to Ned for attending this meeting in honor of his long-time friend and colleague and giving the concluding lecture. Similarly to Marc Feix, he has been one of the pioneers of plasma simulations. In this respect, both of them, together with O. Buneman and J.D. Dawson, were the four members of the scientific committee of the first "Symposium on Computer Simulation of Plasma and Many-Body Problems" which was held at Williamsburg, Virginia, in 1967 — see Part I. Up to now, fifteen conferences on Numerical Simulations of Plasmas have been organized. A very amazing and funny story from this Part is the way in the fifties, Ned Birdsall obtained the values of the ratio $zJ_1(z)/J_0(z)$ (the J_i's are the Bessel functions) for complex values of the argument. The digital computer could not calculate it and, together with a colleague, he obtained the results from analog computations achieved by applying a voltage at the ends of a ... water tank! They needed two days to make one plot. Thirty years later (in 1982), another scientist published a similar plot and he said that a fast personal computer could do it in less than one minute! This made a ratio of about 10^3 between the two durations of calculation and, today, I think that it would be certainly greater than 10^6. Pioneering trials (and errors) are essential to all scientific progress. Someone must propose and try new studies to sort good ideas from bad ones. So it has been with plasma simulations with progress made step by step with computers becoming faster, smaller, cheaper and easier to use and leading, therefore, to applications to more and more complex problems. We are now close to simulating whole plasma reactors employing radio frequency glow discharge as used in 40% of semiconductor processing (a huge industry, in the realm of tens of billions of dollars per year). One might also imagine simulating the plasma physics of an entire fusion reactor before building such a device. Efforts to simulate the present ITER (at some ten billion dollars, an expensive fusion reactor) go in this direction. One guesses that the cost for these simulations should be only a few percent of the total price of such a device. Finally, Ned gives his thoughts about the use of computers in the near and farther future.

For the quality of the contributions and for the large variety of topics that are covered, this book give a good up-to-date picture of the state of the art in these fields. The book does not provide an exhaustive self-contained description of the entire subjects, but rather gives an outline of the most recent and relevant results, exhibits the way to work with analogies between different topics and demonstrates how numerical and analytical approaches can complement each other in such a way that this collection of papers should stimulate graduate students as well as researchers.

Thank you very much, dear Marco, for all that you did.

Paris, Commissariat à l'Energie Atomique, October 1998 S.E. Bouquet

Part I

The World of Marc Feix

Part I

Honouring Marc Feix

Pierre Bertrand

Université Henri Poincaré, Nancy-1, LPMI,
BP 239, F-54506 Vandœuvre-lès-Nancy cedex, France

Abstract. Numerical simulation has long been recognised has an essential tool to study the complex nature of plasma physics. For those who are involved in this field it is essential to identify the pioneering work of some contributors. The purpose of this short paper is to recall the early contributions by Marc Feix to numerical plasma simulation and also to more general problems in other areas of mathematical physics.

1 Foreword

Over 30 years Professor Marc Feix and I have developed a very special relationship. In fact I think of him as my mentor and, indeed, I have the privilege to be his first graduate student following his return from the U.S. in 1967. He is one of the bright stars of my graduate education. I owe him a considerable debt. In this I am not alone. Over the course of his career Marc Feix has been mentor to more than forty other students. From my experience it is clear that spending time with his graduate students is a high priority for him. In fact sometimes I get the feeling that he prefers it to many other duties. His guidance to students is what makes him so special to all of us.

For those who may be familiar with one part of his works and not another, I have summed up in the following sections those areas in which Marc has been a pioneer. Obviously computer simulation is for what he is very famous. With it we have been able to look at nonlinear processes and to study processes where there are many competing phenomena. While Oscar Buneman should be considered as the grandfather of plasma simulation, the fathers are undoubtedly John M. Dawson, Charles K. (Ned) Birdsall and Marc Feix. Sect. 5 will particularly illustrate this point.

If that were all the work of Marc, we would be congratulating him for his accomplishments, but Marc offers much more. Stellar dynamics, information theory, group theory for the equations of mathematical physics, quantum mechanics and statistical physics are other fields in which Marc has proved to be a leader. Many of these topics will be touched upon during this Conference.

It is the purpose of following parts of my talk to state by title Marc's early works. Recent developments of these topics are available in the articles following in this book.

2 Early Work at Collège de France

Marc began his long scientific career at the famous Collège de France in Paris, studying the motion of an electron in a magnetic wiggler (Combe and Feix 1953). Those involved now in free electron lasers will recognise a very up to date problem, but it should be noticed that this work was carried on in the early 50s! The idea was to use accelerated electrons to produce microwaves with relativistic Doppler effect in a magnetic wiggler.

At that time the technology could only afford modest electromagnetic power and Marc turned to neutronics. Under the supervision of Francis Perrin he presented a thesis (Feix 1958) entitled: *A contribution to the theory of fast neutron diffusion.* Forty years later Super Phœnix (and its numerous problems) is still under discussion...

The second thesis Marc submitted to receive the French PhD degree (Docteur ès-Sciences) was devoted to plasma magnetohydrodynamics: the first encounter between Marc and plasma physics and the beginning of an everlasting love story.

3 Plasma Physics

3.1 Linear Plasma Theory

Since the first paper by Landau (Landau 1946) the linear plasma theory was still under consideration. The main problem was to study the propagation of an electric field in a one-dimensional collisionless plasma using the linearised Vlasov-Poisson equations. Performing a Fourier (space) and Laplace (time) transform the solution was easily obtained in the transformed $k-\omega$ variables. The definition of the analytical continuation of the plasma dispersion function $\mathcal{E}(k,\omega)$ allowed one to perform the inverse Laplace transform yielding the now well-known Landau damping[1] phenomena in the k,t variables.

The remaining problem was to perform the k-integration, in order to come back to the x,t variables, using a temporally and spatially localised perturbation as an experimentally realisable initial condition. Marc and his collaborators (Engelmann et al. 1961) showed that around the perturbation centre a zone the size of which increases as $t^{1/2}$ is established in which the amplitude of the electric field, oscillating with plasma frequency, is spatially almost constant and decreases in time as $t^{-1/2}$, thus showing essential characteristics of a diffusion process.

3.2 Numerical Simulation and Microscopic Plasma Theory

In 1959 Oscar Buneman had led the way to particle simulation in a paper (Buneman 1959) concerning the nonlinear evolution of the beam-plasma instability and its saturation, which was not known. This article showed the power of particle simulation.

---

[1] Depending on the shape of the velocity distribution function

In order to elucidate the subtle points of the microscopic theory of a warm one-dimensional plasma, Owen Eldridge and Marc Feix, in a series of famous papers (Eldridge and Feix 1962a, 1962b, 1963), introduced the Lagrangian model of charged sheets. Following the 1D motion of a relatively small number of particles in a Debye length (with $n\lambda_D$ varying from 1/4 to 10) they recovered the thermodynamic properties, together with the correct Fokker-Planck drag and diffusion coefficients for a 1D plasma. At the same time John Dawson provided a similar model (Dawson, 1962). While it was generally thought that a real plasma would require a large number of particles in a Debye sphere[2], this pioneering work showed that pertinent simulations could be performed with far fewer than a million of particles, which would have been impossible to run on the early 60s computers!

The three papers by Eldridge and Feix and the one by Dawson form the very basis of numerical experiments using particles, with an exponentially increasing number of papers on PIC codes stimulated by Ned Birdsall and his co-workers (see for instance the book by Birdsall and Langdon, 1985).

On the other hand Marc early recognised that, besides the Lagrangian approach (particle simulation), there exists an Eulerian approach, in which the nonlinear Vlasov equation is solved numerically. This equation presents a formidable problem even in the simplest case (one dimension has three variables x,v,t). One idea is to use a Fourier-Hermite expansion of the distribution function (Grant and Feix 1967a and 1967b). We briefly outline this method.

Consider the one-dimensional Vlasov-Poisson system for the motion of electrons against a uniform positive neutralising background (in normalised units)

$$\frac{\partial f}{\partial t}+v\frac{\partial f}{\partial x}-E\frac{\partial f}{\partial v}=0$$

$$-\frac{\partial E}{\partial x}=\int_{-\infty}^{\infty} f\,\mathrm{d}v \quad - \quad 1.$$

One makes a double series expansion of the distribution function

$$(2\pi)^{\frac{1}{2}} f=\sum_{r=-\infty}^{\infty}\sum_{s=0}^{\infty} a_{rs}\exp(irk_0x)\,\overline{H}_s(v)\exp\left(-\tfrac{1}{2}v^2\right),$$

where
$$\overline{H}_s(v)=(-1)^s\exp\left(\tfrac{1}{2}v^2\right)\left(\frac{\mathrm{d}}{\mathrm{d}v}\right)^s\exp\left(-\tfrac{1}{2}v^2\right),$$

$$\int_{-\infty}^{\infty}\overline{H}_m(v)\overline{H}_n(v)\exp\left(-v^2\right)\mathrm{d}v=\delta_{mn}(2\pi)^{\frac{1}{2}}n!\ .$$

With a reasonable number of Hermite polynomials (20 to 40) it was possible to recover the Landau damping. Unfortunately a huge number of Hermite polynomials was required to study nonlinear effects. Indeed this idea of Eulerian Vlasov codes was

[2] In 1D this quantity is replaced by the number of particles in a Debye length

too pioneering in the 60s and it is only during the last decade, with the emergence of vector and parallel super computers, that those codes proved to be a useful tool beside the particle codes[3]. Although the methods to solve the Vlasov equation are quite different, the Fourier-Hermite expansion is still used for some specific applications (see for instance Hammet et al. 1993, Parker and Carati 1995, Korn and Schammel 1996).

Another exciting idea was the introduction of the so-called *water bag* model (Bertrand and Feix 1968) which illustrates Marc's claim that numerical simulation must not stay too far from analytical calculations. In the water bag model for a collisionless one dimensional electron plasma, the distribution function $f(x,v,t)$ is a constant A inside a certain region of phase space plane, bounded by two curves (the *contours*) $v_+(x,t)$ and $v_-(x,t)$ and zero elsewhere. Due to Liouville's theorem this is not an approximation but rather a special class of initial conditions and the model remains valid at all time. Instead of solving the Vlasov-Poisson system for the variables x,v,t we have just to follow the motion of the two contours supplemented by Poisson's equation

$$\frac{\partial v_\pm}{\partial t} + v_\pm \frac{\partial v_\pm}{\partial x} = \frac{e}{m} E,$$

$$\frac{\partial E}{\partial x} = \frac{e}{\varepsilon_0}\left[A\left(v_+ - v_-\right) - n_0\right].$$

Introducing the electron density $n(x,t) = A\left(v_+ - v_-\right)$ and the mean (fluid) velocity $V(x,t) = (1/2)\left(v_+ + v_-\right)$ we obtain

$$\frac{\partial n}{\partial t} + \frac{\partial (nV)}{\partial x} = 0,$$

$$\frac{\partial V}{\partial t} + V\frac{\partial V}{\partial x} = \frac{e}{m}E - \frac{1}{mn}\frac{\partial P}{\partial x},$$

$$\frac{\partial E}{\partial x} = \frac{e}{\varepsilon_0}(n - n_0),$$

$$\textit{where} \quad P = \frac{m}{12A^2} n^3.$$

These equations are nothing but the hydrodynamic equations of an adiabatic gas with $\gamma = 3$. This very simple result, which provides a bridge between collective collisionless and fluid models, was the starting point of my PhD thesis (Bertrand 1972).

The *water bag* concept is now very well known and used by a lot of people, but I am not sure that everyone knows who introduced it...

---

[3] See for instance the paper by Ghizzo et al. in the present book (and references therein)

4 First Conference on the Numerical Simulation of Plasmas

In 1967 Marc organized *the Symposium on Computer Simulation of Plasma and Many-Body Problems*, held at the College of William and Mary, Williamsburg, Virginia. The *facsimile* of the first page (from my own library) is reproduced below.

Below is the *facsimile* of the page showing the list of the scientific committee: C.K. Birdsall, O. Buneman, J. Dawson and M.R. Feix!

ACKNOWLEDGEMENTS

The Organizing Committee for the symposium wishes to acknowledge its indebtedness to The College of William and Mary and to the Langley Research Center of the National Aeronautics and Space Administration for their sponsorship. Their joint cooperation turned the conception of this symposium into reality.

The assistance and guidance of the Scientific Advisory Committee is also acknowledged as having played a significant role in providing the guidelines which made the program so successful. The members of this Committee were:

Charles K. Birdsall, Department of Electrical Engineering, University of California, Berkeley, California

Oscar Buneman, Institute for Plasma Research, Stanford University, Stanford, California

John M. Dawson, Plasma Physics Laboratory, Princeton University, Princeton, New Jersey

Marc R. Feix, Physics Department, College of William and Mary, Williamsburg, Virginia, and NASA Langley Research Center, Hampton, Virginia

The Organizing Committee:

Fred R. Crownfield
College of William and Mary, Williamsburg, Virginia

Marc R. Feix
College of William and Mary and NASA Langley Research Center, Hampton, Virginia

Frank Hohl
NASA Langley Research Center, Hampton, Virginia

Gary A. Massel
NASA Langley Research Center, Hampton, Virginia

iv

The plasma community owes Marc Feix a special homage for being the first to establish this conference. Fourteen other Conferences on the Numerical Simulation of Plasmas have been held. The next will take place in Santa Barbara, California, on February 10-12, 1998.

As a final excerpt this is a *facsimile* of the first page of a panel discussion addressing the question of particle codes vs Vlasov codes. Thirty years later it is still under discussion.

Panel Discussion On:

METHODOLOGY IN THE N-BODY PROBLEM:
LAGRANGE VERSUS EULER APPROACH

FIRST PANEL:

Oscar Buneman
John M. Dawson
Marc R. Feix (Moderator)
Roger W. Hockney
Craig G. Smith

Feix: One of the difficulties with most panel discussions is that each of the panel members stands up and gives a 10 minute talk with no audience participation. I hope that this won't be the case here. I would first like to have all the panel members give their philosophy about the computer experiments.

Let me start with a remark which is half serious and half joking and show you a slide which has been brought to my attention by Macon Ellis:

"I think we may have more difficulty in exploring the full limits of the computer than we have had with earlier gadgets. I think there may be more danger in the period of trial and error than there has been with earlier devices. These earlier devices - looms, engines, generators - resisted at critical points human ignorance and stupidity. Overloaded, abused, they stopped work, stalled, broke down, blew up, and there was the end of it. Thus they set clear limits to man's ineptitudes. For the computer the limits, I believe, are not so obvious. Used in ignorance or stupidity, asked a foolish question, it does not collapse, it goes on to answer a fool according to his folly. And the questioner being a fool will go on to act on the reply."[1]

I think there is a lot of truth in this little paragraph, but let us go back to the main discussion!

Oscar Buneman has already given his opinion in his introductory address and I think John Dawson disagrees with this philosophy.

Dawson: I feel that the computer is a very powerful new tool and we are really just beginning to learn how to use it. We must develop new techniques and we must learn what the computer can do and what answers we can get from it. Generally, we get so many answers that we don't know what to do with them so we have to learn how to ask the right questions. I think a computer is suited for two types of use - one is a practical use where we try to simulate real plasma devices and let the computer tell us what they are going to do.

The second use is really to look at fundamental problems, such as the statistical mechanics problem Marc mentioned yesterday, and then we try our

[1]E. E. Morison in "Men, Machines and Modern Times," M. I. T. Press, Cambridge, 1966, p. 91.

407

5 Other Contributions

The following is a nonexhaustive list of other pioneering contributions in various areas in mathematical physics. Some of them are at least as important as those in plasma physics and - I am convinced - so they are for Marc.

5.1 Stellar Dynamics and Astrophysics

A self-gravitating star system (cluster or galaxy) is very similar to a collisionless electron plasma except the sign of the Coulomb force and the lack of neutralising background. Marc's expertise in Vlasov plasma theory and simulation invited him to apply his knowledge to the study of stellar dynamics (Hohl and Feix 1967). As an alternative to Lynden Bell's work he demonstrated a minimum energy principle for water bag equilibrium (Hohl and Feix 1968).

5.2 Information Theory, Statistical Mechanics

The first related paper dealt with the connection between information theory and entropy principle (Engelmann et al. 1960). Then Marc's interest turned to encoding processes of a binary channel with erasing (Mathé 1969).

More recently Marc could apply his knowledge of numerical simulation and modeling to the study of connection networks with a special emphasis on telecommunications applications. This problem opened new issues in statistical mechanics and sorting theory (Bonomi 1980, Bonomi et al. 1983).

5.3 Algebraic-numerical Methods, Invariance, Rescaling

Marc was one of the first physicists to recognise the usefulness of a computer not only in dealing with numerical simulation but also in helping to solve *analytically* equations in mathematical physics. For instance the removal of the secular terms in Krylov-Bogoliubov methods is not an easy task. Instead of trying to obtain an approximation valid for all time, he proposed to keep the secular expansion over a time T such that $\varepsilon T \ll 1$ and to reinitialise numerically after the 'giant step' T. Thus he derived an *algebraic-numerical* method which he could successfully apply to various problems (Kalman and Feix 1969).

On the other hand self-similar transforms are known to simplify the solution of ordinary or partial differential equations, the price to pay being to consider a restricted class of initial and boundary conditions. Marc introduced a new concept 'the quasi-invariance transform', including an arbitrary function of time, allowing the adjustment of initial conditions, simplification of the computational scheme or the taking into account of asymptotic solutions through a transformed finite time. This concept was further reformulated through the rescaling transformation and Lie group symmetry.

This was the starting point (see for instance Burgan et al. 1978) of a series of very important papers and the reader will find in the different articles of the present book various applications of these ideas.

5.4 To Be Continued...

After the early works I have described above, the fields of interest Marc has explored has never ceased to expand.

Among them let me quote:

- Eulerian codes with special emphasis on laser-plasma interaction and stimulated Raman scattering, gyrokinetic simulation of Tokamak plasmas, phase space holes in electrostatic plasmas and self-gravitating stellar systems;
- self-similarity and rescaling (with applications to plasma expansion, classical and quantum harmonic oscillator with variable frequency, shallow water waves, nonlinear oscillators, ...);
- astrophysics and gravitational dynamics;
- integrability properties of Lotka-Volterra systems;
- statistical physics of 1D systems;
- phase space representation of quantum mechanics, the Wigner equation and application to signal theory;
- fuzzy sets and decision theory;
- ...[4]

6 Concluding Remarks

Marc's contribution to physics continues still. I have summarised above the early contributions of Marc, showing the place he should occupy among other mathematical and numerical simulation physicists. Marc's interaction with Physics has been so fruitful: he signed more than 180 scientific publications.

I hope that I have convinced the reader of the pioneering and very creative aspect of his research. As a matter of fact almost all the articles in this book have a paper by Marc as a starting reference. Some of them describe a part of his present work and I am sure some of them will be considered as pioneering ones in the coming decades.

More than fourty graduate students (and I am one of them) shared the honour to have had Marc as a thesis supervisor.

Thank you very much, Marc, for what you brought us!

[4] I know that I have forgotten some topics. Forgive me Marc.

References

Bertrand, P., Feix M.R. (1968): Non Linear Electron Plasma Oscillation: the Water Bag Model. Phys. Lett. A, **28**, 68-70

Bertrand, P. (1972): Contribution à l'étude mathématique de plasmas non collisionnels. Ph.D. thesis, Université de Nancy, France

Birdsall, C.K., Langdon, A.B. (1985): *Plasma Physics via Computer Simulation* (McGraw-Hill, New York)

Buneman, O. (1959): Dissipation of Currents in Ionized Media. Phys. Rev. **115**, 503-517

Bonomi, E., Feix, M.R., Hebuterne G. (1980): Théorie de l'Information. Structure et Modélisation d'un Réseau de Connexion. Annales des Télécommunications **35**, 39-59

Bonomi, E., Lutton, J.L., Feix, M.R. (1983): Three Stages Rearrangeable Connecting Network in the Thermodynamic Limit: Number of Needed Reswitching. IEEE Transactions on Communications **31**, 110-117

Burgan, J.R., Gutierrez, J., Munier, A., Fijalkow, E., Feix, M.R. (1978): Group Transformations in Phase Space Fluids. *Strongly Coupled Plasmas* ed. G. Kalman (Plenum Press, London), 597-643

Combe, R., Feix, M.R., (1953): Mouvement d'un électron dans un onduleur magnétique. C. R. A. S. Paris **237**, 1318-1321

Dawson, J.M., (1962): One-Dimensional Plasma Model. Phys. Fluids **5**, 445-459

Eldridge, O.C., Feix, M.R. (1962a): One-Dimensional Plasma Model at Thermodynamic Equilibrium. Phys. Fluids **5**, 1076-1080

Eldridge, O.C., Feix, M.R. (1962b): Fokker-Planck Coefficients for a One-Dimensional Plasma. Phys. Fluids **5**, 1307-1308

Eldridge, O.C., Feix, M.R. (1963): Numerical Experiments with a Plasma Model. Phys. Fluids **6**, 398-406

Engelmann, F., Feix, M.R., Minardi, E., Oxenius, J. (1960): The problem of Irreversibility within the Framework of a Generalized Entropy Concept. Z. Für Naturforschung **16 A**, 1223-1240

Engelmann, F., Feix, M.R., Minardi, E., Oxenius, J. (1961): On the Diffusion of an Electric Field in a Collisionless Plasma. Il Nuovo Cimento **22**, 1012-1022

Feix, M.R., (1958): A) Contribution à la théorie de la diffusion des neutrons rapides. B) Magnétohydrodynamique des Plasmas. Ph.D. thesis, Université de Paris, France

Grant, F.C., Feix, M.R., (1967): Fourier-Hermite Expansion of the Vlasov Equation in the Linearized Limit. Phys. Fluids **10**, 696-702

Grant, F.C., Feix, M.R., (1967): Transition between Landau and Van Kampen Treatment of the Vlasov Equation. Phys. Fluids **10**, 1356-1357

Hammet, G.W., Beer, M.A., Dorland, W., Cowley, S.C., Smith, S.A., (1993): Developments in the Gyrofluid Approach to Tokamak Turbulence Simulations. Plasma Phys. Control. Fusion **35**, 973-985

Hohl, F., Feix, M.R., (1967): Numerical Experiments with One Dimensional Model for a Self Gravitating System, Ap. J. **147**, 1164-1180

Hohl, F., Feix, M.R., (1968): A Variational Principle for a One Dimensional Stellar System. Ap. J. **151**, 783-778

Kalman, G., Feix, M.R. (1969): *Non linear Effects in Plasma* (Gordon and Breach, New York)

Korn, J., Schammel, H., (1996): Electron holes. Part 1. Immobile Ions. J. Plasma Phys. **56**, 307-337

Landau, L.D., (1946): On the vibration of the Electronic Plasma. J. Phys. U.S.S.R. **10**, 25-37

Mathé, F. (1969): Etude et Simulation sur Ordinateur des Différents Procédés de Codage pour un Canal Binaire avec Effaçage. Thèse 3° cycle, Université de Nancy

Parker, S.E., Carati, D., (1995): Renormalized Dissipation in Plasmas with Finite Collisionality. **75**, 441-444

Part II

Plasma and Gravitation

Plasmas occur in environments ranging from far-reaching voids of interstellar space to very compact devices in the laboratory, from the grandiose displays of the northern and southern lights to to the pyrotechnics of a fluorescent tube in a violent thunderstorm. They are studied in the heavens to seek an answer to the behaviour and properties of the universe and the stars which compose it and in the laboratory to seek yet one more way to satisfy our desire for effortfree energy. The governing equations are about as complex a set of differential equations as could be imagined to describe a physical system: the equations of fluid dynamics, unless one wishes to use the equations of statistical physics for an ensemble of particles; Maxwell's equations; and, in the case of massive plasmas – stars – the gravitational equations of Newton and Einstein.

It comes as no surprise that such a complex system of partial differential equations does not admit a general solution. We must resort to particularism in the mathematical modelling to seek to highlight some specific aspect in an analytic procedure. Even with the high powered computers of today, a complete numerical simulation is not possible and again specific aspects must be modelled. When one is forced to look at parts of the total, there is an added burden, which is to avoid a distortion of the description of the whole in the manner of the committee that reported on the nature of an elephant.

Interaction of a Ultra-High-Intensity Electromagnetic Pulse with an Overdense Plasma

Guy Bonnaud, Erik Lefebvre, and Catherine Toupin

CEA, Centre d'Etudes de Bruyères-le-Châtel, DPTA, BP 12,
F-91680 Bruyères-le-Châtel, France

Abstract. Both one-and-a-half and two-dimensional simulations of a dense neutral plasma impinged upon by a ultra-intense laser pulse are presented. The various scenarii of plasma interaction are reviewed.

Keywords. Laser-plasma interaction, Particle simulation, Ultrashort pulse, Ultrahigh intensity pulse.

1 Introduction

Before the early seventies, the development of laser technology led to a large interest in the interaction of high-flux nanosecond laser pulses with matter. One of the most spectacular effect evidenced at that time was the self-focusing process caused in both atomic or ionized media by thermal, ponderomotive, molecular or semi-conductor effects (Chiao et al. 1964, Kaw et al. 1973, Sodha et al. 1974). Ionization appears when the irradiance is between 10^8 and 10^{14} W/cm^2, depending on the medium impinged upon by the laser wave. Until recently, the highest irradiances of focused beams, i.e. 10^{16} W/cm^2, were achieved in the huge laser facilities developed for inertial confinement fusion (ICF) purposes.

During the present decade, a new generation of compact lasers has been developed, capable of delivering powers in excess of 1 terawatt in subpicosecond pulses. One of these laser types is based on the chirped-pulse technique and works in the infrared region, e.g. 1.053 μm for Nd-glass lasers (Maine et al. 1988, Ferray et al. 1990, Watteau et al. 1992). The irradiance, reached after beam focusing onto a focal spot of approximately 10-wavelength diameter, is currently in excess of 10^{18} W/cm^2. This ultra-high irradiance (UHI) corresponds to an electromagnetic pressure of 300 I_{18} in Mbars, where I_{18} denotes the irradiance in units of 10^{18} W/cm^2. When such a laser beam is incident upon a plasma, the free electrons get a relativistic motion: their quiver momentum, normalized to $m_e c$, is indeed written : $a_0 = 0.85\ (I_{18}\lambda_0^2)^{1/2}$, where m_e denotes the electron mass, c is the velocity of light in vacuum, and λ_0 is the laser wavelength in μm. The electric field amplitude at focus is 27 $I_{18}^{1/2}$ in units of GV/cm; at $I_{18} = 1$, its value is 5 times the atomic field strength in hydrogen atom.

These lasers have given rise to an increasing number of new kinds of experiments, which would have been impossible to conceive of earlier. This is stimulating a great interest in the scientific community largely specialised in laser-plasma interaction in the nanosecond regime, who has to delve into the interaction processes at the picosecond level. Theoretical and experimental reports of clearly labelled phenomena are lacking, and theoreticians naturally turn to simulation codes and large computers to extend their current knowledge to this new realm of physics.

This paper is organized as follows : Section 2 describes the physics underlying the coupling of the laser pulse with a solid-density plasma. Section 3 addresses the model and the computer tools used in this paper. Section 4 is concerned with the results from a one-and-a-half-dimension (1.5-D) model where the incident pulse is assumed to remain planar, whereas Section 5 deals with specific results involved in a two-dimension (2-D) plasma by a finite focal spot. In Section 6, concluding observations are given.

2 The Physics of UHI Laser-Solid Interaction

This new generation of lasers has encompassed two factors: the pulse brevity and the high peak irradiance. The interaction with a gas allows the laser wave to propagate; a large diversity of physical mechanisms can then be triggered, but these lie outside the scope of this paper where we shall limit our discussion to solid matter. The interaction with solid matter rapidly ionizes the solid surface to create a plasma; if the target is thick, the wave is prevented from propagating further into the plasma and is backscattered or absorbed at the solid surface: the high density of free electrons gives rise to a large-amplitude oscillating current and, subsequently, to a strong electromagnetic emission that alters the laser beam propagation. The key parameter is the laser wavelength λ_0: a plasma with electron density above $n_c = 10^{21}/\lambda_0{}^2$ cm^{-3}, with λ_0 in μm, appears to be opaque for the laser wave. The pulse is partially reflected and partially absorbed by the so-called normal skin effect, resulting from the electron-ion collisions inside the wave penetration-depth $c/(\omega_{pe}^2 - \omega_0^2)^{1/2}$; ω_{pe} and ω_0 denote the electron plasma frequency and the incident radial frequency, respectively. The high irradiance allows the target surface to ionize well before the peak of the pulse and the shortness of the laser pulse prevents expansion of the target after ionization; even a 1 keV electron temperature (which is the typical temperature for nanosecond pulses) gives an exponential profile with a characteristic length $L(\mu m) = 0.2\ \tau(ps)$ where τ is the pulse duration. Hence it is possible to achieve high temperature plasmas and at the same time their confinement to densities close to solid density, i.e. 10^{23} - 10^{24} cm^{-3}.

High electron quiver velocities result in reduced collisions and anomalous skin effect (Rae and Burnett 1990, Kato et al. 1991, Andreev et al. 1992, Yang et al. 1995). The combination at the solid surface of the high electromagnetic energy and high density matter would favor the observation of new plasma effects that need to be characterised in more details, but can be classified as thermal (isotropic particle velocity distribution), kinetic (directed energetic particle beams), hydrodynamical (increase of electron and ion density by a shock wave or electromagnetic radiation pressure) or electromagnetic mechanisms (harmonic generation and static fields). Various phenomena can be specifically studied, but strong interplay is the general rule. Particle-in-cell (PIC) simulations have clearly exhibited collisionless absorption via kinetic effects: resonant absorption and vacuum screening effect for oblique incidence, and magnetic **j**×**B** force for normal incidence generate fast electron bursts entering the solid target. Hydrodynamic scenarii have been identified, namely plasma pushing, electrostatic shock and hole boring. High-order harmonic generation has also been evidenced (Gibbon 1996).

The influence of the laser incidence angle on absorption has been interpreted through two effects (Gibbon and Bell 1992, Gibbon 1994, Bulanov et al. 994): (i) the "screening surface effect" and (ii) conventional resonant absorption. The former appears when surface electrons are pulled into the vacuum by the laser wave and, under the subsequent space charge field, return with large velocity into the solid. When the prime mover is the axial electric field originating from an obliquely incident light, this mechanism is coined "vacuum heating" (Brunel 1987, Brunel 1988, Bonnaud et al. 1991, Kato et al. 1993); when the prime mover is the ponderomotive force oscillating at twice the incident frequency, it is termed "**j**×**B** heating" (Kruer and Estabrook 1985, Lefebvre and Bonnaud 1997). Resonant absorption is well known in the context of ICF research: it is characterized by the secular growth of a collective electron oscillation at the reflection point of the laser field when the incident laser wave has oblique incidence (Forslund et al. 1977, Estabrook and Kruer 1978).

Besides these electron acceleration processes, ion motion has also been considered. Electrons are pushed along the laser propagation direction by the axial ponderomotive force at zero-frequency; the space-charge electric field drags ions into the plasma. The natural thermal expansion of the plasma is then inhibited and surface recoil can be observed (Wilks et al. 1992, Lefebvre and Bonnaud 1995). The ion motion can also launch a density bump inside the plasma. Such a bump parts from the surface and can be either a soliton-like perturbation of the density profile or a shock which is associated with an electrostatic field localized within a Debye length at the forefront of the shock (Denavit 1992). Ions can be reflected by this wave and accelerated to high energy.

The previous mechanisms were bound to an axial motion of the charged particles. The finite size of the laser pulse along the radial direction causes the ponderomotive force to expel the electrons from the axis of the laser beam. A hole is then bored inside the plasma (Wilks 1993); the density profile inhomogeneity acts like a lens: it can induce the refraction of the light towards the beam axis and cause self-focusing, increasing the laser irradiance.

3 Model and Simulations

The physics we explore here is characterised by collective relativistic electromagnetic effects in which the magnetic force is as important as the electric field. It can be studied within a non-quantum framework for both the fields and the particles: the quantum physics is confined to the description of the ionization that is the first laser-interaction step to be considered before a large density of free charges is available. The physics is nonlinear for two reasons: the relativistic dynamics of the electrons and the large modulation of the electron and ion densities.

Specific codes are required to tackle those electromagnetic mechanisms that are basically kinetic effects, involving strongly non-Maxwellian distributions, varying rapidly in space and time. The plasma can be reduced to two distributions of electrons and ions, whose charges compose an electrically-neutral medium, and which evolve according to the Vlasov equation with self-consistent fields. The fields evolve according to Maxwell equations, with the plasma charges and currents as sources. The plasma can be considered collisionless since the quiver motion along the laser electric field makes the electron-ion collision frequency much weaker than the electron plasma frequency. The Vlasov equation is solved with a particle-in-cell (PIC) method: the orbits of a large number of free charges with relativistic velocities are calculated in the presence of self-consistent electric and magnetic fields. The phase space, initially unlimited, is explored by a small number of particles in each space mesh. This method requires less computing time for 2-D or 3-D simulations, but also induces a much larger noise level, than a classical discretization scheme.

The lack of time/space hierarchy makes the scenarii difficult to interpret and pure analysis untractable. Numerical computation is therefore required without resorting to any *ab initio* choice of frequencies or wavelengths.

The 1.5-D code EUTERPE has been designed ten years ago and has been used for both ICF and UHI contexts (Bonnaud and Reisse 1986, Bonnaud et al. 1994). At present, it can model oblique incidence at an angle θ, by normal incidence on a plasma drifting at velocity $c \sin(\theta)$ (Bourdier 1983, Lefebvre 1996). The 2-D code MANET has been recently designed. It operates in 2-D cartesian geometry in which the beam and the plasma appear as lines with infinite length along one transverse axis. Both codes can handle linearly polarized light and are fully relativistic. The particles are reflected on the laser-entrance side and reflected (resp. reinjected with initial thermal velocity) on the opposite side when the initial plasma reaches (resp. does not reach) the right boundary of the simulation box. For MANET, periodic conditions have been used along the transverse direction.

The diagnostics are very complete; the most interesting information comes from: particle phase spaces along two axis chosen among the set of 3 (resp. 4) coordinates in 1.5-D (resp. 2-D), lineouts of the density profile, kinetic energy distributions, time evolutions of the kinetic and the various electromagnetic energies, orbits of selected particles versus time, wave-vector/frequency spectra of the electric fields.

4 1.5-D Simulation Results

4.1 Self-induced Transparency

Let us consider a sharp-edged irradiated plasma surface. If the incident irradiance is low enough for the electron motion to be non-relativistic ($a_0 \ll 1$), the electron current dominates the ion current, due to the ion mass inertia. For the so-called critical density n_c, the electron current balances the displacement current. The plasma is opaque and allows for only an evanescent wave to exist inside the plasma: the light interacts only with the plasma surface. High irradiances cause the electron velocity to approach c and the current to saturate, instead of linearly increasing with the wave amplitude. The increase of the inertial electron mass and the subsequent decrease of the effective plasma frequency allows a laser to penetrate a "classically-opaque" plasma: this transformation from surface to volume laser-plasma interaction has been recently evidenced by PIC simulations (Lefebvre and Bonnaud 1995, Sakagami and Mima 1996, Guérin et al. 1996).

The differences between opacity regime and transparency regime are illustrated by the electron orbits displayed in Fig. 1. For the lower irradiance, the laser wave does not penetrate the plasma but pushes its surface, resulting in a large density increase and a large reflectivity; electrons and ions delineate the same plasma surface. For the larger irradiance, the plasma ions remain nearly immobile, while the electrons have long oscillations correlated with the laser wave propagation deeper and deeper in the target. Both pictures show a surface recoiling at steady velocity. When a_0 is increased, this velocity v exhibits (i) a linear increase in the opacity regime according to $v/c = a_0/[2\,(n_e/n_c)\,(m_i/m_e)]^{1/2}$, (2) a jump for a critical laser stength $a_0 \approx 1.65\,(n_e/n_c - 0.5)$ and (3) a rapid increase in the transparency regime. The light reflected by the plasma displays opposite trends due to a Doppler shift from the plasma surface. Contrasting with previous analytical predictions based on fluid arguments, these results stress the importance of both the kinetic effects (in addition to the transverse quiver motion the axial heating causes the plasma frequency to decrease) and the ion motion (mobile ions raise the transparency threshold because of steepened ion density profile).

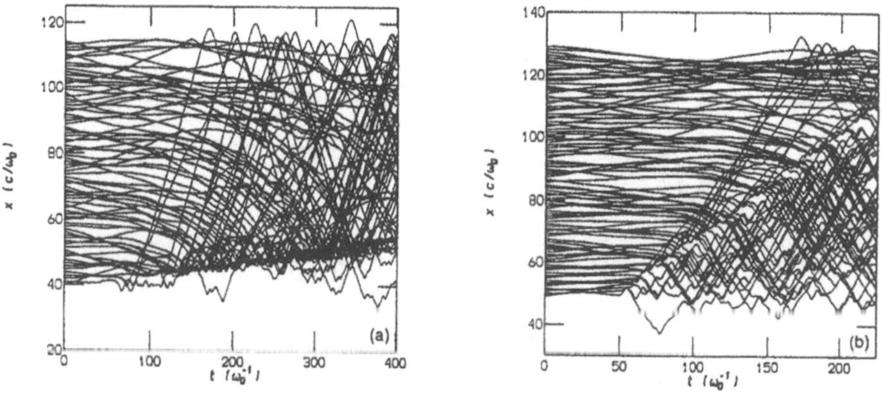

Figure 1: Electron trajectories for $n_e/n_c = 2$, $m_i/Zm_e = 1836$ and (a) $a_0 = 2$ or (b) $a_0 = 4$. The plasma surface is initially located at (a) $x = 40\ c/\omega_0$, (b) $x = 50\ c/\omega_0$. EUTERPE simulations.

4.2 Electron Heating

The irradiation at normal incidence of an overdense plasma causes the acceleration of electrons to relativistic energies. The drag and pull motion of surface electrons driven by the oscillating ponderomotive force and the charge space field cause electrons to exit the plasma and have an excursion in vacuum in front of the plasma surface. They undergo a strong acceleration before returning into the plasma. There, quasi no electric field exists and these electrons are free-streaming deeper and deeper in the plasma until collisional processes slow them down.

1.5-D PIC simulations, performed with fixed ion background, have shown that the fast electron temperature was dependent on both laser irradiance, plasma density and density gradient at the plasma surface (Lefebvre and Bonnaud 1997). If the plasma is sharply-edged, the electron heating originates from the dephasing of the electron orbits in front of the surface; this dephasing is related to the time when they exit the plasma. The denser the plasma, the colder the fast electrons (Fig. 2).

For an extended surface plasma (gentle density gradient), the incident and reflected waves form a nearly standing pattern in front of the target. When the density gradient is smooth enough, the laser reflection point is located inside the plasma and the electrons that are below this turning point interact with this standing wave. It makes them oscillate stochastically with a large longitudinal amplitude (Bauer *et al.* 1995). Their thermal energy increases with the density gradient, and approaches the ponderomotive potential: $\Phi_p = m_e c^2 ((1+a_0^2)^{1/2} - 1)$. Moreover, the density ramp improves the laser-plasma coupling efficiency, which can peak at 30 %.

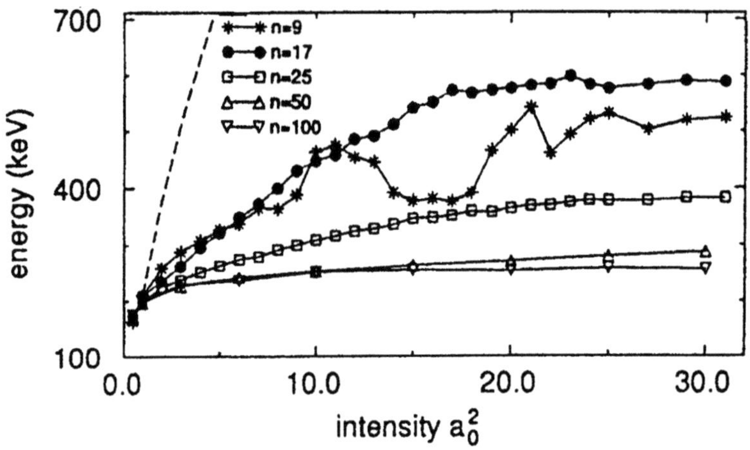

Fig. 2 : Temperature of electrons penetrating the plasma, from kinetic EUTERPE simulations. The dotted line corresponds to the ponderomotive potential.

4.3 Collisionless Shock

The free expansion of a hot plasma in vacuum is well known: the density is given by a self-similar profile $n_e = Zn_i = n_0 \exp(-v/c_s)$, where c_s is the ion-acoustic velocity ($c_s/c = 10^{-3} (ZT_e/A)^{1/2}$, with T_e in keV; Z and A denote the ion charge and mass numbers) and v the plasma velocity ($v = c_s + x/t$ where x is the distance to the initial surface and t the time). When an intense laser wave is present, this expansion can be stopped and the surface pushed if the laser irradiance is large enough (Fig. 1(a)). Momentum and energy exchange between the wave and the plasma leads to the following surface velocity (Lefebvre and Bonnaud 1995), with $n_e(\mathrm{cm}^{-3})$ and R the reflection rate:

$$\frac{v}{c} \approx 0.014 \left(\frac{Z}{A} \frac{R I_{18}}{n_e/10^{21}} \right)^{\frac{1}{2}}$$

which is quite similar to Wilks rule (Wilks et al. 1992).

Deeper in the plasma, a second density front (Fig. 3(a)) can be observed, that characterizes a solitary wave; its velocity is Mc_s. For the largest accessible velocity, i.e., $M_c = 1.6$, the associated electrostatic potential is $e\Phi$ (MeV) = 1.3 T_e. At this velocity, the solitary wave transforms into a shock wave (Denavit 1992). When the thermal ion population at the forefront of this shock is passed through by this wave, ions with kinetic energy (in the shock frame) below $e\Phi$, will be reflected; initially immobile ions can then get a velocity twice the shock front velocity, as if pushed by a piston (Fig. 3(b)). The behaviour of this ion wave is crucially dependent on the effective electron temperature at the shock location. The 1.5-D simulations have evidenced (i) strong shocks (piston) for both high density and high irradiance and (ii) soliton or weak shocks for either low irradiance or high density. A fit of the typical ion kinetic energies is:

$$\mathcal{E}_{ki}\ (\text{MeV}) \approx 5\ Z \frac{(I_{18}\lambda_0^2)^{4/3}}{n_e/n_c}$$

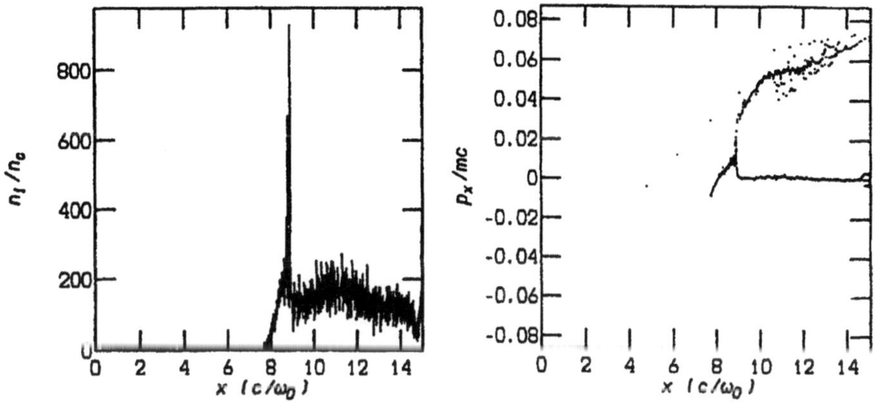

Figure 3 : (left) Ion density profile and (right) ion phase-space for $a_0 = 16$, $n_e/n_c = 100$, $m_i/Zm_e = 1836$, at time 150 ω_0^{-1}. EUTERPE simulation

5 2-D Simulation Results

5.1 Channeling

Along the transverse direction, the ponderomotive force $m_e c^2 \nabla\Phi_p$ causes the electrons to be rapidly expelled out of the light beam; it also acts on the ions but is m_i/m_e times smaller than on the electrons. If the laser beam is maintained inside the plasma over a time that largely exceeds the escaping electron time, an equilibrium between the ponderomotive force and the space-charge force takes place, in which the electrons are stabilized at some distance from the beam center. A large space-charge electric field appears, which largely exceeds the ponderomotive force on the ions and drives the ions out of the beam. A new equilibrium can be found where the ponderomotive force is balanced by the plasma thermal force. Once the electrons are outside the beam, the plasma behaves as an optical fiber, the optical index of which peaks on the beam axis and decreases on the beam periphery. This induces beam guiding and striction and subsequent amplification of the irradiance. In addition to the relativistic transparency discussed in Sec. 4.1, this mechanism can allow the beam to pass trough an overcritical plasma slab, as in Fig. 4.

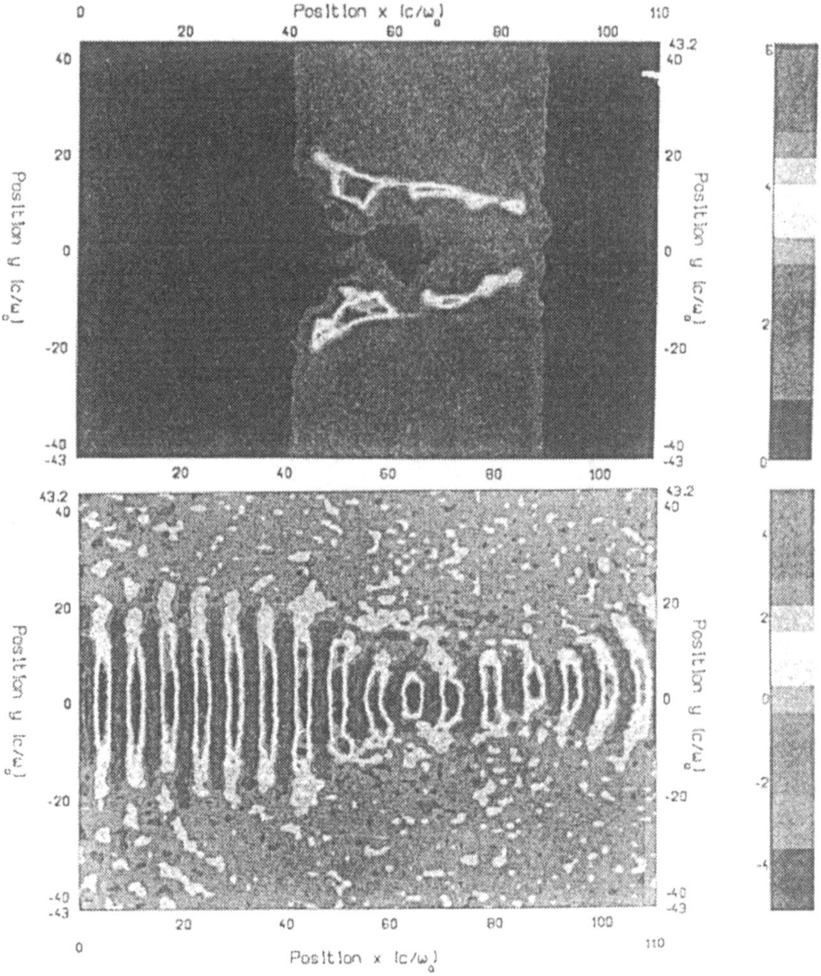

Figure 4 : Density channeling : (top) density map and (bottom) electric field. $a_0 = 4$, $n_e/n_c = 2$, $m_i/Zm_e = 1836$, time = $350\ \omega_0^{-1}$. MANET simulation.

Compared to underdense plasma conditions where a critical beam power is clearly identified, the overdense plasma situation is much more complicated. The beam propagation is bound to the hot electrons which act as a guide for the laser front into the plasma. The high rate of electron heating induces a large magnetic field at the edges of the transparent channel; this field tends to pinch the electron beam and then modify the beam propagation. The most spectacular effect has been revealed in 3-D simulations (Pukhov and Meyer-ter-Vehn 1996) where initial light filaments collapse into one super-intense filament.

5.2 Electron Heating

A probe disk has been defined inside the plasma to count the number of electrons which cross this disk until a given time. The figure below shows the distribution of electrons: a return current can be seen to balance the electrons that enter the target and go straight ahead into the right boundary of the simulation box. The temperature, measured in the range [100-500] keV, is about 500 and 200 keV along the forward and backward directions, respectively.

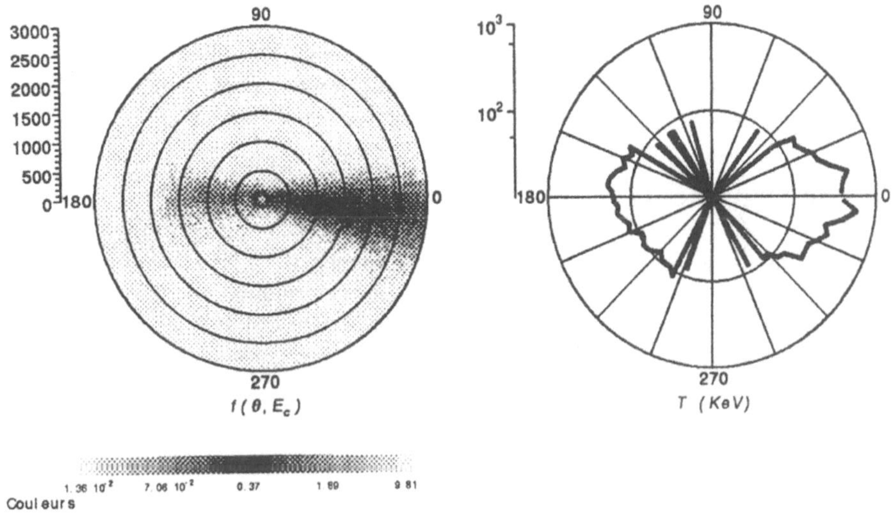

Figure 5 : (left) Polar distribution, as a function of their kinetic energy (radially in keV), of the electrons which crossed a probe-disk; (right) corresponding temperature (in keV). The probe was located in the plasma on the beam axis at 27 c/ω_0 of the surface and had a radius of 3 c/ω_0. $a_0 = 3$, $n_e/n_c = 2$, $m_i/Zm_e = 3672$. The time is 630 ω_0^{-1} . Colors are given on a 5-decade log scale from purple to red. MANET simulation.

6 Conclusions

We have focussed our attention on the coupling mechanims between an overdense collisionless plasma and a ultra-intense laser wave at normal incidence. The physics is novel, highly non-linear, and requires simulation tools to stimulate the physicists' imagination, to check the reality of theoretical speculations, and provide quantitative measurement such as absorption.

Various experiments have evidenced the mechanisms we have commented in Sec. 2: absorption in steep gradients (Kieffer et al. 1989, Teubner et al. 1993, Teubner et al. 1995), x-ray generation (Kmetec et al. 1992, Teubner et al. 1993, Schnürer et al. 1995), reflectivity (Liu and Umstadter 1992, Sauerbrey et al. 1994), MeV electron detection (Malka and Miquel 1996), hole boring (Kodama et al. 1996).

The applications of a short pulse high-intensity laser include the production of energy in fast particles or hard photons. Hot plasmas and high-order harmonics are other possible sources of short x-ray pulses. For instance, the 240th harmonic of Nd:glass laser light, lying in the water-window around 44 Å, could be a very useful diagnostic tool in chemistry or biology.

Another potential application is the fast ignitor, proposed by the Lawrence Livermore Laboratory (Tabak et al. 1995): a picosecond pulse generates kilojoules of MeV electrons that would act as a spark by heating a high-density fuel core compressed independently by multi-nanosecond laser beams . One key process for the success of such an application is the transport of these suprathermal electrons in the dense plasma, which has been only scantly addressed. The understanding of this transport in presence of neutralizing return current and self-consistent magnetic field is an open area for investigation in model and design of simulation tools.

Acknowledgments

The authors are indebted to A. Adolf for designing the first parallel version of MANET, and to J. Chave and P. Bain for their contribution to subsequent code developments. Fruitful discussions are acknowledged with J.M. Rax.

References

Andreev A.A., Gamaly E.G., Novikov V.N., Semakhin A.N., Tikhonchuk T.K. (1992): Heating of a Dense Plasma by an Ultrashort Laser Pulse in the Anomalous Skin-Effect Regime. Sov. Phys. JETP **74**, 963-973.

Bauer D, Mulser P., Steeb W.H. (1995): Relativistic Ponderomotive Force, Uphill Acceleration, and Transition to Chaos. Phys. Rev. Lett. **75**, 4622-4625.

Bonnaud G., Reisse C. (1986): Particle Code Study of the Influence of Non-Monochromacity of Laser Light on Stimulated Raman Scattering in Laser Irradiated Plasmas, Nuclear Fusion **26**, 633-646.

Bonnaud G., Gibbon P., Kindel J., Williams E.A. (1991): Laser Interaction with a Sharp-Edged Overdense Plasma. Laser Part. Beams **9**, 339-354.

Bonnaud G., Teychenné D., Bobin J.L. (1994): Plasma Edge Effect on the Wake Field Induced by a Short Laser Pulse. Europhys. Lett. **26**, 91-96.

Bourdier A. (1983): Oblique Incidence of a Strong Electromagnetic Wave on a Cold Inhomogeneous Electron Plasma. Relativistic Effects. Phys. Fluids **26**, 1804-1807.

Brunel F. (1987): Not-So-Resonant, Resonant Absorption. Phys. Rev. Lett. **59**, 52-55.

Brunel F. (1988): Anomalous Absorption of High Intensity Subpicosecond Laser Pulses. Phys. Fluids **31**, 2714-2719.

Bulanov S.V., Naumova N.M., Pegoraro F. (1994): Interaction of an Ultrashort, Relativistically Strong Laser Pulse with an Overdense Plasma. Phys. Plasmas **1**, 745-757.

Chiao R.Y., Garmire E., Townes C.H. (1964): Self-Trapping of Optical Beams. Phys. Rev. Lett. **13**, 479-482.

Denavit J. (1992): Absorption of High-Intensity Subpicosecond Lasers on Solid Density Targets. Phys. Rev. Lett. **69**, 3052-3055.
Estabrook K.G., Kruer W.L. (1978): Properties of Resonantly Heated Electron Distributions. Phys. Rev. Lett. **40**, 42-45.
Ferray M., Lompré L.A., Gobert O., Mainfray G., Manus C., Sanchez A., Gomes A. (1990): Multiterawatt Picosecond Nd-Glass Laser System at 1053 nm. Optics Comm. **75**, 278-282.
Forslund D.W., Kindel J.M., Lee K. (1977): Theory of Hot-Electron Spectra at High Laser Intensity. Phys. Rev. Lett. **39**, 284-288.
Gibbon P., Bell A.R. (1992): Collisionless Absorption in Sharp-Edged Plasmas. Phys. Rev. Lett. **68**, 1535-1538.
Gibbon P. (1994): Efficient Production of Fast Electrons from Femtosecond Laser Interaction with Solid Targets. Phys. Rev. Lett. **73**, 644-667.
Gibbon P. (1996): Harmonic Generation by Femtosecond Laser-Solid Interaction: a Coherent Water-Window Light Source? Phys. Rev. Lett. **76**, 50-53.
Guérin S., Mora P., Adam J.C., Héron A., Laval G. (1996): Propagation of Ultraintense Laser Pulses through Overdense Plasma Layers. Phys. Plasmas **3**, 2693-2701.
Kato S., Kawakami R., Mima K. (1991): Nonlinear Inverse Bremsstrahlung in Solid-Density Plasmas. Phys. Rev. A **43**, 5560-5567.
Kato S., Bhattacharyya B., Nishiguchi A., Mima K. (1993): Wave Breaking and Absorption Efficiency for Short Pulse *p*-Polarized Laser Light in a Very Steep Density Gradient. Phys. Fluids B **5**, 564-570.
Kaw P., Schmidt G., Wilcox T. (1973): Filamentation and Trapping of Electromagnetic Radiation in Plasmas. Phys. Fluids **16**, 1522-1525.
Kieffer, Audebert P., Chaker M., Matte J.P., Pépin H., Johnston T.W., Maine P., Meyerhofer D., Delettrez J., Strickland D., Bado P., Mourou G. (1989): Short-Pulse Laser Absorption in Very Steep Plasma Density Gradients. Phys. Rev. Lett. **62**, 760-763.
Kmetec J.D., Gordon C.L., Macklin J.J., Lemoff B.E., Brown G.S., Harris S.E. (1992): MeV X-Ray Generation with a Femtosecond Laser. Phys. Rev. Lett. **68**, 1527-1530.
Kodama R., Takahashi K., Tanaka K.A., Tsukamoto M., Hashimoto H., Kato Y. Mima K. (1996): Study of Laser-Hole Boring into Overdense Plasmas. Phys. Rev. Lett. **77**, 4906-4909.
Kruer W.L., Estabrook K. (1985): **JxB** Heating by Very Intense Laser Light. Phys. Fluids **28**, 430-432.
Lefebvre E., Bonnaud G. (1995): Transparency / Opacity of a Solid Target Illuminated by an Ultrahigh-Intensity Laser Pulse. Phys. Rev. Lett. **74**, 2002-2005.
Lefebvre E. (1996): *Mécanismes d'absorption et d'émission dans l'interaction d'une impulsion laser ultra-intense avec une cible solide*. PhD dissertation (in french), (Université Paris-XI, Paris).
Lefebvre E., Bonnaud G. (1997): Nonlinear Electron Heating in Ultrahigh-Intensity-Laser-Plasma Interaction. Phys. Rev. E **55**, 1011-1014.
Liu X., Umstadter D. (1992): Competition between Ponderomotive and Thermal Forces in Short-Scale-Length Laser Plasmas. Phys. Rev. Lett. **69**, 1935-1938.
Maine P., Strickland D., Bado P., Pessot M., Mourou G. (1988): Generation of Ultrahigh Peak Power Pulses by Chirped Pulse Amplification. IEEE J. Quantum Electron. **24**, 398-403.
Malka G., Miquel J.L. (1996): Experimental Confirmation of Ponderomotive-Force Electrons Produced by an Ultrarelativistic Laser Pulse on a Solid Target. Phys. Rev. Lett. **77**, 75-78.
Pukhov A., Meyer-ter-Vehn J. (1996): Relativistic Magnetic Self-Channeling of Light in Near-Critical Plasma - Three-Dimensional Particle-in-Cell Simulation. Phys. Rev. Lett. **76**, 3975-3978.
Rae S.C., Burnett K. (1990): Nonlinear Absorption of Intense Laser Light by Collective Plasma Excitations. Phys. Fluids B **2**, 1015-1020.
Sakagami H., Mima K. (1996): Anomalous Penetration Mechanisms of Very Intense Laser

Pulses into Overdense Plasmas. Phys. Rev. E **54**, 1870-1875.

Sauerbrey R., Fure J., Le Blanc S.P., Van Wonterghem B., Teubner U., Schäfer F.P. (1994): Reflectivity of Laser-Produced Plasmas Generated by a High Intensity Ultrashort Pulse. Phys. Plasmas **1**, 1635-1642.

Schnürer M., Kalashnikov M.P, Nickles P.V., Schlegel T., Sandner W., Demchenko N., Nolte R., Ambrosi P. (1995): Hard X-Ray Emission from Intense Short Pulse Laser Plasmas. Phys. Plasmas **2**, 3106-3110.

Sodha M., Ghatak A., Tripathi V. (1974): *Self-focusing of laser beams in dielectrics, plasmas and semi-conductors* (Tata Mc Graw-Hill: New-Dehli).

Tabak M., Hammer J., Glinsky M.E., Kruer W.L., Wilks S.C., Woodworth J., Campbell E.M., Perry M.D., Mason R.J. (1995): Ignition and High Gain with Ultrapowerful Lasers. Phys. Plasmas **1**, 1626-1634.

Teubner U., Bergmann J., van Wonterghem B., Schäfer F.P., Sauerbrey R. (1993): Angle-Dependent X-Ray Emission and Resonance Absorption in a Laser-Produced Plasma Generated by a High Intensity Ultrashort Pulse. Phys. Rev. Lett. **70**, 794-797.

Teubner U., Wülker C., Theobald W., Förster E. (1995): X-Ray Spectra from High-Intensity Subpicosecond Laser Produced Plasmas. Phys. Plasmas **2**, 972-981.

Watteau J.P., Bonnaud G., Coutant J., Dautray R., Decoster A., Louis-Jacquet M., Ouvry J., Sauteret J., Seznec S., Teychenné D. (1992): Experimental Program on the 20-TW Laser System. Phys. Fluids B **4**, 2217-2223.

Wilks S.C., Kruer W.L., Tabak M., Langdon A.B. (1992): Absorption of Ultra-Intense Laser Pulses. Phys. Rev. Lett. **69**, 1383-1386.

Wilks S.C. (1993): Simulation of Ultraintense Laser Plasma Interactions. Phys. Fluids B **5**, 2603-2608.

Yang T.Y.B., Kruer W.L., More R.M., Langdon A.B. (1995): Absorption of Laser Light in Overdense Plasmas by Sheath Inverse Bremsstrahlung. Phys. Plasmas **2**, 3146-3154.

Plasma Accelerator Simulations Using Semi-Lagrangian 1D and 2D Vlasov Code

Alain Ghizzo, Pierre Bertrand, and Marie-Line Bégué

Université Henri Poincaré, BP 239, 54506 Vandoeuvre-les-Nancy Cedex, France

Abstract. A semi-lagrangian relativistic Vlasov model is described for a hierarchy of applications of increasing complexity, in one and two dimensions. Attention is focused on its accuracy, stability, efficiency properties and implementation facilities on massively parallel computers. Such a code provides good resolution all over phase space, including very low density regions where Particles-In-Cell PIC codes simply run out of particles for which to compute. Recent developments in applying semi-lagrangian Vlasov codes to 1D and 2D problems in an electromagnetic regime (including Forward Raman Scattering and Relativistic Modulational Instability) are then reviewed.

1 Introduction

Since the suggestion of Tajima and Dawson (1979) of particle acceleration by means of plasma waves in 1979, various schemes have been proposed to excite large amplitude electron plasma waves (EPW) (theoretically capable of reaching an electric field of the order of GV/m). Such a wave is of interest as a particle accelerator concept since an electron plasma wave amplitude can largely exceed the breaking limit of the standard metallic cavity based accelerator, which is of the order of $30MV/m$. One of the promising schemes for exciting strong plasma waves remains the Plasma Beatwave Accelerator, which is based on the injection of two electromagnetic waves in a plasma of low density. The difference of frequencies of the two electromagnetic waves is chosen to be equal to the plasma frequency. Under these conditions the beat of these two waves resonantly induces a high-phase velocity longitudinal plasma wave which traps and accelerates electrons to ultra-relativistic energies. A very similar mechanism is involved in the Forward Raman Scattering (FRS) process, in which an incident electromagnetic wave decays into an electron plasma wave and a scattered electromagnetic wave. In the fully relativistic regime (corresponding to the Relativistic Modulational Instability RMI), generally obtained with a large amplitude electromagnetic wave with an irradiance of a few 10^{18} W/cm^2, the electron motion becomes relativistic. Because of the high laser intensities the maximum EPW amplitude can be limited by wavebreaking.

In the present paper examples of FRS and RMI simulations are given using semi-lagrangian 1D or 2D relativistic Vlasov code (Ghizzo et al. 1990;

Johnston et al. 1992; Bertrand et al. 1990; Ghizzo et al. 1996). Such semi-lagrangian (see Staniforth and Coté 1991) codes render possible a detailed examination of the low density regions of the phase space especially the description of the tail phenomena, where only a small number of electrons is involved. In this type of problem "Particles-In-Cell" (PIC) codes suffer from poor statistics. This because the PIC codes lack sufficient simulation particles to display the detailed phase space structure of the distribution function which is often obtained in those regions of phase space where particle and phase velocities are comparable and where trapping or wavebreaking occurs.

The paper is organised as follows. Section 2 describes the basic equations of our 2D Vlasov code. Initial value periodic simulations of FRS at moderate laser intensities (p_{osc}/mc$\sim$ 0.1) are presented in Sect. 3, which also includes a discussion on 2D transverse effects. Attention is focused in Sect. 4 on efficiency properties on parallel computers. Section 5 deals with a 1D periodic model and emphasises the strong and nonlinear wave-particle interaction and plasma wavebreaking at ultrahigh intensities corresponding to an electron quiver momentum $p_{osc}/mc \sim 1$ (related to the complex problem of the RMI). Finally the conclusion is given in Sect. 6.

2 The 2D Semi-Lagrangian Vlasov Code

In FRS the ponderomotive force drives a large amplitude plasma wave along the laser wavevector direction and produces trapped electrons with very high momenta. To model this we consider an infinite homogeneous plasma of density n_o in both x and y directions with a laser wavevector in the x-direction: we specialise to wave propagation in one dimension, i.e. $\mathbf{\nabla} = \hat{e}_x \partial/\partial x$, all field quantities being functions of the space variables, x and y. With the choice of the Coulomb gauge $\mathbf{\nabla A} = 0$ the potential vector $\mathbf{A}$ is in the z direction. Recalling that

$$\mathbf{E} = -\mathbf{\nabla}\Phi - \frac{\partial \mathbf{A}}{\partial t} \tag{1}$$

we have in the perpendicular direction z:

$$E_z = -\frac{\partial A_z}{\partial t} \tag{2}$$

and in the plasma plane:

$$E_x = -\frac{\partial \Phi}{\partial x} \quad \text{and} \quad E_y = -\frac{\partial \Phi}{\partial y}. \tag{3}$$

The electron distribution function $F(x, y, \mathbf{p}, t)$ obeys the relativistic Vlasov equation:

$$\frac{\partial F}{\partial t} + \frac{\mathbf{p}}{m\gamma}\frac{\partial F}{\partial \mathbf{r}} + e\left[\mathbf{E} + \frac{\mathbf{p}}{m\gamma} \times \mathbf{B}\right]\frac{\partial F}{\partial \mathbf{p}} = 0, \tag{4}$$

where $\mathbf{r} = (x, y, 0)$ and $\mathbf{p} = (p_x, p_y, p_z)$ with the Lorentz factor

$$\gamma = \sqrt{1 + \frac{\mathbf{p}^2}{m^2c^2}}. \tag{5}$$

Huge memory requirements are necessary to handle a full 21/2 D model, i.e. five phase space variables x, y, p_x, p_y and p_z as discussed above. Thus let us consider the following class of exact solutions of (4):

$$F(x, y, p_x, p_y, p_z, t) = f(x, y, p_x, p_y, t)\, \delta\left[p_z - P_z(x, y, t)\right]. \tag{6}$$

The reduced 2D distribution function $f(x, y, p_x, p_y, t)$ describing the particle motion in the (x, y) plane satisfies the two-dimensional relativistic Vlasov equation:

$$\frac{\partial f}{\partial t} + \frac{p_x}{m\gamma}\frac{\partial f}{\partial x} + \frac{p_y}{m\gamma}\frac{\partial f}{\partial y} + e\left[E_x - \frac{P_z B_y}{m\gamma}\right]\frac{\partial f}{\partial p_x} + e\left[E_y + \frac{P_z B_x}{m\gamma}\right]\frac{\partial f}{\partial p_x} = 0 \tag{7}$$

with a Lorentz factor of

$$\gamma = \sqrt{1 + \frac{p_x^2 + p_y^2 + P_z^2(x, y, t)}{m^2c^2}}. \tag{8}$$

In this model the transverse momentum effects in the z direction are described by a cold "fluid" model. The transverse momentum $P_z(x, y, t)$ is obtained through the introduction of the canonical generalised momentum $\mathcal{P}_z = P_z + eA_z = 0$, which obeys the fluid equation:

$$\frac{\partial P_z}{\partial t} = eE_z. \tag{9}$$

The plasma self consistent electric field component E_x and E_y are directly given by Poisson's equation according to (3):

$$\Delta\Phi = -\frac{e}{\varepsilon_0}\left[n_e(x, y, t) - n_o\right], \tag{10}$$

where n_o is the homogeneous ion density and $n_e = \int f \mathrm{d}p_x \mathrm{d}p_y$. The electromagnetic field components (E_z, B_x, B_y) obey Maxwell's equations:

$$\frac{\partial B_x}{\partial t} = -\frac{\partial E_z}{\partial y} \tag{11}$$

$$\frac{\partial B_y}{\partial t} = \frac{\partial E_z}{\partial x} \tag{12}$$

$$\frac{\partial E_z}{\partial t} = c^2\left(\frac{\partial B_y}{\partial x} - \frac{\partial B_x}{\partial y}\right) - \frac{J_z}{\varepsilon_o}, \tag{13}$$

where $J_z = en_eP_z/m$. A splitting scheme (Knor and Cheng 1976; Gagné and Shoucri 1977) is then used to integrate the Vlasov equation.

3 2D Vlasov Code : Numerical Simulation of FRS

The FRS instability is a parametric instability involving three waves: the incident electromagnetic wave, here called the "pump" wave (ω_o, k_o), which drives two unstable waves: a scattered electromagnetic wave (ω_s, k_s) and an electron plasma wave (ω_e, k_e). The Raman instability occurs when the usual matching conditions hold:

$$\omega_o(k_o) + \Delta\omega = \omega_s(k_s) + \omega_e(k_e) \text{ and } k_o = k_s + k_e. \tag{14}$$

We assume perfect k-matching (since we have a periodic simulation and match mode numbers exactly), while $\Delta\omega$ contains the mismatch (if any). The matching conditions can be satisfied only if $n_e \leq n_{crit}/4$, where n_{crit} is the critical density above which an electromagnetic wave will not propagate. (A high amplitude electromagnetic wave can, however, propagate through an overdense plasma provided that the electon quiver velocity is close to the velocity of light.) We have performed numerical simulations using periodic boundary conditions in both x and y directions and an initial state consisting of a signifiant electromagnetic pump propagating in the positive x direction and homogeneous in the y direction and a propagating electrostatic plasma wave of modest amplitude, acting as a perturbation from which FRS can grow since a Vlasov code is essentially noiseless. Since the velocities are normalised to c and the frequencies to ω_p, the choice of $k_o(= 2k_s$ or $2k_e$) determines the plasma box length L_x in term of c/ω_p. The plasma was chosen with two electron temperature components, the major (95%) component with a 15 keV temperature in both directions p_x and p_y (high enough for electron Landau damping to subdue the usually rapidly growing but here unwanted backward Stimulated Raman Scattering) and a minor (5%) p_x-component at 100 keV (to enhance wave-particle interaction). With these parameters a good frequency match was obtained by choosing $k_oc/\omega_p = 2.4$, i.e. a box length $L_x = 5.23c/\omega_p$ (we take in the second direction $L_y = 20c/\omega_p$). The normalised pump electric field amplitude is $eE_o/m\omega_pc = 0.28$ which gives a quiver momentum of $p_{osc}/mc \simeq 0.10$ (for a $10.6\mu m - CO_2$ laser. The corresponding intensity is $1.8\ 10^{14} W/cm^2$). The density as compared to the critical density is then $n_e/n_{crit} \simeq 0.15$.

Figure 1 shows the $x - p_x$ phase space representation (in fact we have plotted the function at $y = L_y/2$ and $p_y = 0$) afforded by the Vlasov code. Here shading is used to indicate the relative values of normalised phase space density between 10^{-3} and 10^{-4}. The curves shown here are in good agreement with numerical results obtained directly by a 1D periodic Vlasov code (see Johnston et al. 1992) and seem to indicate a classic acceleration process when the transverse laser profile is chosen to be homogeneous. They clearly exhibit the acceleration of positive velocity particles followed by the trapping and formation of vortices with spiral orbits in the phase space.

To further illustrate the influence of transverse effects on the particle acceleration mechanism, a numerical simulation including a pump light with

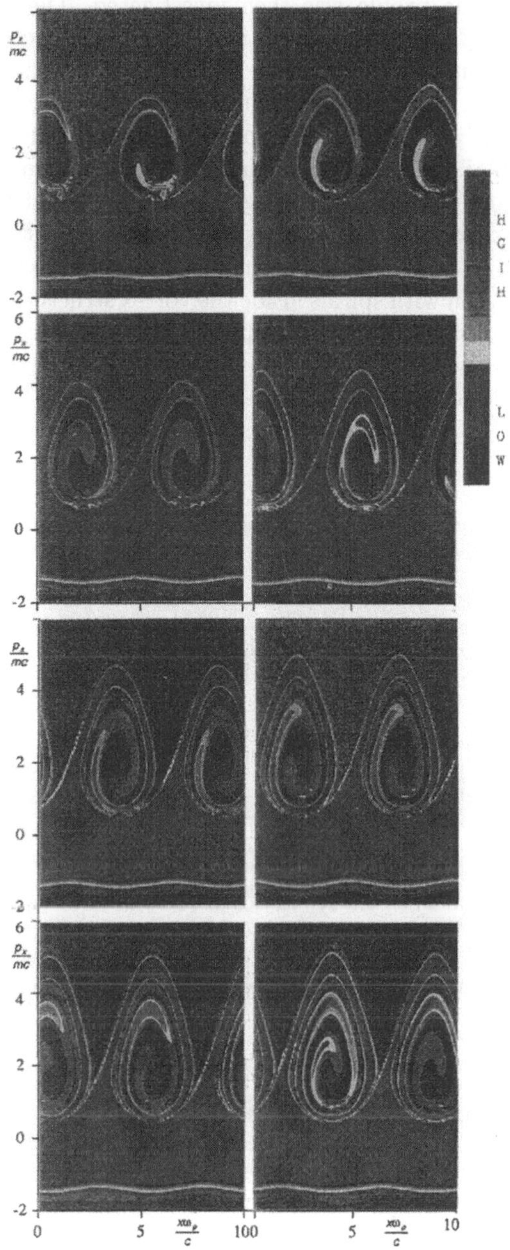

Fig. 1. Figure 1 shows the $x - p_x$ phase space representation afforded by the Vlasov code from time $t\omega_p = 130$ till $t\omega_p = 200$ by steps of $10\omega_p^{-1}$. The sequence clearly exhibits the acceleration of positive particles followed by trapping and the formation of vortices induced by FRS.

a gaussian amplitude profile was then investigated. The E_z amplitude is given in the form $E_{zo}(y) = E_o \exp[-\beta (y - y_o)^2]$. We choose $eE_o/m\omega_p c = 0.35$. This second numerical simulation is conducted with a maximum quiver velocity of $p_{osc}/mc \sim 0.134$ located initially at $y = L_y/2$. The time step used here is $\Delta t\omega_p = 0.05$. Using a phase space grid of $N_x N_y N_{p_x} N_{p_y}$ of $64 \times 32 \times 128 \times 32$, i.e. 8,388,608 "particles", the CPU time on the Cray-C94 computer is 0.56 μs per time step per grid point, i.e. about 8h CPU time up to 300 ω_p^{-1}. The corresponding phase space plots of the distribution function $f(x, y = L_y/2, p_x, p_y = 0)$ are shown in Fig. 2. Similar particle acceleration structures can be seen with spirals inside the vortex structure which implicitly reflect the history of the particles trapped as the wave built up. There is, however, little change in the maximum momentum leading to an acceleration until $mc^2(\gamma - 1) \sim 1.5 MeV$.

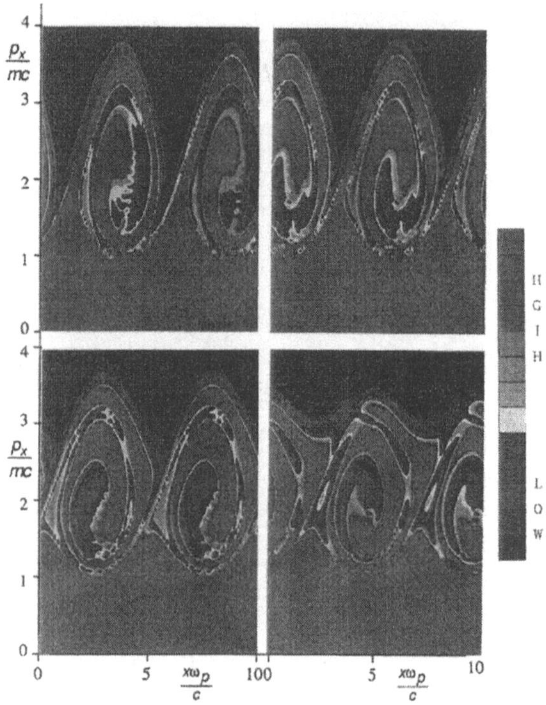

Fig. 2. Figure 2 shows the $x - p_x$ phase space representation in the case of a transverse gaussian profile of the laser light at different times during plasma evolution $t\omega_p = 140, 160, 180, 200$.

4 Numerical Comparison Between Cray C94 and Cray T3D-T3E Vlasov Simulations

All present results were processed on the Cray C94-C98, Cray T3D and Cray T3E computing systems. The vectorised version of the Vlasov code has achieved a data processing rate of 250 Mflops and demonstrates a reasonably high vectorised efficiency. Two grid systems $N_x N_y N_{p_x} N_{p_y}$ of $128 \times 8 \times 128 \times 32$ and then of $128 \times 32 \times 128 \times 32$ were investigated resulting in corresponding numbers of particles of 4,194,306 and 16,777,216 respectively. The CPU time is close to $0.54 \mu s$ per time step per particle on the Cray C94 computer while an early version of the parallelised code gave a value of $24,64 \mu s$ per time step per particle and per processor. A substantial amount of improvement is achieved by optimising the node-to-memory communications which critically affect the parallel efficiency. The optimised version leads then to a CPU time of $9.5 \mu s$ per time step, per particle, per processor. The improvement in numerical efficiency is by factor of three and seems to indicate that one processor of Cray C90 is equivalent to 18 processors on the T3D computer (and equivalent to 14 processors on the T3E computer). Additional efficiency improvement for the present code is still possible by replacing global synchronous message passing calls with several asynchronous calls using the MPI[1] library or the nonstandard SHMEM[2] library of the Cray system. In essence a sustained effort must be maintained to realise the full potential of scalable parallel systems.

5 Numerical 1D Vlasov Simulation of the Relativistic Modulational Instability

As our second example we consider a large amplitude electromagnetic wave (with an irradiance of a few $10^{18} W/cm^2$ or $p_{osc}/mc \sim 1$) in order to handle the difficult problem of the relativistic modulational instability. After focusing, the light intensity is so high that an electron is brought deeply into the relativistic regime by interacting with the wave electromagnetic field. In a plasma such a wave can excite large-amplitude plasma waves by making use of plasma wake fields and laser wake fields. Under these conditions the plasma wave can reach the relativistic wave breaking limit (Dawson 1959). This wave breaking process produces relativistic electrons in the forward direction which are important for heating the ignition spot in the compressed fusion target.

We limit our discussion here to a one-dimensional model, which precludes the two-plasmon instability and the Raman sidescattering instability, but retains forward-backward Raman instabilities and the longitudinal RMI. Here

[1] M.P.I.: Message Passing Interface

[2] SHMEM: SHared MEMory

we choose a couple of parameters $n_e/n_{crit} \sim 0.58$ and $p_{osc}/mc \sim 1$ in order to generate a strong RMI instability.

In the fully relativistic regime perturbation theory cannot be used to take into account nonlinearities of the primary wave (thus the case of an electromagnetic wave with a linear polarisation is more complicated to handle). In order to overcome this difficulty, the nonlinear electromagnetic wave is assumed to be circularly polarised. It is well known that for such a polarisation the amplitudes of the electric field and of the electron quiver velocity remain constant (Akhiezer and Polovin 1956; Guérin et al. 1995).

We start from the 1D relativistic Vlasov model:

$$\frac{\partial f}{\partial t} + \frac{p_x}{m\gamma}\frac{\partial f}{\partial x} + e\left[E_x - \frac{mc^2}{2\gamma}\frac{\partial}{\partial x}\left(a^2\right)\right]\frac{\partial f}{\partial p_x} = 0, \tag{15}$$

where $\gamma = \sqrt{1 + \frac{p_x^2}{m^2c^2} + a^2}$ is the Lorentz factor and $a = eA(x,t)/mc$ is the normalised amplitude of the potential vector $\mathbf{A} = (0, A_y, A_z)$. The Vlasov equation is coupled with Maxwell's equations:

$$\begin{aligned}
\frac{\partial E_y}{\partial t} &= -c^2\frac{\partial B_z}{\partial x} + \omega_p^2 A_y \rho(x,t) \\
\frac{\partial E_z}{\partial t} &= c^2\frac{\partial B_y}{\partial x} + \omega_p^2 A_z \rho(x,t) \\
\frac{\partial B_y}{\partial t} &= \frac{\partial E_z}{\partial x} \\
\frac{\partial B_z}{\partial t} &= -\frac{\partial E_y}{\partial x},
\end{aligned} \tag{16}$$

where $\rho(x,t) = \int \mathrm{d}p_x \; f/m\gamma$. The components of the potential vector are then computed using

$$\frac{\partial A_y}{\partial t} = -E_y \quad \text{and} \quad \frac{\partial A_z}{\partial t} = -E_z. \tag{17}$$

The longitudinal component of the field is obtained by solving Poisson's equation:

$$\frac{\partial E_x}{\partial x} = \frac{e}{\varepsilon_o}\left(n_e(x,t) - n_i\right). \tag{18}$$

We start with an initial homogeneous Maxwellian distribution function with a temperature, $T_e = 5keV$, without a perturbation term. The cutoff in the momentum phase space is $\pm p_{x\,\max}/mc = \pm 3$. The plasma is embedded in a periodic box of length $L_x\omega_p/c = 4\pi$. The initial large amplitude pump wave (ω_o, k_o) is taken as $\mathbf{E} = (0, E_o\cos k_o x, E_o \sin k_o x)$. (ω_o, k_o) satisfy the relativistic dispersion relation of circularly polarised waves

$$\omega_o^2 = \omega_p^2/\gamma_o + k_o^2 c^2,$$

where

$$\gamma_o = \sqrt{1 + \left(\frac{eE_o}{m\omega_p c} \times \frac{\omega_p}{\omega_o}\right)^2} = \sqrt{1 + \frac{p_{osc}^2}{m^2c^2}}.$$

Choosing $k_o c/\omega_p = 2\Delta k = 1$ and $p_{osc}/mc = 1$ we obtain $\omega_o = 1.306\omega_p$. The density as compared to the critical density is then $n_e/n_{crit} \simeq 0.58$. Figure 3 shows the $x - p_x$ phase space representation at different times (from $t\omega_p = 90$ to $t\omega_p = 98$). The large amplitude pump wave generates a strong modulation of the background distribution function, followed by trapping of particles in the plasma wave. The plasma wave, strongly excited by the pump light, tends to break leading to the production of relativistic electrons in the forward direction followed by trapping and the formation of a vortex structure of the whole distribution function. The interaction process observed directly in phase space seems therefore very different from a classic heating of the distribution function and due to the occurence of complex but coherent vortex structures.

6 Conclusion

We have presented several problems, one and two-dimensional, in an electromagnetic regime related to FRS or RMI. It is interesting to note that for the 1D plasma workstations provide today the required memory size and processing speed to handle these 1D plasma semi-lagrangian Vlasov codes. However, the 1D open system requires a higher sampling of the distribution function and the use of vectorial computer may be necessary. Of course 2D problems imply an effort 10^3 or 10^4 times bigger and we reach the level of tomorrow's supercomputer with a few gigawords of memory and the consequent speed. However, the efficiency loss of the Vlasov code for higher dimensions might be compensated by the use of massively parallel computers. To realise the gains in computational power made possible by massively parallel computers generally requires that algorithms used in the traditional serial and vector computers be adapted or redesigned to use optimally a particular parallel architecture (both the processor's specifications and node-to-node communications). Because of the eulerian characteristic of the distribution function, the Vlasov code is probably much easier to realise in such computers since one does not have to cope with the allocation of the particles between processors as Euler elements stay in place. On the other hand its lagrangian characteristic (because we move along the characteristics) allows us to choose a reasonable time step. Both approaches (Particles-In-Cell and Vlasov code) require a more complicated comparison on the interplay between the physical aspect of the model on the one hand and performance and optimisation issues on the other.

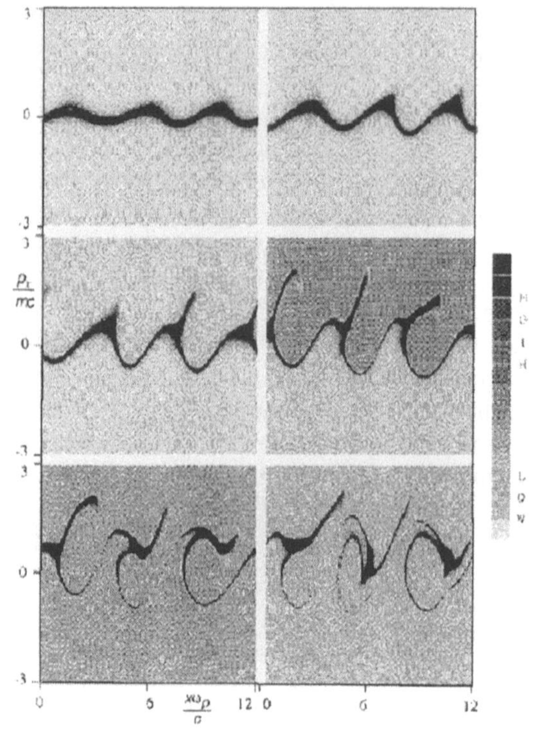

Fig. 3. Figure 3 shows the $x - p_x$ phase space representation of the background distribution function for RMI at different times during plasma evolution $t\omega_p = 88$, 90, 92, 94, 06, 98.

References

Akhiezer, A.I., Polovin, R.V. (1956): Theory of wave motion of an electron plasma. Sov. Phys, JETP **3**, 696–705; see also Kaw, P., Dawson, J. (1970): Relativistic nonlinear propagation of laser beams in clod overdense plasmas. Phys. Fluids **13**, 472–483

Bertrand, P., Ghizzo, A., Johnston, T.W., Shoucri, M., Fijalkow, E., Feix, M.R. (1990): A nonperiodic Euler-Vlasov code for the numerical simulation of laser-plasma beatwave acceleration and Raman Scattering. Phys. Fluids B **2** (5), 1028-1037

Dawson, J.M. (1959): Nonlinear electron oscillations in a cold plasma. Phys. Rev **113**, 383–387; see also, Dawson, J.M., Sheng, Z.M., meyer-ter-Vehn, T. (1997): Relativistic wave breaking in warm plasmas. Phys. Plasmas **4** (2), 493-495

Gagné, R.J., Shoucri, M. (1977): A splitting scheme for the numerical solution of the one-dimensional Vlasov equation. J. Comp. Phys. **24**, 445–449

Ghizzo, A., Bertrand, P., Lebas, J., Johnston, T.W., Shoucri, M. (1996): A hybrid Eulerian Vlasov code. Study of high frequency beatwave experiment in a finite casual system. Phys. Plasmas **3** (2), 650-668

Ghizzo, A., Bertrand, P., Shoucri, M., Johnston, T.W., Fijalkow, E., Feix, M.R. (1990): A Vlasov code for the numerical simulation of stimulated Raman scattering. J. Comp. Phys. **90** (2), 431-457

Guérin, S., Laval, G., Mora, P., Adam, J.C., Héron, A., Bendib, A. (1995): Phys. Plasmas **2** (7), 2807-2814

Johnston, T.W., Bertrand, P., Ghizzo, A., Shoucri, M., Fijalkow, E., Feix, M.R. (1992): Stimulated Raman Scattering: action evolution and particle trapping via Euler-Vlasov fluid simulation. Phys. Fluids B **4** (8), 2523-2537

Knor, G., Cheng, C.G. (1976): Integration of the Vlasov equation in configuration space. J. Comp. Phys. **22**, 330–351

Staniforth, A., Coté, J. (1991): Semi-lagranfian integration schemes for atmospheric models. Monthly Weather Review **119**, 2206-2223

Tajima, T., Dawson, J.M. (1979): Laser electron accelerator. Phys. Rev. Lett. **43**, 267–270

2D Vlasov-Eulerian Simulations: Extension to the Relativistic Case

Eric Fijalkow

Laboratoire de Mathématique, Applications et Physique Mathématique - UMR 6628, Université d'Orléans, UFR des Sciences, BP 6759, F-45067 Orléans Cedex 2, France

Abstract. The splitting schemes of the 1D and 2D Vlasov equations are presented for the electrostatic (with an external magnetic field in 2D) problem. A solution of the split equations by a Flux Balance Method (F.B.) is shown and extended to cases where the transport coefficient depends upon the variable (as in the relativistic Vlasov problem). A short comparison with a 1D spline model is given and a 2D phase space visualisation of an electrostatic simulation is presented.

1 Introduction

Starting in the sixties many works were done to solve numerically the Vlasov equation (Vlasov 1945) on a computer. As examples we mention only a very few of them. Firstly consider the transform methods as presented by Armstrong et al. (1970). These methods are based on direct solutions of the Vlasov equation by very accurate representation of the function by orthogonal polynomials (Fourier series, Hermite polynomials) and calculation of the derivatives by use of the polynomials' derivation properties. Globally the results are very limited, the number of required terms increasing tremendously in time. Finite difference methods (Killeen and Rompel 1966; Telegin 1976) are also used to approximate the phase space derivatives and again the results are quite poor. The time splitting method (Yanenko 1966) first introduced into plasma physics (Cheng and Knorr 1976) by Cheng and Knorr transforms the Vlasov equation to a set of first order differential equations which can be solved formally in some cases or transformed to integral form in others. In the present work we present the general symmetric time splitting scheme as used in our codes, introduce the smoothing scheme of ENO type (Harten et al. 1987) and compare results of the method we call Flux Balance with a formal solution method using interpolation by splines (Shoucri and Gagné 1978; Ghizzo et al. 1993), in one dimension. At that stage we present the problems of generalisation of these methods to the relativistic (and the gyrokinetic) case. The necessary condition to allow the generalisation is given. A two dimensional result, obtained with the flux balance method, which presents a generalisation of the one dimensional case is shown at the end of this paper.

2 The Splitting Scheme

The basic idea of time splitting is as follows.

Given a differential equation of the form

$$\frac{\partial F}{\partial t} + LF = 0; \qquad F(t=0) = F_0, \tag{1}$$

where L is a general differential operator of the form

$$L = \sum_{j=1}^{neqs} L_j \tag{2}$$

and the summation is over the number of equations ($neqs$), the differential equation (1) can be solved by the scheme

$$\begin{aligned} &\frac{\partial F_1}{\partial t} + L_1 F_1 = 0; \; F_1(t=0) = F_0 \\ &\frac{\partial F_2}{\partial t} + L_2 F = 0; \; F_2(t=0) = F_1(t=\Delta t) \\ &\dots \; \dots \\ &\dots \; \dots \\ &\frac{\partial F_{neqs}}{\partial t} + L_{neqs} F_{neqs} = 0; \; F_{neqs}(t=0) = F_{neqs-1}(t=\Delta t). \end{aligned} \tag{3}$$

In this case the solution at $t = \Delta t$ is $F(t = \Delta t) = F_{neqs}(\Delta t)$. In order for the solution not to increase the weight of one of the equations (3) we can take the operator L to be symmetric (as in the Leapfrog method). This choice makes the method at least of second order in time.

Let us now see how this method is used in case of Vlasov equation. In one dimension the equation is

$$\frac{\partial f(t,x,v)}{\partial t} + v\frac{\partial f}{\partial x} + E(t,x)\frac{\partial f}{\partial v} = 0 \tag{4}$$

so that the differential operator is

$$L = v\frac{\partial}{\partial x} + E\frac{\partial}{\partial v} + E\frac{\partial}{\partial v} + v\frac{\partial}{\partial x} \tag{5}$$

and the system to be solved is given by

$$\begin{aligned} &\frac{\partial f_1}{\partial t} + v\frac{\partial f_1}{\partial x} = 0 \qquad f_1(t=t_0) = f(t_0) && (a) \\ &\frac{\partial f_2}{\partial t} + E\frac{\partial f_2}{\partial v} = 0 \qquad f_2(t=t_0) = f_1(t=t_0+\Delta t/2) && (b) \\ &\frac{\partial f_3}{\partial t} + E\frac{\partial f_3}{\partial v} = 0 \qquad f_3(t=t_0+\Delta t/2) = f_2(t=t_0+\Delta t/2) && (c) \\ &\frac{\partial f_4}{\partial t} + v\frac{\partial f_4}{\partial x} = 0 \qquad f_4(t=t_0+\Delta t/2) = f_3(t=t_0+\Delta t) && (d) \\ &\qquad f(t_0+\Delta t) := f(t=t_0+\Delta t) && (e) \end{aligned} \tag{6}$$

In two dimensions the equation is (with a constant external magnetic field)

$$\frac{\partial f}{\partial t}(t,x,y,v_x,v_y)+v\frac{\partial f}{\partial x}+v_y\frac{\partial f}{\partial y}+(E_x-v_yB_z)\frac{\partial f}{\partial v_x}+(E_y+v_xB_z)\frac{\partial f}{\partial v_y}=0 \quad (7)$$

and the differential operator is now

$$L = v_x\frac{\partial}{\partial x} + v_y\frac{\partial}{\partial y} + (E_x - v_yB_z)\frac{\partial}{\partial v_x} + (E_y + v_xB_z)\frac{\partial}{\partial v_y} + \\ (E_y + v_xB_z)\frac{\partial}{\partial v_y} + (E_x - v_yB_z)\frac{\partial}{\partial v_x} + v_y\frac{\partial}{\partial y} + v_x\frac{\partial}{\partial x}. \quad (8)$$

Actually equation (7) splits into eight equations of the same type as those in (6). Two methods can now be used to solve any of (6) or those following from expanding the operator (8). They are formal solution and interpolation or flux balance.

3 The Flux Balance Method

As $E(t,x)$ is given by the Poisson equation

$$\frac{\partial E}{\partial x} = \int_{-\infty}^{+\infty} f(t,x,v)\mathrm{d}v - N_0 \quad (9)$$

and since (6b) and (6c) are shifting $f(t,x,v)$ in v only, and not in x, the electric field does not vary in stages like them. This allows us to solve (6) or the equation following from (8), which are of the general form

$$\frac{\partial f}{\partial t} + a\frac{\partial f}{\partial x} = 0, \quad (10)$$

either by the method of characteristics (Cheng and Knorr 1976) or by a method of flux balance. The formal solution of (10) along its characteristics is obviously $f(t+\Delta t,x) = f(t,x-a\Delta t)$. When solved on a grid, the function $f(t+\Delta t,x)$ is to be calculated by an interpolation method from the previously known values of $f(t,x)$ on the grid. Two methods are mainly used: Fourier interpolation (Izrar et al. 1989; Klimas and Farrel 1994) or splines (Killeen and Rompel 1966; Ghizzo et al. 1990). The Fourier method is simple to implement with the use of a fast Fourier transform (FFT) routine, but is very space consuming in nonperiodic space (as the velocity space) to avoid spurious solutions due to aliasing.

To introduce the flux balance method we write equation (10) in the equivalent form, since a does not depend on x in the case of the Vlasov Electrostatic equation,

$$\frac{\partial f(t,x)}{\partial t} + \frac{\partial}{\partial x}(af(t,x)) = 0. \tag{11}$$

As (11) is of the form of a fluid continuity equation, we can use now the methods of fluid dynamics. Following Harten (1987) we look for a solution to (11) on a grid, assuming the function to be smooth in each elementary cell and replacing $f(t,x)$ by its smoothed approximation

$$\bar{f}(x,t) = \frac{1}{\Delta x}\int_{-\Delta x/2}^{\Delta x/2} f(x+z)\mathrm{d}z. \tag{12}$$

The introduction of (12) into (11) gives for $f(t,x)$ a new equation of the same smoothness as the original one:

$$\frac{\partial f}{\partial t}(t,x) + \frac{1}{\Delta x}[af(t,x+\Delta x/2) - af(t,x-\Delta x/2)] = 0. \tag{13}$$

Integrating this equation over a step Δt we obtain

$$f(t+\Delta t,x) = f(t,x) - \frac{\Delta t}{\Delta x}\int_t^{t+\Delta t} [af(t,x+\Delta x/2) - af(t,x-\Delta x/2)]\mathrm{d}t. \tag{14}$$

Until now we have supposed a to be a constant. In the cases of the relativistic or gyrokinetic Vlasov equation, it is a function of x, t or both, *ie* $a(t,x)$.

In such a case (14) takes the form

$$\begin{aligned} f(t+\Delta t) = f(t,x) - \frac{\Delta t}{\Delta x}\int_t^{t+\Delta t} & [a(t,x+\Delta x/2)f(t,x+\Delta x/2) \\ & -a(t,x-\Delta x/2)f(t,x-\Delta x/2)]\,\mathrm{d}t. \end{aligned} \tag{15}$$

To resolve the terms under the integral we seek a function $X(t_0,t,x)$ such that

$$f(t,x) = f(X(t_0,t,x)\frac{\partial X}{\partial x}(t_0,t,x) \tag{16}$$

and which is a solution of the PDE

$$\frac{\partial X}{\partial t}(t_0,t,x) + a(t,x)\frac{\partial X}{\partial x}(t_0,t,x) = 0 \tag{17}$$

with boundary conditions

$$\begin{aligned} &\frac{\partial X}{\partial t_0}(t_0,t,x) = -a(t_0,t,x) \\ &X(t_0,t_0,x) = x. \end{aligned} \tag{18}$$

Now to simplify we note that we wish to transform integral terms over time, $\int_t^{t+\Delta t} a(t,x)f(t,x)dt$, to integrals over space. We start by introducing (16) into the integral so that

$$\int_t^{t+\Delta t} a(t,x)f(t,x)\mathrm{d}t = \int_t^{t+\Delta t} a(t,x)f(X(to,t,x))\frac{\partial X}{\partial x}\mathrm{d}t. \tag{19}$$

With the use of (17) this term transforms to

$$\int_t^{t+\Delta t} a(t,x)f(t,x)\mathrm{d}t = -\int_t^{t+\Delta t} f(X)\frac{\partial X}{\partial t}\mathrm{d}t.$$

Then after the substitution $y = X$ and remembering that at the beginning of the step $X(t,t,x) = x$

$$\int_t^{t+\Delta t} a(t,x)f(t,x)\mathrm{d}t = -\int_x^{X(t,t+\Delta t)} f(y)\mathrm{d}y. \tag{20}$$

Solving $X(t_0,t,x)$ for a step in time we obtain, starting from (18)

$$\begin{aligned} &X(t,t+\Delta t,x) = x - a(t,x)\Delta t \text{ to first order in } \Delta t \\ &X(t,t+\Delta t,x) = x - a(t+\Delta t, x - a(t,x))\Delta t)\Delta t \text{ to second order.} \end{aligned} \tag{21}$$

Equation (15) now becomes (since for positive $a(t,x)$ $x > X$)

$$\begin{aligned} f(t+\Delta t,x) = f(t,x) - \frac{\Delta t}{\Delta x}\Bigg\{ &\int_{X(t,t+\Delta t,x+\Delta x/2)}^{x+\Delta x/2} [f(t,y+\frac{\Delta x}{2})\mathrm{d}y \\ &- \int_{X(t,t+\Delta t,x-\Delta x/2)}^{x-\Delta x/2} f(t,y-\frac{\Delta x}{2})]\mathrm{d}y\Bigg\}. \end{aligned} \tag{22}$$

The simplest way to represent the terms under the integral sign is by a linear approximation.

In the case of the Vlasov electrostatic equation $a(t,x)$ is a constant so that $X(t,t+\Delta t,x) = x - a\Delta t$.

4 The Vlasov Relativistic Equation

The general Vlasov relativistic equation in 2D is

$$\frac{\partial f}{\partial t}(t,\mathbf{x},\mathbf{p}) + \frac{\mathbf{p}}{\sqrt{1+p^2}}.\nabla_x f(t,\mathbf{x},\mathbf{p}) + \left(\mathbf{E}(t,\mathbf{x}) + \frac{\mathbf{p}\wedge\mathbf{B}}{\sqrt{1+p^2}}\right).\nabla_p f = 0 \tag{23}$$

with $p^2 = p_x^2 + p_y^2$. As $\mathbf{p}$ is not a function of $\mathbf{x}$ we can obviously write

$$\frac{\mathbf{p}}{\sqrt{1+p^i}}.\nabla_x f = \nabla_x(\mathbf{p}f(t,\mathbf{x},\mathbf{p})).$$

Again, as $\mathbf{E}(t,\mathbf{x})$ does not depend upon $\mathbf{p}$, $\mathbf{E}.\nabla_p f = \nabla_p.(\mathbf{E}f)$. The term $\frac{\mathbf{p}\times\mathbf{B}}{\sqrt{1+p^2}}.\nabla_p f$ gives $\nabla.\left(\frac{\mathbf{p}\times\mathbf{B}}{\sqrt{1+p^2}}f - f\nabla_p.\frac{\mathbf{p}\times\mathbf{B}}{\sqrt{1+p^2}}\right)$ and as the rightmost term is null we can rewrite (23) in the form

$$\begin{aligned}&\frac{\partial f}{\partial t}+\frac{\partial}{\partial x}\frac{p_x}{\sqrt{1+p^2}}f+\frac{\partial}{\partial y}\frac{p_y}{\sqrt{1+p^2}}f\\&+\frac{\partial}{\partial p_x}\left\{E_x+\frac{p_yB_z}{\sqrt{1+p^2}}\right\}f+\frac{\partial}{\partial p_y}\left\{E_y-\frac{p_xB_z}{\sqrt{1+p^2}}\right\}f=0.\end{aligned} \tag{24}$$

In the present case the split equations are

$$\begin{aligned}\frac{\partial f}{\partial t}+\frac{\partial}{\partial x}\left(\frac{p_x}{\sqrt{1+p^2}}f\right)&=0 \ (a)\\ \frac{\partial f}{\partial t}+\frac{\partial}{\partial y}\left(\frac{p_y}{\sqrt{1+p^2}}f\right)&=0 \ (b)\\ \frac{\partial f}{\partial t}+\frac{\partial}{\partial p_x}\left\{\left(E_x+\frac{p_yB_z}{\sqrt{1+p^2}}\right)f\right\}&=0 \ (c)\\ \frac{\partial f}{\partial t}+\frac{\partial}{\partial p_y}\left\{\left(E_y-\frac{p_xB_z}{\sqrt{1+p^2}}\right)f\right\}&=0. \ (d)\end{aligned} \tag{25}$$

Now for (25 c, d) the coefficient $a(t,x)$ depends on the variable so that the correction terms given by (21) have to be take into account.

5 Numerical Results

5.1 One Dimensional Comparison

In the following we compare a 1D simulation, which was carried out with a spline code, to the same simulation realised with a Flux Balance code. The initial conditions are a four BGK equilibrium given by

$$f(\epsilon)=\frac{\mu}{\sqrt{2\pi}}\frac{2-2\xi}{3-2\xi}\left(1+\frac{\epsilon}{1-\xi}\right)e^{-\epsilon}. \tag{26}$$

The energy ϵ is given by $\xi = v^2/2+\phi(x)$ with the initial potential

$$\frac{d^2\phi}{dx^2}+\mu\left(\frac{3-2\xi+2\phi}{3-2\xi}\right)e^{-\phi}-1=0. \tag{27}$$

Equation (27) is solved with the boundary conditions $\phi(0) = 0$, $\left.\frac{d\phi}{dx}\right|_{x=0}$ until $\phi(x)$ is null again.

As the BGK equilibria are unstable we perturbed the energy dependent function (26) slightly to see the evolution. The initial condition (Fig. 1) is given by

$$f(t = 0, x, v) = f(\epsilon)(1 + \alpha \cos k.x) \tag{28}$$

with $\alpha = 0.0001$.

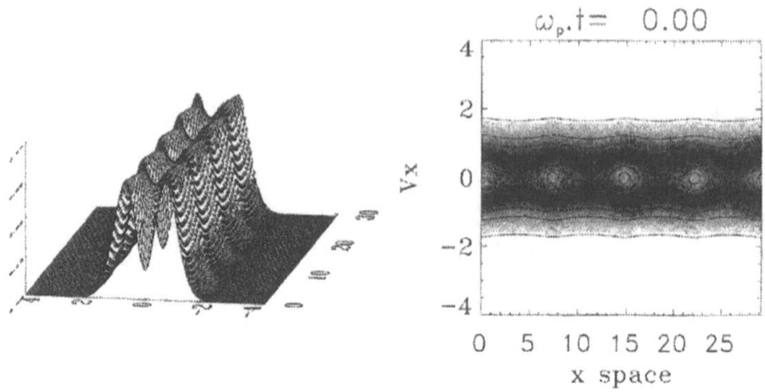

Fig. 1. The BGK initial condition

The time evolution for the two codes is given in Fig. 2. Two time values are given, viz. $t = 150\omega_p^{-1}$ and $t = 210\omega_p^{-1}$. Since the distribution function is given by a cruder representation in the cell in the Flux Balance method, the evolution starts faster for that method (upper part of Fig. 2) at $t = 150\omega_p^{-1}$). At a later time the results are practically the same for the two methods (Fig. 2, $t = 210\omega_p^{-1}$).

5.2 Two Dimensional Simulation

We start the following simulation from a two stream situation, both in x and y, with a perturbation term:

$$\begin{aligned} f(t = 0, x, y, v_x, v_y) = \frac{1}{2\pi} & \exp\left[-((|v_x| - 2)^2 + (|v_y| - 2)^2)/2\right] \\ & \times (1 + 5(v_x^2 + v_y^2))(1 + \text{pert}) \end{aligned} \tag{29}$$

$$\text{pert} = 0.02[\cos(k_0 x)/50 + \cos 2k_0 x + \cos(k_0 y)/50 + \cos 3k_0 y]. \tag{30}$$

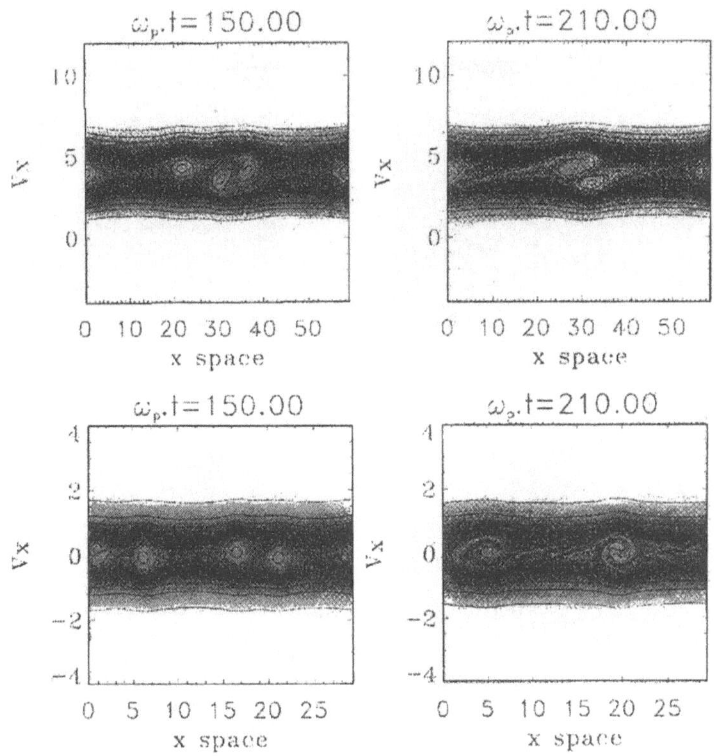

Fig. 2. Time evolution

With these initial conditions the double stream evolves to a two hole BGK equilibrium in x space and a three hole equilibrium in y space (Fig. 3). With time the holes coalesce, directly to one hole in the x direction, to 2 holes in y (Fig. 4) and at the end (Fig. 5) only one hole remains in x and y, exactly as what was found in 1D simulations (Fig. 2).

References

Armstrong, T.P., Harding, R.C., Knorr, G., Montgomery, D. (1970): Solution of Vlasov's Equation by Transform Methods in *Methods in computational Physics* Vol 9, Adler, B., Fernbach, S., Rotenber, M. eds. (Academic Press, New York)

Cheng, C.Z., Knorr, G. (1976): The Integration of Vlasov Equation in Configuration Space. J. Comp. Phys **22**, 330–351

Gagné, R., Shoucri, M. (1977): A Splitting Scheme for the Numerical Solution of the One-Dimensional Vlasov Equation. J. Comp. Phys **24**, 445–449

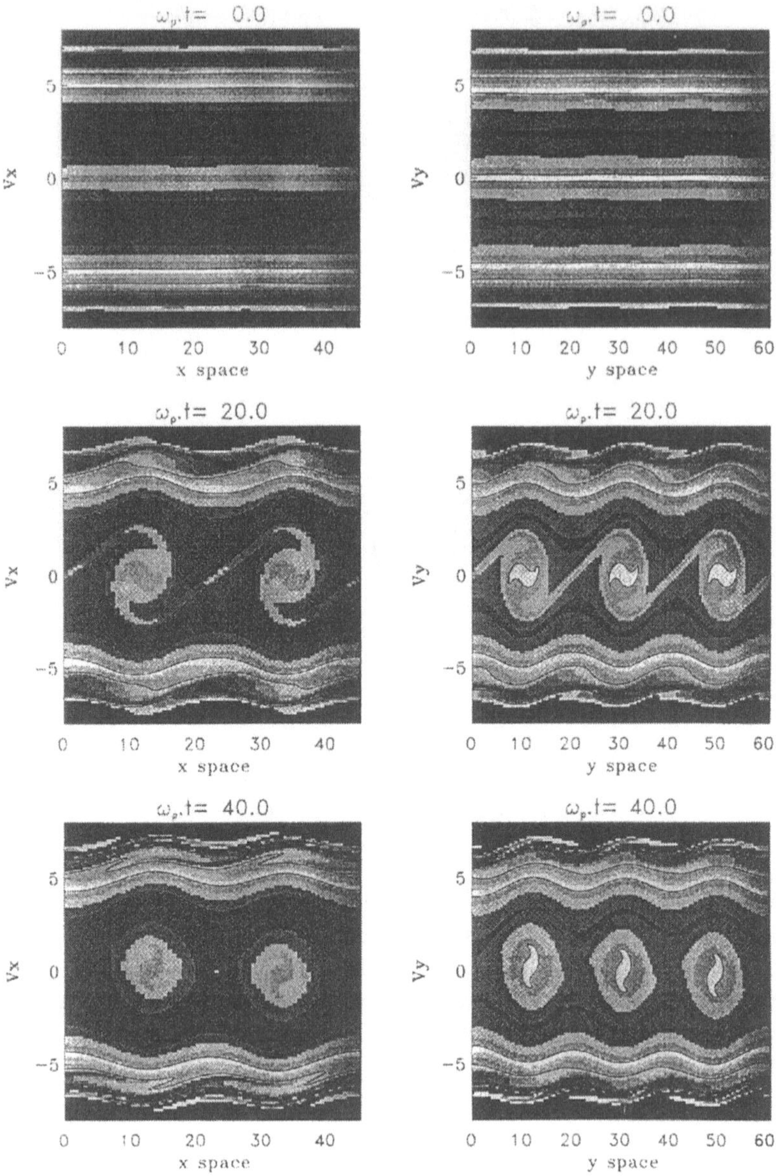

Fig. 3. The double stream evolves into a two hole equilibrium in x space and a three hole equilibrium in y space.

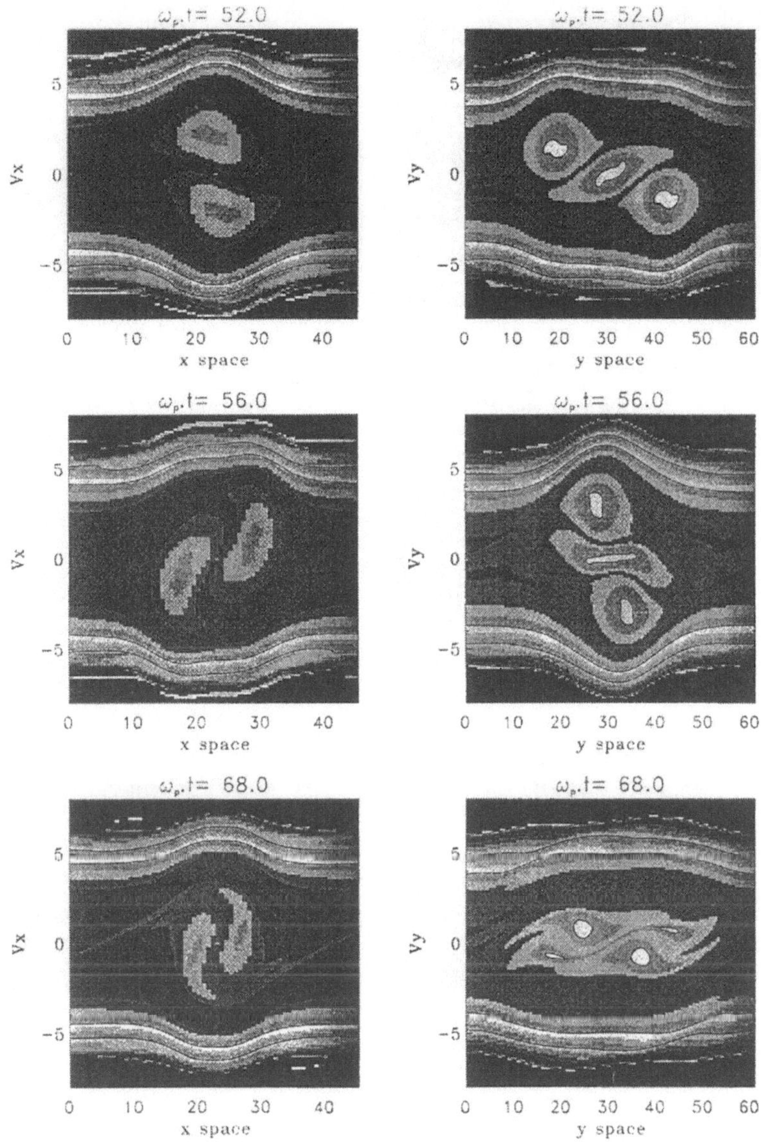

Fig. 4. The three holes on the y direction presented Fig. 3 have coalesced into two holes.

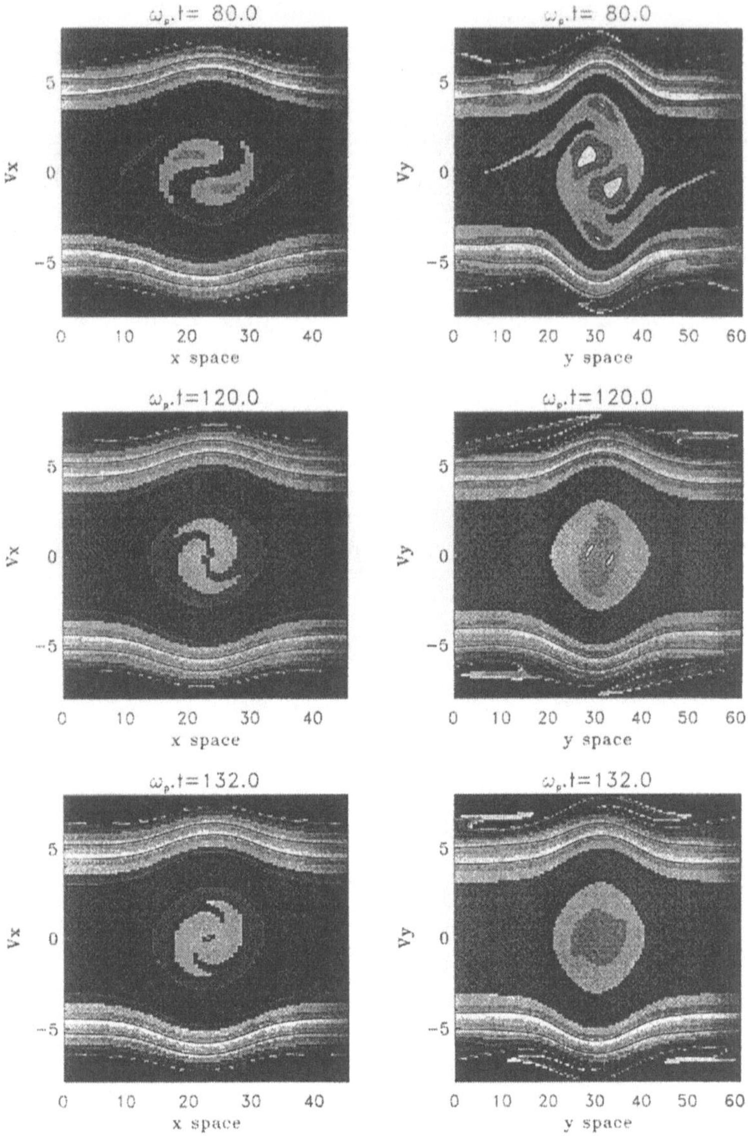

Fig. 5. The three holes on the y direction presented Fig. 3 have coalesced into a single one.

Ghizzo, A., Bertrand, P., Shoucri, M., Fijalkow, E., Feix, M.R (1993): An Eulerian Code for the Study of the Drift-Kinetic Vlasov Equation. J. Comp. Phys. **108**, 105–121

Ghizzo, A., Bertrand, P., Shoucri, M., Johnston, T.W., Fijalkow, E., Feix, M.R. (1990): A Vlasov Code for the Numerical Simulation of Stimulated Raman Scattering. J. Comp. Phys. **90**, 431–457

Harten, A., Engquist, B., Osher, S., Chakravarthy, S.R. (1987): Uniformally High Order Accurate Essentially Non-Oscillatory Schemes III. J. Comp. Phys. **71**, 231–303

Izrar, B., Ghizzo, A., Bertrand, P., Fijalkow, E., Feix, M.R. (1989): Integration of Vlasov Equation by a Fast Fourier Eulerian Code. Comp. Phys. Comm. **52**, 375–382

Killeen, J., Rompel, S.L. (1966): A Computation for Studying the Formation of a Relativistic Electron Layer in Astron. J. Comp. Phys. **1** (1), 29–50

Klimas, A.J., Farrel, W.M. (1994): A Splitting Algorithm for Vlasov Simulation with Filamentation Filtration. J. Comp. Phys. **110**, 150–163

Shoucri, M., Gagné, R. (1978): Splitting Schemes for the Numerical Solution of the Two Dimensional Vlasov Equation. J. Comp. Phys. **27**, 315–322

Telegin, V.I. (1976): A Different Scheme for Vlasov's Equation. Zh. vychisl. Math. mat. Fiz. **16** (5), 1191–1197

Vlasov, A.A. (1945): On the Oscillation properties of an Electron Gas. J. Phys. (USSR) 9, 25–41

Yanenko, N.N. (1971): *The Method of Fractional Steps* (Springer Verlag, New-York)

A Limiting Case of Averaging Compactness Lemmas Motivated by the Kinetic Formulation of Some Classical Systems of Fluid Mechanics

Benoit Perthame[1] and Panagiotis E. Souganidis[2]

[1] Laboratoire d'analyse Numérique, Université P. et M. Curie, 4 Place Jussieu, F-75252 Paris Cedex 05, France et Institut Universitaire de France

[2] Department of Mathematics, University of Wisconsin-Madison, 480 Lincoln Drive, Madison, WI 53706, USA

Abstract. We consider transport equations with a second member which is a full space derivative. This structure coresponds to numerous models of fluid dynamics and to the kinetic formulation of isentropic gas dynamics and, for the case of scalar conservation laws, averaging lemmas prove regularizing effects. We extend the averaging lemmas to second members which are a full space derivative. Our method uses Calderón-Zygmund theory.

1 Introduction

Models of fluid dynamics often lead to transport equations which advection is lead by a combination of the macroscopic velocity ($u(t,x)$ in the sequel, see (3)) and of the kinetic velocity (denoted v). For instance the dynamics of hard rods leads to the equation (see Spohn 1991)

$$\frac{\partial}{\partial t}f + \frac{\partial}{\partial x}\left[v + \frac{a\rho}{1-a\rho}(v-u)\right] f = 0,\ t \geq 0,\ x \in \mathbb{R},\ v \in \mathbb{R}, \tag{1}$$

where a is a positive parameter and we have set

$$\rho(t,x) = \int f(t,x,v)dv, \tag{2}$$

$$\rho u(t,x) = \int v f(t,x,v)dv. \tag{3}$$

Also the dynamics of balls in a potential flow leads to the coupled system

$$\frac{\partial}{\partial t}f + \mathrm{div}_x(H_v f) - \mathrm{div}(H_x f) = 0,\ t \geq 0,\ x \in \mathbb{R}^3,\ v \in \mathbb{R}^3, \tag{4}$$

$$\begin{cases} H(t,x,v) = \frac{1}{2}|v + a\nabla_x\varphi|^2, \\ -\mathrm{div}_x((1+a\rho)\nabla_x\varphi) = \mathrm{div}_x(\rho u),\ t \geq 0,\ x \in \mathbb{R}^3, \end{cases} \tag{5}$$

where we have used the same notations as before (see Russo and Smereka 1996; Herero et al. 1997). Finally a last example is the Hamiltonian formulation of incompressible Euler equations. Then the equation (4) is determined by the Hamiltonian

$$\begin{cases} H(t,x,v) = v \cdot q(t,x), \\ q = \rho u - \nabla\varphi, \\ \Delta_x \varphi = \mathrm{div}_x(\rho u). \end{cases} \tag{6}$$

One checks that the function $q(t,x)$ then satisfies

$$\begin{cases} \mathrm{div}_x q = 0, \\ \partial_t q + \mathrm{div}_x q \otimes q + \nabla_x p = 0, \\ p = \partial_t \varphi + \frac{|q|^2}{2} + q \cdot \nabla_x \varphi \end{cases} \tag{7}$$

(incompressible Euler system). This formalism has been introduced by Kuz'min (1983), Oseledets (1988), and developed by Buttke (1993), Maddocks (1995), Benedetto (1986), Benedetto (1997) and Smereka (1996).

Here we consider a simple structure of equations related to the previous examples. Let the transport equation with a singular second member be

$$\frac{\partial}{\partial t} f + v \cdot \nabla_x f = h \tag{8}$$

$$h(t,x,v) = \sum_{j=1}^{d} \frac{\partial}{\partial x_j} \frac{\partial^k}{\partial v^k} g_j, \tag{9}$$

where k is a multi-index $(k_1, ..., k_d)$ and $\partial_v^k = \partial_{v_1}^{k_1} ... \partial_{v_j}^{k_d}$, $t \geq 0$, $x \in \mathbb{R}^d$, $v \in \mathbb{R}^d$. In a recent work (Perthame and Souganidis 1997) we prove that the averaging compactness lemmas hold for this equation. These results are given in Sect. 2. More precise motivations for the kinetic structure of systems of conservation laws are given in Sect. 3.

2 Averaging Compactness Lemma

The averaging lemmas show that the integrated quantities

$$\overline{\rho}(t,x) = \int f(t,x,v) \varphi(v) dv \tag{10}$$

are more regular than the solutions f of the transport transport itself (for instance (8)).

This phenomenon was discovered by Agoshkov (1984) and Golse, Perthame and Sentis (1985). When the righthand side of the transport equation does not contain derivatives, their final form was given by Golse, Lions, Perthame and Sentis (1988). The case of v derivatives was then treated by R.J. DiPerna and P.L. Lions (1989) (L^2 framework), then by DiPerna, Lions and Meyer

(1991) (L^p framework and fractional derivatives of order less than one in x) and the optimality is discussed in Lions (1995). Also Bézard (1994) obtains similar results to DiPerna et al. (1991) in Sobolev spaces (in place of Besov). Finally Bouchut and Desvillettes (1997) give a result which is local in time and applies to fractional step methods.

A limiting case of velocity averaging was demonstrated by Gérard (1990) who shows that $\overline{\rho}$ is compact in L^2 if the second member, h, of the transport equation (8) is compact in H^{-1}. This is the closest result to ours and we make it more precise.

In Perthame and Souganidis (1997) we show that we can still obtain an estimate in the case (9) of a full x derivative.

Theorem. *Let two functions f, g satisfy (8), (9). For all p, $1 < p < \infty$, $\alpha \in [0, \min(\frac{1}{p}, \frac{1}{p'})[$, and $\varphi \in C^{|k|}(B_R)$, where B_R is the ball of centre 0 and radius R in $\mathbb{R}^d$ and $|k| = k_1 + ... + k_d$, we have*

$$||\overline{\rho}||_{L^p(\mathbb{R}^{d+1})} \leq C||f||_{L^p(Q_R)}^{1-\frac{\alpha}{k+1}} ||g||_{L^p(Q_R)}^{\frac{\alpha}{k+1}}, \tag{11}$$

where we have set $Q_R = \mathbb{R}_x^d \times \mathbb{R}_t \times B_R$, $C = C(d, R, \alpha, p, ||\varphi||_{W^{|k|,\infty}})$.

The way to use this result is given in the following corollary which allows a general framework for compactness:

Corollary. *With the assumptions of the above theorem, if g_n is a compact sequence of $L^p(Q_R)$ and f_n is a bounded sequence of $L^p(Q_R)$, then $\overline{\rho}_n$ is compact in $L^p(\mathbb{R}^{d+1})$.*

One can see in this corollary that the inequality (11) is much more interesting for "g_j small" than the elementary inequality

$$\begin{aligned} ||\overline{\rho}||_{L^p(\mathbb{R}^{d+1})} &\leq \int |\varphi(v)| \, ||f||_{L^p(\mathbb{R}^{d+1})} dv \\ &\leq ||f||_{L^p(Q_R)} ||\varphi||_{L^{p'}(Q_R)}. \end{aligned} \tag{12}$$

Note also that this result extends without difficulty to more general velocities as in Lions, Perthame and Tadmor (1991, 1994) (i.e. $a(v)$ and not v).

Our approach contains a number of variants compared to the classical methods, quoted above for averaging lemmas. We do not introduce cut-off in v to evaluate the integral (10). More effectively we introduce a small perturbation of the equation (8). Also the small parameter for perturbation is not only the Fourier variable, but also a L^p norm. Finally this result can be proved by a combination of the Calderón-Zygmund theory (see Stein 1970, for instance) and of the classical averaging technique in L^2.

3 Compactness in Some Nonlinear Hyperbolic System

One of the motivations for studying the averaging lemmas of Sect. 2 comes from the kinetic formulation of hyperbolic systems with many entropies. We here recall the multidimensional scalar case where averaging lemmas apply and then the case of isentropic gas dynamics where the only compactness argument remains compensated compactness (except for $\gamma = 3$ and the systems of Brenier and Corrias (1996), see Lions et al. (1991)). We refer to Serre (1996) for a general presentation of hyperbolic systems.

We consider a scalar conservation law

$$\frac{\partial u}{\partial t} + \sum_{i=1}^{d} \frac{\partial}{\partial x_i} A_i(u) = 0, \tag{13}$$

with all the entropy inequalities

$$\partial_t S(u) + \sum_{i=1}^{d} \frac{\partial}{\partial x_i} \eta_i(u) \leq 0, \tag{14}$$

for $S(\cdot)$ convex, $\eta_i'(\cdot) = a_i(\cdot) S'(\cdot)$ and we have set

$$a_i = A_i' \in C^1(\mathbb{R}), \quad 1 \leq i \leq d. \tag{15}$$

It is proved in Lions et al. (1991) (and the references therein) that, for an initial data $u^0 \in L^2(\mathbb{R}^d)$, this problem can equivalently be written

$$\frac{\partial \chi}{\partial t} + a(v) \cdot \nabla_x \chi = \frac{\partial m}{\partial v}, \quad t \geq 0,\ x \in \mathbb{R}^d, v \in \mathbb{R}, \tag{16}$$

$$\chi = \chi(v, u(t,x)) = \begin{cases} +1 & \text{for } 0 \leq v \leq u, \\ -1 & \text{for } u \leq v \leq 0, \\ 0 & \text{if not} \end{cases} \tag{17}$$

and

$$m(t,x,v) \text{ is a bounded nonnegative measure.} \tag{18}$$

Using next the local aspect of the hyperbolic equation (13), the fact that the meausres are compactly embedded in negative Sobolev spaces and the averaging compactness lemmas, we obtain that, for all sequences of initial data u_n^0 bounded in $L^1 \cap L^\infty(\mathbb{R}^d)$, the sequence of solutions $u_n(t,x)$ is locally compact in $L^1(\mathbb{R}_t^+ \times \mathbb{R}_x^d)$, under the only nondegenerating condition

$$mes\{v \in \mathbb{R}, |a(v) \cdot \xi - \tau| \leq \varepsilon\} \leq \omega(\varepsilon), \tag{19}$$

for all $(\xi, \tau) \in \mathbb{R}^{d+1}$ with $|\xi|^2 + |\tau|^2 \leq 1$. Here $\omega(\cdot)$ denotes a continuous function such that $\omega(0) = 0$. This generalises a famous earlier result of Tartar (1979) in one dimension.

Now consider the case of isentropic gas dynamics

$$\begin{cases} \partial_t \rho + \partial_x \rho u = 0, \\ \partial_t \rho u = +\partial_x(\rho u^2 + p(\rho)) = 0, \quad t \geq 0,\ x \in \mathbb{R}, \\ p(\rho) = K\rho^\gamma, \quad 1 < \gamma < +\infty,\ K = \frac{(\gamma-1)^2}{4\gamma}. \end{cases} \tag{20}$$

With all the entropy conditions for the so-called *weak entropies* (those compatible with vacuum $\rho = 0$) one can show (Lions et al. 1994) that this problem admits the kinetic formulation

$$\frac{\partial}{\partial t}\chi + \frac{\partial}{\partial x}[\theta v + (1-\theta)u]\chi = \frac{\partial^2 m}{\partial v^2}, \tag{21}$$

$$\begin{cases} \theta = \frac{\gamma-1}{2}, \\ \chi = (\rho^{\gamma-1} - (v-u)^2)_+^\lambda, \quad \gamma = \frac{3-\gamma}{2(\gamma-1)}, \\ m(t,x,v) \text{ is a non positive bounded measure.} \end{cases} \tag{22}$$

These equations, as those of the introduction, display a structure in which the advection is a combination of the microscopic velocity v and of macroscopic velocity $u(t,x)$. Except for the purely kinetic case $\gamma = 3$, one cannot apply the averaging compactness lemmas here in order to prove regularity or compactness (partly because of the term in space derivatives of $u\chi$, but also because for $\gamma = 3$, the function χ has very low regularity).

The only issue for proving compactness of families of solutions to the system (20), and then existence follows, consists in using the compensated compactness arguments of Murat (1978) and Tartar (1979) as Di Perna (1983) did originally. A complete proof of compactness and existence, for all values of γ, can be found in Lions et al. (1996) and uses the formula (21).

References

Agoshkov, V. I. (1984): Spaces of functions with differential-difference characteristics and smoothness of solutions of the transport equation. Soviet Math. Dokl. **29**, 662–666

Buttke, T. F. (1993): Velicity methods: Lagrangian numerical methods which preserve the Hamiltonian structure of incompressible fluid flows in *Vortex flow and related numerical methods*, Beale, J. T., Cottet, G. H., Huberson S. eds.(Kluwer, Dordrecht)

Benedetto, D. (1986): Convergence of velicity blobs method for the Euler Equation. Math. Methods Appl. Sc. **19**, 463–479

Benedetto, D.(1997): A remark on the Hamiltonian structure of the incompresible flows. J. stat. Phys. **79** (to appear)

Bézard, M. (1994): Régularité précisée des moyennes dans les équations de transport. Bull. Soc. Math. France **22**, 29–76

Bouchut, F., Desvillettes, L. (1997): Work in preparation

Brenier, Y., Corrias, L. (1996): A kinetic formulation for multibranch entropy solutions of scalar conservation laws. Ann. Inst. H. Poincaré Analyse non-linéaire

Di Perna, R. (1983): Convergence of the viscosity method for Isentropic Gas Dynamics. Comm. Math. Phys. **91**, 1–30

Di Perna, R.J., Lions, P.-L. (1989): Global weak solutions of Vlasov-Maxwell systems. Comm. Pure Appl. Math. **42**, 729–757

Di Perna, R.J., Lions, P.-L., Meyer, Y. (1991): L^p regularity of velocity averages. Ann. Inst. H. Poincaré Ann. Non Linéaire **8**, 271–287

Gérard, P. (1990): Moyennisation et régularité deux-microlocale. Ann. Sci. Ecole Normale Sup. **4**, 89–121

Golse, F., Perthame, B., Sentis, R. (1985): Un résultat de compacité pour les équations du transport. C. R. Acad. Sci. Paris *Série I 301*, 341–344

Golse, F., Lions, P.-L., Perthame, B., Sentis, R. (1988): Regularity of the moments of the solution of a transport equation. J. Funct. Anal. **76**, 110–125

Herero, H., Lucquin, B., Perthame, B.(1997): On the motion of dispersed balls in a potential flow. (Raport LAN 97021, Université Paris 6)

Kuz'min, G.A. (1983): Ideal incompressible hydrodynamics in terms of the vortex momentum density. Phys. Lett. **96A**, 88–90

Lions, P.-L. (1995): Régularité optimale des moyennes en vitesses. C. R. Acad. Sci. Paris, t.320 Série I, 911–915

Lions, P.-L., Perthame, B., Souganidis, P.E. (1996): Existence of entropy solutions to isentropic gas dynamics system. C.P.A.M. **49**, 599-638

Lions, P.-L., Perthame, B., Tadmor, E. (1994): A kinetic formulation of multidimensional scalar conservation laws and related equations. J. Amer. Math. Soc. **7**, 169–191

Lions, P.-L., Perthame, B., Tadmor, E. (1991): Formulation cinétique des lois de conservation scalaires. C.R. Acad. Sc. Paris, t.312 Série I, 97–102

Maddocks, J.H., Pego, R.L. (1995): An unconstrained Hamiltonian formulation for incompressible fluid flow. Comm. Math. Phys. **170**, 207–217

Murat, F. (1978): Compacité par compensation. Ann. Sc. Norm. Sup. Pisa **5**, 489–507

Oseledets, V.I. (1989): On a new way of writing the Navier-Stokes equation: the Hamiltonian formalism, Comm. Moscow Math. Soc. (1988), translated in Russ. Math. Surveys **44**, 210–211

Perthame, B., Souganidis, P.E. (1987): A limiting case of averaging lemmas. (Rapport LAN, Université Paris 6)

Russo, G., Smereka, P. (1996): Kinetic theory for bubbly flow I : collisionless case. SIAM J. Appl. Math. **562**, 327–357

Serre, D (1996): *Systemes hyperboliques de lois de conservation, Partics I et II.* (Diderot, Paris)

Smereka, P., Vlasov, A. (1996): description of the Euler equation. Nonlinearity **9**, 1361–1386

Spohn, H. (1991): *Large scale dynamics of interacting particles* (Springer-Verlag, Berlin)

Stein, E.M. (1970): *Singular Integrals and Differentiability Properties of Functions* (Princeton University Press, Princeton)

Tartar, L. (1979): Compensated compactness and applications to partial differential equations in *Research Notes in Mathematics, Nonlinear Analysis and Mechanics*, Herriot-Watt Symposium Vol. 4, Knops, R.J. ed. (Pitman, London)

Expansion of a Negative Ion Plasma into Vacuum

Habiba Figua[1], Léonardo G. Garcia[2], Eric Fijalkow[1], and Joao Goedert[2]

[1] Laboratoire de Mathématique, Applications et Physique Mathématique - UMR 6628, Université d'Orléans, UFR des Sciences, BP 6759, F-45067 Orléans Cedex 2, France

[2] Instituto de Física-UFRGS, Caixa Postal 15051, 91500-970 Porto Alegre, RS - Brazil

Abstract. A negative ion plasma expansion into vacuum is simulated by use of a one-dimensional, fully Eulerian code. As for the case of a two species plasma, the time-asymptotic regime is found to be self-similar with the temperature decreasing as t^{-2}. The numerical results exhibit clearly the physically expected effects of mass separation.

1 Introduction

The process of a plasma expansion into vacuum has deserved significant attention over the last decades. This phenomenon was studied not only for a two component plasma (Gurevich et al. 1966; Manfredi et al. 1993), but also for plasmas with three species (El-Zein et al. 1995; Luo and Yu 1992). Particularly, the interest on the plasma expansion process can be justified by its application potential in space physics (Gurevich et al. 1969) and in laboratory experiments (Wagli and Donaldson 1978; Pearlman et al. 1977). In these applications it is common to find large number of negatively charged dust grains that can significantly alter the properties of the plasma. Dusty plasmas are frequently approximated by "negative ion plasmas" (NIP) which consist of positive ions, electrons and negative ions of constant charge. Our main proposal here is to investigate the expansion of a NIP into vacuum by using a one-dimensional Vlasov-Poisson model. The basic equations are treated analytically, by use of a rescaling transformation, and numerically through a one-dimensional Eulerian code. The present work extends the procedure used by Manfredi et al. (1993) in the two component plasma expansion problem.

The organization of the paper is as follows: in section 2 we present the Vlasov-Poisson system and the rescaling transformation used to treat the NIP expansion; numerical results are presented in section 3; the conclusions are given in section 4.

2 Basic Equations and Rescaling

We consider a one-dimensional, collisionless NIP, which is freely expanding into vacuum, experiencing no external fields. The unperturbed plasma densities are $n_0, (1-\epsilon)n_0$ and ϵn_0 for positive ions, electrons and negative ions, respectively. The charge of the positive ions is $+e$ and the electrons and negative ions have charge $-e$. The basic Vlasov-Poisson system is:

$$\frac{\partial f_e}{\partial t} + p\frac{\partial f_e}{\partial x} - E(x,t)\frac{\partial f_e}{\partial p} = 0, \tag{1}$$

$$\frac{\partial f_i}{\partial t} + \frac{p}{M_{ie}}\frac{\partial f_i}{\partial x} + E(x,t)\frac{\partial f_i}{\partial p} = 0, \tag{2}$$

$$\frac{\partial f_n}{\partial t} + \frac{p}{M_{ne}}\frac{\partial f_n}{\partial x} - E(x,t)\frac{\partial f_n}{\partial p} = 0, \tag{3}$$

$$\frac{\partial E(x,t)}{\partial x} = \int_{-\infty}^{+\infty} [f_i - \epsilon f_n - (1-\epsilon) f_e]\, dp. \tag{4}$$

In equations (1-4), for convenience, we have taken $q_i = -q_e = -q_n = m_e = \epsilon_0 = 1$; the suffices e, i and n refer to electrons, positive ions and negative ions, respectively; M_{ie} is the quotient of the mass of the positive ions and electrons, $M_{ie} = m_i/m_e$; $M_{ne} = m_n/m_e$ is the quotient of the mass of the negative ions and electrons, and $\epsilon = n_{0-}/n_0$ is the uniform negative ion concentration. Following the procedure first proposed by Manfredi et al. (1993), we do some preliminary analytical work, by means of rescaling transformations, in order to avoid difficulties concerned with the free boundary conditions. Rescaling methods have been used as an effective tool to obtain solutions of nonlinear problems (Besnard et al. 1983, Mola et al. 1993, Burgan et al. 1983). In the work of Manfredi et al. (1993) this method was used to obtain numerical solutions of the plasma expansion problem over long times. The main goal of the rescaling method is to introduce new space and time variables so that, asymptotically, the evolution of the system is totally absorbed by the transformation. As a consequence, in the new variables, the system is recast into a form where the physics is easier to interpret, at least for large times. In this paper we adopt the same rescaling technique used in the two component plasma expansion problem. A complete description of the method can be found in Manfredi et al. (1993). In the rescaled space, the Vlasov-Poisson system can be written as

$$\frac{\partial F_e}{\partial \theta} + \frac{\Pi}{\mu(\theta)}\frac{\partial F_e}{\partial \xi} - \mu(\theta)\tilde{E}\frac{\partial F_e}{\partial \Pi} = 0, \tag{5}$$

$$\frac{\partial F_i}{\partial \theta} + \frac{\Pi}{M_{ie}\mu(\theta)}\frac{\partial F_i}{\partial \xi} + \mu(\theta)\tilde{E}\frac{\partial F_i}{\partial \Pi} = 0, \tag{6}$$

$$\frac{\partial F_n}{\partial \theta} + \frac{\Pi}{M_{ne}\mu(\theta)}\frac{\partial F_n}{\partial \xi} - \mu(\theta)\tilde{E}\frac{\partial F_n}{\partial \Pi} = 0, \tag{7}$$

$$\frac{\partial \tilde{E}}{\partial \xi} = \int_{-\infty}^{+\infty} [F_{\rm i} - (1-\epsilon)F_{\rm e} - \epsilon F_{\rm n}]\, d\Pi, \tag{8}$$

where $\mu(\theta) = (1 + \Omega\theta/2)^3$. System (5-8) can be easily obtained from (1-4) by transforming (x, p, t) into (ξ, Π, θ) using

$$x = C(t)\xi, \tag{9}$$

$$dt = A(t)^2 d\theta, \tag{10}$$

$$\Pi = [\mu(t)m]\eta, \quad \eta = \frac{d\xi}{d\theta}, \quad \mu(t) = \frac{C^2(t)}{A^2(t)}, \tag{11}$$

$$E(x,t) = \tilde{E}(\xi, \theta), \tag{12}$$

$$f_{\rm k}(x,p,t) = F_{\rm k}(\xi, \Pi, \theta), \qquad \text{k=i, e, n}, \tag{13}$$

with $A^4 = C = 1 + \Omega t$. The advantage of treating numerically equations (5-8) rather than (1-4), is that the system (5-8) describes the evolution of "particles" that have a variable mass, $m_{\rm k}\mu(\theta)$, which increases with time. As a consequence, the distribution $F(\xi, \Pi, \theta)$ is frozen in a finite interval of ξ, facilitating the application of the free boundary conditions.

3 Numerical Results

The numerical analysis of the system (5-8) was performed by means of an Eulerian code, which was slightly modified to incorporate the time-dependent coefficients appearing in the transformed Vlasov equations. In the following plots, we show only that part of the phase space where $\xi > 0$ or $x > 0$. For $\xi < 0$ or $x < 0$, the results are symmetric. Space is measured in units of the electron Debye length, time in units of the electron plasma period (the inverse of the plasma frequency) and velocity in units of the electron thermal velocity. For $t = 0$, the plasma is locally neutral, with uniform distribution from ξ or $x = -6$ to ξ or $x = 6$. In the velocity space, the initial particle distribution is specified by the maxwellian

$$\exp(-m_s v^2/2T_s)\,, \qquad s = \text{i, e, n}$$

where $T_{\rm i}$, $T_{\rm e}$ and $T_{\rm n}$ are the positive ion, electron and the negative ion temperatures, respectively.

In the plots presented in the sequel we have used

$$\Omega = 0.5\,, \quad M_{\rm ie} = 3\,, \quad M_{\rm ne} = 10\,, \quad \epsilon = 0.2\,,$$

$$T_{\rm e} = 1\,, \quad T_{\rm i} = 0.5\,, \quad T_{\rm n} = 0.35\,.$$

The parameter Ω is free and was chosen empirically. Its optimum choice keeps the dimensions of the asymptotic state as close as possible to the dimensions of the initial state. The choice $\epsilon = 0.2$ represents a mixture of negative species with 20% of negative ions and 80% of electrons. In the following graphs that represent distribution functions, dark areas correspond to regions of high density. The electron, as well as the positive and negative ions distribution functions, are shown in Figs. 1, 2 and 3, in the real space (x, v), for several different time values. From these figures we can clearly see an evolution similar to that obtained for a two component plasma expansion (Manfredi et al. 1993). In the early stages, some electrons run ahead the bulk of the plasma, generating an electric field that accelerates a group of positive ions into the vacuum. In the asymptotic regime, i.e., for large times, the evolution becomes self-similar with the distribution functions assuming the form

$$f(x,v,t) = n(x,t)\delta(v - x/t)\,.$$

This is indicated by the formation of a straight line structure in phase space.

The net charge density in the rescaled space,

$$n(\xi) = \int [F_i(\xi,\Pi) - (1-\epsilon)F_e(\xi,\Pi) - \epsilon F_n(\xi,\Pi)]\, d\Pi,$$

is presented in Fig. 4. A formation of alternatively positive and negative layers the dimensions of which go to zero with time is clearly seen. Later we will show that the dimensions of these layers tend to zero faster than t in the rescaled space, indicating that the plasma becomes neutral over a smaller and smaller scale.

Figure 5 shows the negative ion density $n_{\mathrm{n}}(\xi)$ (solid line), positive ion density $n_{\mathrm{i}}(\xi)$ (broken line) and electron density $n_{\mathrm{e}}(\xi)$ (dotted line) in the rescaled space. Again we can notice that charge neutrality is approached asymptotically. Nevertheless this behavior differs slightly from that of an electron-ion plasma. There the electron and ion densities tend to assume the same profile. In the case of a NIP the negative ion density vanishes before the plasma front, indicating that, as a consequence of their bigger inertia, negative ions stay behind the other species in the expansion process.

We now investigate same thermodynamical properties of the NIP expansion. Let us define the local temperature $T(x,t)$, pressure $P(x,t)$ and density $n(x,t)$ by

$$\begin{aligned} kT(x,t) &= m\,\frac{\int (p- <p>)^2 f(x,p,t)dp}{\int f(x,p,t)dp}\,, \\ P(x,t) &= m \int (p- <p>)^2 f(x,p,t)dp\,, \qquad (14) \\ n(x,t) &= \int f(x,p,t)dp\,, \end{aligned}$$

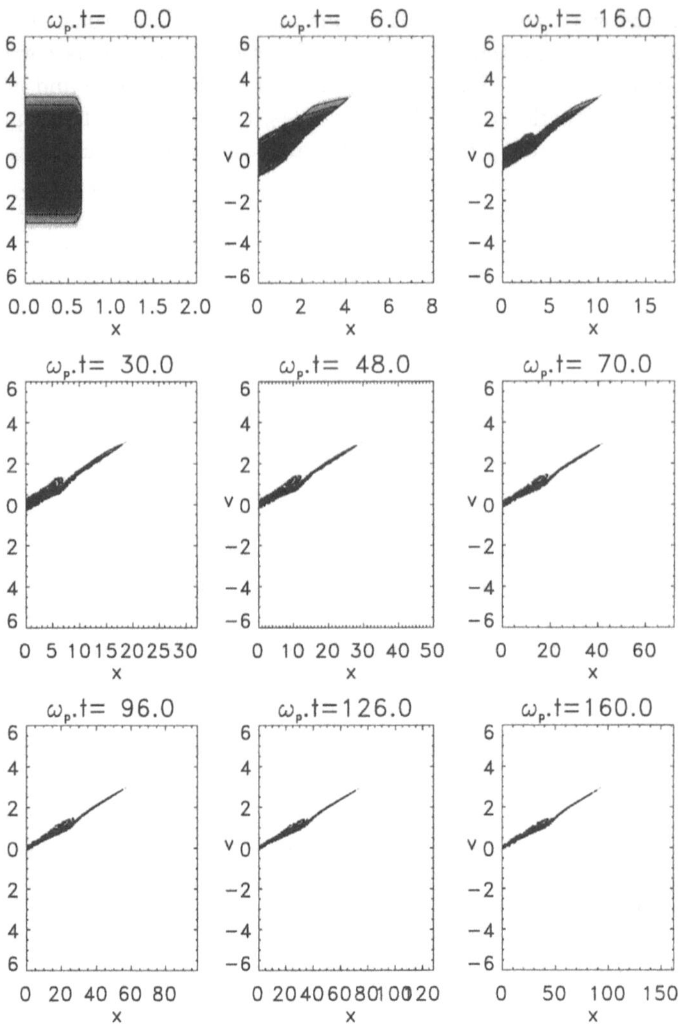

Fig. 1. Electron distribution function in the real phase space (x, v)

where k is the Boltzmann constant.

From the rescaling given in Sec.2, we can define "new" temperature, pressure and density as

$$k\hat{T}(\xi,\theta) = m \frac{\displaystyle\int (\Pi - <\Pi>)^2 F(\xi,\Pi,\theta) d\Pi}{\displaystyle\int F(\xi,\Pi,\theta) d\Pi},$$

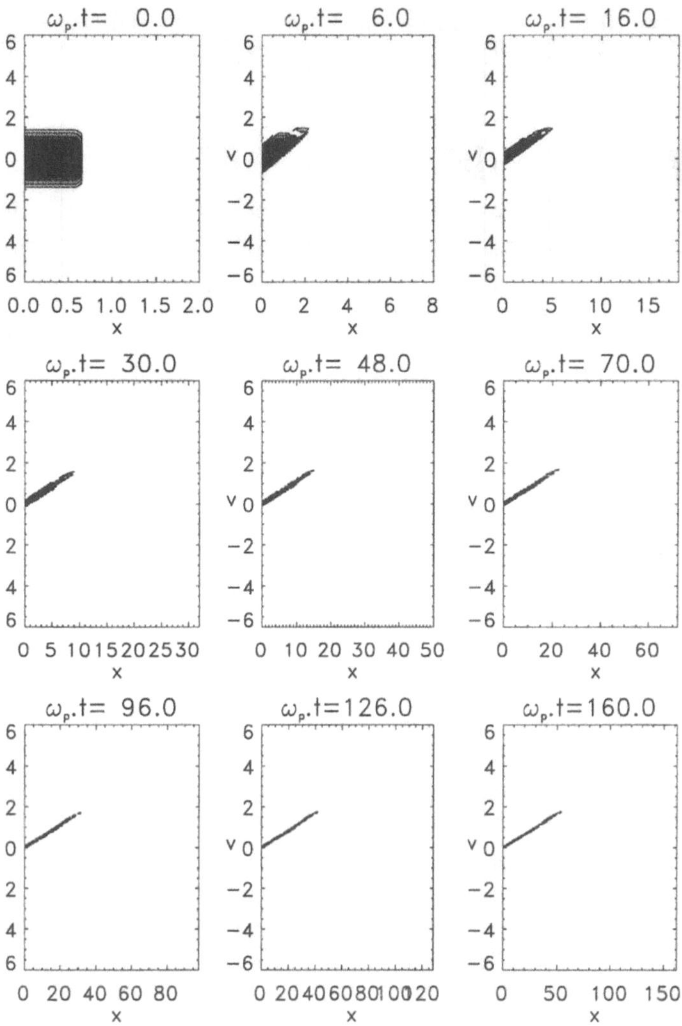

Fig. 2. Ion distribution function in the real phase space (x, v)

$$\hat{P}(\xi,\theta) = m\int(\Pi - <\Pi>)^2 F(\xi,\Pi,\theta)d\Pi\,, \tag{15}$$

$$\hat{n}(\xi,\theta) = \int F(\xi,\Pi,\theta)d\Pi\,.$$

The relationship between the old (eq. 14) and the new (eq. 15) thermodynamical functions is given by

$$T(x,t) = [1/(1+\Omega t)^2]\hat{T}(\xi,\theta)\,,$$

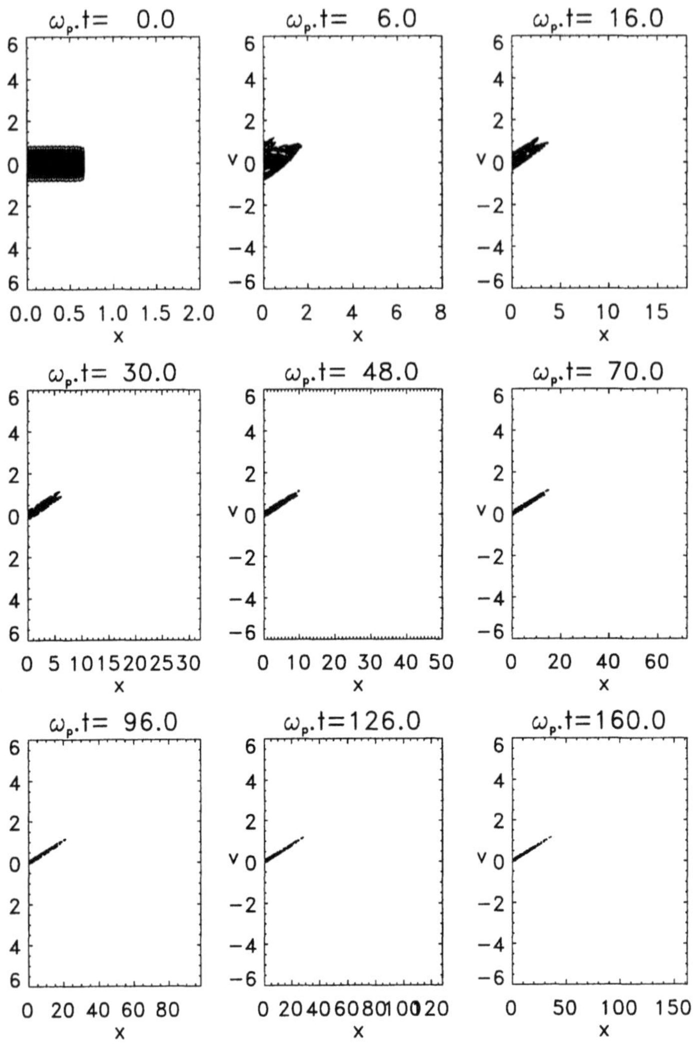

Fig. 3. Neg. ion distribution function in the real phase space (x, v)

$$P(x,t) = [1/(1+\Omega t)^3]\hat{P}(\xi,\theta)\,, \qquad (16)$$
$$n(x,t) = [1/(1+\Omega t)]\hat{n}(\xi,\theta)\,.$$

The asymptotic behavior of the rescaled thermodynamical functions (14) indicates that the NIP has a temperature evolution identical to that of a two component plasma (Manfredi et al. 1993). This can be seen from Figs. 5 and 6, which show that, asymptotically, $\hat{P}$, $\hat{n}$ and $\hat{T}$ do not depend on time anymore ($\hat{T}$ is the ratio between $\hat{P}$ and $\hat{n}$). From eq. (16), it is easily seen

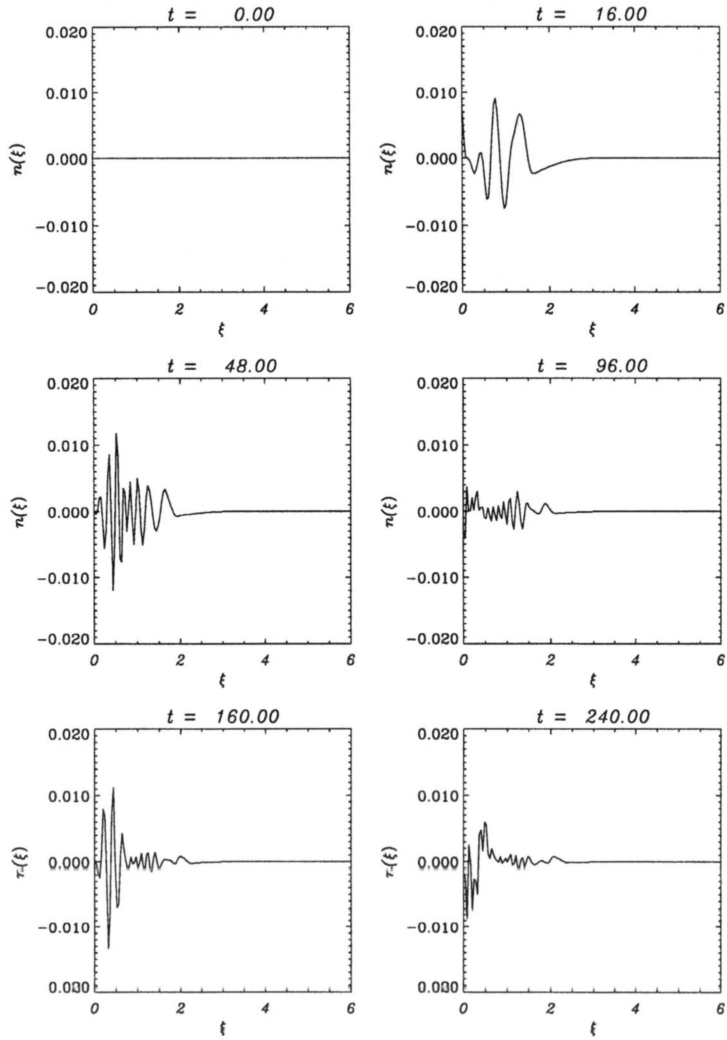

Fig. 4. Net charge density in the rescaled space

that, if $\hat{T}$ does not depend on time asymptotically, then the temperature $T(x,t)$ decreases in time as t^{-2}.

We are now able to show that the dimensions of the charged layers (Fig. 5) tend to zero faster than t in the rescaled space. The Debye length is proportional to $(T/n)^{1/2}$, i.e.,

$$\lambda_{\mathrm{D}} \sim (T/n)^{1/2}.$$

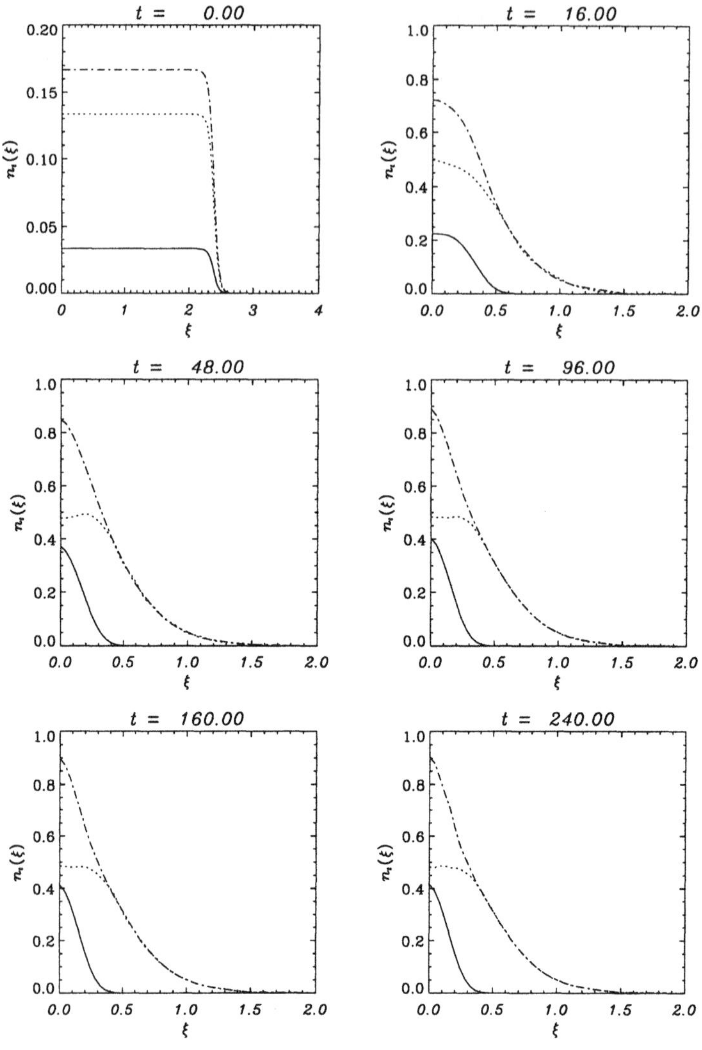

Fig. 5. Number density for electrons (dotted line), ions (broken line) and negative ions (solid line) in the rescaled space

Since the density varies as t^{-1} and the temperature as t^{-2} for large times, this implies that $\lambda_D \sim t^{-1/2}$. As a consequence, in the real space, the Debye length tends to zero, indicating that the plasma goes into a regime of quasineutrality.

The electric potential and kinetic energies are presented in Fig. 7. This plot shows that all kinetic energies are asymptotically constant. However, they oscillate differently, since the electrons, negative ions and positive ions

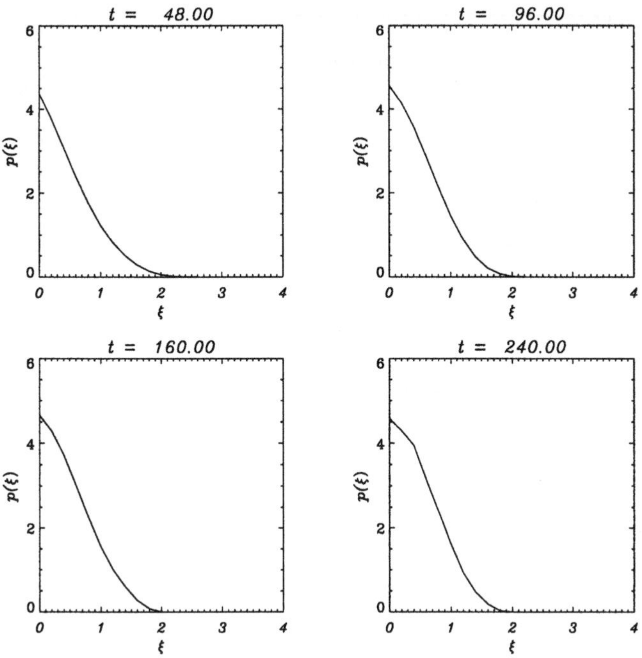

Fig. 6. Negative ions pressure in the rescaled space

have different masses. The potential energy is virtually zero. To allow it to be shown on the same scale as the kinetic energies it is multiplied by a factor of eight.

4 Conclusions

In this paper we have analysed numerically the process of expansion into vacuum of a NIP. Paralleling the analysis of the expansion of a two component plasma, a rescaling transformation of the basic equations was applied in order to deal with the moving boundary conditions. The numerical simulation accounted quite properly for the effects of mass segregation and for the existence of a time-asymptotic ballistic regime. The temperature was shown to follow the same t^{-2} power law as found in the two component plasma. These results attest the adequacy and efficiency of the Eulerian code to treat the expansion process of a negative ion plasma. The modification of some parameters such as initial temperatures, mass ratios, charge and negative ions concentration may have a significant effect on the NIP expansion (Garcia et al. 1997). For example, the rate at which the electrons leave the center of the plasma increases as we increase the proportion of negative ions. These issues are discussed in a separate study in progress.

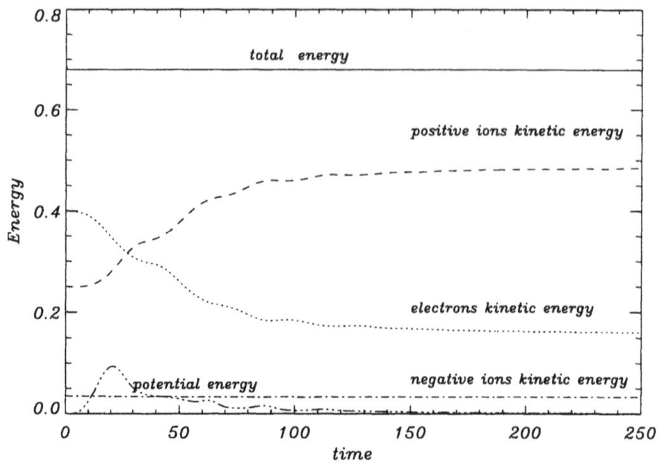

Fig. 7. Evolution of the kinetic and potential energies with time

Acknowledgments

We gratefully acknowledge Dr M. Feix for fruitful discussions and G. Manfredi for discussions and for providing us with the basic code. This work has been partially supported by the Brazilian agencies Conselho Nacional de Desenvolvimento Científico e Tecnológico (CNPq), Fundação de Amparo a Pesquisa do Rio Grande do Sul and Financiadora de Estudos e Projetos (FINEP).

References

Besnard, D., Burgan, J.R., Feix, M.R., Fijalkow, E., Munier, A. (1983): Nonlinear time-dependent anharmonic oscillator: Asymptotic behavior and connected invariants. J. Math. Phys. **5**, 1123–1128

Burgan, J.R., Feix, M.R., Fijalkow, E., Munier, A. (1982): Self-similar and asymptotic solutions for a one dimensional Vlasov beam. J. Plasma Phys. **29** part1, 139–142

El-Zein Amin, A., Kim, H-S., Yi, S., Lonngren, K.E. (1995): Expansion of a negative ion plasma into vacuum. Phys. Plasmas **2**, 1073–1076

Garcia, L.G., Goedert, J., Figua, H., Fijalkow, E., Feix, M.R. (1997): Numerical simulation of a negative ion plasma expansion into vacuum. Phys.Plasmas **4**, 4240–4253

Gurevich, A.V., Pariiskaya, L.V., Pitaievskii, L.P. (1966): Self-similar motion of rarefied plasma. Sov. Phys. JETP **22**, 449–454

Gurevich, A.V., Pitoevskii, L.P., Smirnova, V.V. (1969): Ionospheric aerodynamics. Space Sci. Rev. **9**, 805–871

Luo, H., Yu, M.Y. (1992): Self-similar expansion of dusts in a plasma. Phys. Fluids B **4**, 3066–3069

Manfredi, G., Mola, S., Feix, M.R. (1993): Rescaling methods and plasma expansions into vacuum. Phys. Fluids B **5**, 388–401

Mola, S., Manfredi, G., Feix, M.R. (1993): Expansion of a quantum electron gas. J. Plasma Phys **50**, 145–162

Pearlman, J.S., Thomson, J.J., Max, C.E. (1977): Polarization-dependent absorption of laser radiation incident on dense-plasma planar targets. Phys. Rev. Lett. **38**, 1397–1400

Wagli, P., Donaldson, T.P. (1978): Fast-ion emission and resonance absorption in laser-generated plasma. Phys. Rev. Lett. **40**, 875–879

A Presentation of the Diffusion Velocity Method

Sylvie Mas-Gallic

Centre de Mathématiques Appliquées, Ecole Polytechnique, 91128 Palaiseau Cedex, France
and
LAMI, Université d'Evry-Val-d'Essonne, 4 Bd. F. Mitterand, 91025 Evry Cedex, France

1 Introduction

For a long time particle methods have been extensively used for flow computation of inviscid, incompressible fluid mechanics problems such as the incompressible Euler equations (see e.g. the survey of Leonard 1980). They have been much used and studied in plasma physics problems such as the Vlasov equation (see e.g. the survey of Harlow 1964).

In fact, since they contain very few numerical diffusion effects, are easy to use and quite accurate, these methods are very well suited to purely convective problems. It was thus a natural idea to try to extend them to nonpurely convective effects and, in the case of diffusion effects, this was first done by addition of a random walk to to the movement of the particles (see Chorin 1973, for the Navier-Stokes equation and Grmela et al. 1980 and Hockney and Eastwood 1981, for the case of plasma physics problems). The next method to appear followed from a complete change of philosophy in the sense that particles had not only their position changing in time but also their weights. In this kind of method, called particle strength exchange (PSE), the particles are no longer independent and the information travels thanks to the time evolution of the coefficients (see Degond and Mas-Gallic 1989 and also Fishlov 1990).

Another method arose more recently: the method of diffusion velocity. Although this method appeared for the first time in the case of the Fokker-Planck equation, I mainly focus my attention on the case of Navier-Stokes equation.

After the pioneering work of Fronteau et al. (1984) based on the introduction of Fick's law to compute approximately the solution of a model Fokker-Planck equation, Degond and Mustieles (1990) have adapted the method, given a different way to compute the flux and obtained interesting results on a general advection-diffusion equation as well as on a two dimensional Fokker-Planck equation.

Later different teams worked on the method and applied it to the Navier-Stokes equations in two or three dimensions for an incompressible fluid.

Strickland et al. (1996) used the classical theory (Chorin 1973) to construct the scheme though the methods of Rivolaen (1994), Huberson et al. (1997a), Huberson et al. (1997b) introduced a local decomposition of the vorticity to construct their method. Let us make precise that, even if we simply consider for what follows problems on the whole space, in the case of a problem with some boundary conditions it is also possible to adapt the method.

An outline of the paper is as follows. The method is, for simplicity, presented firstly in the case of the one dimensional heat equation. Then we consider the 2-d incompressible Navier-Stokes model written in the vorticity-velocity formulation and present the extension of the method. Finally the difficulty of the three-dimensional case is discussed for the Navier-Stokes system.

2 The One-Dimensional Case

The basic idea of the diffusion velocity method is to transform a diffusion equation into an advection equation by the application of Fick's law. In one dimension this is, at least formally, easily done as follows: consider the heat equation with initial condition on the whole space

$$\frac{\partial \omega}{\partial t} - \nu \frac{\partial^2 \omega}{\partial x^2} = 0, \quad \omega(.,0) = \omega_0.$$

Then, assuming that the solution ω is always different from zero (the problem of vanishing ω can be overcome), we observe that this equation can be rewritten as

$$\frac{\partial \omega}{\partial t} + \frac{\partial}{\partial x}(a\omega) = 0, \quad \omega(.,0) = \omega_0,$$

where we have set $a(x) = -\nu \dfrac{\partial \omega(x)}{\partial x} / \omega(x)$. In order to define the particle method we introduce the usual small parameter, $\epsilon > 0$, of particle methods, a smoothing function ζ and the scaled function $\zeta_\epsilon : \zeta_\epsilon(x) = \epsilon^{-1}\zeta(x/\epsilon)$ and we take as velocity for the particles the *smoothed* function a_ϵ defined by

$$a_\epsilon(x) = -\nu \frac{\omega * \zeta'_\epsilon(x)}{\omega * \zeta_\epsilon(x)},$$

where we denote by $*$ the convolution product. We choose a quadrature rule with weighted nodes $(w_p^0, x_p^0)_p$. The initial condition is approximated in the sense of measures by the following linear combination of Dirac measures $\omega_0^h(x) = \sum_p \alpha_p^0 \delta(x - x_p^0) \approx \omega_0(x)$, where $\alpha_p^0 \approx w_p^0 \omega_0(x_p^0)$, and the particle approximation ω^h of ω at time t is thus

$$\omega^h(x,t) = \sum_p \alpha_p^0 \delta(x - X_p(t)),$$

where the X_ps are solution of the characteristic curves of a_ϵ:

$$\frac{dX_p}{dt}(t) = a_\epsilon(X_p(t)), \quad X_p(0) = x_p^0.$$

We now make come comments on this one dimensional case before focusing on the two dimensional Navier-Stokes equation. Firstly the only computations which have to be done in this method are the particle positions. The second remark is that in the case of a one dimensional advection diffusion equation the velocity would be directly obtained by addition of the velocity given in the equation and the diffusion velocity which is already smoothed. Lastly, although this method has been known for a long time, it seems that no analysis of it has been done yet, but this is work currently in progress.

3 The Case of Higher Dimensions

We consider the two dimensional Navier-Stokes equation for an incompressible fluid. Written in velocity, u, and vorticity, ω, variables this system is

$$\begin{cases} \dfrac{\partial \omega}{\partial t} + \nabla(\omega u) - \nu \Delta \omega = 0, \\ \nabla . u = 0, \quad \nabla \times u = \omega, \\ \omega(.,0) = \omega_0, \end{cases}$$

where ω is a scalar function. We observe that, thanks to Biot-Savart's law, it is classical to rewrite equivalently this system as follows

$$\begin{cases} \dfrac{\partial \omega}{\partial t} + \nabla(\omega u) - \nu \Delta \omega = 0, \\ u = K * \omega, \\ \omega(.,0) = \omega_0. \end{cases}$$

Then, in order to extend the diffusion velocity method to this case, we mimic what was done in the one dimensional case, assume first that the vorticity is never equal to zero (notice that as in the one dimensional case this difficulty can be overcome), rewrite the evolution equation of ω which is an advection-diffusion equation as a purely transport equation and obtain the following system

$$\begin{cases} \dfrac{\partial \omega}{\partial t} + \nabla(\omega(u + u_d)) = 0, \\ u = K * \omega, \quad u_d = -\nu \dfrac{\text{grad } \omega}{\omega}, \\ \omega(.,0) = \omega_0. \end{cases}$$

We have thus introduced a new velocity which is the diffusion velocity and the system will be easily discretised the usual way by introduction of the usual small parameter, $\epsilon > 0$, of particle methods, a smoothing function ζ and the scaled function $\zeta_\epsilon : \zeta_\epsilon(x) = \epsilon^{-2}\zeta(x/\epsilon)$. The particle velocity is the *smoothed* function u_ϵ defined by :

$$u_\epsilon(x) = u * \zeta_\epsilon(x) - \nu \frac{\omega * \text{ grad } \zeta_\epsilon(x)}{\omega * \zeta_\epsilon(x)} = \omega * K_\epsilon(x) - \nu \frac{\omega * \text{ grad } \zeta_\epsilon(x)}{\omega * \zeta_\epsilon(x)}.$$

We choose a quadrature rule with weighted nodes $(w_p^0, x_p^0)_p$. The initial condition is approximated in the sense of measures by the following linear combination on Dirac measures $\omega_0^h(x) = \sum_p \alpha_p^0 \delta(x - x_p^0) \approx \omega_0(x)$, where $\alpha_p^0 \approx w_p^0 \omega_0(x_p^0)$, and the particle approximation ω^h of ω at time t is

$$\omega^h(x,t) = \sum_p \alpha_p^0 \delta(x - X_p(t)),$$

where the X_ps are solution of the characteristic curves of u_ϵ:

$$\frac{dX_p}{dt}(t) = u_\epsilon(X_p(t)), \quad X_p(0) = x_p^0.$$

We now make a short overview of the three dimensional case. We can write the Navier-Stokes system of equations for an incompresible fluid in velocity and vorticity variables. We find a system in which both quantities u and ω are vectors so that the extension of the previous method is no longer straightforward. Of course it is no longer possible to do in three dimensions what was done in both one and two dimensions which is to divide by ω, a vector now. Actually we could try to apply directly the previous method on each component of the vorticity, but this would lead to the definition of a diffusion velocity for each component of the vorticity and this, of course, is not reasonable.

This important remark is the basis of the different works of Strickland et al. (1996) on one hand and of Rivolaen (1994), Huberson et al. (1997a), Huberson et al. (1997b) on the other hand. In fact, though they used quite different approaches to extend the diffusion velocity method to the three dimensional case, the main idea is to try to define a local diffusion velocity.

The Navier-Stokes system in velocity, u, and vorticity, ω, variables is

$$\begin{cases} \dfrac{\partial \omega}{\partial t} + \nabla(u \otimes \omega) - (\omega.\nabla)u - \nu \Delta \omega = 0, \\ \nabla.u = 0, \quad \nabla \times u = \omega, \\ \omega(.,0) = \omega_0. \end{cases}$$

Huberson et al. (1997a) introduced a local decomposition of the vorticity field which, fundamentally, leads them back to the scalar case. They define the diffusion velocity u_ν by

$$u_\nu = -\nu \frac{(\nabla \otimes \omega).\omega}{|\omega|^2}$$

and obtain an equation of transport for the vorticity of the following form

$$\frac{\partial \omega}{\partial t} + \nabla((u + u_\nu) \otimes \omega) - (\omega.\nabla)u - \omega_\nu \times \omega = 0.$$

The work of Strickland et al. (1996) is different since they assume that some of the terms in the decomposition of the vorticity are negligible. This assumption is actually not clear from a theoretical point of view, but, since they restrict themselves to the two dimensional case for the computations, their results are good.

As a conclusion we can say that this method seems promising for computation of two and three dimensional flows of viscous incompressible fluids as well as for numerical resolution of plasma physics equations.

Figures

In the following are found several figures obtained with the velocity diffusion method by the different teams previously mentioned.

Figure 1 is extracted from Degond and Mustieles (1990) in which the authors show the exact solution (solid line) and the numerical solution (dashed line) of the heat equation respectively at time $t = 0.2$ and $t = 0.6$, with a regularisation parameter $\epsilon = 0.2$ and the number of particles $N = 100$.

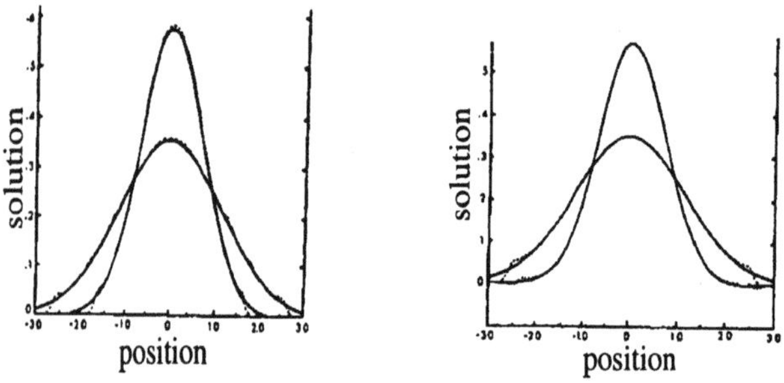

Fig. 1. Left: Order 2 B-spline cut-off, time step $\Delta t = 0.01$. Right: Super gaussian cut-off, time step $\Delta t = 0.02$

The two sets of figures in Figs. 3 and 4 represent the convection field approximation for a number of particles $N = 32$ and two different cut off functions. In the left figures the approximation of $u(x) = \exp[-x^2]/\sqrt{\pi}$ is plotted; the central figure is the approximation of u' and the right figure is the approximation of $u'(x)/u(x)$. We note that a larger parameter ϵ is needed for u' than for u because the differentiation of the cut off ζ_ϵ lowers the accuracy of the quadrature rule error.

References

Chorin, A.J. (1973): Numerical study of slightly viscous flows. J. Fluid Mech. **57**, 785–796

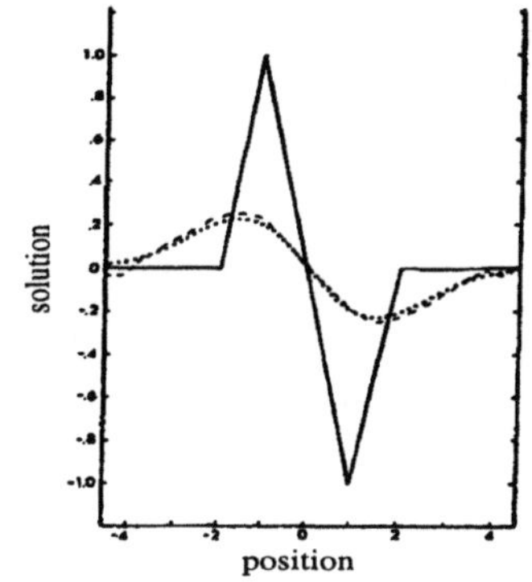

Fig. 2. Heat equation solution with non constant sign, $N = 100$ particles, $\epsilon = 0.3$, $\Delta t = 0.01$. Initial condition (solid line). Exact solution (dotted line). Numerical solution (dashed line) (Degond and Mustieles 1990)

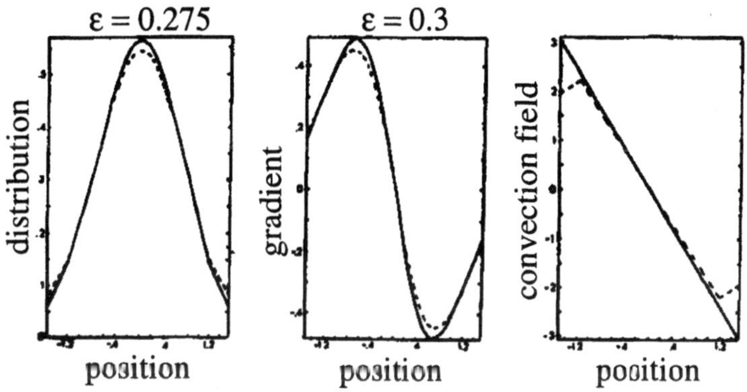

Fig. 3. Order 2 B-spline cut off (Degond and Mustieles 1990)

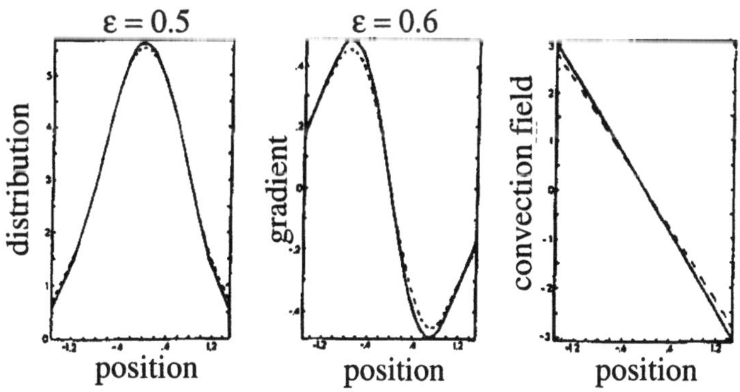

Fig. 4. Super gaussian cut off (Degond and Mustieles 1990)

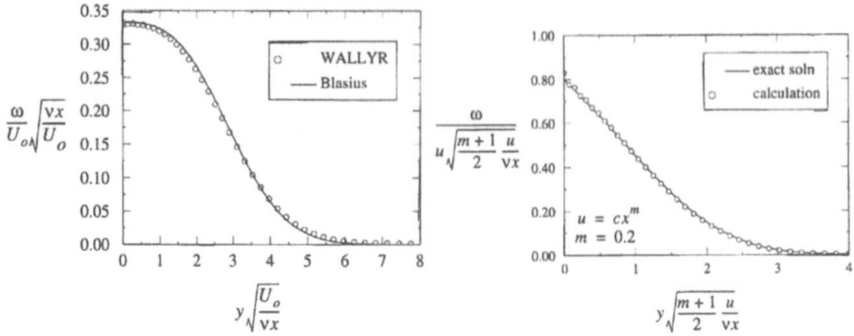

Fig. 5. Exact and computed solutions for flow past a flat plate (left) and past a wedge (right), from Strickland et al. (1996)

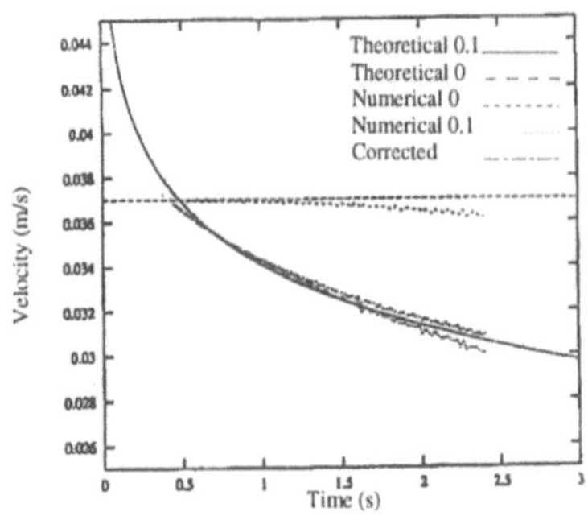

Fig. 6. Induced velocity by a vortex ring (Huberson et al. 1997a)

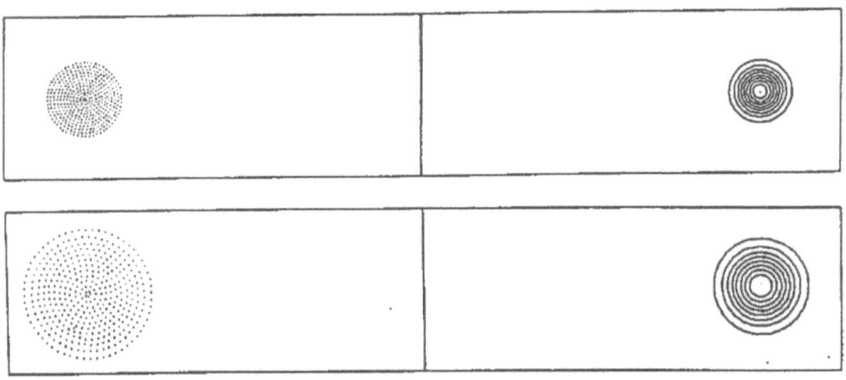

Fig. 7. Position particles (left figures) and vorticity isovalues (right figures) at time $t = 0.4$ (top.figures) and $t = 2.4$ (bottom figures) (Huberson et al. 1997a)

Degond, P., Mas-Gallic, S. (1989): The weighted particle method for convection-diffusion equation. Math. Comp. **53**, 485–525

Degond, P., Mustieles, F.J. (1990): Approximation of diffusion equations by deterministic convections of particles. SIAM J. on Scientific and Statistical Computing **11**, 293–310

Fishelov, D. (1990): A new vortex scheme for viscous flows. J. Comp. Phys. **86**, 211–224

Fronteau, J., Combis, P. (1984): A Lie-Admissible Method of Integration of Fokker-Planck Equations with Non-Linear Coefficients (Exact and Numerical Solutions). Hadronic J. **7**, 911-930

Grmela, M., Fronteau, J., Tellez-Arenas, A. (1980): Inverse liouville problem. Hadronic J. **3**, 1209–1241

Harlow, F.H. (1964): *Methods in computational physics.*, Alder, B., Fernbach, S. and Rotenberg, M. ed. (Academic Press, San Diego) **3**

Hockney, R.W., Eastwood, J.W. (1981): Computed simulations using particles, (Mc Graw-Hill, New York)

Huberson, S., Rivolaen, E. (1997): Numerical simulation of axisymmetric viscous flows by means of particle method. (submitted)

Huberson, S., Rivolaen, E., Hauville, F. (1994): Simulation numérique des équations de Navier-Stokes 3D par une méthode particulaire. C.R. Acad. Sci. Paris **324**, Série II*b*, 543–549

Leonard, A. (1980): Vortex methods for flow simulation. J. Comp. Phys. **37**, 289–335

Rivolaen, E. (1994): PhD Thesis, Le Havre

Strickland, J.H., Kempka, S.N., Wolfe, W.P. (1996): Viscous diffusion of vorticity using the diffusion velocity concept *in ESAIM Proceedings, Vortex flow and related numerical methods II* **1**, 135–151

Finite-Difference Schemes for the Diffusion Equation

Giovanni Manfredi[1] and Maurizio Ottaviani[2]

[1] UKAEA Fusion, Culham, OX14 3DB, United Kingdom (UKAEA/Euratom Fusion Association) **
[2] JET Joint Undertaking, Abingdon, OX14 3EA, United Kingdom

Abstract. The Crank-Nicolson scheme is widely used to solve numerically the diffusion equation, because of its good stability properties. It is, however, ill-behaved when large time-steps are used: the short wave-lengths may happen to be *less* damped than the long ones. A detailed analysis of this flaw is performed and an alternative scheme is proposed, which removes this difficulty while preserving all other good properties (unconditional stability and second-order accuracy).

1 Introduction

The diffusion equation plays an important role in many branches of physics, in fields as different as fluid and plasma dynamics, thermodynamics and neutron transport. From the point of view of numerical analysis, diffusive terms are often necessary when solving partial differential equations in order to control the numerical noise at small wave-lengths. Indeed, although a fine resolution is needed to reproduce accurately the larger scales, small scales do not contain any significant physical information. If, however, such small scales are not damped away by some diffusive mechanism, in the long run they can corrupt the entire solution, even at long wave-lengths.

There is a vast literature of numerical methods for the diffusion equation, which we make no attempt to review even partially here (Marchuk 1982; Morton and Mayer 1994; Richtmyer and Morton 1967; Press et al. 1992; Crank 1975; Roache 1972), (Morton and Mayer 1994 provide a list of fourteen finite-difference methods). According to the time-stepping technique used, these schemes can be either explicit or implicit. Explicit schemes are simpler and computationally faster, but often impose a very restrictive upper bound on the maximum time-step, beyond which they are unstable. Implicit schemes are often unconditionally stable, but require the inversion of a (tridiagonal, in one dimension) matrix, which is more time consuming, although not prohibitive.

The scope of this communication is to point out a weakness of a class of widely used implicit, finite-difference techniques, which seems to be generally overlooked. A prototype of such techniques is the Crank-Nicolson (CN)

** Present address: Dublin Institute for Advanced Studies, Ireland scheme (Crank and Nicolson 1947), possibly the most popular of all finite-difference schemes for the diffusion equation. The problem with the CN scheme is that it is extremely ill-behaved for large values of the time-step. By the expression "ill-behaved" we mean that, although the results are of course accurate (to second order) for small time-steps, taking large time-steps radically changes the nature of the equation we are trying to solve. For example, in the limit of very large time-steps, a monochromatic wave is no longer damped (as it should be, according to a diffusion process), but travels with constant phase velocity $\lambda/(2\Delta t)$ (where λ is the wave-length and Δt the time-step). Another, more serious, shortcoming is that, when Δt exceeds a certain value, short wave-lengths can be *less* strongly damped than some long wave-lengths. Bearing in mind that diffusive terms are devised to get rid of the small scale noise, this result can have disastrous consequences.

After recognizing the source of this incorrect behaviour, we propose a class of schemes that does not suffer from this flaw, being "well-behaved" in the limit of large time-steps. These schemes are still second order accurate and unconditionally stable, and only involve the inversion of tridiagonal matrices (just like the CN scheme), but require two matrix inversions rather than one. Therefore, although they are computationally more expensive than CN by a factor of two, they allow the use of an arbitrarily large time-step *without changing the diffusive nature of the equation.*

From the previous discussion it is apparent that "good behaviour" is a crucial property (albeit loosely defined) for many numerical schemes and should be mentioned together with the more familiar notions of accuracy and convergence. Accuracy means that the discrete model closely approximates (in a quantitative sense) the original differential equation when the time-step and grid size are sufficiently small. Good behaviour means that, far away from this limit, the discrete model still preserves qualitatively the most relevant features of the original equation. These notions will become clearer in the following sections, where we work a concrete example.

The remaining material of this paper is organized as follows. In Sect. 2 we briefly review a few schemes for the diffusion equation, then identify the problem with the CN technique and propose an alternative, well-behaved scheme. Section 3 presents a more detailed stability analysis of both CN and our technique, including a numerical example of how CN fails when our scheme succeeds. Finally, in Sect. 4 we draw our conclusions and discuss more extensively the concept of good behaviour of a numerical scheme.

2 Properties of Some Finite-Difference Schemes

Our objective is to solve numerically the diffusion equation

$$\frac{\partial F}{\partial t} = -\hat{D}F, \tag{1}$$

where $\hat{D} = -\sigma\, \partial^2/\partial x^2$ is the diffusion operator (σ is the diffusion coefficient), the discrete approximation of which to second order in the grid size is

$$(\hat{D}F)_j = -\sigma\,(F_{j-1} - 2F_j + F_{j+1})/\Delta x^2 + O(\Delta x^3). \tag{2}$$

We use this notation because, with suitable boundary conditions, $\hat{D}$ is a positive, hermitian operator (i.e. all of its eigenvalues are real and positive), which, for all practical purposes can be treated as a real positive number. Equation (1) can be solved formally from $t_n = n\Delta t$ to t_{n+1}, yielding

$$F^{n+1} = \mathrm{e}^{-\Delta t \hat{D}} F^n, \tag{3}$$

where $F^n = F(t_n)$. Most numerical schemes can be formally represented as an approximation to some order of the exponential appearing in (3). This exponential is the evolution operator for our problem. Thus we have, respectively for the explicit first-order, implicit first-order and Crank-Nicolson schemes:

$$\exp(-\Delta t\, \hat{D}) = 1 - \Delta t\, \hat{D} + O(\Delta t^2) \tag{4a}$$

$$\exp(-\Delta t\, \hat{D}) = (1 + \Delta t\, \hat{D})^{-1} + O(\Delta t^2) \tag{4b}$$

$$\exp(-\Delta t\, \hat{D}) = \frac{1 - \Delta t\, \hat{D}/2}{1 + \Delta t\, \hat{D}/2} + O(\Delta t^3). \tag{4c}$$

The explicit scheme (4a) is unstable for $\sigma\Delta t/\Delta x^2 > 1/2$, whereas the implicit and the CN schemes (4b-c) are always stable. This can be checked by verifying that the approximate evolution operators have all eigenvalues smaller than unity in absolute value (heuristically, this is obvious by treating $\hat{D}$ as a real positive number). However, the implicit first-order scheme is also well-behaved since, for large time-steps, its evolution operator goes to zero without changing sign, just like the exact exponential. On the contrary the CN evolution operator approaches, for large time-steps, the negative of the identity operator, thus giving rise to unphysical oscillations while suppressing all damping. For instance a sinusoidal wave would, in this limit, simply change sign at each time-step, which is the same as traveling with a phase velocity equal to $\lambda/(2\Delta t)$. This is the paradox first mentioned in the introduction.

The CN scheme is often considered as an ideal choice for the diffusion equation, since it displays some key good properties: (a) it is second order accurate both in space and in time; (b) it is unconditionally stable; and (c) it requires the inversion of a tridiagonal matrix, which can be performed exactly in $O(N)$ operations, N being the number of grid points (Morton and Mayer 1994; Crank 1975). However, as we have seen, when the time-step exceeds a certain value (which will be specified in the next section), the CN scheme is ill-behaved, and gives qualitatively incorrect results.

The problem discussed above does not seem to be mentioned explicitly in the literature. Roache (1972) rather briefly suggests that "large Δt will cause some Fourier modes to overshoot". He then correctly recognizes that "the CN second-order method is less accurate than the first-order implicit method for large enough Δt". He wrongly generalizes this fact to suggest that, for large time-steps, first-order schemes are more accurate than second-order schemes. The results presented in the next paragraph prove that this is not generally true. The authors of *Numerical Recipes* (Press et al. 1992) also acknowledge that small scale structures are not damped for the CN scheme in the limit of large time-steps, but do not give a detailed analysis of why it is so. They also incorrectly suggest that the only possible solution is to turn to a first-order implicit method.

We now show that it is possible to devise a scheme which, while preserving most or all of the CN advantages, is also well-behaved for $\Delta t \to \infty$. The Padé approximation to the exponential appearing in (3) is

$$\exp(-\Delta t\ \hat{D}) = (1 + \Delta t\ \hat{D} + \Delta t^2\ \hat{D}^2/2)^{-1} + O(\Delta t^3). \tag{5}$$

The scheme resulting from this approximation is still second order accurate and unconditionally stable and also well-behaved, but, due to the term $\hat{D}^2$, requires the inversion of a non-tridiagonal matrix. This property may not be crucial since in more than one dimension the matrix is more complicated even for the CN scheme. However, the multi-dimensional diffusion equation can always be reduced to a sequence of one-dimensional equations by the time splitting technique (Yanenko 1971) and, therefore, tridiagonality can be an important issue.

Let us write the formal solution corresponding to (5)

$$\hat{M} F^{n+1} = F^n, \tag{6}$$

where $\hat{M}^{-1}$ is given by the right hand-side of (5). One could split the matrix $M = M_1 \cdot M_2$, where $M_{1,2}$ are linear in $\hat{D}$ (and therefore tridiagonal). The solution is then obtained in two steps

$$\hat{M}_1 F^* = F^n \tag{7a}$$

$$\hat{M}_2 F^{n+1} = F^*. \tag{7b}$$

Unfortunately $\hat{M}$ can only be factored by introducing complex coefficients $\hat{M} = (1 + a_+ \Delta t \hat{D})(1 + a_- \Delta t \hat{D})$, with $a_\pm = (1 \pm \mathrm{i})/2$, which would unduly complicate the scheme. We can try a more general approximation of the evolution operator:

$$\exp(-\Delta t\ \hat{D}) \simeq \frac{1 + \alpha \Delta t\ \hat{D}}{(1 + \beta_+ \Delta t \hat{D})(1 + \beta_- \Delta t \hat{D})}. \tag{8}$$

We call $g(z) = (1 + \alpha z)/(1 + \beta_+ z)(1 + \beta_- z)$ the *response function*. By expanding both sides of (8) in a Taylor series, we obtain

$$1 - z + z^2/2 - z^3/6\ =$$

$$1+(\alpha-\beta_+-\beta_-)z+[\beta_+^2+\beta_-^2+\beta_+\beta_- - \alpha(\beta_++\beta_-)]z^2 - \quad (9)$$
$$[\beta_+^3+\beta_-^3+\beta_+\beta_-^2+\beta_+^2\beta_- - \alpha(\beta_+^2+\beta_-^2+\beta_+\beta_-)]z^3+O(z^4),$$

where $z = \Delta t\hat{D}$. Matching the coefficients at first and second order we can express $\beta_\pm$ as a function of α

$$\beta_\pm = \frac{1+\alpha \pm \sqrt{\alpha^2-2\alpha-1}}{2}. \quad (10)$$

The truncation error $E(\alpha)$ is the modulus of the coefficient of the third order term in (9). After some algebra, it can be shown that $E(\alpha) = \alpha/2 + 1/6$.

We have thus obtained a family of numerical schemes parameterized by the real number α. However, some conditions must be satisfied for the scheme resulting from (8) to be meaningful. First of all it must have $\beta_\pm \geq 0$ so that the response function $g(z)$ remains finite. This condition implies $\alpha \geq -1/2$. Note that, for $\alpha = -1/2$ (yielding $\beta_- = 0, \beta_+ = 1/2$), we recover the Crank-Nicolson scheme. In order to have real coefficients the argument of the square root in (10) must be non-negative, which requires $\alpha \leq 1-\sqrt{2}$ or $\alpha \geq 1+\sqrt{2}$. Furthermore, if $\alpha \geq 0$, the response function is positive and decreases monotonically for positive values of z. The special case $\alpha = 0$ yields the Padé approximation (5), as can easily be checked. Finally, by taking $\alpha = -1/3$, we obtain $E = 0$ and the scheme is third order accurate in time.

According to our definition the numerical scheme is well-behaved if the response function $g(z)$ qualitatively preserves some of the relevant properties of the exact exponential *for all values of z*. What is meant by 'relevant properties' depends on the problem under study: in our case it is crucial that $g(z)$ be positive and decreasing for $z > 0$ since this property guarantees that smaller wave-lengths be more damped. If, in addition, we want all coefficients to be real, we need $\alpha \geq 1+\sqrt{2}$. The minimum truncation error compatible with this choice is obtained when taking the equality sign in the previous expression. This case will be analyzed in detail in the remainder of the article. Other choices are possible, however. For example, if reducing the truncation error is important, the Padé approximation ($\alpha = 0$) is more convenient, although complex numbers must be used in that case.

With our choice ($\alpha = 1+\sqrt{2}$) we obtain $\beta_\pm \equiv \beta = 1 + 1/\sqrt{2}$ and the scheme can be written in a two-step form (each involving the inversion of the same tridiagonal matrix)

$$(1+\beta\Delta t\hat{D})F^* = (1+\alpha\Delta t\hat{D})F^n \quad (11a)$$

$$(1+\beta\Delta t\hat{D})F^{n+1} = F^*. \quad (11b)$$

This is the scheme that we propose in order to circumvent the bad properties of the CN scheme. It is second-order accurate in space and time, unconditionally stable and well-behaved for large time-steps.

3 Stability Analysis and a Numerical Example

The CN scheme can be written explicitly as follows

$$-\frac{s}{2}F_{j-1}^{n+1} + (1+s)F_j^{n+1} - \frac{s}{2}F_{j+1}^{n+1} = \frac{s}{2}F_{j-1}^{n} + (1-s)F_j^{n} + \frac{s}{2}F_{j+1}^{n}, \quad (12)$$

where $s = \sigma \Delta t/\Delta x^2$. Stability can be checked with the Von Neumann method by taking $F_j^n = F^n \exp(ikx_j) = \exp(i\xi j), \xi = k\Delta x$. Substitution into (12) yields

$$\frac{F^{n+1}}{F^n} = \frac{1 - s(1-\cos\xi)}{1 + s(1-\cos\xi)} \equiv G, \quad (13)$$

where $G(\xi)$ is the so-called amplification factor. The scheme is unconditionally stable because $|G| < 1$ for every value of s and ξ. In the limit $\xi \ll 1$ (which simply means that we have enough points to describe every wave-length) the amplification factor becomes

$$G = \frac{1 - s\xi^2/2}{1 + s\xi^2/2}. \quad (14)$$

A fundamental property of G to guarantee the good behaviour of the numerical scheme is that its modulus be monotonically decreasing with wave-number: this implies that the damping increases with increasing wave-number. However, from the plot of $|G(\xi)|$ for a given (large) value of s (Fig. 1), we see that this is not true for the CN technique. Indeed, to the right of its zero $\xi = \sqrt{2/s}$, $|G(\xi)|$ increases with wave-number. Let us study the efficiency of the scheme for different values of s. Obviously the well-behaved region is the one for which $0 < \xi < \sqrt{2/s}$. If $s \ll 1$, then the zero of the function $G(\xi)$ will be much larger than unity: the restriction $\xi < 1$ then means that we are exploiting only a small fraction of this region. When $s \simeq 1$, the scheme is well-behaved for all waves satisfying $\xi < 1$. However, we have not gained much with respect to the explicit scheme, the upper bound of which for stability is $s < 1/2$. When $s \gg 1$, $\sqrt{2/s} \ll 1$ and waves with $\xi > \sqrt{2/s}$ are still sufficiently sampled by the spatial grid, but, for these waves, the damping *decreases* with wave-number, giving completely incorrect results. We must therefore conclude that the interesting region $s > 1$ (i.e. $\Delta t > \Delta x^2/\sigma$), which, if accessible, would free us from the restriction of explicit schemes, is still forbidden for the CN method.

How serious the above restriction is in practice will of course be determined by the physical problem under consideration. When studying two-dimensional fluid turbulence, for example, the time-step is generally determined by other terms in the equations, typically the convective terms, and must obey $\Delta t \lesssim \Delta x/u$, where $u = O(1)$ is a typical velocity of the flow. On the other hand, since the physical Reynolds number is generally much larger than what can be afforded in the computation, the diffusion coefficient is chosen mainly for numerical reasons. In order for wave-lengths comparable

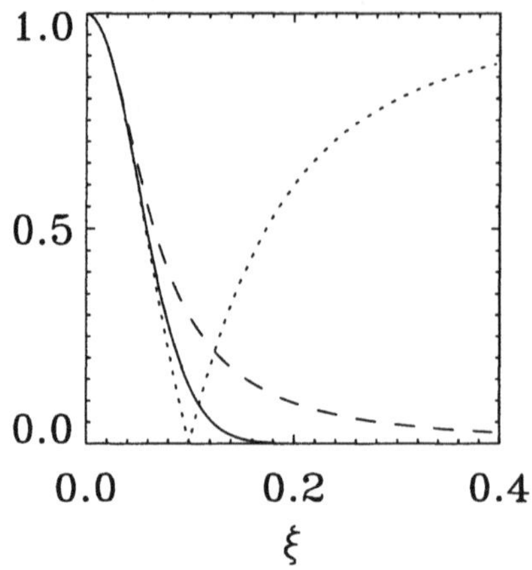

Fig. 1. Modulus of the amplification factor $G(\xi)$ as a function of $\xi = k\Delta x$ for the exact solution (solid line), the Crank-Nicolson scheme (dotted line) and our proposed scheme (dashed line). The curves are plotted for $s = 200$. For the CN case, the curve has a zero at $\xi = (2/s)^{1/2}$.

to the grid size to be damped quickly, one must have $\sigma \simeq u\Delta x^2/a$, where $a = O(1)$ is a macroscopic length scale. The stability parameter then becomes $s \simeq \Delta x/a \ll 1$, which ensures stability even for explicit schemes. We conclude that, in two-dimensional turbulence, the use of a more costly implicit scheme to treat the diffusive terms is hardly justifiable.

In other physical situations the restriction can be less severe. Let us consider a purely diffusive process, in which we only want to isolate a few dominant modes allowed by the boundary conditions, which are all long wavelength modes. This imposes a less strict bound on Δt. Suppose that we want to treat correctly all modes in the range $k_0 < k < k_1$, where the fundamental mode is $k_0 = 2\pi/L$. The maximum mode present is $k_{max} = 2\pi/\Delta x$. It is crucial to require that all modes larger than k_1 should be more damped than k_1 itself. The maximum k that satisfies this condition can be found by solving the algebraic equation $|G(\xi_1)| = |G(\xi_{max})|$, where we recall that $\xi = k\Delta x$. The result, expressed as an upper bound for the time-step, is

$$\sigma\Delta t < \frac{2}{k_1 k_{max}} = \frac{\Delta x}{\pi\, k_1}. \tag{15}$$

Thus, for this very particular case, Δt scales as Δx instead of Δx^2, as for the explicit scheme. However, for large values of k_1, the above upper limit can still be very stringent: in the next paragraph it will be shown that, with our technique, all restrictions on the time-steps are lifted.

Turning to the scheme that we propose (11) we have

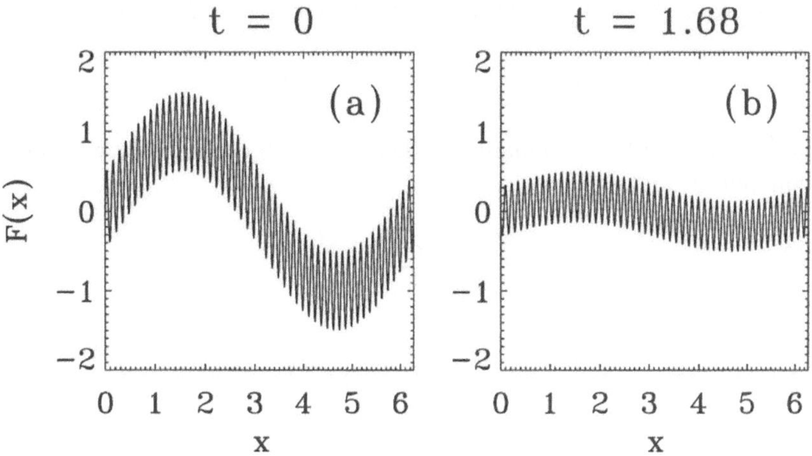

Fig. 2. Plot of $F(x)$ at different times (a) $t = 0$, (b) $t = 1.68$, from a numerical solution with the Crank-Nicolson scheme. The long wave-length is more strongly damped than the short wave-length

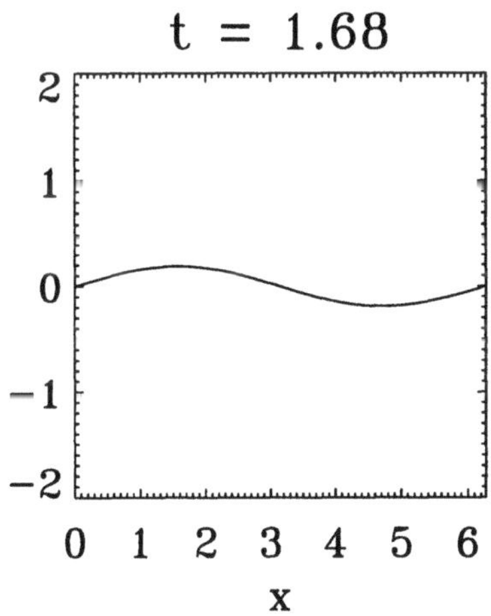

Fig. 3. Plot of $F(x)$ at $t = 1.68$, from a numerical solution with our proposed scheme. The initial condition is the same as shown in Fig. 2a. Now the short wave-length is rapidly damped

$$-\beta s F^*_{j-1} + (1+2\beta s)F^*_j - \beta s F^*_{j+1} = -\alpha s F^n_{j-1} + (1+2\alpha s)F^n_j - \alpha s F^n_{j+1} \quad (16a)$$

$$-\beta s F^{n+1}_{j-1} + (1+2\beta s)F^{n+1}_j - \beta s F^{n+1}_{j+1} = F^*_j. \quad (16b)$$

Using again the Von Neumann method to investigate stability we find that the amplification factor is

$$G = \frac{1 + 2\alpha s(1 - \cos\xi)}{[1 + 2\beta s(1 - \cos\xi)]^2}. \quad (17)$$

It can easily be verified that $|G| < 1$ for any value of ξ and s, if α and β are those of our choice ($\alpha = 1 + \sqrt{2}, \beta = 1 + 1/\sqrt{2}$): this is sufficient to ensure stability. The plot of $|G(\xi)|$, in the case $s \gg 1$, is given in Fig. 1. Note that, in contrast to the CN scheme, the amplification factor decreases monotonically to zero. It is obvious that, even for very large values of s, the property that short wave-lengths be *more* damped than long ones is preserved for the entire spectrum. The scheme is thus, according to our definition, well-behaved.

We now illustrate the results obtained above with a numerical example. We solve the diffusion equation (1), in a periodic domain of extension $L = 2\pi$, with diffusion coefficient $\sigma = 1$. The initial condition is the sum of two waves with very different wave-lengths:

$$F(x, t = 0) = A_1 \sin(k_1 x) + A_2 \sin(k_2 x), \quad (18)$$

with $A_1 = 1, A_2 = 0.5, k_1 = 1, k_2 = 50$. For the numerical solution we take $N = 500, \Delta x = 0.0126$, so that $k\Delta x < 1$ for both waves. Normally, the second wave should decay much faster, leaving only the large scale perturbation. However, for the CN scheme, if we choose the time-step so that $|G(k_1)| = |G(k_2)|$, then the two waves will decay with the same rate. For a still larger time-step the first wave will decay faster than the second one, yielding completely incorrect results. The upper bound for Δt – given by (15) with k_1, k_2 replacing k_1, k_{max} – is in this case $\Delta t_{lim} = 0.04$.

We present a numerical solution with $\Delta t = 0.08$, first using the standard CN scheme. Fig. 2 shows $F(x)$ at two different times: as expected, the large scale wave decays faster. Using the scheme of (11),(16) completely eliminates the problem, as is apparent from Fig. 3, and the small wave-length is now correctly damped faster. Fig. 4 shows the amplitude of the Fourier coefficient of each wave for the two cases. The damping of the first wave ($\gamma_1 = -\sigma k_1^2$) is reproduced accurately in both cases, since $|\gamma_1|\Delta t \ll 1$. However, the CN scheme grossly underestimates the damping rate γ_2 of the second wave, as is apparent from Fig. 2. Moreover, spurious oscillations appear, which are due to the fact that $G(\xi)$ is negative for the value of $\xi = k\Delta x$ corresponding to k_2. Our scheme also underestimates γ_2, but automatically preserves the crucial relation $|\gamma_2| > |\gamma_1|$.

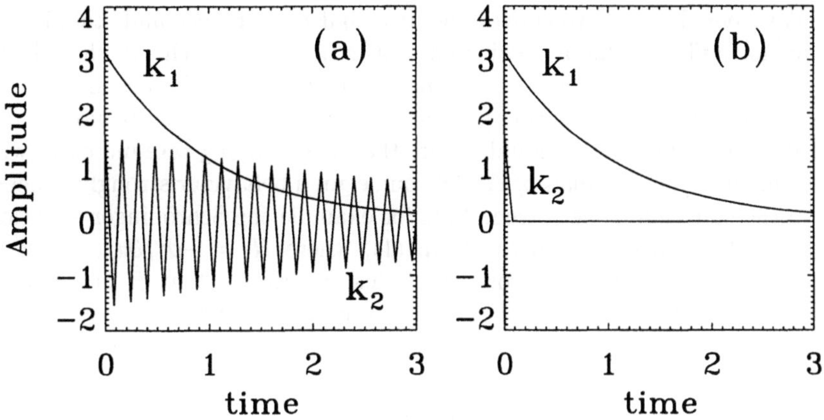

Fig. 4. Time evolution of the Fourier coefficient of both waves for (a) the Crank-Nicolson scheme and (b) our alternative scheme. The damping of the small wave-number k_1 is well reproduced in both cases, but the large wave-number k_2 is poorly treated by the Crank-Nicolson scheme

4 Conclusion

We have identified a flaw occurring in a class of finite-difference, implicit schemes for the diffusion equation, the prototype of which is the Crank-Nicolson scheme. Although the CN is stable for arbitrary values of the time-step, in practice it gives poor results for $\Delta t > \Delta x^2/\sigma$. In particular, short wave-lengths may happen to be less damped than long ones. For problems of fluid turbulence, the constraint on the size of the time-step is virtually as severe as that of explicit schemes ($\Delta t < \Delta x^2/\sigma$), although, for flows at high Reynolds numbers, the convective terms impose a time-step which always falls in the stability region. For purely diffusive problems, the constraint is somewhat less severe (Δt scales as Δx, instead of Δx^2). The possibility of using an even larger time-step is in any case ruled out.

The technique that we propose to overcome this restriction is based on a Padé-like approximation of the exact evolution operator for the diffusion equation. This allows us to construct a new scheme that preserves all the good properties of CN, while avoiding the above shortcoming. For this scheme, short wave-lengths are always more damped than long ones. The only price to pay is that the computation time *per time-step* is roughly twice that of CN, but now a much larger time-step can be used so that the total computation time will in fact be much shorter. We note that our choice, expressed by (8), is not the only one possible. The denominator on the right hand-side does not need to be a perfect square, although this slightly simplifies the resulting scheme since the same matrix is inverted twice.

We have also tried to convey the idea that convergence and stability are often not sufficient for a good numerical scheme. The scheme should also preserve qualitatively some key properties of the original equation over all the range of the discretization parameters: it should be "well-behaved". For the diffusion equation it should ensure that the damping rate increases with wave-number *for all values of* Δt. Of course good behaviour strongly depends upon the equation to be solved. Take for example the Schrödinger equation, obtained by replacing $\hat{D}$ with $i\hat{H}$ in (1), where $\hat{H} = -\partial^2/\partial x^2 + V(x)$ is the Hamiltonian. The key property of this equation is that $\hat{H}$ is a unitary operator, i.e. the integral of $|F(x)|^2$ is conserved in time. It is easy to show that the CN scheme provides the only discrete approximation to $\hat{H}$ which is both unitary and second order accurate. Both schemes (5) and (11) introduce some numerical damping that violates unitarity. The CN scheme is thus well-behaved for the Schrödinger equation, although not for the diffusion equation.

This work was partially funded by the Commission of the European Communities under contract ERBCHBICT941009.

References

Crank, J. (1975): *The Mathematics of Diffusion* (Clarendon, Oxford)

Crank, J., Nicolson, P. (1947): A practical method for numerical evaluation of solutions of partial differential equations of the heat conduction type. Proc. Cambridge Philos. Soc. **43**, 50–67

Hobson, J.M., Wood, N., Mason, P.J. (1996): A new finite difference diffusion scheme. J. Comput. Phys. **125**, 16–25

Marchuk, G.I. (1982): *Methods of Numerical Mathematics* (Springer, Berlin, Heidelberg)

Morton, K.W., Mayers, D.F. (1994): *Numerical Solution of Partial Differential Equations* (Cambridge University Press, Cambridge), 25–27

Press, W.H., Teukolsky, S.A., Vetterling, W.T., Flannery, B.P. (1992): *Numerical Recipes* (Cambridge University Press, Cambridge), 839–840

Richtmyer, R.D., Morton, K.W. (1967): *Difference Methods for Initial Value Problems* (Wiley, New York), 186–188

Roache, P.J. (1972): *Computational Fluid Dynamics*, (Hermosa, Albuquerque), 85–86

Yanenko, N.N. (1971): *The Method of Fractional Steps* (Springer, Berlin, Heidelberg)

1, 2, 3 Plasmas!

Jean Clérouin

Commissariat à l'Energie Atomique, 94195 Villeneuve St Georges Cedex, France.

Abstract. Playing with the geometrical dimensionality of plasmas models gives us the opportunity to explore models with different properties. In particular, a lower dimensionality allows us to avoid a quantum description of interactions between oppositely charged particles, totally in one dimension and partially in two dimensions. We will describe some models; from the One and Two Component Plasma in one and two dimensions to the "true" three dimensional plasma combined with a Thomas-Fermi description of the electrons.

1 Introduction

1.1 Potentials

The electrostatic potential at a point r due to a charge e placed at the origin is given by the solution of Poisson's equation in d dimensions

$$\nabla^2 \varphi(r) = -C_d e \delta(r),$$

where $r = |\mathbf{r}|$ and C_d=2, 2π, 4π in 1, 2 and 3 dimensions. The solutions are

$$\varphi(r) = -er \qquad (d=1)$$

$$\varphi(r) = -e \log(r/L) \qquad (d=2)$$

$$\varphi(r) = -e / r. \qquad (d=3)$$

While the functional form of $\varphi(r)$ depends on dimensionality, its Fourier transform exhibits the k^{-2} behaviour which follows directly from Poisson's equation :

$$\tilde{\varphi}(k) = \int \exp\left[i\, \mathbf{k}.\mathbf{r}\right] \varphi(r)\, \mathrm{d}^d r = \frac{C_d e}{k^2}$$

It is also possible to consider d-dimensional plasma where a geometric dimension d does not coincide with the Coulomb dimension d'. For example point charges in two dimensions are $(d,d') = (2,3)$ plasmas (or shortly, 2-3 plasmas) and point charges in one dimension are 1-3 plasmas. Those plasmas, as shown by Lotte and Feix using rescaling methods (Lotte and Feix 1984), are not governed by the same plasma frequency at high and low temperature, but they correspond to real models such as electrons trapped on the surface of helium (Grimes and Adams 1979) or ions in a magnetic trap (Dubin 1996).

1.2 Graininess Parameter

A usual way to characterise a plasma is to define a graininess parameter g in such way that, if we perform the splitting experiment which consists in cutting in two parts the particles ($e \to e/2$, $m \to m/2$, $n \to 2n$, $v \to v$, $T \to T/2$), this parameter will be also divided by 2. Another requirement is that the plasma frequency $\omega_p^2 = C_d ne^2/m$ and the Debye length $\lambda_D^2 = kT/ne^2$ will be preserved in the limit $g \to 0$ (Vlasov limit) where only collective effects exist. The collective character of the plasma is then characterized by $g = 1/n\lambda_D^d$, which is nothing else that the inverse of the number of particles in the Debye cube.

It is also convenient to define a coupling parameter which is the ratio of the electrostatic energy to the mean kinetic energy. In d dimensions we have:

$$\Gamma^2 = e^4 / \left(\varepsilon_0^2 n^{4/d-2} kT\right)$$

which becomes :

$$\Gamma = e^2/nkT = g^2 \qquad (d=1)$$

$$\Gamma = e^2/kT = \frac{1}{2\pi} g \qquad (d=2)$$

$$\Gamma = e^2 n^{1/3}/kT = \frac{1}{4\pi} g^{2/3} . \qquad (d=3)$$

It is interesting to note that the coupling Γ does not depend on the density in 2 dimensions (2D), is proportional to $n^{1/3}$ in 3D and inversely proportional to n in 1D.

1.3 Dense Plasmas and Simulation

If we consider a system of N particles in d dimensions enclosed in a box of side L with a density n, the total number of particles can be written as

$$N = nL^d = n\lambda_D^d\left(\frac{L}{\lambda_D}\right)^d = \frac{1}{g}\left(\frac{L}{\lambda_D}\right)^d .$$

Two cases must be considered:

a) we are interested in simulating a Vlasov plasma : g must be very small (e.g. 10^{-2}) and the box length must be much bigger than the Debye length (e.g. L/λ_D=100). We need 10^4 particles in 1D, 10^6 in 2D and 10^8 in 3D.

b) we are interested in strongly coupled plasmas. In this case $g \geq 1$, and the Debye length is also much smaller (of the same order than the inter-particle spacing e.g. L/λ_D=20). The number of particles needed is then 20, 4.10^2 and $8\ 10^3$, in 1D, 2D and 3D respectively.

This simple argument indicates that the simulation of dense plasmas is much easier than Vlasov plasmas. Even in 3D a few hundreds of particles is enough to approach the thermodynamic limit and a simple scheme of simulation based on the solution of the N body problem is efficient. Unfortunately, due to the slow decay of the Coulomb potential, all interactions between particles must be treated (N^2). For periodic boundary conditions Ewald summations must be used in 2D and 3D (for details see the review by Hansen (Hansen 1986). On the other hand, Vlasov plasmas can be simulated by so called "particle codes" using millions of pseudo particles and using Poisson solvers on a grid, thus ignoring all individual interactions.
In the following we focus on the simulation of strongly coupled plasmas in d dimensions.

2 One Component Plasmas

When the electronic density is high enough, all levels are filled to the Fermi level and electrons cannot exchange momentum with the ions. The electron gas is said to be fully degenerate. The electrons can be viewed as a fixed neutralising continuum and the system reduces to the ionic component. This model, known as the One Component Plasma (OCP), can be appropriate in extreme situations such as in interior of Jovian planets, stars and white dwarfs, liquid metals or Inertial Confinement pellets. Due to its simplicity, the OCP is also a very good model to test statistical mechanics.

2.1 One Dimension

The one dimensional OCP has been investigated by Marc Feix (Feix 1976). The system is made up by infinite charged planes constrained to move in one dimension x. The trajectories are a succession of sinusoids, and the simulation code reduces to a book keeping of crossings times. At low temperature each particle is trapped in its own potential well and oscillates without energy exchange with its neighbors. This is an example of a non ergodic system.

2.2 Two Dimensions

The two dimensional OCP has been more studied due to its similarity to real systems such as electrons trapped on helium ($1/r$ interaction) surfaces or vortex in superfluids ($\ln(r)$ interaction). The equation of state has been first calculated by Calinon (Calinon, Choquard et al. 1980). Moreover the nature of the transition between the fluid phase and the ordered phase was a great subject of discussion in the light of the Kosterlitz-Thouless theory (Kosterlitz and Thouless 1973), which was predicting the existence of a intermediate hexatic phase with an exponential decay for translational order, and algebraic for orientational order. The search for an evidence of the hexatic phase stimulated a great number of simulations, using various boundary conditions, without definitive conclusion on the nature of the transition (Caillol et al. 1982)

A carefully study of the system below the melting temperature revealed the existence of cooperative premelting phenomena, such as loops or permutations of six or more particles or gliding rows which were responsible for the rise of the entropy (Choquard and Clérouin 1983). Because the same kind of phenomena were observed for hard disks systems (Alder et al. 1963), we concluded the universality of cooperative premelting phenomena. Recently the same kind of permutation has been observed in high temperature materials (Doan and Adda 1987) and also between magnetic vortex in high T_c compounds (Yates et al. 1995).

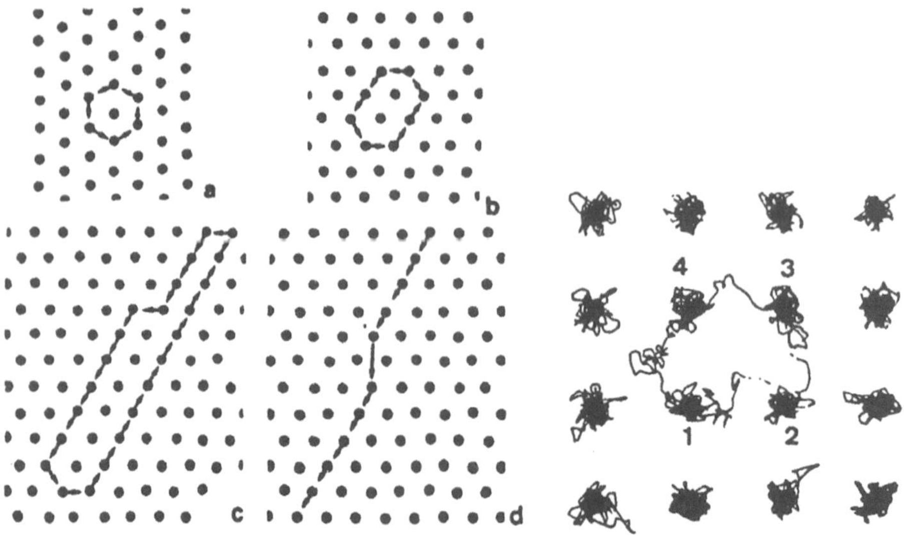

Figure 1: Cooperative motions observed below the melting of the 2D- OCP (left), and cooperative motions observed in a high temperature solid (right) (Doan and Adda 1987).

2.3 Three Dimensions

The 3D OCP has been widely studied and is now considered as the starting point of any "real" dense plasma model. We just give a bibliographic reference of a review on this subject (Baus and Hansen 1980).

3 Two Component Plasmas

3.1 One Dimension

The 1D Coulomb potential, being regular at the origin, it is easy to consider plasmas of opposite charges, "ions" and "electrons", in a purely classical frame. At high temperature the system behaves as a N particle plasma and at low temperature, or higher coupling parameter, pairs of oppositely charged particles are forming. The equation of state PV/NkT changes from 1 at the kinetic limit, $g \to 0$, to 1/2 in the strongly coupled regime with all paired particles (Feix 1976). This behavior reproduces one the most important features encountered in "real" laser plasmas when a sudden rise in the temperature yields a strong augmentation of the pressure due to ionization (More 1983).

3.2 Two Dimensions

In 2D the stability of oppositely charged particles is only marginal. The system is stable only for a coupling $\Gamma = e^2/kT < 2$. For higher values of Γ a short range correction must be added to the $\ln(r/L)$ potential. It can be a repulsive short range part (hard cores) or a regularized parabolic core, corresponding to smeared out ions (Thomson model).

The first model, Coulomb potential plus hard cores, with the ions fixed on a lattice allowed us to show off a continuous delocalisation transition between a dielectric state with all "ions" and "electrons" paired, and a conducting state (Clérouin and Hansen 1985). This transition, which is also a Kosterlitz-Thouless transition, can be evidenced by the critical onset of the electric conductivity in the vicinity of the transition (Clérouin et al. 1987). A second kind of transition appeared with the soft core model. When the radius σ of the soft cores are of the same order of magnitude than the mean ionic disk radius defined as $a_i=(4\pi n)^{-1/2}$, the electrons are again delocalized. At low temperature, their equilibrium position changes from the center of core disks to some equilibrium position between "ions" (Clérouin et al. 1987).

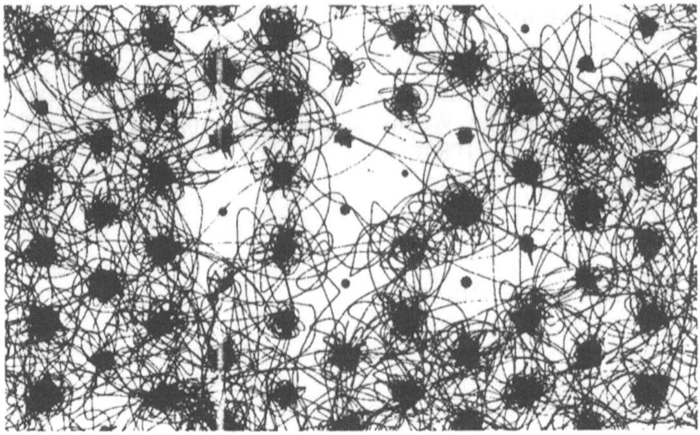

Figure 2: Electron and ion trajectories for the hard core model in two dimensions, at moderate coupling.

This transition is very reminiscent of the so called pressure ionization in real plasmas, where a simple criterion, given by D. More (More 1983) predicts that electrons are ionized as soon the mean electronic distance a_e is of the same order as the Bohr radius $r_s=a_e/a_B\approx 1$. Here σ plays the role of a "two-dimensional Bohr" radius and can be understood as a Heisenberg smearing due to the incertainity of their positions. In the (ρ,T) plane this two dimensional model reproduces two very important process of real plasmas: thermal and pressure ionization leading to a very rich phase diagram (Choquard et al. 1989).

The soft core model with mobile ions is even more exciting. Some exploratory runs indicated a very complex behavior starting from a gas of neutral paired "molecules" and finishing at high temperature to a "ions" and "electrons" plasma with all intermediate states consisting of mixtures of charged and neutral particles. Unfortunately this study has not gone further and we just mention a theoretical paper on the succession of phase transition occuring in this model (Gallavoti and Nicolo 1985).

3.3 Three Dimensions

In 3 dimensions the simulation of point particles with opposite charges appears more difficult. First the validity domain of such a model is limited because one has to consider a region where, simultaneously, the electronic De Broglie length must be smaller than the ionic sphere radius ($\Lambda_e/a < 1$), the plasma must be fully ionized ($kT > 1$ Ryd) and the coupling parameter must be greater than unity to allow for molecular dynamics simulations. The second problem is the validity of approximating diffraction and symmetry effects by a two body effective potential, such as the one given by Deutsch (Minoo et al. 1981). Moreover the electron to ion mass ratio makes the simulation very slow which yields results with poor statistics. In spite of these difficulties, interesting results on the electrical conductivity have been obtained by Hansen (Hansen and McDonald 1981).

These difficulties are calling for a more physical description of electrons, with the electronic density treated as a continuous variable, including some quantum physics, such as the Pauli principle to ensure the stability of the system.

4 Ions and Continuous Electronic Density

The density functional theory (Parr and Yang 1989) gives a fundamental basis to deal with the electronic density in the external field of the ions. This theory states that the density which minimises the total energy of the system corresponds to the fundamental level of the system. The coupling of this theory with molecular dynamics has been achieved by Car and Parrinello (Car and Parrinello 1985) who proposed to use a fictitious Lagrangian to control at the same time the evolution of the electronic density and the ionic motion. The trick is that, after a first electronic minimisation, it is not necessary to diagonalise repeatedly the Hamiltonian at each ionic time step. The evolution of the electronic density is computed "on the fly" by introducing new degrees of freedom in the Lagrangian which guarantees that the density will stay close to the Born-Oppenheimer surface during the simulation (for a synthetic review of the method see Remler and Madden 1990).
The method is based on the electronic density, but the expression of the kinetic energy of the electron gas requires orbitals, which in the genuine method are the variables controlled by the Lagrangian. The cost of the method scales as the cube of the number of orbitals, which forbids the simulation of large systems. To introduce this method in the field of plasmas physics, we suggested the use of some approximated form of the electron kinetic energy. The simplest one being the Thomas-Fermi approximation, we used the fictitious Lagrangian frame of Car and Parrinello to combine a Thomas-Fermi description for the electrons with a classical molecular dynamics for the ions.

4.1 A One Dimensional Prototype

Thomas Fermi Molecular Dynamics (TFMD) was first tested on a one dimensional model (Clérouin et al. 1990; Clérouin et al. 1991) The system is made up of ions in 1D immersed in an electronic continuum of density $\rho(x)$ ensuring the global neutrality. In one dimension the Fermi energy is :

$$K_F = \frac{h^2}{32m}\int_{-a}^{a}\rho^3(x)dx \ .$$

The Lagrangian of the whole system is written

$$L = \int \frac{1}{2}\mu\dot{\rho}^2(x)dx + \frac{1}{2}\sum_I M_I \dot{X}_i^2 - E\left[\rho(x), X_I\right] + \Lambda\left(\int \rho(x) - N\right),$$

where the first two terms are, respectively, the fake kinetic energy associated with the electronic density and the real ionic kinetic energy. The total energy is the sum of the Fermi energy K_F (which acts here as a potential term) and of the usual electrostatic contributions U_{ee} ,U_{ei} and U_{ii} . The last term Λ is a Langrange multiplier to ensure global neutrality. By taking the derivatives of the Lagrangian with respect to the density and the ionic positions, we get the equation of evolution of the electronic density :

$$\mu\ddot{\rho}(x,t) = \frac{\partial E}{\partial \rho(x,t)} + \Lambda$$

and of the ions :

$$M_I \ddot{X}_I = -\frac{\partial E}{\partial X_I(t)} \; .$$

After a first electronic minimization, the two dynamics are started and, provided the electronic masses μ are small compared to the ionic ones (but not so small than 1/1852), the electronic density will follow adiabatically the ionic motion. In fig.3 we have drawn the electronic density for a random ionic distribution, after the first minimisation.

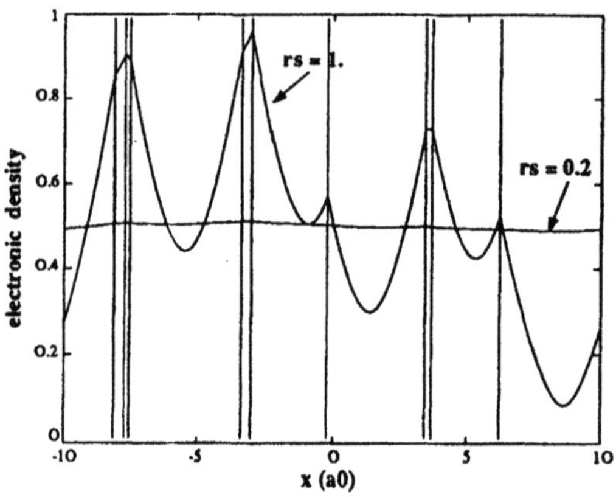

Figure 3: Electronic density for a random configuration of the ions (vertical lines) and for two electronic densities r_s=0.2 and r_s=1.

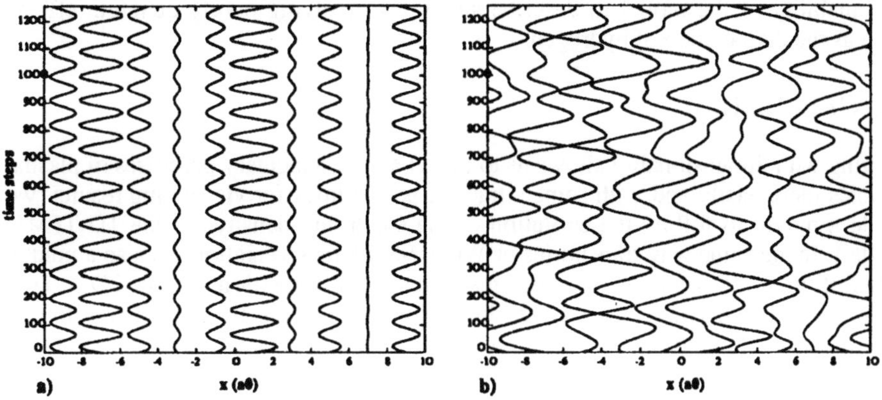

Figure 4: x versus time for the 1D TFMD model, for r_S=0.2 (OCP like) and r_S=1. and for the same coupling Γ=7 and initial conditions.

At high density $r_s=a_e/a_B=0.2$, the electronic density is almost constant and the ionic interactions reduce to the bare coulomb potential of the OCP. At lower density, $r_s=1$, the electrons are now strongly polarized by the ions, screening their interactions.This screening has a dramatic effect on the periodic structure of the 1D-OCP. When $r_s=0.2$, the ions are oscillating independently and the system remains trapped as in the OCP case. When the electronic density is lowered, $r_s=1$, the ionic interactions are more screened and consequently weaker. The crystal melts. In some way the 1D-plasma recovers its lost ergodicity through the screening of interactions (fig. 4).

4.2 Full 3D Model

The three dimensional model (Clérouin et al. 1992) is more complicated because of the number grid points needed to define accurately the electronic density, and also due to the computation of the electric field. A very efficient Poisson solver using 3D Fast Fourier transforms on at least 32^3 grid points is necessary. It now possible to compute the evolution of systems containing 250 ions over hundreds of plasmas periods and for an arbitrary electronic degeneracy. We must also mention that all Thomas-Fermi functionals can be easily extended to finite electronic temperatures, which is much more difficult for quantum models.
To test this model we have first quantified the effect of screening on static and dynamical properties of a dense hydrogen plasma, by comparing our results with linear screening theories and the OCP model (Zérah et al. 1992). The equation of state of this system has been found also in good agreement with quantum calculations (Penman et al. 1995). Static properties and transport coefficients such as diffusion are also comparing favourably with Car-Parrinello, Thigth Binding or Quantum Monte-Carlo calculations for a much lower computational effort (Clérouin and Bernard 1997) in the dense regime. However, the price to pay for a simplified density functional is

the unability of the model to describe situations involving purely quantum effects such as the dissociation of hydrogen.

5 Conclusion

Playing with dimensionality allows us to model some elementary processes of plasma physics more easily and with some attractive possibilities for visualizing results (we made a movie to follow the evolution of the electronic density for the 1D-TFMD model). In the same way trajectories or Wigner-Seitz representation are also a unique possibility of 2D models (e.g. the metal-insulator transition of the 2D - Two Component Plasmas). At least lower dimensionality models are a good way to test new ideas in a simple manner (e.g.: the 1D-TFMD model) and to design prototypes for 3D models.
We would not finish without mentioning the extraordinary experiment of D. Dubin (Dubin 1996) where a few hundreds of ions are trapped by a magnetic field and cooled to cryogenic temperatures. Depending upon the choice of parameters, the ions arrange themself in 3D-ellipsoid, or in a 2D-lattice (a 2-3 plasmas) and even in a 1D-chain (i.e., 1-3 plasmas). This last configuration is unstable because the ions have a tendancy to rearrange along planes and hence to restore the proper coulombic dimensionality of 1-1 plasmas.

References

Alder, B. J., Hoover, W. G., Wainwright, T. E. (1963): Cooperative motion of hard disks leading to melting. Phys. Rev. Lett. **11**, 241-243

Baus, M., Hansen, J. P. (1980): Statistical Mechanics of Simple Coulomb Systems. Physics Reports **59**, 1-94

Caillol, J. M., Levesque, D., Weis, J. J., Hansen, J. P. (1982): A Monte Carlo study of the classical two-dimensional one-component plasma. J. Statistical Physics **28**, 324-348

Calinon, R., Choquard, P., Jamin, E., Navet, M. (1980): On the equation of state of the 2-D Wigner model, in *Ordering in Two Dimensions* (Sinha, Elsevier North Holland), 317-322

Car, R., Parrinello, M. (1985): Unified Approach for Molecular Dynamics and Density-Functional Theory. Phys. Rev. Lett. **55**, 2471-2474

Choquard, P., Clérouin, J. (1983): Cooperative Phenomena below Melting of the One-Component Two-Dimensional Plasma. Phys. Rev. Lett. **50**, 2086-2089

Choquard, P., Piller, B., Rentsch, R., Clérouin, J., Hansen, J. P. (1989): Ionization and phase diagram of classical Thomson atoms on a triangular lattice. Phys. Rev. A **40**, 931-945

Clérouin, J., Bernard, S. (1997): Dense hydrogen plasma: a comparaison between models. Phys. Rev. E **56**, 3534-3539

Clérouin, J., Hansen, J. P. (1985): Delocalization Transition of a Classical Two-Dimensional Coulomb Lattice. Phys. Rev. Lett. **54**, 2277-2280

Clérouin, J., Hansen, J. P., Piller, B. (1987): On the Dielectric Phase of the Periodic Two-Dimensional Coulomb Gas: Optical Modes, Delocalization and Melting. Europhysics Lett. **4**, 771-776

Clérouin, J., Hansen, J. P., Piller, B. (1987): Two-dimensional classical electron gas in a periodic field: Delocalization and dielectric-plasma transition. Phys. Rev. B **36**, 2793-2810

Clérouin, J., Pollock, E. L., Zérah, G. (1992): Thomas-Fermi Molecular Dynamics. Phys. Rev. A **46**, 5130-5137

Clérouin, J., Zérah, G., Benisti, D., Hansen, J. P. (1990): Plasma Simulations Using the Car-Parrinello Method. Europhysics Lett. **13**, 685-690

Clérouin, J., Zérah, G., Hansen, J. P. (1991): Plasma Simulations Using the Car-Parrinello Method, Proceedings of *Computer Simulation in Material Science* conference, Aussois France (M. Meyer and V. Pontikis ed., Kluwer Academic)

Doan, N. V., Adda, Y. (1987): Nouveau mécanisme de diffusion dans les solides à haute température. Philosophical Magazine A **56**, 269-283

Dubin, D. H. E. (1996): Correlation effects on static and dynamic properties of nonneutral plasmas, Proceedings of *Physics of Strongly Coupled Plasmas* conference, Binz/Rügen Germany (W. Kraeft and M. Schlanges ed., World Scientific)

Feix, M. R. (1976): Computer Experiments in One-Dimensional Plasmas, Proceedings of *Strongly Coupled Plasmas* conference, Orleans France (C. G. Kalman ed., Plenum, New York)

Gallavoti, G., Nicolo, F. (1985): The "Screening Phase Transitions" in the Two-Dimensional Coulomb Gas. J. Statistical Physics **39**, 133-152

Grimes, C. C., Adams, G. (1979): Evidence for a Liquid-to-Crystal Phase Transition in a Classical, Two-Dimensional Sheet of Electrons. Phys. Rev. Lett. **42**, 795

Hansen, J. P. (1986): Molecular-Dynamics Simulation of Coulomb Systems in Two and Three Dimensions, Proceedings of *Molecular Dynamics Simulations of Statistical-Mechanical Sytems* conference, Varenna Italy (G. Ciccotti and W. G. Hoover ed., Soc. Italiana di Fisica Bologna)

Hansen, J. P., McDonald, I. R. (1981): Microscopic simulation of a strongly coupled hydrogen plasma. Phys. Rev. A **28**, 2041-2059

Kosterlitz, J. M., Thouless, D. J. (1973): Ordering, metastability and phase transitions in two-dimensional systems. J. Phys. **C6**, 1181-1203

Lotte, P., Feix, M. R. (1984): Plasma models and rescaling methods. J. Plasma Physics **31**, 141-151

Minoo, H., Gombert, M. M., Deutsch, C. (1981): Temperature-dependent Coulomb interactions in hydrogenic systems. Phys. Rev. A **23**, 924-943

More, D. (1983): *Atomic Processes in High-Density Plasmas* (Plenum, New-York)

Parr, R. G., Yang, W. (1989): *Density-Functional Theory of Atoms and Molecules* (Oxford Science, Oxford)

Penman, J. I., Clérouin, J., Zérah, G. (1995): Equation of state of a hydrogen plasma by density functional molecular dynamics. Phys. Rev. E **51**, R5224-R5227

Remler, D. K., Madden, P. (1990): Molecular Dynamics wihout effective potentials via the Car-Parrinello approach. Mol. Phys. **70**, 921-966

Yates, K., Newman, D. J., DeGroot, P. A. J. (1995): Effect of pin density on planar vortex mobility. Phys. Rev. B **52**, R13149-R13151

Zérah, G., Clérouin, J., Pollock, E. L. (1992): Thomas-Fermi Molecular-Dynamics, Linear Screening, and Mean-Field Theories of Plasmas. Phys. Rev. Lett. **69**, 446-449

The Notion of Integrability for Time-Dependent Hamiltonian Systems: Illustration from the Relativistic Motion of a Charged Particle in an Electromagnetic Field

Alain Bourdier and Serge Bouquet

Commissariat à l'Energie Atomique, Centre de Bruyères-le-Châtel, DPTA/PPE, BP 12, F-91680 Bruyères-le-Châtel, France

Abtract. It is shown that "Liouville's theorem" on integrability still holds in the case of time-dependent Hamiltonian systems; when they have n independent possibly time-dependent invariants in involution the solution can be found by quadratures and no chaos can take place. This is applied to the relativistic motion of a charged particle in a homogeneous constant magnetic field and a transverse circularly polarized homogeneous electric field.

Keywords. Integrability, Relativistic Hamiltonian, Electromagnetism.

1 Introduction

One might have expected that knowledge of 2n independent invariants would always be required for obtaining the general solution for an autonomous Hamiltonian system with n degrees of freedom. This is suggested by the fact that Hamilton's equations are a set of 2n first order equations whose general solution can be specified by 2n independent invariants. However, from a result due to Bour and often attributed to Liouville[1], the existence of n independent, time independent invariants in involution is sufficient for deriving a complete solution to the problem.[2,3] Actually, Liouville extended this result on integrability to non autonomous systems by showing that n independent, possibly time-dependent invariants in involution can be used to obtain n additional invariants.[1,4] However, he did not construct the 2n invariants as n canonically conjugate pairs.[5]

An autonomous Hamiltonian system is called completely integrable if it possesses n independent, time-independent invariants in involution.[2,3,6,7] The solution for the motion of a completely integrable system can be expressed in terms of canonical action-angle variables.[2,3,6,7] Unfortunately, this does not mean that the equations of motion can be integrated analytically. Nevertheless, the solution of the equations always exists and is unique for specified initial data. Moreover, when the motion is completely integrable, all the Lyapunov exponents one can compute equal zero. Chaotic trajectories cannot fill a non zero measure phase-space volume.

A general time-dependent Hamiltonian system with n degrees of freedom is usually considered to be equivalent to a particular autonomous Hamiltonian system in an extended phase space with n + 1 degrees-of-freedom.[7] In Sec. 2, the Hamiltonian in the extended phase space is introduced carefully. We show that the equivalent system is completely integrable if the original, non autonomous system possesses n independent, possibly time-dependent invariants in involution. It can be proved that it is possible to construct 2(n + 1) invariants as n+1 canonically conjugate pairs.[5] The demonstration for the case when n = 1 is given by one method in Section 2.1 and by a second method in Section 2.2. On the other hand, it was shown by Kozlov and Kolesnikov that the solution of the original non-autonomous system can be found by quadratures.[2] This result has been shown explicitly in the case of a one-degree-of-freedom Hamiltonian system for which some invariant is specified. A second invariant is expressed in terms of quadratures which permits one to derive formally the solution as a function of the two constants.[8,9,10] This is outlined at the beginning of Section 2. In addition it can be shown, by using the definition of a Lyapunov exponent, that no chaos can take place in the case of integrable systems.[11] We have decided to call a time-dependent Hamiltonian system integrable if it possesses n independent, possibly time-dependent invariants in involution. This is an extension of the first definition of integrability. In this sense, we say that "Liouville's theorem" on integrability still holds in the case of time-dependent Hamiltonian systems.

We illustrate the concept of integrability of time-dependent systems in terms of an example which concerns the motion of a charged particle in a constant homogeneous magnetic field and transverse circularly polarised electric field. This problem has already been explored by other authors.[12,13,14] It has been shown recently how the Hamiltonian formalism brings enlightenment to their discussions.[15] The integrability of this two-degree-of-freedom system can be shown by finding two independent constants in involution, one obtained by using Noether's theorem.[2,3,16,17] Alternatively the system can be reduced to a one-dimensional problem which can be solved by quadratures in two different ways and, because of the extension of the definition of integrability we have made, shown to be integrable.[15] Two canonically conjugate invariants are obtained simply by integrating Hamilton's equations. Canonical transformations[2,3,6,7,15,17] permit us to take the two first invariants as two new conjugated variables and, consequently, we can reduce the system to a time-dependent Hamiltonian system with one degree-of-freedom possessing one invariant. It is therefore integrable in the new "Liouville sense". Then, as we already know, a second integral can be derived which permits to give the solution in terms of quadratures. Another approach by quadratures exists. It consists in showing that on the one hand the solution can be written in terms of the energy of the particle and on the second hand the energy is shown to be solution of an integrable differential equation.

2 Integrability of Time-Dependent Hamiltonian Systems

In this section, we study the integrability of a n-degrees-of-freedom time-dependent Hamiltonian system. However, in order to make the approach easier, we begin with a one-degree-of-freedom time-dependent Hamiltonian system.

2.1 Integration by Quadratures of One-Degree-of-Freedom Time-Dependent Systems

This is an outline of what has been presented in Refs. 8, 9, 10. We consider one-dimensional time-dependent problems where the Hamiltonian H(Q, P, t) is a function of canonically conjugate variables (Q, P) and time t. It must be noted that H is not a constant of motion as it depends explicitly on time. The existence of a first integral I(Q, P, t) is assumed. The formal inversion of I with respect to P provides

$$P = G(Q, I, t), \tag{1}$$

where G is the reciprocal function of I. Using equation (1) in one of Hamilton's equations, we get

$$\dot{Q} = h(Q, I, t), \tag{2}$$

where $h(Q, I, t)$, is a specified function. Equation (2) can be considered as a first-order differential equation with parameter I. In order to integrate equation (2) we look for an integrating factor $\alpha(Q,I,t)$ in such a way that $\alpha dQ - \alpha h dt$ is a total differential form. As a consequence, the derivative of the coefficient of dQ with respect to t must equal the derivative of the coefficient of dt with respect to Q. This provides $\alpha_t = -(\alpha h)_Q$. We obtain, this way, the following partial differential equation

$$\alpha_t + \alpha_Q h + \alpha h_Q = 0, \tag{3}$$

where the subscript stands for the partial derivative. One can show that a solution to equation (3) is[8,9]

$$\alpha = G_I(Q, I, t) = \left[I_P(Q, P, t)\right]^{-1}, \tag{4}$$

where $G_I = \partial G/\partial I$ and $I_P = \partial I/\partial P$. Then, to integrate equation (2), we note that, since $G_I dQ - G_I h dt$ is a total differential form, it can be interpreted as the total derivative of a first integral J(Q,I,t) such that $dJ = G_I dQ - G_I h dt$ (no term is proportional to dI as I is a first integral). Under these circumstances, one must have

$$J_Q(Q, I, t) = G_I(Q, I, t), \quad J_t(Q, I, t) = -G_I(Q, I, t)\, h(Q, I, t). \tag{5}$$

Integration leads to

$$J(Q, I, t) = \int_0^Q \frac{dQ'}{I_P\left[Q'; P = G(Q', I, t); t\right]} - \int_0^t \frac{H_P\left[0; P = G(0, I, t'); t'\right]}{I_P\left[0; P = G(0, I, t'); t'\right]}\, dt'. \tag{6}$$

This way, a second invariant is expressed in terms of quadratures.[8,9,10] As announced in the introduction I and J are canonically conjugate, i.e., [J, I] = 1, where [A, B] stands for the Poisson bracket of A with B.[5,10] Equation (6) shows that Q can be derived formally as a function of time and as a function of two arbitrary constants of motion I and J. This is in good agreement with the work of Kozlov and Kolesnikov, who proved that in such a case the solution can be found with quadratures.[2] In that sense our system is integrable.

2.2 Integrability of Time-Dependent Systems with n Degrees-of-Freedom

Let us consider an autonomous Hamiltonian $\boldsymbol{H}$ (q_j, p_j, j = 1, n+1) with n + 1 degrees-of-freedom of the form

$$\boldsymbol{H}\left(q_j, p_j; j = 1, n+1\right) = p_{n+1} + H\left(q_i, p_i, q_{n+1}; i = 1, n\right), \tag{7}$$

where $H\left(q_i, p_i, q_{n+1}; i=1, n\right)$ is an arbitrary function of its arguments. The reason for this specific form will become apparent further. In such a case, Hamilton's equations are (τ is the time associated with this system)

$$\frac{dq_1}{d\tau} = \frac{\partial \boldsymbol{H}}{\partial p_1} = \frac{\partial H}{\partial p_1}, \quad \frac{dp_1}{d\tau} = -\frac{\partial \boldsymbol{H}}{\partial q_1} = -\frac{\partial H}{\partial q_1}, \tag{8a,b}$$

$$\frac{dq_j}{d\tau} = \frac{\partial \boldsymbol{H}}{\partial p_j} = \frac{\partial H}{\partial p_j}, \quad \frac{dp_j}{d\tau} = -\frac{\partial \boldsymbol{H}}{\partial q_j} = -\frac{\partial H}{\partial q_j}, \tag{8c,d}$$

$$\frac{dq_{n+1}}{d\tau} = \frac{\partial \boldsymbol{H}}{\partial p_{n+1}} = 1, \quad \frac{dp_{n+1}}{d\tau} = -\frac{\partial \boldsymbol{H}}{\partial q_{n+1}} = -\frac{\partial H}{\partial q_{n+1}}. \tag{8e,f}$$

Since this system is autonomous, $\boldsymbol{H}$ is conserved

$$\boldsymbol{H} = I_1, \tag{9}$$

where I_1 is a constant. Equation (8e) can be readily integrated to give

$$q_{n+1} - \tau = I_{n+1}, \tag{10}$$

where I_{n+1} is a constant. Thus two invariants are $\boldsymbol{H}$ and $q_{n+1} - \tau$. They are not in involution. Instead they are canonically conjugate

$$\left[q_{n+1} - \tau, \boldsymbol{H}\right] = 1. \tag{11}$$

One concludes that in the case n=1, two invariants can be derived for which the Poisson bracket is unity. This is in good agreement with the previous section.

Let us now replace the variable q_{n+1} by t (we will see below that t plays the role of a time) and one can write the equations of motion (Eqs. 8) in the following way

$$\frac{dq_1}{dt} = \frac{\partial H(q_i, p_i, t)}{\partial p_1}, \quad \frac{dp_1}{dt} = -\frac{\partial H(q_i, p_i, t)}{\partial q_1}, \tag{12a,b}$$

$$\frac{dq_i}{dt} = \frac{\partial H(q_i, p_i, t)}{\partial p_i}, \quad \frac{dp_i}{dt} = -\frac{\partial H(q_i, p_i, t)}{\partial q_i}, \tag{12c,d}$$

$$\frac{dp_{n+1}}{dt} = -\frac{\partial H(q_i, p_i, t)}{\partial t}. \tag{12e}$$

The first 2n equations are those of an n-dimensional non-autonomous system for which $H(q_i, p_i, t)$ is the time-dependent Hamiltonian. Finally, since $dH/dt = \partial H/\partial t$, the last equation (Eq. 12e) can be written as

$$\frac{dp_{n+1}}{dt} = -\frac{dH}{dt}. \tag{13}$$

By taking equations (7) and (9) into account, its solution can be written as

$$p_{n+1} = -H + I_1. \tag{14}$$

This equation, which is a consequence of Eq.(12e), is just Eq.(7) where I_1 stands for $\boldsymbol{H}$. The difference between the autonomous Hamiltonian function $\boldsymbol{H}$ and its value I_1 must be emphasised. The value of $\boldsymbol{H}$ is an arbitrary constant along the solution trajectories where I_1 can chosen to be zero. This result leads to an important remark concerning what has appeared in the literature about the extended phase space. When an n-dimensional time-dependent Hamiltonian $\widetilde{H}(q_i, p_i, t)$ $(i = 1,n)$ is considered in Ref.(7), the authors let $q_{n+1} = t$ and $p_{n+1} = -\widetilde{H}$, and take $\widetilde{\mathbf{H}}(q_i, p_i, q_{n+1}, p_{n+1}) = \widetilde{H} + p_{n+1} \equiv 0$ as a new Hamiltonian in the extended phase space. However, it is usually (incorrectly) omitted that this value is valid only along a solution trajectory. In addition, this is misleading as $\widetilde{\mathbf{H}}$ is identical to zero and cannot be considered as an arbitrary constant of motion. $\boldsymbol{H}$ can be chosen as a new Hamiltonian and is an arbitrary constant if one lets

$$p_{n+1} = -\widetilde{H} + I_1. \tag{15}$$

Let us consider again equations (12) and assume $H(q_i,p_i,t)$ has n independent first integrals, possibly time-dependent, $J_i(q_i, p_i, t)$ $(i=1,n)$ in involution. We have

$$\frac{dJ_i}{dt} = \frac{\partial J_i}{\partial t} + \sum_{i=1}^{n}\left(\frac{\partial H}{\partial p_i}\frac{\partial J_i}{\partial q_i} - \frac{\partial H}{\partial q_i}\frac{\partial J_i}{\partial p_i}\right) = 0. \tag{16}$$

Then

$$\left[J_i,\boldsymbol{H}\right] = \sum_{i=1}^{n}\left(\frac{\partial J_i}{\partial q_i}\frac{\partial \boldsymbol{H}}{\partial p_i} - \frac{\partial J_i}{\partial p_i}\frac{\partial \boldsymbol{H}}{\partial p_i}\right) - \frac{\partial \boldsymbol{H}}{\partial t}\frac{\partial J_i}{\partial p_{n+1}} + \frac{\partial \boldsymbol{H}}{\partial p_{n+1}}\frac{\partial J_i}{\partial t}$$

$$= \sum_{i=1}^{n}\left(\frac{\partial J_i}{\partial q_i}\frac{\partial H}{\partial p_i} - \frac{\partial J_i}{\partial p_i}\frac{\partial H}{\partial q_i}\right) + \frac{\partial J_i}{\partial t}\times 1 = 0 \ , \tag{17}$$

and because $\partial J_i/\partial \tau = 0$ as well as $\left[J_i,\boldsymbol{H}\right] = 0$, the integrals J_i are also invariants for the system associated with $\boldsymbol{H}$ $(q_i, p_i, q_{n+1}, p_{n+1})$. The system with (n+1) degrees-of-freedom has (n+1) independent invariants in involution and is therefore completely integrable ($\boldsymbol{H}$ is independent of the J_i because it is the only invariant to depend upon p_{n+1}). We have shown that the time-dependent system associated with H which is integrable in the sense the solution can be found by quadratures (according to Kozlov and Kolesnikov) and is equivalent to a completely integrable system with an additional degree of freedom.

The solution for the motion of the system with (n+1) degrees-of-freedom can be expressed in terms of canonical action-angle variables. The solution for the motion of the system with n degrees-of-freedom can then be found by using the transformation to the system with n degrees-of-freedom. It should be noted that being able to transform the non-autonomous system to an integrable autonomous system does not imply that the solution of the non autonomous system can be expressed simply or that it is geometrically simple. An example of that for a linear, one-degree-of-freedom, time-dependent system was given by A. Salat and J. Tataronis.[18]

If the system is completely integrable in the extended phase space, the numerically calculated Lyapunov exponent corresponding to any trajectory equals zero. In this space one can consider couples of two trajectories with the same initial time τ and the same initial p_{n+1}. The Lyapunov exponents are given by $\sigma_{Ext} = \lim (1/\tau)\ln (D(\tau)/D(0))$ in the limit when $\tau \to \infty$ and $D(0) \to 0$, where $D(\tau)$ measures the distance between the two trajectories at a time τ.[6,7] In the same way the corresponding Lyapunov exponents can be calculated in the space of origin by taking the limits of $\sigma_{Orig} = \lim (1/t)\ln (d(t)/d(0))$ when $t \to \infty$ and $d(0) \to 0$, where $d(t)$ represents the distance between the two trajectories at a time t. It has been shown that, in this case, $D(\tau) > d(t)$ and $D(0) = d(0)$.[11] Then $\sigma_{Ext} > \sigma_{Orig}$, and one can conclude that all the Lyapunov exponents equal zero in the space of origin when they equal zero in the extended phase-space. As a consequence, in the case of time-dependent Hamiltonian problems with n independent constants of motion (possibly time-dependent) in involution, only a set of measure zero can contain chaotic trajectories.

3 Application to the Dynamics of a Charged Particle in a Constant Homogeneous Magnetic Field and a Transverse Homogeneous Rotating Electric Field

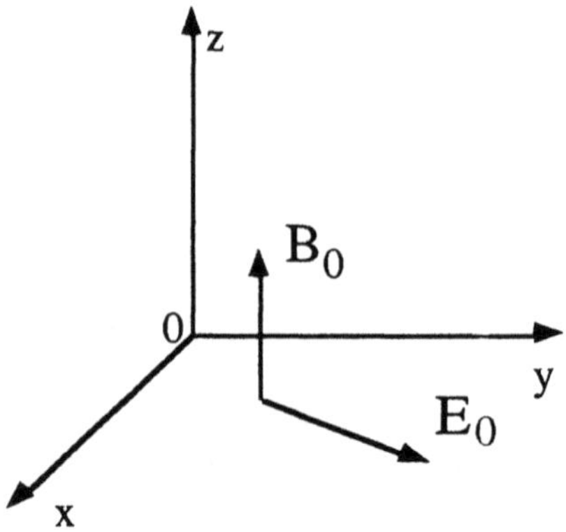

Fig. 1 : The coordinate system

The constant magnetic field $\mathbf{B}_0$ is assumed to be along the z-axis (Fig.1), and the electric field has the components

$$\begin{aligned} E_x &= E_0 \cos \omega_0 t, \\ E_y &= E_0 \sin \omega_0 t, \\ E_z &= 0, \end{aligned} \tag{18}$$

where E_0 and ω_0 are constants. The following gauge is chosen for the electromagnetic field

$$\mathbf{A} = -\left(\frac{B_0}{2} y + \frac{E_0}{\omega_0} \sin \omega_0 t\right) \hat{\mathbf{e}}_x + \left(\frac{B_0}{2} x + \frac{E_0}{\omega_0} \cos \omega_0 t\right) \hat{\mathbf{e}}_y. \tag{19}$$

As a consequence of Maxwell's equations, the electric field and the total magnetic field cannot in fact be constant along the z-axis. Still, for long wavelengths, the variation of the magnetic field can be neglected in a region where the electric field is maximum. The treatment given here applies in this regime. The motion of the charged particle is assumed to be in the x-y plane. Its relativistic Hamiltonian expressed in MKSA units is

$$H = \left[\left(P_x - \frac{eE_0}{\omega_0} \sin \omega_0 t - \frac{eB_0}{2} y\right)^2 c^2 + \left(P_y + \frac{eE_0}{\omega_0} \cos \omega_0 t + \frac{eB_0}{2} x\right)^2 c^2 + m^2 c^4\right]^{1/2}, \tag{20}$$

where $-e$ and m are the charge and rest mass of the particle.

3.1 First Demonstration of the Integrability of the Problem

The Hamilton equations allow us to find two constants of motin readily[15]

$$C_1 = P_x + \frac{eB_0}{2} y, \qquad C_2 = P_y - \frac{eB_0}{2} x. \tag{21}$$

Another constant of motion can be obtained by using Noether's theorem.[2,3,15,16] The result is[15]

$$C_3 = yP_x - xP_y + H/\omega_0. \tag{22}$$

It can be noted that the first two constants are canonically conjugate

$$\left[C_1, \frac{C_2}{eB_0}\right] = 1. \tag{23}$$

where now C_1 and C_2 must be though to be functions of their arguments, i.e., the right hand side of Eqs.(21). Among these three constants of motion, one cannot find two constants in involution. This becomes obvious from the relations

$$[C_1, C_3] = C_2 \ , \qquad [C_2, C_3] = - C_1. \tag{24}$$

Another constant of motion is given by

$$C_4 = C_1^2 + C_2^2. \tag{25}$$

It satisfies the relation

$$[C_4, C_3] = 2C_1[C_1, C_3] + 2C_2[C_2, C_3] = 0. \tag{26}$$

As two constants are independent and in involution, the system is integrable according to the definition given in the introduction.

3.2 Reduction to a One-Dimensional Problem, Second Demonstration of the Integrability of the Problem

What follows is a part of what has been done in Refs.15(a,b).

Written in terms of the following dimensionless variables and parameters

$$\hat{x} = x\frac{\omega_0}{c},\ \hat{y} = y\frac{\omega_0}{c},\ \hat{P}_{x,y} = \frac{P_{x,y}}{mc},\ \hat{t} = \omega_0 t,\ \hat{H} = \gamma = \frac{H}{mc^2},\ a = \frac{eE_0}{mc\omega_0},\ \Omega_0 = \frac{eB_0}{m\omega_0},$$

the Hamiltonian is

$$\hat{H} = \left[\left(\hat{P}_x - a\sin\hat{t} - \frac{\Omega_0}{2}\hat{y}\right)^2 + \left(\hat{P}_y + a\cos\hat{t} + \frac{\Omega_0}{2}\hat{x}\right)^2 + 1\right]^{1/2} . \tag{27}$$

In these variables, the constants of motion corresponding to C_1 and C_2 are

$$\hat{C}_1 = \hat{P}_x + \frac{\Omega_0}{2}\hat{y}\ , \quad \hat{C}_2 = \hat{P}_y - \frac{\Omega_0}{2}\hat{x}. \tag{28}$$

They satisfy

$$\left[\widehat{C}_1, \widehat{C}_2\right] = \Omega_0. \tag{29}$$

By using this property we can show that the system can be described by a single degree-of-freedom time-dependent Hamiltonian. To do so, let us choose the two constants $\widehat{C}_1$ and $\widehat{C}_2$ (one must be normalised by Ω_0) as new conjugate momentum and coordinate. We choose a first canonical transformation, $\left(\hat{x}, \hat{y}, \widehat{P}_x, \widehat{P}_y\right) \to \left(\tilde{x}, \tilde{y}, \widetilde{P}_x, \widetilde{P}_y\right)$, defined by the following Type 2 generating function: $F_2 = \left(\widetilde{P}_x - \Omega_0/2\,\hat{y}\right)\hat{x} + \widetilde{P}_y\hat{y}$. [2,3,6,7,15] Then a second canonical transformation, $\left(\tilde{x}, \tilde{y}, \widetilde{P}_x, \widetilde{P}_y\right) \to \left(Q_1, Q_2, P_1, P_2\right)$, generated by $F_2 = \left(P_2 + \Omega_0\tilde{x}\right)\tilde{y} + P_1\left(\tilde{x} + P_2/\Omega_0\right)$ is introduced. The resulting transformation, which is the product of the two transformations, is given by

$$\begin{gathered}\hat{x} = Q_1 - P_2/\Omega_0,\ \hat{y} = Q_2 - P_1/\Omega_0, \\ \widehat{P}_x = 1/2\left(\Omega_0 Q_2 + P_1\right),\ \widehat{P}_y = 1/2\left(\Omega_0 Q_1 + P_2\right).\end{gathered} \tag{30}$$

In terms of these variables one has

$$Q_2 = \frac{\widehat{C}_1}{\Omega_0}, \qquad P_2 = \widehat{C}_2, \tag{31}$$

and the Hamiltonian is

$$\overline{H} = \left[\left(P_1 - a\sin\hat{t}\right)^2 + \left(\Omega_0 Q_1 + a\cos\hat{t}\right)^2 + 1\right]^{1/2}. \tag{32}$$

As expected P_2 and Q_2 are cyclic variables since they do not appear in the Hamiltonian. This Hamiltonian depends on time and has one degree-of-freedom. In terms of the new variables, the constant C_3 becomes the following constant

$$K = \overline{H} - \frac{P_1^2}{2\Omega_0} - \frac{\Omega_0}{2}Q_1^2. \tag{33}$$

The fact the problem can be reduced a one-degree-of-freedom system can help to predict the aspect of trajectories easily.[15] As we have one constant of motion, we can conclude that the system is integrable. This is a second way to show the integrability of (20). Finally, according to Section 2.1, we can conclude that a second constant of motion of (32) can be obtained by quadratures.

3.3 Another way to Solve the Problem

The equations of Hamilton derived from Eq. (32) are

$$\dot{P}_1 = -\frac{\Omega_0}{\gamma}\left(\Omega_0 Q_1 + a\cos\hat{t}\right), \quad \dot{Q}_1 = \frac{1}{\gamma}\left(P_1 - a\sin\hat{t}\right). \tag{34}$$

Introducing the variables

$$\overline{Q}_1 = Q_1 + {}^{a}/_{\Omega_0} \cos\hat{t}, \ \overline{P}_1 = P_1 - a\sin\hat{t}, \tag{35}$$

and the complex quantity $Z = \overline{P}_1 + i\,\Omega_0\overline{Q}_1$, taking into account that $\overline{H} = \gamma = \sqrt{1 + |Z|^2}$, the equations of Hamilton (Eqs.(34)) are equivalent to the following equation

$$\dot{Z} = \frac{i\Omega_0 Z}{\sqrt{1 + |Z|^2}} - a\exp(i\hat{t}), \tag{36}$$

which is the equation of a nonlinear oscillator under the action of an external force. Formally, the solution of this equation can be written

$$Z = A_0 \exp i\left[\sigma(\hat{t}) + \delta\right] - a\int_0^{\hat{t}} \exp i\left[\sigma(\hat{t}) - \sigma(\tau) + \tau\right] d\tau, \tag{37}$$

with

$$\sigma(\hat{t}) = \Omega_0 \int_0^{\hat{t}} d\tau\gamma^{-1}(\tau), \tag{38}$$

where A_0 and δ are real constants. Then

$$P_1 = A_0\cos\left[\sigma(\hat{t}) + \delta\right] + a\sin\hat{t} - a\int_0^{\hat{t}} \cos\left[\sigma(\hat{t}) - \sigma(\tau) + \tau\right] d\tau,$$

$$Q_1 = \frac{A_0}{\Omega_0}\sin\left[\sigma(\hat{t}) + \delta\right] - \frac{a}{\Omega_0}\cos\hat{t} - \frac{a}{\Omega_0}\int_0^{\hat{t}} \sin\left[\sigma(\hat{t}) - \sigma(\tau) + \tau\right] d\tau. \tag{39}$$

The quantities A_0 and δ are determined so that at $\hat{t} = 0$, $A_0^2 = \gamma_0^2 - 1 = \hat{p}_{x0}^{\,2} + \hat{p}_{y0}^{\,2}$ and $\tan\delta = \hat{p}_{y0}/\hat{p}_{x0}$ ($\hat{p} = p/mc$, p is the momentum of the particle). The subscript 0 appended to variables γ and p refers to their initial values.

We now derive an equation for γ. Taking the time derivative of Eq.(32) with respect to time and using Eqs. (39), we obtain

$$\gamma\,\dot{\gamma} = -a\left\{A_0\cos\left[\sigma(\hat{t})-\hat{t}+\delta\right] - a\int_0^{\hat{t}}\cos\left[\sigma(\hat{t})-\sigma(\tau)+\tau-\hat{t}\right]d\tau\right\}. \tag{40}$$

This equation, multiplied by $\Omega_0/\gamma - 1$ and integrated between 0 and $\hat{t}$, leads to

$$\Omega_0(\gamma-\gamma_0) - \frac{(\gamma^2-\gamma_0^2)}{2} - aA_0\sin\delta = -a\left\{A_0\sin\left[\sigma(\hat{t})-\hat{t}+\delta\right]\right.$$

$$\left. - a\int_0^{\hat{t}}\sin\left[\sigma(\hat{t})-\sigma(\tau)+\tau-\hat{t}\right]d\tau\right\}. \tag{41}$$

Then Eq.(40) is differentiated with respect to time and Eq.(41), multiplied by $\Omega_0/\gamma - 1$, is added to it. The resulting equation is multiplied by $\gamma\,\dot{\gamma}$ and integrated between 0 and $\hat{t}$. In this way, the following differential equation for the energy is derived

$$(\dot{\gamma})^2 + \frac{\gamma^2}{4} - \Omega_0\gamma + R_0 - \frac{\Gamma_0}{\gamma} - \frac{K_0}{\gamma^2} = 0 \tag{42}$$

with

$$R_0 = aA_0\sin\delta + \Omega_0^2 + \Omega_0\gamma_0 - \frac{\gamma_0^2}{2} - a^2, \tag{43}$$

$$\Gamma_0 = 2\Omega_0 aA_0\sin\delta + 2\Omega_0^2\gamma_0 - \Omega_0\gamma_0^2 \tag{44}$$

and

$$K_0 = a^2A_0^2\cos^2\delta - \Gamma_0\gamma_0 + R_0\gamma_0^2 - \Omega_0\gamma_0^3 + \frac{\gamma_0^4}{4}. \tag{45}$$

This result is in agreement with the one obtained by Roberts and Buchsbaum.[12] Equation (42) describes a motion in a one-dimensional potential. It admits a solution which gives time in terms of a sum of elliptic integrals of the first and third type.[19] Hence, the fact that an analytical solution of Eq. (42) exists permits us, with the help of Eqs.(39), to express Q_1 and P_1 in terms of quadratures.

4 Conclusion

For time-dependent problems with n-degrees-of-freedom, the Hamiltonian in the extended phase space is carefully introduced. In this space, when one has n independent, possibly time-dependent constants of motion in involution in the original space, one has n + 1 independent invariants in involution. This allows us to conclude that, in this case, the system is completely integrable in the extended phase space. The n first integrals permit us to solve the system by quadratures and no finite part of the original space can be filled by chaotic trajectories. It is shown explicitly in the case of a one-dimensional problem that, if one has one invariant, one can derive a second invariant which is canonically conjugate to the first one and which gives the solution in terms of quadratures. In that sense, we say that Liouville's theorem can be applied in its usual way. This generalised Liouville's theorem was applied to the relativistic motion of a charged particle in a constant homogeneous magnetic field and a transverse rotating electric field. The integrability of this problem was shown by using two different ways. Noether's theorem was used to find a constant of motion for the system. An invariant is derived which is independant and in involution with the other one. This is sufficient to prove the integrability as this problem has two-degrees-of-freedom. Then, using canonical transformations, we reduced this problem to a time-dependent one with a single degree-of-freedom. This system has a constant of motion which is the one found previously by using Noether's theorem, and expressed with the new variables.This was the second way to show that this problem is integrable. In this case we know that a second integral can be found which permits us to solve the system by quadratures. Another approach by quadratures was given. It consists in deriving an equation for the energy which is integrable. As the coordinates are shown to be expressible as quadratures containing the energy, it proves a second time the solution of the system can be found by quadratures.

Acknowledgments

The authors wish to thank Professor H.R. Lewis for very valuable discussions.

References

[1]Liouville, J. (1855): Sur l'Intégration des Equations Différentielles de la Dynamique, Journal de Mathématiques **20**, 137-138

[2]Arnold, V.I. (1988): *Dynamical Systems* III (Springer-Verlag)

[3]Rasband, S.N. (1983): *Dynamics* (John Wiley & Sons, New York)

[4]Whittaker, E.T. (1952): *A Treatise on the Analytical Dynamics of Particles and Rigid Bodies* (Cambridge at the University Press)

[5]Lewis, H.R., Bouquet, S. (1997): Hamilton's Principal Function and Integration by Quadratures for an N-Degree of freedom Non-Autonomous System Given N Invariants in Involution, submitted to Phys. Lett. A

[6]Ott, E. (1993): *Chaos in Dynamical Systems* (Cambridge University Press)

[7]Lichtenberg, A.J., Liebermann, M.A. (1983): *Regular and Stochastic Motion* (Springer-Verlag, New York)

[8]Bouquet, S., Dewisme, A. (1992): *Nonlinear Evolution Equations and Dynamical Systems* (Proceedings Needs'91), edited by M. Boiti, L. Martina and F. Pempinelli (World Scientific, Singapore), 395-403

[9]Dewisme, A., Bouquet, S. (1993): First Integrals and Symmetries of Time-Dependent Hamiltonian Systems, J. Math. Phys. **34**(3), 997-1006

[10]Bouquet, S., Lewis, H.R. (1996): A second Invariant for One-Degree-of-Freedom, Time-Dependent Hamiltonians given a First Invariant, J. Math. Physics, **37**(11), 5509-5517

[11]Bouquet, S., Bourdier, A. (1997): accepted for publication in Phys. Rev. E

[12]Roberts, C.S., Buchsbaum, S.J. (1964): Motion of a Charged Particle in a Constant Magnetic Field and a Transverse Electromagnetic Wave Propagating along the Field, Phys. Rev. A **135**, 381-389

[13]Hakkenberg, A., Weenink, M. P. H. (1964): The Solution of the Relativistic Equation of Motion of a Charged Particle in a Homogeneous Magnetic and Rotating Electric Field Perpendicular to each other, Physica **30**, 2147-2152

[14]Jory, H.R., Trivelpiece, A.W. (1968): Charged-Particle Motion in Large-Amplitude Electromagnetic Fields, J. Appl. Phys. **39**, 3053-3060

[15](a)Bourdier, A., Valentini, M. Valat, J. (1996): Hamiltonian Structure of Motion and Dynamics of a Particle in a Constant Homogeneous Magnetic Field and a Transverse Electric Field, Phys. Lett. A **215**, 219-228

(b)Bourdier, A., Valentini, M, Valat, J. (1996): Dynamics of a Relativistic Charged Particle in a Constant Homogeneous Magnetic Field and a Transverse Homogeneous Rotating Electric Field, Phys. Rev. E **54**, 5681-5691

[16]Jones, R. D. (1981): Constants of Motion in a Helical Magnetic Field, Phys. Fl. **24**(3), 564-565

[17]Goldstein, H. (1980): *Classical Mechanics*, 2nd ed. (Addison - Wesley, New York)

[18]Salat, A., Tataronis, J. (1982): Stochasticity and Order in a Linear Quasiperiodic Differential Equation, *Long Time prediction in Dynamics* (John Wiley & Sons, New York), 213-226

[19]Byrd, P.F., Friedman, M.D. (1954): *Handbook of Elliptic Integrals for Engineers and Physicists* (Springer-Verlag, Berlin)

Stochastisation of Phases in Four-Wave Interaction

Bertrand Lefebvre[1], Volodia Krasnosel'skikh[1], and Thierry Dudok de Wit[2]

[1] Laboratoire de Physique et Chimie de l'Environnement, CNRS, 3A av de la Recherche Scientifique, 45071 Orléans Cedex 2, France
[2] Centre de Physique Théorique, CNRS et Université de Provence, Luminy case 907, 13288 Marseille Cedex 9, France

Abstract. We study the statistical properties of nonlinear wave-wave interaction in a system consisting of three Langmuir waves and one ion-sound wave. Such a system describes either a cascade of two decay instabilties or a modified decay instability. We pay a particular attention to the phases which in some cases exhibit a chaotic synchronization, showing a transition from three-wave to four-wave interaction.

1 Introduction

High frequency electrostatic turbulence of Langmuir waves in a plasma is supposed to be excited by an energy source such as an unstable particle distribution function or by parametric instability of electromagnetic waves. It is often encountered in various applications such as ionospheric heating (Fejer et al. 1991), laser-plasma interactions (Kruer 1989) or electron beam propagation in the solar wind (Hospodarsky and Gurnett 1995).

Since the elementary entity of such turbulence is supposed to be linear waves, it is always required that the energy contained in these waves is much lower than the thermal one : $W/(nkT) \ll 1$. Moreover, there is a conventional distinction between weak and strong regimes of Langmuir turbulence.

Weak turbulence is supposed to be quasi-homogeneous and it is assumed that the characteristic time of the non-linear wave-wave interactions is much longer than any time scale which characterizes the behaviour of the waves in the linear approximation, in particular the characteristic dispersion time

$$\tau_{NL}^{-1} \ll \frac{\partial^2 \omega(\mathbf{k})}{\partial k_i \partial k_j} k_i k_j . \tag{1}$$

It is also assumed that the frequencies of the waves are larger than all characteristic growth and damping rates.

There exists a conventional statistical approach to such a type of turbulence called the weak turbulence approximation. Assuming that the phases of the waves are independent random variables with Gaussian probability densities one can write, for example, the second order moment of the complex variable a_k as $\langle a_k a_{k'}^* \rangle = n_k \delta(k-k')$. After this ensemble averaging a simplified description of the turbulence is obtained in terms of a kinetic equation for the numbers of quanta n_k (Zakharov et al. 1985). It is worth mentioning the implicit assumption that the ensemble average can be replaced by a time average and that the phase correlation time is much smaller than the characteristic time of variation of the amplitudes. Indeed the n_ks are supposed to vary slowly with respect to the stochastization time. In this weak turbulence approximation, three-wave interactions (e.g. decay instabilities) are prevalent under certain conditions (see a more detailed discussion of this topic in Zakharov et al. 1985 and Musher et al. 1995).

Strong turbulence occurs when

$$\tau_{NL}^{-1} \approx \frac{\partial^2 \omega(\mathbf{k})}{\partial k_i \partial k_j} k_i k_j. \tag{2}$$

In this case four-wave processes such as modulational instabilities dominate and are responsible for the creation of spatial inhomogeneities (solitons in 1 or 2D or collapses in 2 or 3D).

Both weak and strong Langmuir turbulence are well-described by the Zakharov equations (Zakharov 1972). Simulations of a pump wave propagating in the ionosphere showed that both the Zakharov equations and the kinetic equations for the n_ks predict a singular k-spectrum consisting of jets, implying the presence of a countable number of waves (Hanssen et al. 1993). However, there are some important differences which appear in these predictions when the amplitude of the pump wave becomes sufficiently large, since the k-spectrum predicted by the Zakharov equations presents many fewer peaks than the spectrum predicted by the weak turbulence approximation.

It should be stressed that the difference between these two approaches lies in the description of the phases of the waves. As a consequence the probability densities (especially for the phases) should be key quantities for describing high-frequency wave turbulence. These probability densities give a statistical description of the plasma and result from the complex behaviour which is described by dynamical equations.

These equations of motion can be cast into the general framework of Hamiltonian equations (Zakharov et al. 1985). Moreover, the small parameter $W/(nkT)$ allows one to expand the Hamiltonian $\mathcal{H}(a_k, a_k^*)$ in the variables a_k and a_k^* :

$$\mathcal{H} = \mathcal{H}_0 + \mathcal{H}_1 + \mathcal{H}_2 + ... \tag{3}$$

The first order term $\mathcal{H}_0$ describes the linear waves, without any interaction. In the limiting case of a infinite number of waves, the system can thus be seen as a perfect gas of plasmons. At the first order correction $\mathcal{H}_1$ adds the possibility for having three-wave interactions. At the next order $\mathcal{H}_2$ is related to four-wave interactions and so on.

The three-wave interaction in the absence of a growth rate or damping is known to be integrable by the inverse scattering method. With a growth rate and damping rates, a detuning of the resonance conditions has to be introduced in order to avoid some divergences (Vyshkind and Rabinovich

1976; Wersinger et al. 1980). This detuning parameter is supposed to take into account the broadening of the resonance provided by some neglected phenomena, such as other waves participating into the turbulence or perturbations due to higher order interactions.

We have choosen to study the simplest system which can describe two types of physical processes, namely, decay instability (first order of non-linearity) and modified decay instability (or modulational instability). In the following we present numerical simulations (in 1D) of these two situations, paying particular attention to the probability densities associated with such ensembles. We commence with some remarks about the methods used to describe probability densities in wave turbulence and on the way they are related to the theoretical description of the wavefield.

2 An Experimental Approach to the Estimation of Probability Densities

The conventional way to describe rather complex turbulent phenomena is in term of probabilities in spite of the deterministic origin of the phenemna, thus regarding the measured variables as random variables. Their statistical description is provided by a probability density $p(x)$, which can also be described in the Fourier space by the introduction of the *characteristic function* $G(\theta) = \langle \exp i\theta x \rangle$ ($\langle\rangle$ denotes ensemble average). If for any $q > 0$ the absolute moment $\langle |x|^q \rangle$ exists (which is not true for example for Lévy processes), then one can expand $G(\theta)$ as an infinite series:

$$G(\theta) = \sum_{q=0}^{n} \frac{(i\theta)^q}{q!} \langle x^q \rangle , \tag{4}$$

where $\langle x^q \rangle$ is the qth order moment of the random variable (in the following it is supposed that $\langle x \rangle = 0$). One can also write

$$\ln G(\theta) = \sum_{q=0}^{\infty} \frac{(i\theta)^q}{q!} C_q \quad \text{with} \quad C_q = (-i)^q \left. \frac{\mathrm{d}^q \ln G(\theta)}{\mathrm{d}\theta^q} \right|_{\theta=0} . \tag{5}$$

The quantities C_q are called *cumulants*, or semi-invariants, and the set $\{C_q\}$ uniquely determines the probability density p (provided the series (3) exists). For example, in the case of a Gaussian probability density, C_1 is the mean, C_2 the variance and $C_q = 0$ for $q > 2$. As a consequence knowledge of the cumulants is useful for describing probability densities that weakly depart from a Gaussian, that is, when the series (4) converges rapidly. Cumulants have some other nice properties which make them useful quantities with which to work (Monin and Yaglom 1971).

All these considerations can be quite straightforwardly be extended to a field $\mathbf{x} \in \mathbb{R}^n$. In this a case the cumulants

$$C_{q_1,\ldots,q_n} = (-i)^K \left. \frac{\partial^K \ln G(\theta\ )}{\partial\theta_1^{q_1}\ldots\partial\theta_n^{q_n}} \right|_{\theta\ =0} \quad \text{with } K = q_1 + \ldots + q_n \tag{6}$$

have the property that, if a component of $\mathbf{x}$ is independent from others, then they must vanish (say the component θ_1 is statistically independent from the others, then $G(\theta\ ) = G'(\theta_1)G''(\theta_2, \ldots., \theta_n)$ and therefore the cumulants defined by (6) vanish). This property is generally used to detect nonlinear wave interactions in a wavefield. From a waveform, $x(t)$, one can construct a random field $\mathbf{x}(t)$ of $\mathbb{R}^n$ taking $x\,(t)$, $x(t+\tau_1)$, ..., $x(t+\tau_{n-1})$. For example, the second order cumulant of $\mathbf{x}(t)$, $= (x(t), x(t+\tau))$, is simply the correlation function and by a Fourier transform the power spectrum is obtained (assuming ergodicity):

$$S(\omega) = \int_{-\infty}^{+\infty} \langle x(t)x(t+\tau)\rangle\, e^{i\omega\tau} \mathrm{d}\tau. \tag{7}$$

Following the same procedure one can introduce the higher order spectra (see Brillinger 1965). By means of a Fourier transform of the third order cumulant one obtains the bispectrum

$$B(\omega_1, \omega_2) = \int \langle x(t)x(t+\tau_1)x(t+\tau_2)\rangle\, e^{i\omega\,\cdot\,\tau} \mathrm{d}\tau_1 \mathrm{d}\tau_2. \tag{8}$$

In terms of the Fourier components a_ω of $x(t)$ the bispectrum appears as an average over time intervals of

$$a^*_{-\omega_0} a_{\omega_1} a_{\omega_2}, \tag{9}$$

where $\omega_0 = \omega_1 + \omega_2$. It follows from the properties of the cumulants that. if the phases φ_j of a_{ω_j} are statistically independent random variables, then the bispectrum vanishes. Moreover, the bispectrum is a maximum (in modulus) when relations between the phases are met, viz. $\varphi_0 = -\varphi_1 - \varphi_2$. This allows a normalisation such that the bispectrum is equal to 1 when this condition is satisfied. The normalised bispectrum $b(\omega_1, \omega_2)$ is called the *bicoherence* and is such that $|b|^2 \in [0, 1]$.

As a consequence the bispectrum is useful to investigate the type of three-wave interaction contained in the $\mathcal{H}_1$ of the introduction. Indeed, in this kind of interaction, the waves are supposed to satisfy the same type of resonance conditions and it has been explained that the relations between the phases and their behaviour are determinant for the interaction.

Analogously the trispectrum $T(\omega_1, \omega_2, \omega_3)$ (and the *tricoherence*) can be constructed from the fourth order cumulant and can be used to detect four-wave interactions (Kravtchenko-Berejnoi et al. 1994).

The approach of higher order spectra has therefore the great advantage to be similar to the Hamiltonian description of wave turbulence. We explore in greater detail these similarities between the statistical approach and the dynamical one, investigating the behaviour of a simple system which can describe both three- and four-wave interactions.

3 A Model for Four-Wave Interactions

The description of (weak and strong) Langmuir turbulence relies on a system written by V.E. Zakharov (Zakharov 1972). The first equation describes the propagation of Langmuir waves in a plasma with slow density variations, and has been averaged over the plasma frequency by making the assumption that this is much higher than any other frequency of the system. The second one describes ion-sound waves, with a source term corresponding to the ponderomotive force. They are in one dimension

$$\left(i\frac{\partial}{\partial t} + \frac{3V_{the}^2}{2\omega_{pe}}\frac{\partial^2}{\partial x^2} + i\nu\right)\widetilde{E} = \frac{\omega_{pe}}{2n_0}n\,\widetilde{E} \tag{10}$$

$$\left(\frac{\partial^2}{\partial t^2} + \nu_0\frac{\partial}{\partial t} - c_s^2\frac{\partial^2}{\partial x^2}\right)n = \frac{\epsilon_0}{4M}\frac{\partial^2}{\partial x^2}\left|\widetilde{E}\right|^2 . \tag{11}$$

Here the damping (or growth) rate ν has been introduced under the form of some integral operator depending on k. A dissipation of the wave energy is made possible by Landau damping while an energy input may, for example, come from an electron beam. Similar equations can be written in the presence of a weak magnetic field (Krasnosel'skikh and Sotnikov 1977).

We assume a singular spectrum consisting of three Langmuir waves (modes 1 to 3) and a single ion-sound (mode 0). Furthermore we restrict ourselves to the homogeneous case in which the interaction only affects the temporal evolution of the waves:

$$\widetilde{E} = \sum_{j=1}^{3} a_j(t)\times\exp i(k_jx-\omega_jt) + c.c., \quad 0<\omega_1<\omega_2<\omega_3 \tag{12}$$

$$n = a_0(t)\times\exp i(k_0x-\Omega_0t) + c.c. \tag{13}$$

Assuming the following exact resonance conditions to be met,

$$\omega_0+\omega_1-\omega_2=0, \quad k_0+k_1-k_2=0 \tag{14}$$

$$\omega_0+\omega_2-\omega_3=0, \quad k_0+k_2-k_3=0, \tag{15}$$

we introduce these solutions into the Zakharov system. After a convenient normalization is made, a set of equations which is standard for wave-wave interactions is obtained:

$$\left(\frac{\mathrm{d}^2}{\mathrm{d}t^2} + \nu_0\frac{\mathrm{d}}{\mathrm{d}t} - 2i\frac{\mathrm{d}}{\mathrm{d}t}\right)a_0 = -a_1^*a_2 - a_2^*a_3 + i\nu_0a_0 \tag{16}$$

$$i\frac{\mathrm{d}}{\mathrm{d}t}a_1 = a_0^*a_2 - i\nu_1a_1 \tag{17}$$

$$i\frac{\mathrm{d}}{\mathrm{d}t}a_2 = a_0a_1 + a_0^*a_3 - i\nu_2a_2 \tag{18}$$

$$i\frac{\mathrm{d}}{\mathrm{d}t}a_3 = a_0a_2 - i\nu_3a_3 . \tag{19}$$

This system describes two distinct cases of four-wave interactions:

- When an energy source feeds the highest wavelength mode (i.e. $\nu_3 < 0$), then the system describes an energy cascade toward lower k by means of two decay instabilities ($3 \to 2 + 0$ and $2 \to 1 + 0$).
- If the energy source now makes mode a_2 unstable, then an energy flux occurs toward both its Stokes and anti-Stokes satellites through density fluctuations ($2 + 2 \to 1 + 3$). The modulational and modified decay instabilities belong to this class of four-wave interaction.

These two cases are supposed to correspond to two different regimes of Langmuir turbulence: the inverse energy cascade of the first case is reminiscent of weak turbulence, whereas the four-wave interaction of the second case is supposed to prevail in strong turbulence.

Instead of using a_js, one can introduce the amplitude and the phase of the waves by writing $a_j = u_j \exp(-i\varphi_j)$. The phases appear in the equations in terms of the two differences $\Phi = \varphi_0 + \varphi_1 - \varphi_2$ and $\Psi = \varphi_0 + \varphi_2 - \varphi_3$, which respectively correspond to the resonant interactions (3-4). This reduces formally the dimension of the system from 10 to 8. Such a set of equations constitutes a dissipative dynamical system (provided $\nu_0 + \nu_1 + \nu_2 + \nu_3 > 0$). A simple numerical integration (fourth order Runge-Kutta) allows us study its behaviour.

3.1 Simulations of a Decay Instability

We consider the case in which the shortest wavelength mode (mode 3) is unstable (we write $\nu_3 = -\gamma < 0$). Depending upon the parameters, the systems exhibits many different regimes with attractors the simplest example of which are fixed points. The system (16–19) clearly has a fixed point at the origin, which is always unstable. There exists another fixed point the stability of which depends upon the parameters [1], and its loss of stability is associated with the birth of a limit cycle (periodic motion) though a classical Poincaré-Andronov-Hopf bifurcation.

Although they describe wave-wave interactions, the probability densities associated with some attractors such as fixed points or limit cycles are trivially simple. The situation is more complicated for strange attractors, which may carry (uncountably) many measures. Thus in the experiments there is a selection process of one particular probability measure which is known as the *physical measure* (Eckmann and Ruelle 1985). However, for a low dimensional attractor the corresponding probability density is expected to be much different from a Gaussian, as we observe below.

Attracting limit cycles can be replaced by more complicated attracting sets when the parameters are further varied. Although we only explored a small part of the parameter space (there are four-independent parameters),

[1] This fixed point is defined provided $\nu_1 > \gamma$. The phases Φ^s and Ψ^s at the fixed point are both equal to $\pi/2$.

intermittencies appear to be by far the most frequent scenario of transition to chaos in our system (Fig. 1).

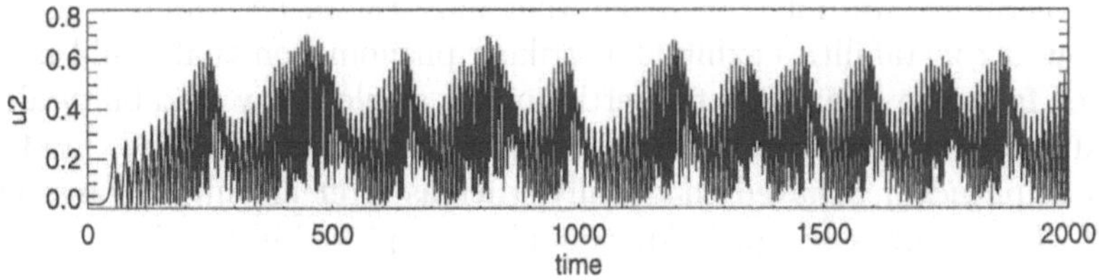

Fig. 1. Intermittent behaviour of the amplitude $u_2(t)$ near the bifurcation for the parameters $\nu_0 = 0.1$, $\nu_1 = 0.08, \nu_2 = 0.03$, $\nu_3 = -0.06$ (equations have been normalized so that $\Omega_s = 1$)

The qualitative features of the observed time series (no growth of a subharmonic mode) and the distribution of the duration of the quiescent periods between two intermittent bursts strongly suggest that we are dealing with type II intermittencies (subcritical Poincare-Andronov-Hopf bifurcation) (Pomeau and Manneville 1980). Note that most intermittent regimes observed so far in three-wave processes belonged to the type I[2] (Meunier et al. 1982). If for example one diminishes ν_2 starting from the parameters choosen in Fig. 1, the quiescent periods become shorter and shorter until they disappear in the chaotic motion.

An indication of the effective number of degrees of freedom involved in the motion is given by the so-called generalized or Rényi dimensions. We computed the information dimension D_1 and the correlation dimension D_2[3] (Hentschel and Proccacia 1983) of the attractor as reconstructed from the time series of the amplitude $u_j(t)$ of one the four modes. Typically, for each mode we obtain

$$3 < D_2 < D_1 \leq 3.5$$

which is considerably lower than the initial number of degrees of freedom. Note that the well-studied single three-wave interaction considered in Vyshkind and Rabinovich (1976), Wersinger et al. (1980), Meunier et al. (1982) and Hugues and Proctor (1990) had a phase space of dimension three only (due to a particular choice of parameters). Thus it appears that a cascade consisting only of two three-wave interactions has typically a more complex behaviour

[2] Indeed, at least four dimensions are required for type II intermittencies, whereas this three-wave interaction had 3 degrees of freedom.

[3] Roughly speaking the information dimension D_1 describes how fast one gains information on the state of the system when the accuracy of measurements ε is decreased (for small ε). The correlation dimension D_2 tells how the probability that two points of the attractor are in the same ball of radius r scales ($\mathcal{P}(r) \propto r^{D_2}$, for small r).

than a single three-wave interaction (in the sense that more degrees of freedom are involved). This higher complexity can not be taken into account by any parameter such as the detuning parameter described in the introduction.

Furthermore, we must point out that our system describing a cascade of two decay instabilities exhibits a striking phenomenon that could not be deduced from the dynamical properties of the single decay instability alone. Indeed, we observed chaotic regimes where the phases Φ and Ψ (and the waves amplitudes) behaved chaotically whereas after an initial transient Φ and Ψ where rigorously equal. An example is displayed in Fig. 2.

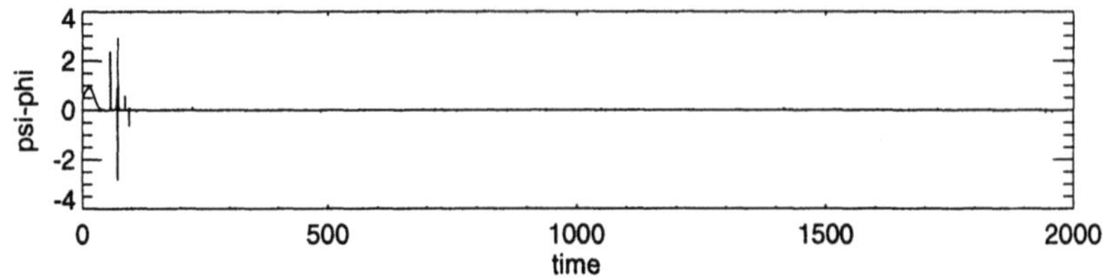

Fig. 2. Although the phases behave chaotically, $\Psi(t) - \Phi(t)$ is constant for $\nu_0 = 0.1$, $\nu_1 = 0.08, \nu_2 = 0.01$, $\nu_3 = -0.06$

This phenomenon, known as chaotic synchronization (Fujisaka and Yamada 1983, Afraimovich et al. 1986, Pecora and Carroll 1990), means that

$$\Psi - \Phi = 2\varphi_2 - \varphi_1 - \varphi_3 = const \tag{20}$$

which is a phase relation for the four-wave interaction with resonance condition

$$2\omega_2 - \omega_1 - \omega_3 = 0, \quad 2k_2 - k_1 - k_3 = 0. \tag{21}$$

This chaotic synchronization can now be interpreted as a transition from a three-wave to a four-wave process (Krasnosel'skikh et al. 1997). Let us check the signature of this phenomenon on the physical measure by comparing it with a nonsynchronized chaotic motion. In the latter case the joint probability density $p(\Phi, \Psi)$ vanishes for some (Φ, Ψ) and is already higher around $\Phi = \Psi$ (Fig. 3a). For the chaotic synchronization it is simply of the form $p(\Phi, \Psi) = p'(\Phi)\delta(\Phi - \Psi)$ (Fig. 3b).

The tricoherence calculated in the case of the chaotic synchronization is as expected nearly 1. The bicoherence is much lower ($b(\omega_1, \omega_2) \approx 0.23$), indicating that the correlation time for the four wave interaction is short while it is nearly infinite for the four-wave interaction. On the other hand, in case (a) of Fig. 3, the correlation time for the three-wave interaction is higher than for the four-wave interaction ($b(\omega_1, \omega_2) \approx 0.44$ and $t(\omega_1, \omega_2, \omega_3) \approx 0.36$). Furthermore the chaotic synchronization appears to be resistant to the introduction of external Gaussian white noise (or at least with a bandwidth much

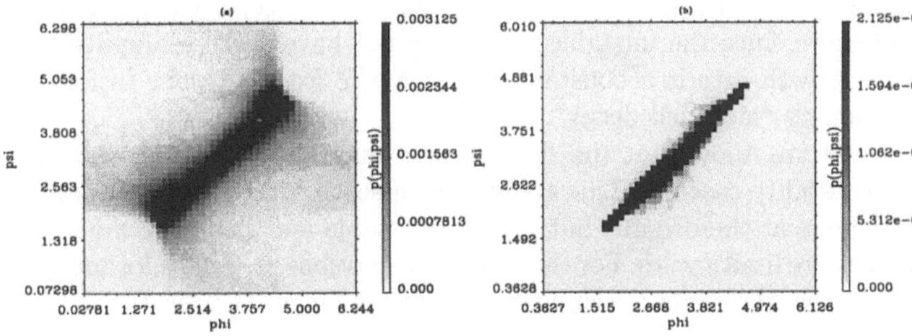

Fig. 3. Joint probability densities $p(\Phi, \Psi)$: (a) for a nonsynchronised chaotic motion ($\nu_0 = 0.07, \nu_1 = 0.05, \nu_2 = 0.03$, $\nu_3 = -0.06$); (b) at the chaotic synchronisation (same parameters as Fig. 2)

higher than that of the electric field) into the eqs. (16)-(19) (Fig. 4 displays an exemple with very strong noise).

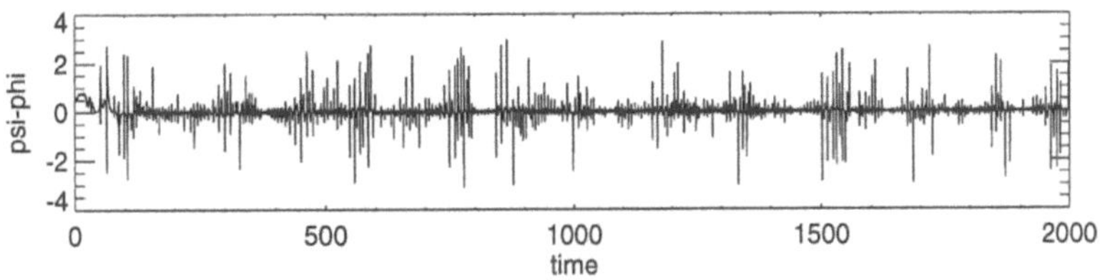

Fig. 4. $\Psi - \Phi$ at the chaotic synchronization in the presence of external Gaussian white noise with standard deviation $\sigma = 0.05$

Finally one may reasonably expect that the chaotic synchronisation would cause a reduction of the dimension of the attractor since on the contrary two degrees of freedom "stick together". However, this turns out not to be case since the generalised dimensions slightly increase. A possible explanation is that the variables we used are not generic (*i.e.* do not contains all the information about the measure on the attractor), which is plausible because of the quite high dimension of the phase space.

3.2 Simulations of a Modified Decay Instability

We now consider the case where the mode 2 is unstable ($\nu_2 = -\gamma < 0$, $\nu_0, \nu_1, \nu_3 > 0$) and generates its Stokes and anti-Stokes satellites 1 and 3. Considering the unstable wave as a pump wave (containing enough energy to keep a constant amplitude) one can distinguish between several limits of this four-wave interaction according to the energy and the wavelength of the perturbations, viz. these limits are modulational instability, modified decay instability etc. (Zakharov et al. 1985). Such a clear separation is not made possible here since the unstable wave does not have a given amplitude, but only its growth rate is a constant. For instance we shall refer to four-wave interaction as "modified decay".

Simulations show that the dynamics is much different from that of the decay instability cascade. This system possesses another fixed point different from the one at the origin[4], but it is never stable and therefore the simplest attractors are limit cycles. For these periodic motions as well as for more complicated ones, the characteristic time scales are longer than in the preceding case for parameters of the same order.

In the limit when the second fixed point is not defined ($\nu_3 \to \nu_1$) the solutions often diverge. However, this divergence disappears when slight detunings of the resonances are introduced into (14-15).

At the exact resonance and when $\nu_3 < \nu_1$, some chaotic behaviour is observed. It is again qualitatively different from what we previously observed and looks like a slow evolution near the fixed point at the origin from time to time interrupted by an exponential growth ("bursts") and followed by an oscillatory relaxation (close to a limit cycle) (see Fig. 5). Such behaviour is reminiscent of the so-called on-off intermittency (Platt et al. 1993), except for the oscillations after the bursts.

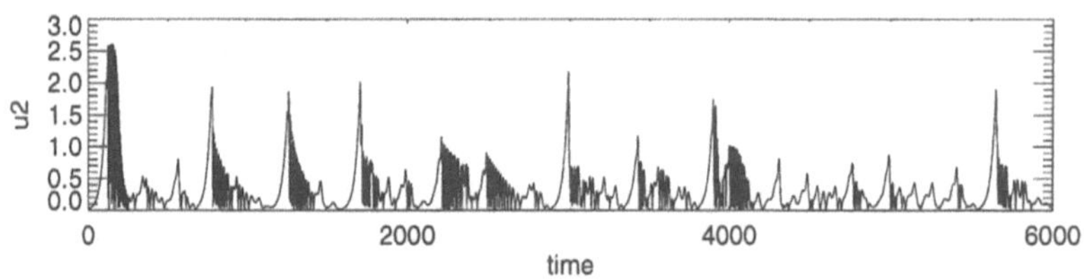

Fig. 5. Amplitude $u_2(t)$ for the parameters $\nu_0 = 0.1$, $\nu_1 = 0.05$, $\nu_2 = -0.04$, $\nu_3 = 0.03$

On-off intermittencies appear more strikingly from observing $\Psi - \Phi$ which spends much time near π[5] as can be seen on the joint probability density $p(\Phi, \Psi)$ (Fig. 6).

In this case the bicoherence as well as the tricoherence are of the same order and rather high (0.83). The estimation of the fractal dimension is more difficult than in the case of decay instability because of the two distinct time-scales involved: the period of the oscillations in the relaxation and the interval between bursts. A calculation gives a correlation dimension

$$D_2 \approx 4.2 \tag{22}$$

[4] Provided $\nu_1 < \nu_3$. The phases are $\Phi^s = -\Psi^s = \pi/2$.

[5] To be related to their values at the fixed point : $\Psi^{ms} - \Phi^{ms} = \pi$ (cf note 4)

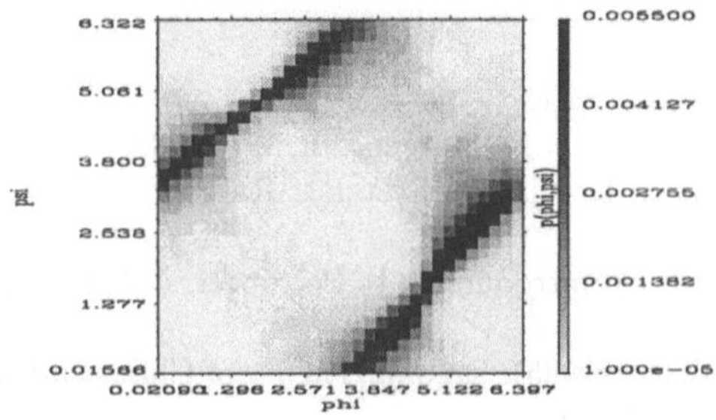

Fig. 6. Joint probability density $p(\Phi, \Psi)$ for the same parameters as in Fig. 5

which suggests that the dynamics of the modified decay is more complex than that of the cascade decay instability, in the sense that more degrees of freedom are excited.

4 Conclusion

These preliminary results show that our system describing the interaction of four waves as a cascade of two decay instabilities as well as a modified decay instability generates probability densities that depart from the Gaussian probability density assumed in the weak turbulence approximation. Such a result was expected because of the relatively low dimension of the phase space.

This system exhibited some nontrivial behaviour such as the chaotic synchronization between phases occurring in a cascade of two decay instabilities. It implies that the correlation time for the phases in the four-wave interaction is much higher than that for the three-wave interactions, whereas this inequality is reversed in other typical chaotic behaviour of this system. The estimations of these correlation times were done using the rather powerful techniques of higher order spectra. One also can note that on nonchaotic attractors these correlation time are not defined and in that sense both four-wave and three-wave interactions are present.

Generally speaking the probability measures appeared to be concentrated on low-dimensional sets of the phase space, thus the departure from Gaussian or uniform distributions is not so surprising. The calculation of the Rényi dimensions showed that rather typically the probability measures on strange attractors have a rather complex structure (multifractal). Furthermore this dimension allows a quantitative distinction between the decay and modified decay instabilities. Finally the probability densities corresponding to the observed attractors (chaotic as well as non-chaotic) do not correspond either to weak or strong turbulence, but instead to some state that could be called "moderate turbulence" which can be characterised by the correlation time for the phases corresponding to the three-wave and the four-wave interactions.

References

Afraimovich, V.S., Verichev, N.N., Rabinovich, M.I. (1986): Stochastic Synchronization of Oscillations in Dissipative Systems. Radiophys. Quantum Electron. **29**, 795–803

Brillinger, D.R. (1965): An Introduction to Polyspectra. Ann. Math. Stat. **36**, 1351–1374

Eckmann, J.P., Ruelle, D. (1985): Ergodic Theory of Chaos and Strange Attractors. Rev. Mod. Phys. **57**, 617–656

Fejer, J.A., Sulzer, M.P., Djuth, F.T. (1991): Height Dependance of the Observed Spectrum of Radar Backscatter from HF-induced Ionospheric Langmuir Turbulence. J. Geophys. Res., **96**, 15985–16008

Fujisaka, H., Yamada, T. (1983): Stability Theory of Synchronized Motion in Coupled-Oscillator Systems. Progr. Theor. Phys. **69**, 32–47

Hanssen, A., Mjolhus, E., Dubois, D.F., Rose, H.A. (1993): Numerical Test of the Weak Turbulence Approximation to Ionospheric Langmuir Turbulence. J. Geophys. Res. **97** A8, 12073–12091

Hentschel, H.G.E, Proccacia, I. (1983): The Infinite Number of Generalized Dimensions of Fractals and Strange Attractors. Physica **8**D, 435–444

Hospodarsky, G.B., Gurnett, D.A. (1995): Beat-Type Langmuir Wave Emissions Associated with a Type III Solar Radio Burst: Evidence of Parametric Decay. Geophys. Res. Lett. **22** 10, 1161–1164

Hugues, D.W., Proctor, M.R.E. (1990): Chaos and the Effect of Noise in a Model of Three-Wave Mode Coupling. Physica D **46**, 163–176

Krasnosel'skikh, V.V., Lefebvre, B., Dudok de Wit, T., Mourenas, D. (1997): Synchronization vs. Stochastization of Phases in Four-Wave Interaction. Physica Scripta **57**, 246–252

Krasnosel'skikh, V.V., Sotnikov, V.I. (1977): Plasma-Wave Collapse in a Magnetic Field. Sov. J. Plasma Phys. **3**(4), 491–495

Kravtchenko-Berejnoi, V., Lefeuvre, F., Krasnossel'skikh, V., Lagoutte, D. (1994): On the Use of Tricoherent Analysis to Detect Non-Linear Wave-Wave Interactions. Signal Processing **42**, 291–309

Kruer, W.L. (1989): *Laser-Plasma Interaction* (Pergamon Press, London)

Meunier, C., Bussac, M.N., Laval, G. (1982): Intermittency at the Onset of Stochasticity in Nonlinear Resonant Coupling Processes. Physica **4**D, 236–243

Monin, A.S., Yaglom, A.M. (1971): *Statistical Fluid Mechanics* (MIT Press, Cambridge USA)

Musher, S.L., Rubenchik, A.M., Zakharov, V.E. (1995): Weak Langmuir Turbulence. Phys. Rep. **252**, 177–274

Pecora, L.M., Carroll, T.L. (1990): Synchronization in Chaotic Systems. Phys. Rev. Lett. **64**(8), 821–824

Platt, N., Spiegel, E.A., Tresser, C. (1993): On-Off Intermittency: a Mechanism for Bursting. Phys. Rev. Lett. **70**(3), 279–282

Pomeau, Y., Manneville, P. (1980): Intermittent Transition to Turbulence in Dissipative Dynamical Systems. Comm. Math. Phys., **74**, 189–197

Vyshkind, S.Ya, Rabinovich, M.I. (1976): The Phase Stochastization Mechanism and the Structure of Wave Turbulence in Dissipative Media. Sov. Phys. JETP **44** No. 2, 292–299

Wersinger, J.-M., Finn, J.M., Ott, E. (1980): Bifurcation and "Strange" Behaviour in Instability Saturation by Nonlinear Three-Wave Mode Coupling. Phys. Fluids **23**(6), 1142–1154

Zakharov, V.E. (1972): Collapse of Langmuir Waves. Soviet Phys. JETP **35**(5), 908–914

Zakharov, V.E., Musher, S.L., Rubenchik, A.M (1985): Hamiltonian Approach to the Description of Non-Linear Plasma Phenomena. Phys. Rep. **129**, 285–366

Light Bullets Propagation in Saturating Nonlinear Optical Medium

Vladimir Skarka[1], Vazha Berezhiani[1,2], and Richard Miklaszewski[3]

[1] Laboratoire POMA, Université d'Angers, 2, boulevard Lavoisier, 49045 Angers, cedex 1, France
[2] Institute of Physics, The Georgian Academy of Science, Tbilisi 380077, Georgia
[3] Institute of Plasma Physics and Laser Microfusion, 00-908 Warsaw, street Hery 23, P.O. Box 49, Poland

Abstract. The simultaneous balance of nonlinear spatial self-focusing with diffraction and the compensation of dispersion with nonlinear temporal self-focusing, in a saturating nonlinear medium, gives stable spatiotemporal solitons – light bullets. An analytical approach for arbitrary saturating nonlinearity is established in order to investigate the dynamics of such light bullets. The analytically predicted exceptional robustness of light bullets, which can be generated over a large range of values of the parameters, is confirmed by numerical simulations.

1 Introduction

There is a growing interest in spatiotemporal solitons, which are characterised by a balance of diffraction and dispersion with respectively nonlinear spatial and temporal self-focusing. The dynamics of such localised structures is governed by a (3+1)-dimensional nonlinear Schrodinger equation (NSE) containing one propagation dimension and three "transverse" dimensions (McLeod et al. 1995).

The materials used in optical systems are usually of Kerr type. Consequently the dynamics of the electromagnetic (EM) pulses is described by an NSE with cubic nonlinearity (CNSE). It is well-known that only solutions of (1+1)-dimensional CNSE are stable (Zakharov and Shabat 1972). These exact analytical solutions obtained by the inverse-scattering method correspond to one-dimension, either temporal or spatial, solitons. However, NSEs with saturating nonlinearities have in some circumstances stable soliton solutions even for two or three transverse dimensions (Vakhitov and Kolokolov 1973). Such solitons, completely localised in three space and one time dimensions, are called "light bullets" (Silberberg 1990). Light bullets are of both fundamental and technological interest since they satisfy the best requirements for digital optical logic (McLeod et al. 1995). Given that materials currently used in optical systems exhibit weak saturation effects, there is an increasing interest for those materials which can sustain stable propagation of solitons limiting the growth of the nonlinear optical index $\delta n(I)$ (where I is intensity).

For higher dimensional NSEs there are no exact analytical methods to derive nonstationary solutions. One often has to rely on computer simulations in order to investigate the solutions of such equations.

General dynamical properties of nonstationary solutions are rather complex, making analytical approximations highly desirable. To describe the dynamics of localised solutions of the NSE various approximation schemes like the paraxial ray theory (Akhmanov et al. 1968), the moment theory (Lam et al. 1977) and the variational approach (Anderson 1981) have been developed.

In the present work we study the dynamics of the light bullets in saturating media, combining the variational method with numerical simulations. A general analytical approach for every transverse dimension, $D = 1, 2, 3$, and an arbitrary saturating nonlinearity is given in Sect. 1. This general approach will be applied to nonlinear materials with negative fourth-order optical index in order to study the conditions under which the propagation of light bullets is stable (Sect. 3). With these analytical results as a guiding line for numerical simulations the studies of exact soliton dynamics are performed in Sect. 4.

2 General Approach Based on the Variational Principle

The interaction of a finite amplitude optical pulse with a nonlinear bulk medium gives rise to a slowly varying optical field envelope $\mathcal{E}$ corresponding to a quasi-monochromatic electric field with carrying frequency ω. The equation for pulse propagation along the z axis is

$$2ik\left(\frac{\partial\mathcal{E}}{\partial z}+\frac{1}{v_g}\frac{\partial\mathcal{E}}{\partial t}\right)+\Delta_\perp\mathcal{E}-k\frac{d^2k}{d\omega^2}\frac{\partial^2\mathcal{E}}{\partial t^2}+2k^2\frac{\delta n(|\mathcal{E}|^2)}{n_o}\mathcal{E}=0, \qquad (1)$$

where v_g is the group velocity of the pulse, n_0 is linear optical index, $k = n_o\omega/c$ is the wave vector and $\Delta_\perp = (\partial^2/\partial x^2 + \partial^2/\partial y^2)$ is the 2D Laplacian describing beam diffraction. In what follows we assume that the optical medium exhibits an anomalous group velocity dispersion $d^2k/d\omega^2 < 0$.

Introducing a "moving" coordinate $\tau = t - z/v_g$ and making a self-evident renormalization of the variables we can rewritte (1) in dimensionless form as

$$i\frac{\partial E}{\partial z}+\frac{\partial^2 E}{\partial\tau^2}+\Delta_\perp E+f(|E|^2)E=0, \qquad (2)$$

where E is the slowly varying amplitude of the electric field $\mathcal{E}$ properly redefined corresponding to the nonlinearity under consideration.

Equation (2) describes spatial and temporal self-focusing of electromagnetic pulse propagating under the combined effects of diffraction and dispersion.

We look for a "spherical" symmetric distribution of the fields. Note that the comoving coordinate, τ, can be treated on an equal footing as a spatial coordinate. In terms of the radial variable $r = (x^2 + y^2 + \tau^2)^{1/2}$ (2) can be rewritten as

$$i\frac{\partial E}{\partial z} + \frac{1}{r^{D-1}}\frac{\partial}{\partial r}\left(r^{D-1}\frac{\partial E}{\partial r}\right) + f(|E|^2)E = 0, \tag{3}$$

where the transverse dimension, D, can be $1, 2$, or 3.

In this paper we use the variational approach. The Lagrangian density corresponding to (3) is

$$L = -r^{D-1}\left|\frac{\partial E}{\partial r}\right|^2 + \frac{i}{2}r^{D-1}\left(E^*\frac{\partial E}{\partial z} - E\frac{\partial E^*}{\partial z}\right) + r^{D-1}F(|E|^2), \tag{4}$$

where the asterisk denotes complex conjugate and

$$F(u) = \int_0^u f(u')\mathrm{d}u' \ . \tag{5}$$

An appropriate variation of the Lagrangian, $\delta L/\delta E^* = 0$, yields (3) as the Euler-Lagrange equation.

In the optimisation procedure the first variation of the corresponding functional must vanish on a set of suitably chosen trial functions. As trial functions we use Gaussian shaped pulses. Thus we assume that the evolution of the pulse can be characterised by the trial function:

$$E = A(z)\exp\left[-\frac{r^2}{2a(z)^2} + ir^2b(z) + i\phi(z)\right] \tag{6}$$

with amplitude A, beam "radius" a, wave front curvature b and phase ϕ as unknown functions of the propagation coordinate, z, which will be furthermore used in order to make the variational functional an extremum.

The set of Euler-Lagrange equations can be derived under the condition that the variation with respect to each of the unknown functions should be zero

$$A^2a^D = A_0^2a_0^D \tag{7}$$

$$\frac{\mathrm{d}^2a}{\mathrm{d}z^2} = \frac{4}{a^3} - \frac{2}{a}\left[K'(A^2) - \frac{K(A^2)}{A^2}\right] \tag{8}$$

$$\frac{\mathrm{d}\phi}{dz} = -\frac{D}{a^2} + \frac{1}{4}\left[(D+2)K'(A^2) - D\frac{K(A^2)}{A^2}\right] \tag{9}$$

$$b = \frac{1}{4a}\frac{\mathrm{d}a}{\mathrm{d}z}. \tag{10}$$

Here A_0 and a_0 are respectively the initial amplitude and the initial "radius" of the EM beam at $z = 0$ and $K(u) = 4/\Gamma\,[D/2] \int_0^\infty \mathrm{d}p\, p^{D-1} F\left(ue^{-p^2}\right)$, where Γ is the Gamma function.

Equation (7) is nothing but a statement of the fact that during the EM pulse evolution its "energy", $N = A^2 a^D$, is conserved. With (7) the integration of (8) gives the Hamiltonian, H, defined by

$$\frac{1}{2}\left(\frac{\mathrm{d}a}{\mathrm{d}z}\right)^2 + V(a) = H = V(a_0), \tag{11}$$

where

$$V(a) = \frac{2}{a^2} - \frac{2a^D}{Da_0^D A_0^2} K\left(\frac{A_0^2 a_0^D}{a^D}\right) \tag{12}$$

plays the role of an effective potential for the evolution of the radius a. We have assumed that the initial beam has a plane front (or zero curvature) $(\mathrm{d}a/\mathrm{d}z|_{z=0} = 0 = b(0))$.

Using the analogy with a particle in a potential well we can acquire a deeper physical understanding of light beam dynamics. Choosing the initial beam radius a_0 to be equal to the equilibrium radius a_e we obtain a stationary solution of (8) if $\partial V/\partial a|_{a=a_e} = 0$. Note that $-\partial V/\partial a$ is equal to the right hand side of (8). The equilibrium radius of the beam is readily found to be

$$a_e^2 = 2\left[K'(A_0^2) - \frac{K(A_0^2)}{A_0^2}\right]^{-1}, \tag{13}$$

The stability of the equilibrium solution, $a_0 = a_e$, can be checked by studying the behaviour of small amplitude disturbances around this equilibrium. Linearising (8) around the equilibrium solution ($a = a_e + \delta a$, $a_e >> \delta a$) we obtain

$$\frac{\mathrm{d}^2}{\mathrm{d}z^2}\delta a + \Omega^2 \delta a = 0, \tag{14}$$

where

$$\Omega^2 = \frac{\partial^2 V}{\partial a^2}|_{a=a_0} = \frac{1}{a_0^2}\left[\frac{4+2D}{a_0^2} - DA_0^2 K''(A_0^2)\right]. \tag{15}$$

If Ω^2 is positive, the equilibrium solution is stable.

In deriving the system (7)-(15) we did not use an explicit form for the functions $f(|A|^2)$; thus the established formalism can be applied to the NSE with an arbitrary nonlinear term.

For the saturating nonlinearity $V(a)|_{a\to 0} \to \infty$, which means that the beam radius is bounded from below. Consequently the so-called collapse does not occur. The beam is either trapped and forms an oscillating waveguide or diffracts monotonically. For the oscillating beam radius there are two distinct modes of behavior (see for instance Fig. 1):

(i) When $a_0 > a_e$, the slope of potential is positive, $\partial V/\partial a|_{a=a_e} > 0$, and the beam radius initially contracts until it reaches the smallest value given by the turning point $a_-(< a_e)$.
(ii) If $a_0 < a_e$, the slope is negative, $\partial V/\partial a|_{a=a_e} < 0$, and the beam radius initially increases. Further evolution depends upon the value of constant Hamiltonian H (see (11)).

If $H < 0$, the beam never diffracts and enters a self-trapped regime.

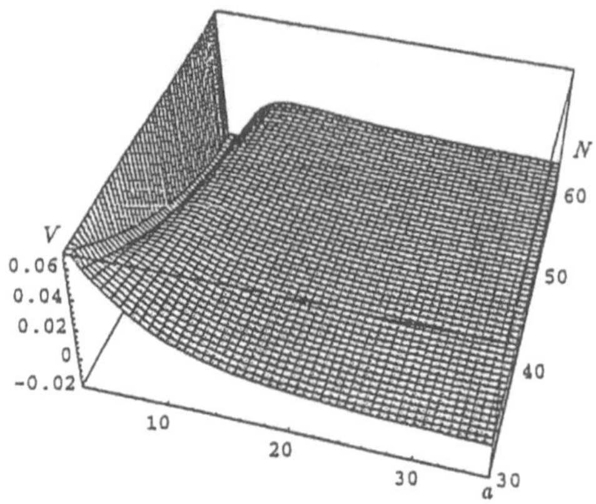

Fig. 1. Diffraction for subcritical energy $N < N_c$ and self-guiding in the supercritical case.

3 Saturating Nonlinearity with Negative Fourth-Order Optical Index

In nonlinear optics several types of saturating nonlinearities are discussed (Karlsson 1992; Berezhiani and Mahajan 1995). Recently a large nonresonant nonlinearity with negative fourth-order index variation has been found in polydiacetylene *para*-toluene sulfonate (PTS) (Lawrence et al. 1994).

In the subsequent analysis we consider a nonlinear term of the following form:

$$f(|E|^2) = |E|^2 - |E|^4 \ . \tag{16}$$

For PTS the second term in the right hand side of (16) can be of the same order as the first one (Wright et al. 1995).

In order to study the conditions under which solitons can propagate in such a material we apply the established general formalism to the saturating nonlinearity given by (16). For the function K one obtains

$$K(A^2) = \alpha A^4 - \beta A^6, \tag{17}$$

where $\alpha = 2^{(-D/2)}$ and $\beta = 2 \cdot 3^{(-D/2-1)}$. The equilibrium radius of the beam, obtained from (13),

$$a_e^2 = \frac{2}{\alpha A_0^2 - 2\beta A_0^4} \tag{18}$$

is plotted in the (a_0^2, A_0^2)-plane (see curve e in Fig. 2). In the region that lies above this curve the beam radius oscillates with initial focusing.

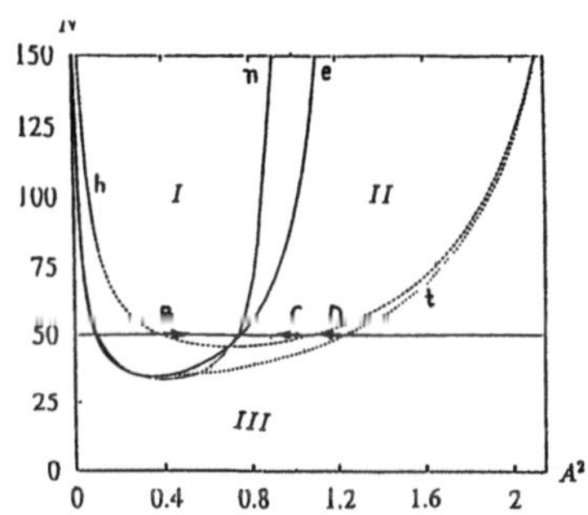

Fig. 2. Energy N as a function of A^2 for the 3D case. Equilibrium curve e (solid line) together with the numerically obtained equilibrium line n (dotted line), curve h, and trapping curve t (both dashed). I: focusing region, II: defocusing region and III: diffraction region.

The equilibrium solution is stable ($\Omega^2 > 0$), i.e. it is a soliton, if the condition

$$A_0^2 > \frac{\alpha}{4\beta}\frac{D-2}{D-1} \tag{19}$$

is satisfied. For 1D and 2D cases the equilibrium solution is stable for arbitrary beam amplitude since the inequality (19) is always satisfied. For the 3D case the solution is stable only if $A_0 > (\alpha/8\beta)^{1/2}(\approx 0.58)$. Thus the small amplitude ($A_0 < 0.58$) equilibrium solutions are unstable.

As we mentioned above, if the Hamiltonian is negative ($H < 0$), then the beam always forms an oscillating waveguide. From (11) and (12) one obtains the beam radius corresponding to the zero Hamiltonian

$$a_H^2 = \frac{D}{\alpha A_0^2 - \beta A_0^4} \tag{20}$$

which is plotted as the curve h in Fig. 2. It can be seen that, if $a_0^2 > a_H^2$, for given A_0^2, then $H < 0$.

The particularity of the 3D case is that $a_e > a_H$ only if the initial amplitude is sufficiently big, $A_0^2 > (\alpha/4\beta)$. The self-trapping of the beam is, however, possible even for a positive Hamiltonian. Indeed, if $A_0^2 < \alpha/4\beta$, then $a_e < a_H$ and, consequently, the equilibrium solution corresponding to the soliton appears in the range of $H > 0$ also. In what follows we consider only the 3D case.

In order to have better insight into the dynamical behavior of the pulse and the graphical presentation in terms of experimentally relevant quantities, we plot in Fig. 2 the conserved energy, N, as a function of the squared amplitude of the pulse, A^2. The solid line corresponds to the equilibrium and the dashed one to the zero Hamiltonian. The light bullet is trapped in an oscillating waveguide provided its energy N is bigger than the critical value given by $N_c = 35.3$. Note that the equilibrium is stable, i.e. $dN/dA^2 > 0$, if $A^2 > 0.34$. Thus for every given $N > N_c$ we have two equilibria. However, only the stable one plays the role of an attractor for each of the points initially located in the regions (I) and (II). This can be easily seen from Fig. 1 by analysing the shape of potential V as a function of radius a and energy N, viz.

$$V(a, N) = \frac{2}{a^2} - \frac{2\alpha N}{3a^3} + \frac{2\beta N^2}{3a^6} \tag{21}$$

which is obtained from (7), (11) and (12).

The front lines (Fig. 1) correspond to the sections with constant energy N less than the critical value, N_c, and the beam diffracts monotonously.

In the supercritical case, $N > N_c$, the potential has two extrema, a minimum and a maximum. Therefore, depending upon $H = V(a_0, N)$, the beam will be either trapped or will diffract. The beam with initial radius corresponding to the point lieing on the potential curve somewhere below the maximum will be always trapped creating a light bullet with oscillating radius. If it starts from the negative slope of the potential, the beam will first diffract (region II in Fig. 2); if on the positive slope, it will first focus (region

I). Only if the initial conditions are chosen to be on the stable part of the curve e will the light bullet have constant (equilibrium) radius since it is already at the bottom of the potential well. However, if the initial radius is so small that it corresponds to a potential greater than its extremum (maximum), the beam can diffract forever. For $N = N_c$ the potential curve is broken at the point where minimum and maximum coincide which corresponds to the tangent point between stable and unstable equilibrium on curve e of Fig. 2.

The trapping line t in Fig. 2 asymptotically approaches the h of the zero Hamiltonian for large amplitude. Above the trapping line (regions I and II in Fig. 2) the light bullets can be created even for a positive Hamiltonian. Below this line (region III) monotonic diffraction occurs.

The dynamics of pulsating light bullets clearly follows from Figs. 1 and 2. From the variational method it can be predicted that for a constant energy N (a horizontal line in Fig. 2) all beams initially in region I, or equivalently on the positive slope, will form light bullets pulsating around the equilibrium radius with initial focusing till the turning point in region II, which can be exactly computed from known initial conditions. For initial points in region III the beam will always diffract even though it crosses the equilibrium line. For instance, if it is initially on the right of the trapping line in Fig. 2, the diffracting beam amplitude will decrease from its maximum value, crossing region II and overshooting the stable equilibrium as well as the unstable one.

The main conclusion which follows from present analyses is the existence of the stable equilibrium which plays a role of attractor provided $N > N_c$. For a given energy, N, the beam will be self-trapped for a wide range of values of the parameters: not only in the domain of the negative Hamiltonian but also in a part of the region where $H > 0$. Such a self-trapped beam generates a light bullet which will pulsate near the stable equilibrium.

4 Numerical Simulation of the Behaviour of the Light Bullets

The main limitation of various approaches like the paraxial, momentum or variational one is that they are valid only in the aberrationless approximation, i.e. they are unable to account for structural changes in the beam shape.

It is also obvious that in the case of big saturating nonlinearities the fundamental, i.e. the ground state soliton profile can be quite different from the Gaussian one. Such aspects of the beam dynamics are better delineated by numerical simulations although the guiding lines for simulation are still provided by approximate analytical approaches.

In this section we present the results of numerical simulations of (2) for an initially Gaussian shaped beam in the 3D case. The simulation data obtained can be qualitatively understood and interpreted with better physical insight using the analytical results of the previous section.

The numerical simulations give the critical energy for the beam trapping to be $N_c = 34.04$. This is the minimum of the dotted line n in Fig. 2, which corresponds to the exact numerical solution of (2) for the ground state soliton. Such a soliton with maximal amplitude, A_m, is stable provided $A_m^2 > 0.41$. For a large amplitude soliton the results of the variational approach (based on a Gaussian trial function) deviate markedly from exact numerical solutions.

Numerical simulations show that, if an initial profile of the beam is close to the stable equilibrium one (the exact numerical solution), the beam quickly attains the profile of a ground state soliton and propagates for a long distance without distortion of its shape. Even if the initial beam is Gaussian shaped, i.e. in a domain of parameters corresponding to the profile quite far from the equilibrium one, this beam will, nevertheless, either focus or defocus to the ground state equilibrium and exhibit damped oscillations around it. The pulsations are damped due to the appearance of a radiation spectrum. This is shown in Fig. 3, where the field intensity ($I = |E|^2$) distribution is plotted versus radius, r, and the propagating coordinate, z, for the initially Gaussian shaped beam $|E(r,0)| = A_0 \exp[-r^2/2a_0^2]$. In this case the parameters of the beam (A_0 and a_0) are chosen to give the energy $N = A_0^2 a_0^3 = 50$ (the horizontal line in Fig. 2). Although the initial amplitude of the beam (points B in Fig. 2) is quite different from the amplitude of the corresponding equilibrium soliton, light bullets exhibit damped pulsations near the equilibrium. If we starting from the initial point B (in domain I of Fig. 2), the amplitude first increases in Fig. 3, which corresponds to an initial focusing as predicted analytically. Then the equilibrium amplitude is a bit overshot in order to be reached immediately. The initial condition B corresponds to the negative Hamiltonian. The other initial point tested, D, belongs to the domain, II, with initial defocusing and with positive Hamiltonian. Also for higher energy the pulse dynamics remains the same. For instance, for $N = 150$, the light bullet evolves toward the stable equilibrium exhibiting damped pulsations (see Fig. 4 and specially Fig. 5 where the damping is more obvious).

It can be concluded that, as predicted by our analytical approach, the exceptional robustness of light bullets which can be generated for a wide range of values of the parameters, even far from stable equilibrium, is confirmed by numerical simulations. Due to their complete localisation and robustness, light bullets appear to be the best candidate for carrying the information that has to be treated in all-optical logic circuits.

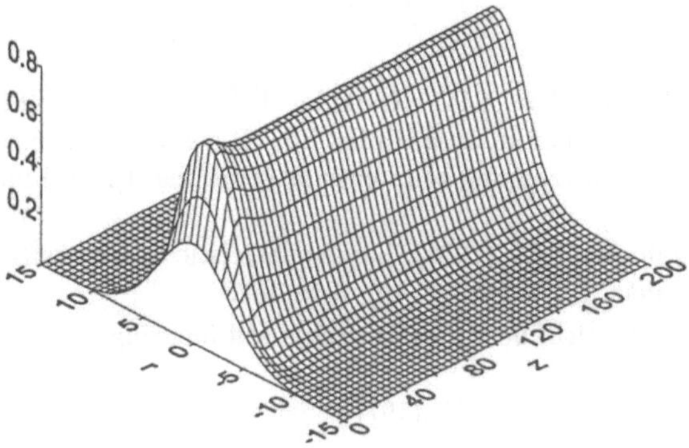

Fig. 3. Generation of a stable light bullet with initial focusing for energy $N = 50$ and initial intensity $A_0^2 = 0.4$.

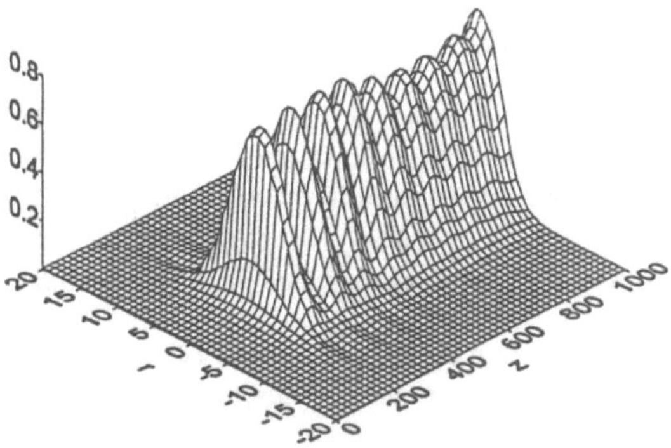

Fig. 4. Formation of pulsating light bullet for $N = 150$ and for initial intensity $A_0^2 = 0.01$.

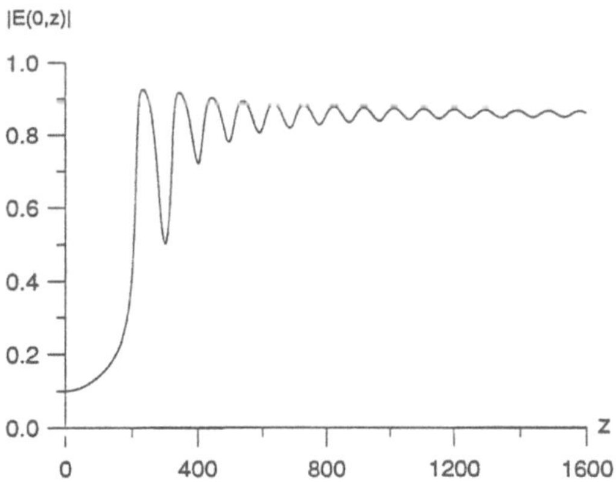

Fig. 5. Dynamics of the pick amplitude of beams $|E(0, z)|$ for $N = 150$ and $A_0^2 = 0.01$.

References

Akhmanov, S.A., Sukharukov, A.P., Khokhlov, R.V. (1968): Self-focusing and diffraction of light in a nonlinear medium. Sov. Phys. Usp. **10**, 609–634

Anderson, D. (1981): Self-trapping of intense laser beams. Phys. Fluids **24**, 2345–2347

Berezhiani, V.I., Mahajan, S.M. (1995): Large relativistic density pulses in electron-positron-ion plasmas. Phys. Rev. E **52**, 1968–1979

Karlsson, M. (1992): Optical beams in saturable self-focusing media. Phys. Rev. A **46**, 2726–2734

Lam, J.F., Lippman, B., Tappert, F. (1977): Self-trapped laser beams in plasma. Phys. Fluids **20**, 1176–1179

Lawrence, B.L., Torrueles, W.E., Cha, M., Sundheimer, M.L., Stegeman, G.I., Meth, J., Etemad, S., Baker, G. (1994): Identification and role of two-photon excited states in a π-conjugated polimer. Phys. Rev. Lett. **73**, 597–600

McLeod, R., Wagner, K., Blair, S. (1995): (3+1)-dimensional optical soliton dragging logic. Phys. Rev. A **52**, 3254–3278

Silberberg, Y. (1990): Collapse of optical pulses. Opt. Lett. **15**, 1282–1284

Vakhitov, N.G., Kolokolov, A.A. (1973): Stationary solutions of the wave equation in the medium with nonlinearity saturation. Sov. Radiophys. **9**, 262–270

Wright, E.M., Lawrence, B.L., Torruellas, W., Stegeman, G.(1995): Stable self-trapping and ring formation in polydiacetylene para *para*-toluene sulfonate. Opt. Lett. **20**, 2481–2483

Zakharov, V.E., Shabat, A.B. (1972): Exact theory of two-dimensional self-focusing and one-dimensional self-modulation of waves in nonlinear media. Sov. Phys. JETP **34**, 62–69

The Dynamics of Some Quantum Open Systems with Short-Range Nonlinearities

Francis Nier

Centre de Mathématiques, URA-CNRS 169, Ecole Polytechnique, F-91128 Palaiseau Cedex, France

Abstract. We propose a functional framework for some nonlinear dynamics, where the nonlinearities are (very) short-range perturbation of a given Hamiltonian. In this framework one can prove the global in time existence and uniqueness of a solution for the time dependent problem and the existence of steady states of the form which can be specified. This problem arises from the modeling of semiconductors and covers different applications.

1 Introduction

The starting point of this analysis was the work of Jona-Lasinio et al. (1995) and Presilla and Sjöstrand (1996) in which they showed with a reduced model the interaction of resonances with nonlinear effects due to Coulombian repulsion of electrons in resonant tunneling diodes. Before analysing the validity of their reduced model via a small parameter asymptotic expansion, it seemed useful to understand the functional framework in which such nonlinear problems are well-posed. The problem studied by Jona-Lasinio, Presilla and Sjöstrand is one-dimensional, but the approach that we propose here holds in a general situation and covers other applications. For example it works also for the description of the dynamics in quantum wave-guides connected to an electronic device where nonlinear phenomena take place. The steady states of this latter problem have been studied in Ben Abdallah (1996). A precise description of these applications can be found in Nier (1997).

We are interested in the quantum dynamics of mixed states which is governed by the quantum Liouville (or Von Neumann) equation. For a given time-dependent self-adjoint Hamiltonian, $H(t)$, acting on a Hilbert space $\mathcal{H}$, the equation is

$$\begin{cases} i\partial_t \rho(t) = [H(t), \rho(t)] & t \in \mathbb{R} \\ \rho(t=0) = \rho_{\mathrm{ini}} \ . \end{cases} \tag{1}$$

The nonlinear problem which we consider is then set by taking a Hamiltonian which depends on $\rho(t)$ as

$$H(t) = H_0 + V[\rho(t)] \ . \tag{2}$$

An example of ρ-dependent perturbation with $\mathcal{H} = L^2(\Omega)$ can be given by calculating the particle density in a bounded region $\omega \subset \Omega$:

$$\int_{\omega} n(x)\varphi(x)dx = \mathrm{Tr}[\rho\varphi(\mathrm{x})] \quad \forall\varphi \in \mathcal{C}_0^{\infty}(\omega) \tag{3}$$

and by taking for $V[\rho]$ the solution of the Poisson equation:

$$\begin{cases} -\Delta V = n & \text{in } \omega \\ V\Big|_{\partial\omega} = 0 \ . \end{cases}$$

In such applications the reference Hamiltonian H_0 is $-\Delta + V_0$, where $-\Delta$ is considered with boundary conditions on $\partial\Omega$ and V_0 is a fixed background potential.

We close this introduction by specifying the kind of solution to (1) and (2) that we seek. We wish to describe an open system with inflow "boundary conditions" at infinity. This has two consequences: Firstly the solution $\rho(t)$ cannot be trace class as a usual state and this density matrix will only be locally trace class. Secondly a natural generalisation of the notion of non-homogeneous boundary conditions consists in looking for a solution in some affine space. As in the classical case good inflow boundary conditions are those which determine uniquely up to bound states (or semi-bound states) a steady state ρ_0 for the reference Hamiltonian, H_0, that is a locally trace class operator ρ_0 so that $[H_0, \rho_0] = 0$. Thus the boundary conditions can be represented by a choice of ρ_0 commuting with H_0 and the affine space will be $\rho_0 + \widetilde{J}$, where $\widetilde{J}$ is a vector space of weighted trace class operators the weight of which ensures that the difference $\rho - \rho_0 \in \widetilde{J}$ is small in incoming regions. The incoming and outgoing regions in the phase space are distinguished by introducing an observable A which has a positive commutator with H_0. This observable A is usually called a conjugate operator for H_0 in the framework of Mourre theory (see Mourre 1981; Cycon et al. 1987; Amrein et al. 1996). The incoming region corresponds to the negative spectrum of the self-adjoint operator, A, while the outgoing region corresponds to the positive spectrum of A. In the standard case for which $H_0 = -\Delta$ in $\mathbb{R}^n$ one takes the generator of dilatations $A = \frac{1}{2}(x.D + D.x)$ which is the (Weyl-) quantisation of the function on the phase space (x, ξ).

2 Assumptions and Definitions

Next we briefly describe the main assumptions on which our analysis relies and introduce some definitions. Their detailed form can be found in Nier (1997). These assumptions are written in an abstract form in order to fit with different applications, but can be considered in a first reading with $\mathcal{H} = L^2(\mathbb{R}^n)$, $H_0 = -\Delta$, $A = \frac{1}{2}(x.D + D.x)$ or A equal to a vector field which coincides with the generator of dilatations out of some compact subset of $\mathbb{R}^n$, $V = V(x)$ and $R = (1 + |x|)$.

Let $\mathcal{H}$ be a separable Hilbert space and let $\mathcal{L}(\mathcal{H})$ and $\mathcal{L}^p(\mathcal{H})$, $1 \leq p \leq \infty$, be respectively the spaces of bounded operators and the pth order Schatten spaces on $\mathcal{H}$.

Hypothesis 1. Existence of a conjugate operator We consider a reference Hamiltonian H_0 which is self-adjoint with domain $D(H_0)$, is bounded from below and admits a conjugate operator.

There exists a self-adjoint operator A with $D(A) \cap D(H_0)$ dense in $D(H_0)$ such that

$$(H_0 + i)^{-1} \mathrm{ad}_{\mathrm{A}}^{\mathrm{k}} \mathrm{H}_0 \in \mathcal{L}(\mathcal{H}), \quad \forall \mathrm{k} \in \mathbb{N}, \tag{4}$$

where ad^{k} denotes the kth order commutator. Moreover for some discrete set, $\tau \subset \mathbb{R}$, the Mourre estimate

$$\chi(H_0)\,[H_0, iA]\,\chi(H_0) \geq c_\chi \chi(H_0)^2 + K_\chi \tag{5}$$

holds with $c_\chi > 0$, K_χ compact on $\mathcal{H}$ as soon as $\chi \in C_0^\infty(\mathbb{R} \setminus \tau)$.

Hypothesis 2. Energy localization. We assume that the reference steady state ρ_0 is localized on the energy axis away from the set of thresholds τ : there exists $\chi_0 \in C_0^\infty(\mathbb{R} \setminus \tau)$ and $\rho_0 = \chi(H_0)\rho_0$.

Hypothesis 3. Short-range condition for the perturbation V. The perturbation V belongs to some Banach space E for which the mapping,

$$E \ni V \to R^{s_0+1+\varepsilon_0} V \in \mathcal{L}(\mathcal{H}), \tag{6}$$

is continuous for some $\varepsilon > 0$, where R is a self-adjoint operator such that

$$R^{-s_0} \chi(H_0 + V) \in \mathcal{L}^1(\mathcal{H}), \quad \forall \chi \in C_0^\infty(\mathbb{R}), \forall V \in E, \tag{7}$$

and

$$R^{-1} A (H_0 + i)^{-1} \in \mathcal{L}(\mathcal{H}) \ . \tag{8}$$

From (6) and (7) we infer that the perturbation $V \in E$ satisfies

$$\chi(H) V \chi(H_0) \in \mathcal{L}^1(\mathcal{H})$$

and Birman's theorem (see Reed and Simon 1975) yields the existence of the wave operators

$$W_\pm := \lim_{t \to \pm\infty} e^{itH} e^{-itH_0}$$

and their asymptotic completeness.

Some definitions:

a) *For $V \in E$, ρ_V denotes the steady state for $H = H_0 + V$ given by $\rho_V := W_- \rho_0 W_-^*$.*

b) *If χ_0 is the energy cut-off introduced in Hypothesis 2, A_0 denotes the localized conjugate operator for H_0, $A_0 := \chi(H_0) A \chi(H_0)$.*

c) *The space $\widetilde{J}$ is defined as the intersection $F_\mu(A_0)^{-1}\mathcal{L}^1(\mathcal{H})F_\mu(A_0)^{-1} \cap \mathcal{L}(\mathcal{H})$, where F_μ is a smooth function which is equal to 1 on $\mathbb{R}^-$ and decays like $\langle x\rangle^{-\mu}$, $\mu > 1$, on $\mathbb{R}^+$. It is a Banach space with the norm*

$$\|\rho\|_0 = \|F_\mu(A_0)\rho F_\mu(A_0)\|_{\mathcal{L}^1} + \|\rho\| \ .$$

With this norm one naturally associates the distance d_0 on the affine space $\rho_0 + \widetilde{J}$.

d) *The vector space $\widetilde{J}$ is also endowed with a coarser topology which is the projective topology associated to the embedding $\widetilde{J} \to F_\mu(A_0)^{-1}\mathcal{L}^1(\mathcal{H})F_\mu(A_0)^{-1}$ and $\widetilde{J} \to \mathcal{L}(\mathcal{H})$, where these two spaces are respectively considered with the norm topology and the strong operator topology. This topology, which is also defined on $\rho_0 + \widetilde{J}$ by translation, is denoted θ.*

We can now specify the type of nonlinearity which is to be considered.

Hypothesis 4. Properties of $V[\rho]$ in the nonlinear case. The mapping $\rho_0 + \widetilde{J} \ni \rho \to V[\rho] \in E$ is assumed to be globally Lipschitz when $\rho_0 + \widetilde{J}$ is endowed with the distance d_0 and continuous on any bounded set of $\rho_0 + \widetilde{J}$ endowed with the topology induced by θ.

The above assumption will be sufficient for proving the well-posedness of the evolution problem (1)(2). The existence of steady states for the nonlinear problem is obtained by being more specific on the functional framework and by strengthening a little the above assumption.

Hypothesis 4'. For the steady states problem, one considers the case $\mathcal{L}(\mathcal{H}) = L^2(\Omega)$, where Ω is a smooth enough unbounded domain of $\mathbb{R}^n$ and where H_0 is a bounded perturbation of the Laplace operator on Ω with homogeneous boundary conditions. The Banach space E, where the perturbation V lies, is assumed to be a subspace of

$$\left\{u \in L^\infty(\Omega;\mathbb{R}),\ \langle x\rangle_\Omega^{n+1+\varepsilon} V \in L^\infty(\Omega;R)\right\}$$

with $\langle x\rangle_\Omega := \langle x\rangle\, 1_\Omega(x)$. One considers the particle density on a bounded sub-domain ω of Ω defined like in (3) as a bounded measure. For a given $\rho \in \rho_0 + \widetilde{J}$ it is denoted by $n[\rho]$ and the potential $V[\rho]$ is supposed to be given by

$$V[\rho] = L\left[n[\rho]\right] + V_1\,, \tag{9}$$

where V_1 is a fixed element of E and L is a linear continuous operator from $\mathcal{M}_b(\overline{\Omega_b})$ to E. Moreover the operator L is assumed to be positivity preserving and compact from $L^\infty(\Omega_b)$ to E.

3 Results

The first result is about the linear problem, where the perturbation $V \in E$ does not depend on time. It relies on the Sigal-Soffer strong propagation estimates (see Derezinski and Gérard 1997; Jensen 1985 and Sigal and Soffer 1988) for the propagator e^{-itH_0} which are derived from *Hypothesis 1*. It says that the affine space $\rho_0 + \widetilde{J}$ provides a good extension to the quantum case of the classical notion of inflow boundary conditions.

Theorem 3.1 *For any $V \in E$ the operator ρ_V belongs to $\rho_0 + (\widetilde{J}, \|.\|_0)$ and the only solutions $\rho \in \rho_0 + \widetilde{J}$ to $[H_0 + V, \rho] = 0$ are the bounded operators of the form*

$$\rho = \rho_V + \sum_{k \in \mathbb{N}} a_k \, (\psi_k, .) \, \psi_k, \tag{10}$$

where the ψ_ks are normalized eigenvectors of H_0+V and the sequence $(a_k)_{k\in\mathbb{N}}$ belongs to $\ell^1(\mathbb{C})$.

Once the linear problem is well understood, the result about the nonlinear Cauchy problem (1) and (2) is proved with the standard argument: a) Local existence and uniqueness: The mapping which associates to $\rho_1(t)$, $t \in [0,T]$, the solution $\rho_2(t)$ to (1) with $V(t) = V[\rho_2(t)]$ is proved to be a contraction in some complete affine space for T small enough; b) The global existence is then obtained with a sublinear estimate and Gronwall lemma.

Theorem 3.2 **a)** *For any $s \in \mathbb{R}$ and any fixed $\rho(s) \in \rho_0 + \widetilde{J}$, there exists a unique element $\rho \in C^0(\mathbb{R}; \rho_0 + \widetilde{J}, \theta)$ such that*

$$\rho(t) = U[\rho](t,s)\rho(s)U[\rho](t,s)^*, \quad \forall t \in \mathbb{R}, \tag{11}$$

where $U[\rho](t,s)$ is the unitary propagator associated with the time-dependent hamiltonian $H(t) = H_0 + V[\rho(t)]$, or equivalently so that

$$i\partial_t \rho(t) = [H_0 + V[\rho(t)]] \, \rho(t) \tag{12}$$

in terms of quadratic forms defined on $D(H_0)$.

b) *The solution of the above nonlinear problem satisfies $\rho(t) = \alpha(t-s)\rho(s)$, where $\alpha(t)$ group of homeomorphisms on $(\rho_0+\widetilde{J}, d_0)$ which is θ-continuous with respect to $t \in \mathbb{R}$. Moreover this group leaves invariant the convex sets*

$$\left(\rho_0 + \widetilde{J}\right)_R = \left\{\rho \in \rho_0 + \widetilde{J}, \; \rho^* = \rho\right\} \tag{13}$$

$$\left(\rho_0 + \widetilde{J}\right)_+ = \left\{\rho \in \rho_0 + \widetilde{J}, \; \rho \geq 0\right\} \; . \tag{14}$$

The group property stated in part b) comes from the fact that the nonlinear system (1, 2) is indeed an autonomous differential system. This result is indeed completed with a local in time stability result with respect to perturbations of the initial data and of the equation (by adding a nonhomogeneous term to (1)). It can also be extended to the case of time-dependent "boundary conditions".

The existence of nonlinear steady states is also obtained with standard techniques: basically the Leray-Schauder fixed point theorem. Indeed for this last result it is not necessary to avoid the threshold energies and Hypothesis 2 can be forgotten.

Theorem 3.3 *When Hypothesis 4 is replaced with Hypothesis 4' and* $\rho_0 \geq 0$ *is assumed, there exists a solution* $(\rho, V) \in \mathcal{L}(\mathcal{H}) \times E$ *to*

$$\begin{cases} [H_0 + V, \rho] = 0, \\ V = V[\rho] \end{cases} \tag{15}$$

with the conditions

$$\rho \geq 0 \tag{16}$$

$$1_{K_0}(H_0 + V)\rho = \rho \ . \tag{17}$$

Moreover such a solution exists with the condition $\rho \in \left(\rho_0 + \widetilde{J}\right)_+$ *and more precisely with*

$$\rho = \rho_V + \sum_{i=1}^{N} \lambda_k (\psi_k, .)\psi_k, \tag{18}$$

where $(\psi_k)_{k\in\{1\dots N\}}$ *is a finite orthonormal sequence of eigenvectors of* $H_0 + V$ *and the real numbers* λ_k*,* $k \in \{1 \dots N\}$*, all belong to* $[0, ||\rho_0||]$.

References

Amrein, W., Boutet de Monvel, A., Georgescu, V. (1996): *C_0-Groups, Commutator Methods and Spectral Theory of N-Body Hamiltonians* (Progress in Mathematics **135**, Birkhäuser, Basel-Boston-Berlin)

Ben Abdallah, N. (1996): On a Multidimensional Schrödinger-Poisson Scattering Model for Semiconductors. (Preprint 96.4, UMR-CNRS 9974, Univ P. Sabatier, 118 route de Narbonne, F31062 Toulouse Cedex, France)

Cycon, H.L., Froese, R.G., Kirsch, W., Simon, B. (1987): *Schrödinger Operators with Application to Quantum Mechanics and Global Geometry* (Texts and Monographs in Physics. Springer-Verlag, Berlin, Heidelberg)

Derezinski, J., Gérard, C. (1997): *Asymptotic Completeness of Classical and Quantum N-Particles Systems* (Texts and Monographs in Physics. Springer Verlag, Berlin, Heidelberg)

Jensen, A.(1985): Propagation estimates for Schrödinger-type operators. Trans. Am. Math. Soc. **291** (1), 129–144

Jona-Lasinio, G., Presilla, C., Sjöstrand, J.(1995): On Schrödinger Equations with Concentrated Nonlinearities. Ann. Phys. **240** (1), 1–21

Mourre, E.(1981): Absence of Singular Continuous Spectrum for Certain Self-Adjoint Operators. Commun. Math. Phys. **78**, 391–408

Nier, F.(1997): The Dynamics of some Quantum Open Systems with Short-Range Nonlinearities. (Prépublication n° 1161 du Centre de Mathématiques, URA-CNRS 169, Ecole Polytechnique, F91128 Palaiseau Cedex, France, soumis.)

Presilla, C., Sjöstrand, J.(1996): Equilibrium and nonequilibrium transport properties in biased resonant tunneling heterostructures. (Preprint 1133 CMAT-Ecole Polytechnique, URA-CNRS 169, F91128 Palaiseau, France)

Reed, M., Simon, B.(1975): *Methods of Modern Mathematical Physics.* (Academic Press, New York, San Francisco)

Sigal, I.M., Soffer, A.(1988): Local Decay and Propagation Estimates for Time Dependent and Time Independent Hamiltonians. (Preprint Princeton University, Princeton)

Global Dynamics of Fusion Plasmas

Hans Wilhelmsson

Chalmers University of Technology, 412 96 Göteborg, Sweden

Abstract. A self-consistent treatment of reaction-diffusion processes is surveyed. Dynamic response to heating processes as well as the phenomenon of profile consistency is analysed.

The nonlinear evolution of diffusion-reaction systems is of considerable interest for describing new physical situations, for example the state of a burning fusion plasma in which deuterium and tritium ions react to produce alpha particles and fast neutrons under the simultaneous influence of diffusion effects.

The diffusion-reaction equations also offer a fruitful area of research from a mathematical point of view (Kadomtsev 1992; Smoller 1983).

The equation of nonlinear heat conduction with a nonlinear source term has numerous applications in science. Analytical solutions to such an equation cannot be obtained for arbitrary initial conditions, as can be shown by a Painlev test (Jancel and Wilhelmsson 1991). One therefore has to develop special analytic methods or rely on numerical solutions. A new technique of analysis based on a central expansion in the spatial variable has been developed with the aim of studying bounded states of a fusion plasma (Wilhelmsson et al. 1992; Wilhelmsson and Le Roux 1993). The technique considerably simplifies the dynamics of the plasma as well as the determination of equilibria. For cases of practical interest where there are simultaneous effects of alpha particle heating, external heating by neutral beam or radio frequency radiation, ohmic heating and losses by bremsstrahlung explicit analytic results are obtained. The analytic results are complementary to those of extensive direct computer calculations and useful as check-points for interpretation.

The technique amounts to introducing several useful dynamic variables which characterise the profile of a fusion plasma and its behaviour in space and time. Such variables are (i): a central value A, (ii): a width L and (iii): a form function η, which all depend only on time, but which enter the power series expansion in terms of the spatial coordinate.

The reaction-diffusion PDE is accordingly transformed into three ordinary coupled nonlinear differential equations in only time, which can be studied e.g. by phase-plane analysis, by simplified computer analysis or which occasionally can be given exact solutions by a suitable transformation of the time variable. The technique can be generalised to consider coupled reaction-diffusion equations, describing the simultaneous evolution in space and time of a fusion plasma with due regard to given spatial boundary conditions and taking into account the influence of external perturbations, describing e.g.

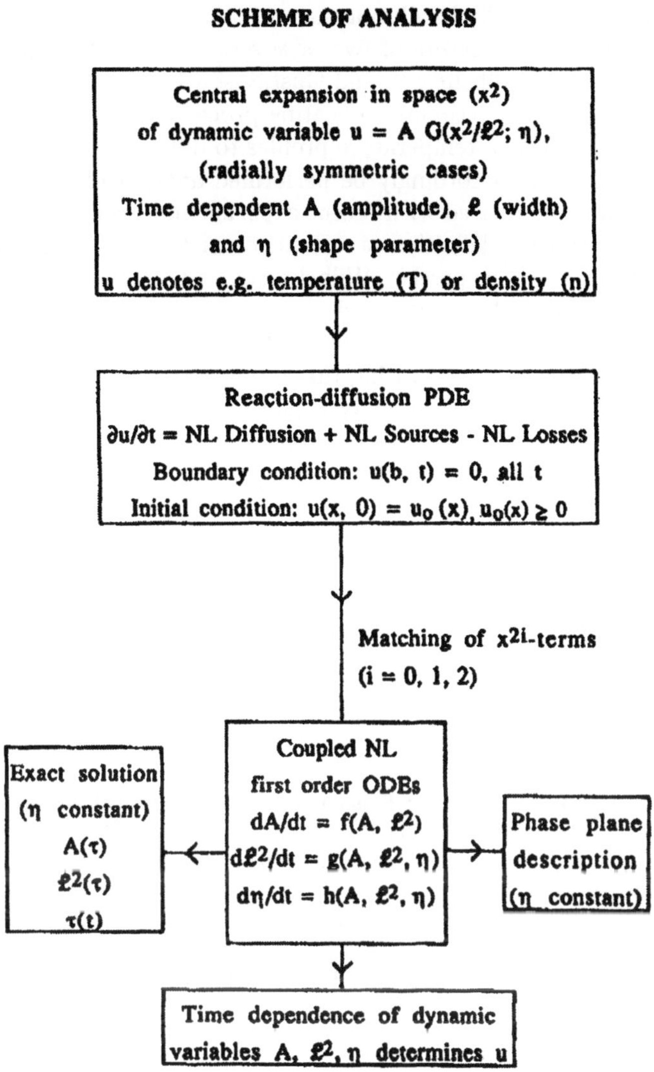

Fig. 1. Schematic outline of the analysis discussed in this paper.

the effects of a periodically oscillating source, of a mixture of two or several oscillating sources or a mixture of two or several oscillating sources having different frequencies (Wilhelmsson and Gustavsson 1994).

Studies of a dynamic response to heating processes and, in particular, of the rigidity of the plasma temperature profiles to different types of localised additional heating can accordingly be performed analytically, providing results which can be compared with recent experimental results (Cirant et al. 1994). The quantitative definition of "profile consistency" (Coppi 1980) can accordingly be made more precise (Wilhelmsson et al. 1996).

The short summary here presented of some work that extended through quite some years in various laboratories in Sweden, France, Italy and elsewhere has the purpose of celebrating the important rle that Marc Feix has played in physics, mathematical analysis and simulations to scientists of all ages all over the world. I would like to express my most sincere appreciation and thanks for your inspiration and friendship.

References

Cirant, S. et al. (1994): IAEA Conference on Controlled Nuclear Fusion and Plasma Physics, Sevilla, paper IAEA/CN/60/G-P-G Sevilla

Coppi, B., (1980): Non classical transport and the principle of profile consistency. Comments Plasma. Phys. Contr. Fusion **5**, 261–268

Jancel, R., Wilhelmsson, H. (1991): Double beam free electron laser. Physica Scripta **44**, 603–605

Kadomtsev, B.B. (1992): Tokamak Plasma: a complex physical system (Plasma Physics Series, Institute of Physics Publishing, Bristol and Philadelphia)

Smoller, J. (1983): Shock waves and reaction-Diffusion Equations, Part II (Springer-Verlag, New-York)

Wilhelmsson, H. (1988): Simultaneous diffusion and reaction processes in plasma dynamics. Phys. Rev. A **38**, 1482–1489

Wilhelmsson, H., Etticher, B., Cairns, R.A., Le Roux, M.N. (1992): Evolution of temperature profiles in a fusion reactor plasma. Physica Scripta **45**, 184–187

Wilhelmsson, H., Lazzaro, E., Cirant, S. (1996): Sensitivity of fusion plasma temperature profiles to localized and distributed heat sources. Physica Scripta **54**, 385–391

Wilhelmsson, H., Le Roux, M.N. (1993): Self-consistent treatment of transport in tokomak palsmas. Physica Scripta **48**, 735–757

Wilhelmsson, H., Gustavsson, H.G. (1994): Dynamic response treatment of a burning fusion plasma. Physica Scripta **49**, 715–723

The Jeans' Criterion and the Gravitational Instability

Serge Bouquet

Commissariat à l'Energie Atomique, Centre de Bruyères-le-Châtel, DPTA/PPE, BP 12, F-91680 Bruyères-le-Châtel, France

Abstract. In this paper, we derive a Jeans' criterion for the dynamical spherical collapse of a gravitational polytropic medium with constant γ. In opposition to the classical (static configuration) Jeans' criterion where a threshold for stability appears, it is shown that the stability of the dynamical solution depends on the value of γ. For γ=4/3, the existence of a threshold is also obtained. However, for $\gamma \neq 4/3$, the system is always unstable with an instability growing according to a time-power, instead of the usual exponential law. In addition, two types of instabilities are obtained: the large wavelengths are unstable from the very beginning of the evolution, whereas the small ones, produce oscillations on a finite range of time and, finally, the system becomes unstable (oscillations stop). In this stage the pressure can be neglected. Since the collapse is unstable, we examine the fragmentation problem. The dynamical Jeans' criterion suggests a cascade in which the mass M and the radius R of the pieces are connected by a relation of the type $M \propto R^2$.

Keywords. Gravitational collapse, Jeans' instability, Self-similar solutions.

1 Introduction

The Jeans' criterion provides the maximum size, R_J, for a gaseous mass to be stable against its own gravitation. For $R>R_J$ the gravitational system may develop disturbances increasing with time. The maximum radius R_J was given by Jeans (1929) as

$$R_J = \left(\frac{\pi}{\rho_0 G}\right)^{1/2} c_0 = 2\pi \frac{c_0}{\Omega_{0J}} , \tag{1}$$

where $\Omega_{0J} = (4\pi\rho_0 G)^{1/2}$ is the so-called Jeans' frequency and c_0 and ρ_0 are respectively the initial sound velocity and the initial density (G is the gravitational constant). The derivation of (1) is based on the uniformity of the density and pressure throughout the gas. The starting point of Jeans' calculation then assumes a linearization about the static solution. However, as noticed by Bonnor (1957), this assumption is questionnable because, for a uniform pressure medium, no gradient exists. It is therefore impossible to balance initially the gravitation and a static solution cannot exist. Bonnor overcame the difficulty in the following way: in an infinite gaseous mass, no preferred direction for the gravitational field can be chosen and it must be zero. As, however, Jeans' calculations and applications deal with finite mass systems, the self-consistency is not clear.

In this paper we study the gravitational instability of a gaseous mass collapsing under its own gravitational field, i.e., in opposition to Jeans' work, we relax the static solution as initial state and we study the dynamics in accordance with the initial conditions.

2 Lagrangian Analytical Solution for a Collapsing Homogeneous Medium

Consider a homogeneous spherical collapsing medium. The equation of motion is $d^2r/dt^2=-GM/r^2$, where M is the mass inside the sphere of radius r. Assuming no crossing of shells, a first integration provides the following relation between the velocity v and the position r

$$\frac{1}{2}v^2 = \frac{GM}{r} + C\,, \tag{2}$$

where C is a constant of integration.

The $C=0$ solution (parabolic motion with particles at rest at infinity) is easily obtained and we are going to see that this simplification is justified from the properties of the solutions for $C>0$. Taking $C=0$, we get

$$r(t) = \left(\frac{9GM}{2}\right)^{1/3} \left(t_c - t\right)^{2/3}\quad , \tag{3}$$

where t_c is a constant of integration. For $t \rightarrow -\infty$, the shell is at infinity and for $t=t_c$, the shell has collapsed into a singularity at the centre. During the collapse the Eulerian velocity $v(r,t)$ is linear with respect to the position according to

$$v(r,t) = \frac{dr}{dt} = -\frac{2}{3}\frac{r}{t_c - t}\,. \tag{4}$$

We now examine the properties of the solutions obtained for $C>0$ (hyperbolic motion). On the introduction of the new variable θ defined as $\mathrm{sh}^2\theta = Cr/GM$, $\theta \geq 0$, the integration of equation (2) gives

$$\frac{1}{2}\,\mathrm{sh}2\theta - \theta = \frac{C\sqrt{2C}}{GM}\,(t_c - t)\,. \tag{5}$$

This parametrization differs basically from that given initially by Mestel (1965) and developed later by Penston (1969) where the particles are initially at rest at a finite distance, a, from the centre. Consequently, C is negative (it is given by $-GM/a$). It corresponds to an elliptic motion and one obtains $\sin\theta$ instead of our $\mathrm{sh}\theta$. In our case, the velocity of any particle can never be zero at a finite distance of the centre. Moreover, for $t \to -\infty$, θ is going to infinity and from our relations between r, θ and t, we get (for $\theta \to +\infty$, $\mathrm{sh}2\theta \propto 2\mathrm{sh}^2\theta$)

$$r(t) \approx \sqrt{2C}\,(t_c - t)\,. \tag{6}$$

The difference between the $C=0$ solution and the general one (Eq.[5]) is that the shell comes from infinity linearly with respect to t in equation (6), while it moves according to $(-t)^{2/3}$ in equation (3). As a consequence, when the shell is at infinity, the velocity $v(t)$ approaches the constant value $-\sqrt{2C}$ and goes to zero with C as $\sqrt{C}$. Although, the solutions given by equations (3) and (5) describe quite different initial stages for the collapse, we prove the important result that the two solutions converge towards the same limit. In the asymptotic stage, i.e., for $t \to t_c$, $\theta \to 0$, we can easily obtain the relation between r and t. Taking the first and third orders of Taylor's series in the relation between r and θ, we get $\theta^2 = Cr/GM$ and (5) becomes

$$\frac{2}{3}\theta^3 = \frac{C\sqrt{2C}}{GM}\,(t_c - t)\,. \tag{7}$$

Eliminating θ between the two relations, we obtain precisely the equation (3) obtained for $C=0$. As a main result we have proved that any solution, obtained for $C>0$ (hyperbolic case), converges as the time elapses towards the special $C=0$ solution (parabolic case). Consequently, the solution given by equation (3) is an attractor. It is surprising and very interesting that this solution which is the more physical one (particles at rest at infinity) possesses the mathematical property of being an attractor.

3 Eulerian Analytical Solution

In opposition to Jeans' assumption we consider a gaseous mass evolving dynamically from the beginning. Since, in any case, equation (4) describes the collapse, it is worthwhile to study the stability of this special solution. This analysis will be performed in terms of the Eulerian coordinates.

3.1 The Equations of the Model

The evolution of the system is governed by the usual hydrodynamical equations written in spherical geometry: the Euler equation for the velocity $v(r,t)$, the Poisson equation for the gravitational field $g(r,t)$, the continuity equation for the density $\rho(r,t)$ and, finally, the polytropic equation which connects the pressure and the density $p = K\rho^\gamma$, where γ is the polytropic constant and K a constant in space and time.

3.2 The Self-Similar Transformation

From equation (3) the Lagrangian position of a particle is proportionnal to a t-power-law which is typical of a self-similar behavior (Bouquet *et al* 1985). It is therefore very interesting to introduce a scaling which absorbs most of the time-dependence of the evolution. This scaling was initially used by Bouquet *et al* (1985) and was again fruitfully applied by Blottiau *et al* (1988) where the new dependent variables $\hat{\rho}$, $\hat{p}$, $\hat{v}$ and $\hat{g}$ have been introduced according to

$$\rho(r,t) = (1+\Omega t)^{-2}\,\hat{\rho}(\hat{r},\hat{t})\,, \qquad p(r,t) = (1+\Omega t)^{-2\gamma}\,\hat{p}(\hat{r},\hat{t})\,, \tag{8}$$

$$v(r,t) = (1+\Omega t)^{-1/3}\left[\hat{v}(\hat{r},\hat{t}) + \frac{2}{3}\Omega\hat{r}\right], \qquad g(r,t) = (1+\Omega t)^{-4/3}\,\hat{g}(\hat{r},\hat{t})\,, \tag{9}$$

and where the new independant variables $\hat{r}$ and $\hat{t}$ were given by

$$\hat{r} = r(1+\Omega t)^{-2/3}\,, \qquad \hat{t} = \Omega^{-1}\mathrm{Log}(1+\Omega t). \tag{10}$$

At the initial time $t = \hat{t} = 0$ and it is worthwhile to note that in this transformation, all the quantities with and without " ^ " coïncide except for the velocity where the translation $2\Omega\,\hat{r}/3$ occurs. The free prameter Ω will be determined later on (see Eq. [12]). Equations (10) give a correspondance between the "usual" or "physical" (r,t)-space-time and the "new" or "rescaled" one $(\hat{r},\hat{t})$. The quantities $\hat{\rho}$, $\hat{p}$, $\hat{v}$ and $\hat{g}$ correspond respectively to the density, the pressure, the velocity and the gravitational field in the rescaled space. In addition, on the substitution of equations (8)-(10) into the initial ones, three of the rescaled equations are invariant, i.e., Poisson, continuity and polytropic equations keep the same form, but the equation of motion is now

$$\frac{\partial\hat{v}}{\partial\hat{t}} + \hat{v}\,\frac{\partial\hat{v}}{\partial\hat{t}} = -\,\frac{1}{3}\Omega\hat{v} + \frac{2}{9}\,\Omega^{2}\hat{r} + \hat{g} - \exp\left(\frac{2}{3}(4-3\gamma)\Omega\hat{t}\right)\frac{1}{\hat{\rho}}\,\frac{\partial\hat{p}}{\partial\hat{r}}\,. \tag{11}$$

The new Euler equation contains two additional terms as well as an explicitly $\hat{t}$-dependant coefficient in front of the pressure gradient. In the case of a uniform density it disappears, but, as soon as any inhomogeneity appears, this term concentrates the difficulty. In spite of it, an analytical treatment is given in Sect. 4. We now try to identify the solution given by equation (3) in terms of the new variables $\hat{r}$ and $\hat{t}$. In opposition to the developments given in Sect. 1, we can now look for a static solution ($\hat{v} = 0$) in the new space without "violating" the Physics. We take a uniform constant density $\hat{\rho}_0$ which corresponds to the initial density ρ_0 in the initial space at t=0. Now the point is that we drop all the terms in equation (11) except the third and the fourth ones of the R.H.S. The gravitation $\hat{g}$ is therefore balanced by the linear field $\frac{2}{9}\Omega^2\hat{r}$ and we obtain the value of Ω (we take the minus sign since we have a collapse) which is proportional to the Jeans' frequency

$$\Omega = -\sqrt{3/2}\ \Omega_{0J} = -\sqrt{3/2}\ \sqrt{4\pi\rho_0 G}. \tag{12}$$

On the other hand, the expresion for the mass provides $\hat{r} = (3M/4\pi\rho_0)^{1/3}$ and using the first equation (10), we obtain

$$r(t) = \left(\frac{9GM}{2}\right)^{1/3}\left(\sqrt{3/2}\ \Omega_{0J} - t\right)^{2/3}. \tag{13}$$

This expression is identical to equation (3) and we deduce that the final instant for the collapse is (remember Eq.[12])

$$t_c = -1/\Omega. \tag{14}$$

As a consequence the static solution in the new space corresponds to the dynamical one given by equation (3). The entire evolution has been absorbed by the analytical transformation (Eqs [8-10]). Coming back to the the (r,t)-space, we obtain the Eulerian functions:

$$v(r,t) = \frac{2}{3}\frac{\Omega r}{1+\Omega t}, \quad \rho(t) = \frac{\rho_0}{(1+\Omega t)^2}. \tag{15}$$

4 Stability Analysis

The stability analysis of any dynamical evolution is a difficult problem because, in opposition to the case of an equilibrium, additional terms with time dependent coefficients appear in the linearized equations. For instance, it is very difficult to carry out the stability analysis of the solution given by equation (15) in the usual (r,t)-space. But, this solution is an equilibrium in the $(\hat{r},\hat{t})$-space and it is therefore possible to perform a classical stability analysis. The stability in the ordinary space results from the inverse transformation obtained from equations (8)-(10). In the rescaled space the static solution corresponding to evolution (3) is given by

$$\hat{v}_0 = 0\ ,\ \hat{\rho}_0 = \rho_0\ ,\ \hat{g}_0 = -(2/9)\ \Omega^2\hat{r}\ ,\ \hat{p}_0 = p_0\ , \tag{16}$$

where p_0 is obtained from the polytropic equation. If perturbations would not appear, the system would evolve ideally according to equations (15) up to the central singularity and it would stay infinitely static in the rescaled space (this property comes out from the fact that $\hat{t} \to +\infty$ as $t \to -1/\Omega$). As soon as perturbations appear, the collapse deviates from equation (15) and it becomes important to know whether this initial perturbation causes large deviations from the unperturbed evolution. The stability analysis will be performed in the new space and a Jeans' criterion will be obtained. We point out that it is completely exact because the linearisation is made about a static state in the rescaled space. Using the inverse transformation from $(\hat{r},\hat{t})$ to (r,t) a Jeans' criterion valid for the dynamical collapse is derived.

4.1 The Linear Perturbation Equation

We consider small deviations about the static solution given by equations (16)

$$\hat{x} = \hat{x}_0 + \delta\hat{x}(\hat{r},\hat{t}) \, , \tag{17}$$

where $\hat{x}$ stands for the velocity, the density, the gravitational field or the pressure. Substituting equation (17) into equation (11); neglecting second order terms and finally eliminating, respectively, the pressure, the velocity and the gravitational perturbation from the polytropic, continuity and Poisson equations, we obtain the equation for the density perturbation (for further details, see Blottiau *et al* 1988)

$$\left(\frac{\partial^2}{\partial\hat{t}^2} + \frac{1}{3}\,\Omega\,\frac{\partial}{\partial\hat{t}} - \Omega_{0J}^2\right)\delta\hat{\rho} = c_0^2\, e^{\frac{2}{3}(4-3\gamma)\Omega\hat{t}}\left(\frac{\partial^2}{\partial\hat{r}^2} + \frac{2}{\hat{r}}\frac{\partial}{\partial\hat{r}}\right)\delta\hat{\rho} \, , \tag{18}$$

where c_0 satisfies $c_0^2 = \gamma p_0/\rho_0 = \gamma K\rho_0^{\gamma-1} = \gamma K\hat{\rho}_0^{\gamma-1}$. Equation (18) describes the propagation of sound waves inside a spherically collapsing gravitational medium. Seeking for a spherical wave for the spatial dependance, we write $\delta\hat{\rho}(\hat{r},\hat{t})$ in the form

$$\delta\hat{\rho}(\hat{r},\hat{t}) = A(\hat{t})\,\sin(\widehat{kr})\,/\,(\widehat{kr}), \tag{19}$$

where $A(\hat{t})$ is the amplitude of the wave. Substituting Eq.(19) in Eq.(18), we obtain

$$\frac{d^2A}{d\hat{t}^2} + \frac{\Omega}{3}\frac{dA}{d\hat{t}} + \left(\hat{k}^2 c_0^2\, e^{\frac{2}{3}(4-3\gamma)\Omega\hat{t}} - \Omega_{0J}^2\right)A = 0 \, . \tag{20}$$

This equation has been solved analytically for γ=4/3 and studied numerically for other values of γ by Blottiau *et al* (1988). In this paper we derive the general analytical solution for $1 \leq \gamma < 4/3$.

4.2 Analytical Solution for $\gamma \neq 4/3$

We introduce the new variables, x and y, given by

$$x = x_0 \exp\left[(4\text{-}3\gamma)\ \Omega\hat{t}/\ 3\right],\ y = x_0^{\mu}\ A(\hat{t}) \exp(\Omega\hat{t}/\ 6)\ , \tag{21}$$

where

$$\mu = 1\ /\left[2\ (4 - 3\gamma)\right],\ x_0 = x_0(\hat{k}) = 3\hat{k}c_0\ /\left[(4 - 3\gamma)\left|\Omega\right|\right] = \sqrt{6}\,\hat{k}c_0\ /\left[(4 - 3\gamma)\Omega_{0J}\right]. \tag{22}$$

The evolution equation (20) becomes

$$x^2\,\frac{d^2y}{dx^2} + x\,\frac{dy}{dx} + (x^2 - \lambda^2)\ y = 0\ , \tag{23}$$

with $\lambda = 5\mu$. This is a Bessel equation and its general solution is

$$y(x) = a_\lambda\ J_\lambda(x) + b_\lambda\ N_\lambda(x)\ , \tag{24}$$

where J_λ and N_λ are respectively the Bessel functions of first and second kind of order λ (a_λ and b_λ are arbitrary constants). If λ is not an integer (24) is no longer valid and we must replace N_λ with $J_{-\lambda}$ (Schwartz 1979). In order to simplify the calculation we restrict our study to values of γ making λ integer. Consequently, we shall not work with the functions $J_{-\lambda}$. This assumption does not limit the physical properties because we may expect them to change continuously with the parameter γ.

From the relation between λ and μ, the smallest admissible value of λ is 3, for which $\gamma = 1.05$. More generally, for $\lambda = \{3, 4, 5, 6,\ldots,p,\ldots, +\infty\}$, we obtain $\gamma = \{1.05, 1.13, 1.17, 1.19,\ldots, (4/3)\text{-}5/(6p),\ldots,4/3\}$. It is interesting to notice that $\gamma = 4/3$ appears as a limiting, but excluded, case. On the other hand, since Ω is negative, the abcissa x decreases (see Eq.[21]) from the initial value x_0 to 0 as the time elapses ($\hat{t}$ goes from 0 to $+\infty$ as t goes from 0 to $-1/\Omega$). As a consequence for $\hat{t} \to +\infty$, $y(x)$ diverges like $N_\lambda(x)$ [the function $J_\lambda(x)$ goes to zero asymptotically like x^λ]

$$y(x) \approx N_\lambda(x) \approx x^{-\lambda},\ x \to 0. \tag{25}$$

In addition it comes out from equations (21) that the variation of the amplitude $A(\hat{t})$ is given by

$$A(\hat{t}) \approx y(x) \exp(-\Omega\,\hat{t}\ /\ 6) \approx \exp(-\Omega\,\hat{t}) \approx \exp[(3/2)^{1/2}\Omega_{0J}\,\hat{t}]. \tag{26}$$

The fourth term of expression (26) comes from equations (21) and (25). The density perturbation increases exponentially with $\hat{t}$ (for $\hat{t}$ large enough) and, surprisingly, regardless of the value of γ. This asymptotic instability was already obtained by

Blottiau *et al* (1988) and it corresponds to the case of a purely gravitational system (Hunter 1967). This limiting case may be recovered from equation (20) in which the term proportional to c_0^2 governing the propagation of sound waves (and containing the single γ-dependence) is neglected. Now, the question is whether this instability develops from the beginning $\hat{t}=0$, or, in contrast, does the system evolves "quietly" over a long period with perturbations coming very late?

4.3 Onset of the Instability in the Rescaled Space

In equation (24) the instability is induced only by the term $b_\lambda N_\lambda(x)$ which is singular at $x=0$ [we drop consequently the regular contribution $J_\lambda(x)$]. At $\hat{t}=0$, the initial value, x_0, of x is given by equation (22) which is linear in $\hat{k}$. If $\hat{k}$ is large enough, x_0 is larger than the first zero, x_λ, of $N_\lambda(x)$. If one remembers that x decreases from x_0 to 0 as $\hat{t}$ increases from 0 to $+\infty$, and since $N_\lambda(x)$ oscillates over the range $[x_0, x_\lambda]$, we see that the function $y(x)$ oscillates also over this range. These oscillations correspond to the phase preceeding the real instability described by equation (25) which occurs as soon as x goes through the first zero x_λ. The larger x_0-x_λ is, the longer the oscillating phase is. This oscillating phase does not stop provided the inequality $x \geq x_\lambda$ is satisfied and the divergence (25) does not happens. This relation is of special interest because it will provide the Jeans' criterion in the rescaled space. The first zero of the Neumann function $N_\lambda(x)$ is given by Abramowitz and Stegun (1964) in the analytical form $x_\lambda=\lambda\ (1 + 0.93\ \lambda^{-2/3} + 0.24\ \lambda^{-4/3} +\ldots)$. With the combination of equations (21), (22) and this expansion, where we neglect the term $\lambda^{-4/3}$, the condition $x \geq x_\lambda$ is

$$\hat{k} \geq \hat{k}_c \exp\left[-(4\text{-}3\gamma)\Omega\hat{t}\,/\,3\right], \tag{27}$$

where $\hat{k}_c=\alpha k_{0J}$ with $\alpha=(25/24)^{1/2}[1 + 0.93(2(4\text{-}3\gamma)/5)^{2/3}+\ldots]$ and $k_{0J}=\Omega_{0J}/c_0$ is the initial wave number given by Eq.(1). Equation (27) provides the Jeans' criterion for stability in the rescaled space. It must be noted that the new time $\hat{t}$ appears explicitly.

For γ=4/3 the time dependence vanishes, the term in bracket in α is unity and we obtain a threshold in the new space where $\hat{R}_J=(24/25)^{1/2}(2\pi)/k_{0J}$ defines a new Jeans' radius which is very close to the one given by Eq.(1) since in the new space the configuration is static. For $\gamma\neq$4/3 equation (27) leads to the following: for a wave number, $\hat{k}$, initially larger than $\hat{k}_c$ (the perturbation is therefore initially stable), we can calculate the time, $\hat{t}_{ins}$, after which the phase of time-oscillations stops and the exponential instability begins to grow. This time is given by the equality in equation (27), $\hat{t}_{ins} = \left[3/(|\Omega|\,(4\text{-}3\gamma))\right] \log\left(\hat{k}\,/\,\hat{k}_c\right)$. On the other hand, if $\hat{k}$ is such that $\hat{k}<\hat{k}_c$, i.e., $x_0 < x_\lambda$, the perturbation is initially unstable with no oscillations.

4.4 Onset of the Instability in the Physical Space

We now recast our results in the physical space. In absence of any perturbation the configuration would collapse homologously as given by equations (15). The radius decreases like $r=r_0(1+\Omega t)^{2/3}$. Similarly, from the definition of the Jeans' frequency, the sound speed and the time-variation of the density (Eq.[15]) we obtain $\Omega_J(t)=\Omega_{0J}(1+\Omega t)$ and $c(t)=c_0(1+\Omega t)^{\gamma-1}$. It must be pointed out that the sound speed increases more slowly than the fluid velocity which varies as $(1+\Omega t)^{-1}$. Especially for $\gamma=1$ (isothermal case) $c=cste=c_0$.

Now, if we introduce a disturbance of eigenmode $\hat{k}$ in the rescaled space, the configuration is perturbed around the **static state**, while in the physical space the configuration experiences perturbations around the **dynamical solution**. The initial wave number, k_0, in the physical space (r,t) is given by $k_0=\hat{k}$. During the collapse, the number of spatial periods in the perturbation remains constant at any time [because it is invariant in the static space $(\hat{r},\hat{t})$]. Since any length rescales according to $(1+\Omega t)^{2/3}$, the wave number becomes $k(t)=k_0(1+\Omega t)^{-2/3}$. Substituting these results in equation (27) and using the connection between $\hat{t}$ and t we obtain the **Jeans' criterion** in the physical (r,t)-space

$$k(t) \geq k_c(t) \equiv \alpha\, k_{0J}\,(1+\Omega t)^{\gamma-2}. \tag{28}$$

This equation differs from the original Jeans' formulation, since it contains the factor α and the additional γ(and t)-dependence. The criterion given by equation (28) distinguishes between unstable $[k(t)<k_c(t)]$ and stable modes $[k(t)>k_c(t)]$. Nevertheless, as the collapse proceeds, since $k_c(t)$ increases as $(1+\Omega t)^{\gamma-2}$, while $k(t)$ varies more slowly according to $(1+\Omega t)^{-2/3}$ **any small wavelength becomes ultimately unstable.** The special case $\gamma=4/3$ does not obey the same conclusion because $k(t)$ and $k_c(t)$ have the same time rescaling (the stability of any mode $k(t)$ at time t is therefore the stability of the mode k_0 at time t=0). For $\gamma\neq 4/3$ we can obtain the time, t_{ins}, at which oscillations of an initially stable mode, k_0, stop. This condition is $k(t)=k_c(t)$ and we obtain

$$t_{ins} = t_c\left[1 - \left(k_c(0)\,/\,k_0\right)^{3/(4-3\gamma)}\right]. \tag{29}$$

Obviously we could have also obtained t_{ins} directly from $\hat{t}_{ins}$ derived in the rescaled space. At this instant the gravitational term dominates in the perturbation equation (see Sect.4.1). The perturbations cannot longer propagate and they are trapped at the sonic point (see the proof in Blottiau *et al* 1988). Any local mass condensation therefore attracts gravitationally the surrounding matter. This is the starting point for the non linear phase and the formation of the seed leading to high dense shells. This was shown numerically for γ=1.20. From equation (29) we conclude the larger the difference 4/3-γ is, the sooner an initial stable mode k_0 becomes unstable. In the limit $\gamma\to 4/3$ we obtain $t_{ins}\to t_c$, in accordance with the previous considerations.

Let us derive the density contrast $\delta\rho(r,t)/\rho(r,t)$. From the Poisson equation and

Eq.(19), $\delta\rho(r,t)/\rho(r,t) = \hat{\delta\rho}(\hat{r},\hat{t})/\hat{\rho}(\hat{r},\hat{t}) = [\delta\rho(r,0)/\rho(r,0)][A(\hat{t})/A(0)]$ along a trajectory. Introducing equations (21) in the last expression and using $\hat{t}$=Log$(1+\Omega t)/\Omega$ we obtain

$$\frac{\delta\rho(r,t)}{\rho(r,t)} = \frac{\delta\rho(r,0)}{\rho(r,0)}\,\frac{x_0^{-\mu}}{A(0)}\left(1+\Omega t\right)^{-1/6} N_\lambda\!\left[x_0\left(1+\Omega t\right)^{(4-3\gamma)/3}\right]. \tag{30}$$

For the sake of clarity we analyse this general formula in two cases $x_0 \ll x_\lambda$ and $x_0 \gg x_\lambda$. We begin with the oscillations ($x_0 \gg x_\lambda$). The function $N_\lambda(x)$ oscillates with amplitude decreasing as $x^{-1/2} \approx (1+\Omega t)^{-(4-3\gamma)/6}$. Since $\rho \approx (1+\Omega t)^{-2}$, we obtain from equation (30)

$$\delta\rho/\rho \approx (\delta\rho/\rho\,)_0\rho^{\,(5-3\gamma)/12} \text{ or } \delta\rho \approx (t_c - t)^{(3\gamma-17)/6}. \tag{31}$$

As soon as $\delta\rho$ is of order of ρ, the linear model is no longer valid. If we label by $(\delta\rho/\rho)_0$ the value of the density contrast at t=0, the value of the density making $\delta\rho \approx \rho$ is given by

$$\rho = \rho_0(\delta\rho/\rho)_0^{\,12/(3\gamma-5)}. \tag{32}$$

This result generalises Hunter's expression (1962), $\rho=\rho_0(\delta\rho/\rho)_0^{-2}$, valid for $\gamma = 1$ without pressure. If for example we take $(\delta\rho/\rho)_0=10^{-2}$, for $\gamma \to 4/3$, we obtain $\rho=\rho_0 10^{24}$ which is a notable compression rate corresponding to a very large range of application of the linear perturbation theory. The time, t_{nl}, from which the system leaves the linear phase is derived from equations (14) and (32) and we obtain $t_{nl} = t_c\,[1 - (\delta\rho/\rho)_0^{6/(3\gamma-5)}]$. In the former case, $\rho=\rho_0 10^{24}$, one gets $(t_{nl} - t_c)/t_c \approx (\rho_0/\rho\,)^{1/2} = 10^{-12}$. The linear phase stops therefore at the very final stage of the collapse and the evolution is nearly entirely governed by the self similar solution (Eq.[13]).

We now examine the case $x \ll x_\lambda$ for which $N_\lambda(x) \approx x^{-\lambda}$. From equation (30), we obtain $\delta\rho/\rho \approx (\delta\rho/\rho)_0\rho^{\,1/2}$ or $\delta\rho \approx (t_c - t)^{-3}$. As discussed at the end of Sect.4.2, this result corresponds to a free-fall collapse (Hunter 1967). In this limit we conclude that, although pressure is included, the system finally forgets it and behaves as a free-falling sphere for large wavelength.

4 Astrophysical Considerations and Conclusion

Before developing astrophysical aspects, we recall the main results. A Jeans' criterion, valid for a collapsing gravitational polytrope (with $1 \le \gamma < 4/3$), is derived. It comes from the stability analysis of the solution describing the collapse of the system, under its own gravitation and pressure. This new criterion generalises the classical Jeans' ones to dynamics and is called "**the dynamical Jeans' criterion**". Similarly to the Jeans' criterion it is shown that the dynamical solution for γ=4/3 exhibits stable and unstable modes. In contrast configurations for γ<4/3 are unstable. Nevertheless, this instability does not have the classical behavior :

i) the perturbation $\delta\rho$ **does not increase exponentially**. It varies according to a time power, given by $\delta\rho \approx (t_c - t)^{-n}$, where t_c is the final instant of the singular collapse and where n is a positive exponent given in **ii)**,

ii) although the dynamical criterion does not contain stable modes for γ<4/3, it distinguishes between two kinds of instabilities according to the value of the wavelength λ. For λ such that $\lambda \geq \lambda_c \equiv 2\pi/k_c$, where k_c is given by equation (28), the value of the exponent in **i)** is n=3. This result does not depend on the value of γ. In contradistinction, a short wavelength, $\lambda \leq \lambda_c$, produces **time oscillations**. The amplitude of the envelop increases with an exponent depending on γ given by $n = (17-3\gamma)/6$. Let us compare the two results. The perturbation given by the first equation increases always more rapidly than the one derived in the second one, because $-3<(3\gamma-17)/6$. As a consequence the limit of validity of the linear theory, $\delta\rho \approx \rho$, will be reached after a long time for small wavelengths and the system will experience **very high compression** before entering the non linear phase. This property is explained as follows. The first result in **ii)** was initially given by Hunter (1967) in the case of gravitational systems with no pressure. In our case, the evolution of a small wavelength perturbation is very slow since it is submitted to the influence of the pressure-gradient which produces time oscillations and delays the instant at which a local condensation will attract gravitationally the neighbouring particles.

iii) although the dynamical criterion given by equation (28) keeps formally invariant when the time elapses, it contains time-dependent quantities. As a consequence, the frontier λ_c used in **ii)** is a **function of time**. According to the definition of k_c we get $\lambda_c(t)=\lambda_J[(t_c - t)/t_c]^{2-\gamma}$, where λ_J is the Jeans' wavelength at the initial time, given by $\lambda_J=(2\pi/k_{0J})/\alpha$. This definition slightly differs from the classical one by the factor α^{-1} which is close to unity. On the other hand, if at t=0 the wavelentgh of the perturbation is λ_0, it is given, at t, by$\lambda(t)=\lambda_0[(t_c - t)/t_c]^{2/3}$. Since γ<4/3, we deduce that $\lambda_c(t)$ shrinks more quickly than $\lambda(t)$ and we conclude that for $\lambda_0>\lambda_J$, we obtain $\lambda(t)>\lambda_c(t)$ at any time, and the perturbation will give rise to the evolution given by the first case of **ii)**. For$\lambda_0<\lambda_J$, we start with an oscillation phase and as soon as the inequality is no longer valid ($t=t_{ins}$) the evolution is given by the second case of **ii)**. After $t=t_{ins}$ the gravitation is definitely stronger than the pressure gradient and this is the starting point for the formation of high density regions from the local condensations. We conclude that, in any case, the effects of the pressure gradients are finally forgotten. Their role concerns only the delay of the final phase which is always purely gravitational.

We examine now the difficult question of fragmentation. The rough application of the Jeans' criterion (Eq.[1]) to the collapse gives $R_J \approx R^{3/2}$. We deduce that R_J decreases more quickly than R and it is expected that smaller and smaller (but more and more dense) configurations become unstable. The exact calculation shows that this approach is correct since the dynamical criterion differs very slightly from equation (1). We have therefore a "cascade scheme" where, starting with a primary mass M_0 with dimension R_0, it becomes unstable and splits up into secondary masses M_1 with dimension R_1, each of them splitting up too. From the considered model we cannot deduce when the fragmentation cascade stops. To obtain this information additional physics in the model is necessary (radiation, nuclear fusion...). The dynamical Jeans' criterion does not contain explicitly neither the mass nor the radius of the system: it depends only on the density. One expects a scaling law for the fragmentation. As soon as a polytropic configuration becomes unstable, its mass and radius are respectively given by (it is deduced from Eq.[28])

$$M = \frac{4\,\pi^{5/2}\,\gamma^{3/2}}{3\,\alpha^3} G^{-3/2}\,P^{3/2}\,\rho^{-2}\,, \qquad R = \frac{\sqrt{\gamma\pi}}{\alpha} G^{-1/2}\,P^{1/2}\,\rho^{-1}\,, \tag{33}$$

where P and ρ are respectively the uniform pressure and density. It is interesting to note that the exponents entering in equations (33) are dependent on γ only in the ratio $\gamma^{1/2}/\alpha$. Assuming that the different regions of the interstellar medium supports the same pressure equilibrium we eliminate ρ between the two equations (33) and we obtain

$$M = \frac{4\,\pi^{3/2}\,\gamma^{1/2}}{3\,\alpha} G^{-1/2}\,P^{1/2}\,R^2\,. \tag{34}$$

We recover the mass-radius relation for molecular clouds, firstly pointed out by Larson (1981) and derived by Chièze (1987) in the case of an evolution described by a series of hydrostatic equilibriums. This relation is therefore extended to the dynamical problem. For γ=1 and P=5.10^{-13}cgs, equation (34) writes $M/M_{soleil} = 64\ R^2$ (where R is expressed in parsec). Conversely, equation (34) may be used to deduce the value of the interstellar pressure. Let us come back to the cascade scheme. For symmetry considerations, a primary mass M_0 will split into secondary masses, M_1, roughly identical. The number of these subsystems is probably two. As a matter of fact, a larger number of secondary masses would imply a smaller radius R_0, corresponding to short wavelengths in the initial configuration. Since the smaller λ is, the later the instability appears, we deduce that the first instability comes out for the greatest admissible value of λ which corresponds to $M_1=M_0/2$. From equation (33) the radius R_1 is $R_1=(1/2)^{1/2}R_0$. After n stages the mass and the size of the fragments are respectively given by $M_n=M_0/2^n$, $R_n=(1/2)^{n/2}R_0$. Finally equation (34) is based on the assumption of pressure equilibrium. If we suppose the existence of regions evolving adiabatically, then the parameter K is the constant of the problem. Under these conditions, equations (33) become respectively

$$M = 4\pi^{5/2}\gamma^{3/2}\,G^{-3/2}K^{3/2}\rho^{(3\gamma-4)/2}/(3\alpha^3)\,, \quad R = \sqrt{\gamma\pi}\,G^{-1/2}K^{1/2}\rho^{(\gamma-2)/2}/\alpha\,. \tag{35}$$

These relations depend upon γ. The mass-radius relation is $M \approx R^{(4-3\gamma)/(2-\gamma)}$. The exponent is always smaller than 1. The maximun value is obtained for $\gamma = 1$.

References

Abramovitz, M., Stegun, I.A. (1964): *Handbook of Mathematical Functions* (Dover: New York)

Blottiau, P., Bouquet, S., Chièze, J.P. (1988): An Asymptotic Self-Similar Solution for the Gravitational Collapse. Astron. Astrophys. **207**, 24-36

Bonnor, W.B. (1957): Jeans' Formula for Gravitational Instability. M.N.R.A.S **116**, 104-117

Bouquet, S., Feix, M.R., Fijalkow, E., Munier, A. (1985): Density Bifurcation in a Homogeneous Isotropic Collapsing Star. Ap.J., **293**, 494-503

Chièze, J.P. (1987): The Fragmentation of Molecular clouds. Astron. Astrophys. **171**, 225-232

Hunter, C. (1962): The Instability of the Collapse of a Self-Gravitating Gas Cloud. Ap. J., **136**, 594-608

Hunter, C. (1967): Galactic Structure, in *Relativity. Theory and Astrophysics* **Vol. 2** ed. J. Ehlers (American Mathematical Society: Providence)

Jeans, J.H. (1929): *Astronomy and Cosmology* (Cambridge University Press: Cambridge)

Larson, R.B. (1981): Turbulence and Star Formations in Molecular Clouds. M.N.R.A.S **194**, 809-826

Mestel, M.L. (1965): Problem of Star formation - I. Quart. J. R. A. S. **6**, 161-198

Penston, M.V. (1969): Dynamics of Self-Gravitating Gaseous Spheres. M.N.R.A.S **144**, 425-448

Schwartz, L. (1979): *Méthode Mathématique pour les Sciences Physiques* (Hermann: Paris)

Recent Progress in Studies of One-Dimensional Gravitating Systems

Bruce N. Miller, Kenneth R. Yawn, and V. Paige Youngkins

Physics Department, Box 298840, Texas Christian University, Fort Worth, TX 76129, USA

Abstract. We review the results of recent investigations of the dynamics of one-dimensional gravitational systems carried out at Texas Christian University. Two types of systems are considered: (1) parallel mass sheets of infinite spatial extent and (2) concentric, spherically symmetric, irrotational mass shells confined between two rigid spherical barriers. By studying the kinetic energy equipartition of a sheet system consisting of two distinct populations with a fixed mass ratio, we show that the system approaches equilibrium on a long time scale. We show that the system of spherical shells has stronger ergodic properties than the planar sheets and that it approaches equilibrium on a much shorter time scale. In addition, for sufficiently small inner barrier radius, it undergoes a phase transition to a state with a greatly condensed core as the energy passes through its phase equilibrium value.

1 Introduction

Because the evolution of gravitating systems such as galaxies and globular clusters (fortunately) takes place on longer time scales than the historical records of our species, they cannot be investigated by direct observation. Astrophysicists have developed theoretical models for different stages in the evolution of these systems, but, in the absence of accurate measurements and controls, they are extremely difficult to test and improve. It would seem that computer simulation would provide the ideal environment for testing theories of gravitational evolution. However, due to complicating details associated with the properties of the two body interaction potential, such as binary formation and escape, the simulation of the gravitational N body problem is notoriously difficult.

In an attempt to sidestep the inherent difficulties of the full, three-dimensional, gravitational N body problem, astronomers and physicists have constructed idealized, one-dimensional, models to test their theories in which the sole interaction is Newtonian gravity. The two systems which have been studied extensively in the past few decades are (1) the system of parallel, planar, mass sheets and (2) the system of concentric, spherically symmetric, "rotating", mass shells. These systems have an enormous computational advantage over the three-dimensional parent system in that their motion can be simulated by event driven algorithms which avoid the tedious, stepwise, numerical integration of the equations of motion. Thus their evolution can accurately be followed over comparatively long time scales with present hardware.

The central question of gravitational dynamics is the time required for a system to reach an equilibrium state from an arbitrary initial configuration. Our entry into this arena commenced with a test of the conjecture by early investigators of the parallel sheet system that the correct time scale was on the order of $N^2 t_c$, where N is the system population and t_c is a typical system crossing time (Hohl 1967). By applying careful, nonparametric, statistical tests to the simulation results we concluded that this was false (Wright et al. 1982). Rather, following a rapid, "violent", virialisation period lasting about $20t_c$, we observed that the system entered a steady, nonequilibrium state for the remainder of the simulation (about $2N^2 t_c$). There followed a fifteen year period in which a number of investigations failed to resolve this issue. However, it was learned that the characteristics of the metastable state following the initial virialisation depend on the system's virial ratio at the initial time (Reidl and Miller 1987), that time correlation functions decay very slowly (Miller and Reidl 1990) and, at least for $N < 11$, stable regions exist in the phase space so the system is not ergodic (Reidl and Miller 1992). Additional claims concerning relaxation to equilibrium were made by other groups, but they were not substantiated and we do not repeat them here. In a later study (Reidl and Miller 1993) the short time Lyapunov exponents associated with perturbations of two unstable periodic orbits were computed from a simulation with $N = 11$. By extrapolating to longer times it was estimated that they converged to each other in about 107 characteristic times, thus setting a lower limit on the system relaxation time which is much greater than that predicted earlier.

If the planar sheet system is to serve as a useful test model, the system dynamics must be sufficiently chaotic (say ergodic and mixing) that the time averages of observables converge to their statistical predictions. One method for testing this property is to construct a system with sheets having two different masses. In equilibrium the kinetic energy of each species would be equal and the heavy sheets should be concentrated in the centre. Previous attempts to observe equipartition in simulations were not successful, probably due to slow hardware. Very recently we were able to observe both mass segregation and the equipartition of kinetic energy in a system of 64 sheets equally divided into two groups with a mass ratio of 3 : 1. We believe that this is the first conclusive demonstration of macroscopic relaxation in this system. The details of our investigation are described in Sect. 2.

The statistical mechanics of gravitating systems is notoriously difficult and unconventional, for example the partition function of a gravitating system diverges. Moreover, the predictions of the microcanonical ensemble, in which a negative heat capacity is possible, differ from those of the canonical ensemble, in which it is always positive. Michael Kiessling (1989) predicted that a phase transition occurs in a purely gravitational three-dimensional system with spherical symmetry (see also Stahl et al. 1995). The transition was missed by earlier authors because they did not admit the possibility of a delta function singularity in the radial density at the system centre. In order to determine if a phase transition could be realised dynamically, we decided to investigate the behaviour of a system of concentric, irrotational, mass shells. In a system consisting of three planar mass sheets Froeschle and Scheidecker (1975) found that only 4% of the phase space was actually occupied by chaotic orbits. In contrast, by constructing a Poincaré surface of section for a system of two spherical mass shells (Miller and Youngkins 1997), we found that the strength of the ergodic properties depends on the energy (which scales out in the planar system) and is, in general, greater than for the former model. Figure 1 shows a specific surface of section for the two shell system. Except for the presence of two stable islands, a single chaotic orbit occupies almost all of the phase space. This suggests that the spherical shell system may reach equilibrium more quickly and may provide a more viable testing environment than the sheet system.

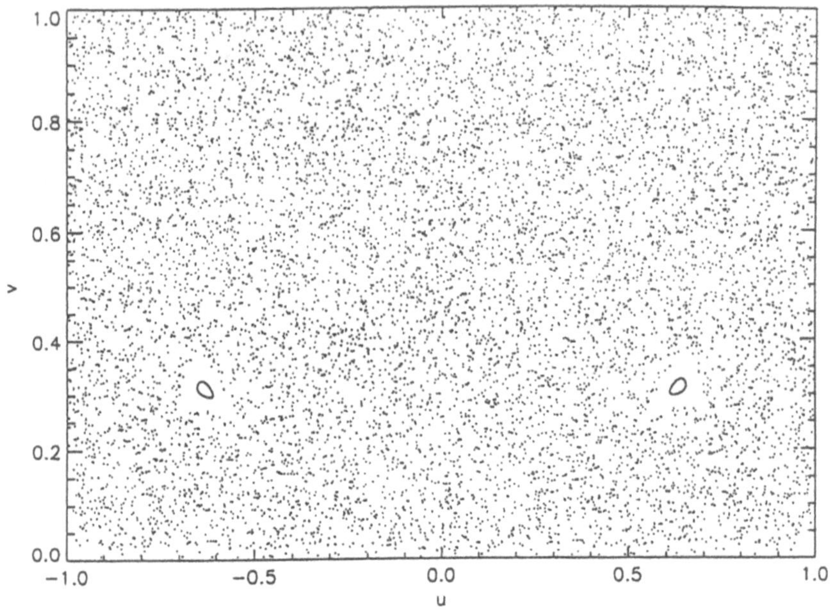

Fig. 1. Sample Poincaré surface of the two shell system.

In the Vlasov limit for the spherical shell system, theory predicts that a phase transition between a quasi-uniform density distribution and a distribution with a concentration of mass at the origin occurs for a specific range of system parameters. We explored how the system behaves with increasing population both far from the transition and near the transition. For system parameter values far from the transition point we determined the Vlasov density distribution for the spherical shell system and compared it to the average particle distribution from simulations of systems with 4, 16 and 64 shells. We found that, as the number of shells in the system increases, the simulation distribution approaches the predicted Vlasov distribution and, even for only 64 particles, the agreement is quite good. In each instance of 4, 16 and 64 particles, the final distribution from the simulation remains the same when the initial conditions are drastically changed. For all simulations the time required for the shell system to converge to the final distribution is significantly smaller than in the parallel sheet system. This robust convergence also suggests that the spherical shell system may be a more useful model for testing astrophysical theories concerning the evolution of stellar systems.

In the Vlasov limit theory predicts that the system undergoes a phase transition in which the central density of the system jumps. Near the phase transition we studied how the finite number of particles in the simulation affects the properties of the phase transition. It is predicted that the finite number of particles will cause the jump to be rounded over a range of values (Binder 1987). In addition the region of transition will be shifted. The amount of rounding and shifting depends upon the population and decreases as the population increases. We studied this dependence by comparing the Vlasov prediction and the simulation results for different numbers of particles. As was expected, small N systems show a significant rounding and shifting of the transition point. An increase in N results in a sharper transition and less shifting. The details of this study are described in Sect. 3 below.

2 The One-Dimensional Gravitating Sheet System

The one-dimensional gravitating sheet system (OGS) is a collection of N planar sheets of constant mass density, infinite in the x and y directions, which move along the z direction under their mutual gravitational attraction. The resulting gravitational field felt by each sheet is uniform. No collisional forces between sheets are considered. Therefore the sheets merely pass through (cross) one another. When two sheets cross, each experiences a discrete jump in its acceleration while its velocity remains a continuous function of time. Between crossings the sheets simply undergo uniform acceleration produced by the inhomogeneity of the mass distribution. This system can also be considered as a collection of point masses confined to move on a line, each with mass m_i, where i labels the mass type of the particle considered. The system members are ordered and labeled for tracking purposes. The acceleration experienced by the jth particle from the left depends only on the total mass of particles to the right and left and is given by

$$150A_j = 2\pi G(M_R - M_L), \tag{1}$$

where M_R (M_L) is the total mass of particles to the right (left) of the jth particle, respectively, and G is the universal gravitational constant. Units are used such that the total system mass, $M, = 1$, $2\pi G = 1$ and position (x), velocity (v) and time (t) are dimensionless. It is customary to define the characteristic period of oscillation of a particle in the system as $t_c = (G\rho_0/\pi)^{-1/2}$, where ρ_0 is the equilibrium mass density evaluated at the origin. In these units, $t_c \approx 2\pi t$.

The single particle distribution functions for the discrete single mass-component system were derived by Rybicki (1971) in both the canonical and microcanonical ensembles. In the Vlasov limit ($N \to \infty$, M held constant) these distributions converge to

$$f_c(v, x) = \frac{1}{2}\pi^{-\frac{1}{2}} \exp(-v^2)\mathrm{sech}^2(x). \quad (2)$$

In this limit the resulting distribution in m-space is a continuous fluid (no discrete particles) that obeys the collisionless Boltzmann equation. The Vlasov limit is often a useful, simplifying approximation for gravitational systems with a large number of system members. However, these results apply only to the single mass component system.

2.1 Recent Investigations on Equipartition of Energy

Equipartition of energy has long been used as a marker for dynamical systems that can approach equilibrium. Several recent papers have used the idea of equipartition of the total energy to investigate the approach to equilibrium of a single mass OGS (Tsuchiya et al. 1996; Yawn and Miller 1997). They track the evolution of the system using a measure (Δ) of the deviation of the average energy per particle from the theoretical equipartition value given by the virial theorem. After an initial short period of microscopic relaxation, Δ tends to decrease steadily until a large peak appears (Fig. 2).

This peak was originally thought to indicate the onset of equilibrium (Tsuchiya et al. 1996). Follow-up investigations showed that these peaks were also seen in systems known to be nonergodic (Yawn and Miller 1997). Thus they were then seen as evidence that the system gets hung-up for long times in restricted (sticky) regions of the phase space (Yawn and Miller 1997). The measure of Δ was therefore not considered convincing evidence that the system indeed approached a state of thermal equilibrium. These findings were our initial motivation for studying the two mass-component OGS.

2.2 Dynamical Simulations of a Two Mass Component System

Recently we investigated equipartition of kinetic energy and mass segregation in a two mass-component one-dimensional system looking for macroscopic evidence of a trend toward relaxation and equilibrium. Equipartition of kinetic energy and mass segregation are seen in three-dimensional simulations of globular clusters as macroscopic evidence of equilibrium. Mass segregation occurs as a consequence of equipartition. As kinetic energy begins to be

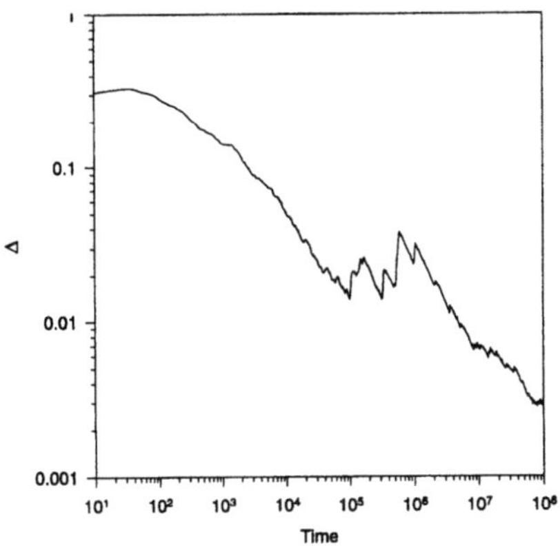

Fig. 2. Δ vs time for a 64 particle OGS. (All units are dimensionless).

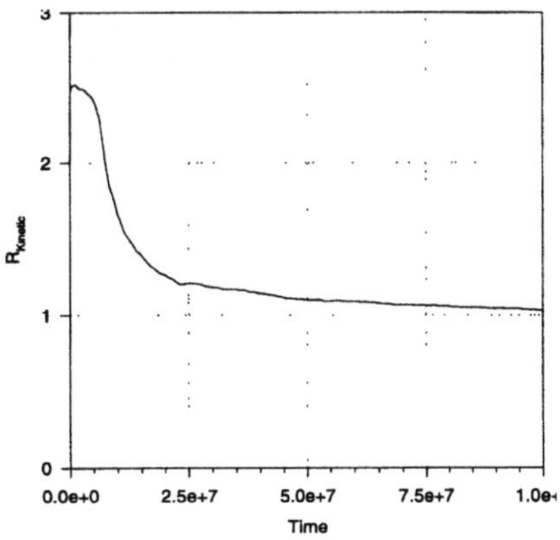

Fig. 3. $R_{Kinetic}$ vs time for the two mass component system. (All units are dimensionless).

partitioned equally between the heavy and light masses, the heavy masses lose velocity on average and move into orbits much closer to the centre of mass while the light masses gain velocity and move away from the centre of mass. This causes a measurable separation in the average positions of the two mass types. Equipartition in the OGS was studied some years ago by several groups (Hohl and Campbell 1968; Severne et al. 1994). Neither of these studies was conclusive due to limited computational capabilities and suspect initial conditions.

In the dynamical simulations initial conditions for the 64 particles were obtained by uniformly sampling a rectangle in μ-space. This distribution has been shown to approximate a Vlasov distribution after the initial period of violent relaxation (Tsuchiya et al. 1994). There were 32 particles of each mass type with the mass of the heavy particle equal to three times the mass of the light, $m_H = 3m_L$. Therefore the ratio of the average kinetic energy of the two mass types was approximately three at the beginning of the simulation. The systems were initialised with a virial ratio of approximately one and total energy of approximately 0.75 in our units. Dynamical simulations of the systems were run for a time of $1 \times 10^8\tau (\approx 1.6 \times 10^7 t_c)$. The average kinetic energy, potential energy, total energy and position of each mass type were tracked throughout the course of the simulation.

Equipartition of Kinetic Energy. The ratio of the long time average of the kinetic energy of the two mass types, $R_{Kinetic}$ vs time, is shown in Fig. 3. After an initial period of approximately $6 \times 10^6\tau$, during which $R_{Kinetic}$ is approximately three, it begins to decrease steadily and approaches one after a very long time indicating equipartition of kinetic energy.

Mass Segregation. The average distance from the centre of mass of the system of each mass type was tracked as a function of time and is shown in Fig. 4. D_H indicates the average distance of the heavy particles and D_L indicates the average distance for the light particles. Figure 5 shows a μ-space

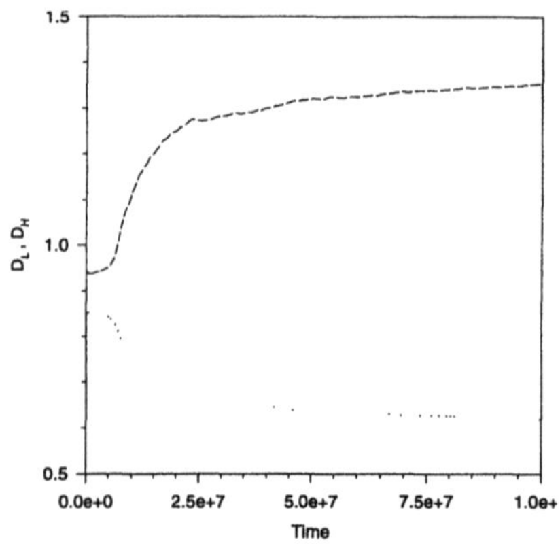

Fig. 4. D_H and D_L vs time. (All units are dimensionless).

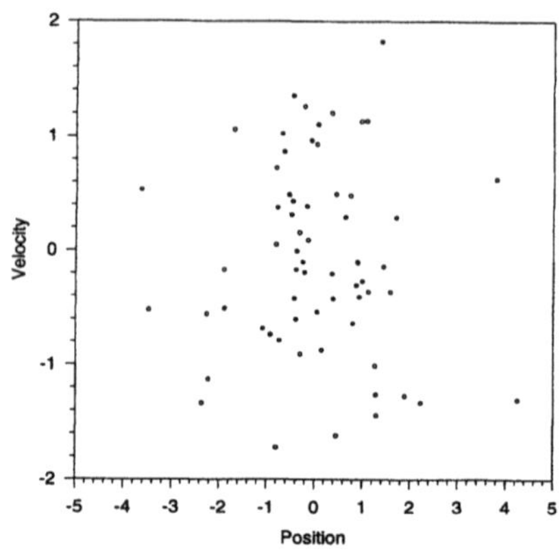

Fig. 5. μ-space plot of the final system conditions (All units are dimensionless).

plot of the system at the end of the run. Mass segregation is clearly visible in the plot.

After the initial period, during which D_H and D_L are approximately constant, the beginning of mass segregation is seen as the averages move apart to their steady state positions. Note the longer relaxation time for the positions (mass segregation) as compared to that of the energy. The longer apparent correlations in position were previously seen by Miller and Reidl (1990).

Backward Averaging. Beginning at $t = 10^7\tau$ and for the remainder of the run, the average kinetic energy of each mass type was computed for consecutive time segments of $6 \times 10^4\tau$. In this way we were able to construct mean values by backward averaging in time from the end of the run (Fig. 3). Doing so provided a means of studying the system properties in the absence of the initial large and long lasting fluctuations. Figure 6 shows clearly that $R_{Kinetic}$ does indeed converge to unity in a finite time.

Time Correlations. Using the data gathered in the $6 \times 10^4\tau$ windows, time correlations for $R_{Kinetic}$ were calculated for a period up to $t = 6.2 \times 10^6\tau$ and are shown in Fig. 7. Large correlations are seen to decay over a period of approximately $t = 1 \times 10^6\tau$, while much smaller correlations persist for very long times. This is again another indicator that very long modes of relaxation exist in this system that are not seen in higher dimensional gravitational systems.

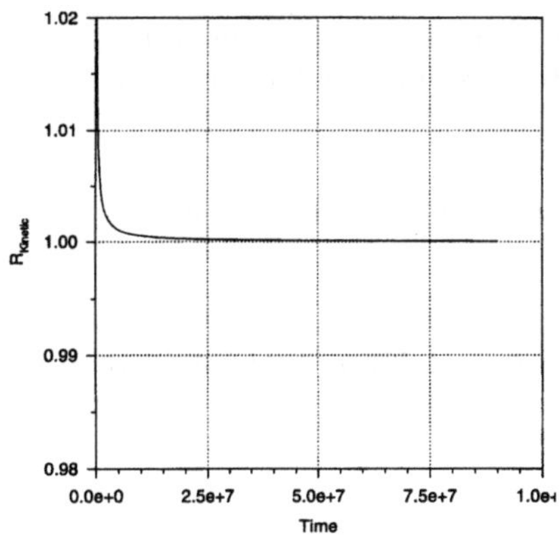

Fig. 6. Backward average of $R_{Kinetic}$ vs time (All units are dimensionless).

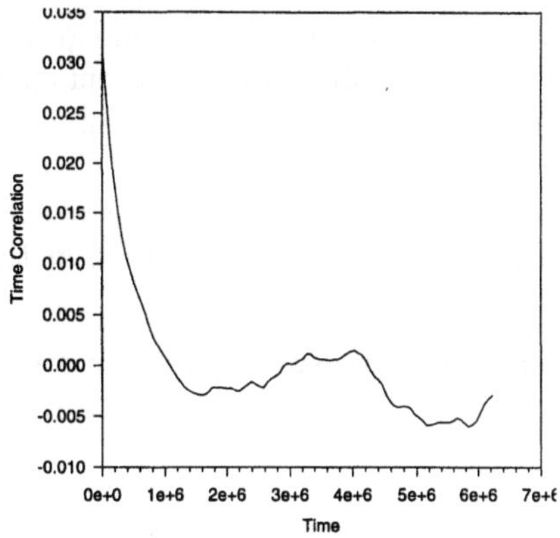

Fig. 7. Time correlations using the short time window data (All units are dimensionless).

3 The Spherical Shell Model

3.1 Description of the System

We consider a system of concentric spherical mass shells of equal mass and uniform mass density. The motion of each shell is purely radial; there is no rotational motion. The shells move under their mutual and self gravitation and the energy of the system is conserved. The system is contained in a spherical box with a reflecting outer boundary. In addition there is a reflecting inner boundary around the origin of the system to avoid numerical difficulties that arise from the singularity of the gravitational force. As the shells move, they can intersect and pass through one another. When a pair of shell trajectories cross, there is a discontinuity in the amount of mass contained within a given shell's radius which results in a discontinuity in each shell's acceleration. Between crossings the acceleration of a shell is continuous and its energy is conserved.

We have developed an algorithm which simulates the evolution of the system without numerically integrating the equations of motion. With the use of the conservation of energy of each shell between crossings it is possible to solve for the time that a shell will reach a specified radius. The event-driven algorithm used in this study is based on the ability to calculate the time a shell reaches a specific radius. There are three possible events for each shell: collision with the boundary, turning point (only possible for a shell with negative energy and positive velocity) and crossing with another shell. The basic strategy for the algorithm is to determine the times at which these events occur for all shells and then sort these times to find the next event. The appropriate initial conditions are updated and the process is repeated. This algorithm for following the evolution of the system has many advantages over one that numerically integrates the equations of motion. Firstly the efficiency of solving only for events allows the system to be followed for a much longer time. In addition the truncation error associated with any numerical integration scheme is avoided. Computer round off error is also reduced since it is introduced only at events instead of at every time increment, as is the case with numerical integration methods.

3.2 Comparison of Vlasov and Simulation Density Distributions far from the Phase Transition

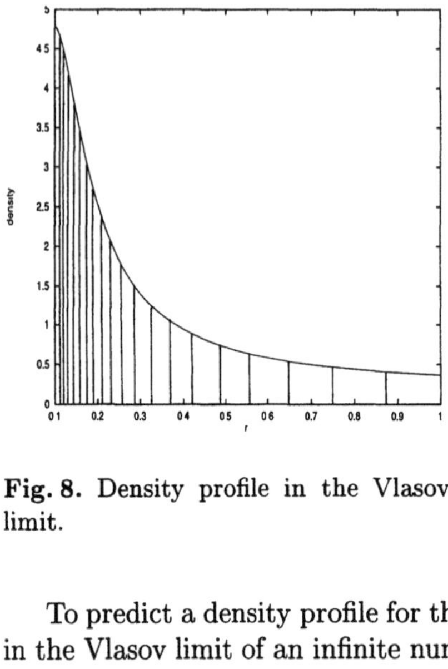

Fig. 8. Density profile in the Vlasov limit.

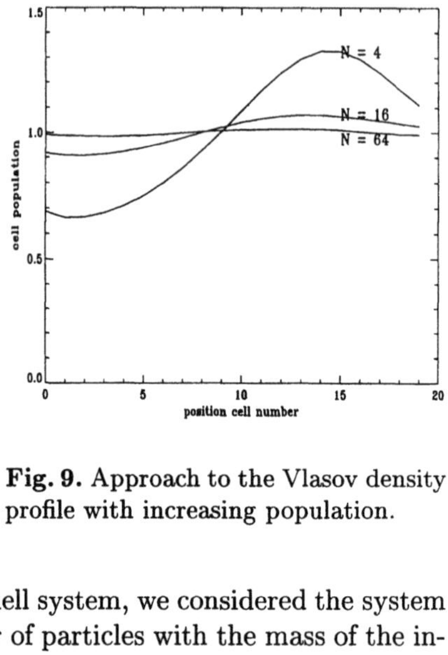

Fig. 9. Approach to the Vlasov density profile with increasing population.

To predict a density profile for the shell system, we considered the system in the Vlasov limit of an infinite number of particles with the mass of the individual particles approaching zero. In this approximation the motion of the system is described by a simple fluid flow. To determine how well this compares to a dynamical simulation with a finite number of particles, we divided the space between the inner and outer boundaries into twentry cells of equal probability, i.e., the area under the Vlasov density curve is the same for each cell (see Fig. 8). The initial positions and velocities for the simulation were assigned according to the total energy and inner boundary radius specified for the Vlasov distribution. The shells' positions were equally spaced between the inner and outer boundary and the difference between the total system energy and their total potential energy was divided equally by assigning equal velocities to all shells. The system was sampled at equal time intervals (every 0.1 time unit in our system of units) and the number of particles was accumulated in the appropriate cell according to the position of the shell.

Figure 9 shows the simulation results for the system with 4, 16 and 64 shells. The horizontal line indicating a population of one for all cells represents the Vlasov density distribution. The distribution for the 4 particle system shows a marked population deficiency in the inner cells and a surplus in the outer cells compared with the Vlasov density. The 16 particle case also shows a similar structure, but the difference between simulation and Vlasov prediction is relatively smaller. For 64 particles the density distribution shows excellent agreement with the Vlasov prediction. This agreement is surprising since the Vlasov theory assumes that the number of particles is infinite.

The relaxation time scales for the parallel sheet and spherical shell systems are significantly different. It has been shown by simulation that the sheet system may reach equilibrium on a time scale of 1×10^7 characteristic times (one characteristic time for the sheet system is the time that a sheet takes to traverse the system), if it does so at all. A commonly used characteristic time for spherical systems is the amount of time that a cold system (no kinetic energy) takes to collapse and then expand back to its original distribution. We found that, in all instances, the simulation converged to the final distribution in less than 4×10^5 characteristic times, about two orders of magnitude fewer than that required for the sheet system in terms of its characteristic time unit.

3.3 Comparison of Vlasov Predictions with Simulations

As discussed in the previous section, we may determine a Vlasov density distribution for a given total energy and inner barrier radius. Here we fix the inner barrier radius and calculate the entropy for each distribution in a range of energies. Figure 10 shows a plot of entropy versus energy at an inner barrier radius of 0.001. Between the energies of around 2.4 to 3, the function is multivalued. However, it is the maximum entropy curve which is significant. The maximum entropy curve exhibits a "kink" where the slope (and therefore, temperature) changes abruptly and a phase transition occurs. In Fig. 11 the virial ratio is calculated at each point along the maximum entropy curve and jumps at an energy of 2.7. To compare this to the simulation, we simulate systems with 16, 32 and 64 particles at energies in the neighbourhood of the transition energy and calculate an average virial ratio (Fig. 12). We see that, as expected, as the population increases, the region over which the transition is smeared and the shifting of this region is decreased.

The numerical experiments described above demonstrate that there is still much to learn from one-dimensional gravitating systems. They finally demonstrate that equilibrium averages are obtained in the planar sheet system on a very long time scale (on the order of 10^7 crossing times). Our earlier work suggests that there are "sticky" regions in the phase space surrounding tori associated with stable periodic orbits. We conjecture that, as the population of the system increases, some of the periodic orbits become unstable, but small stable regions remain. The remaining tori trap the system

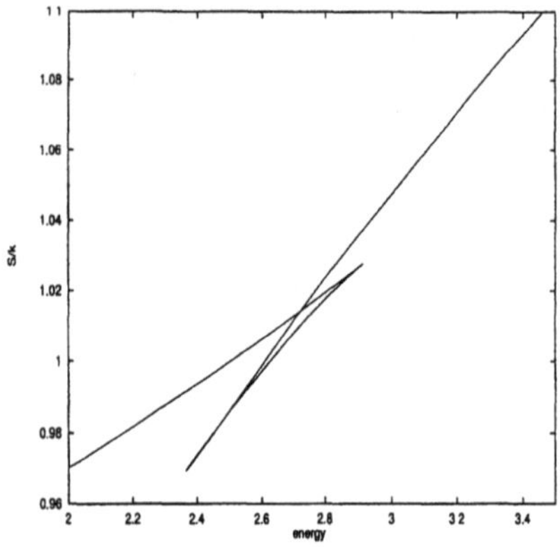

Fig. 10. Entropy versus energy showing the phase transition.

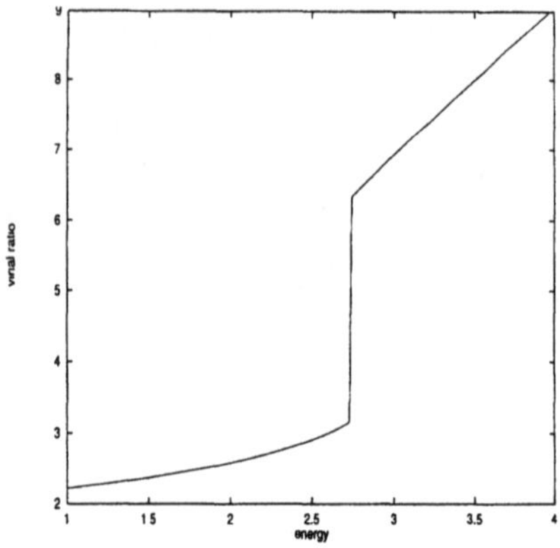

Fig. 11. Discontinuity in the virial ratio as predicted by mean field (Vlasov) theory.

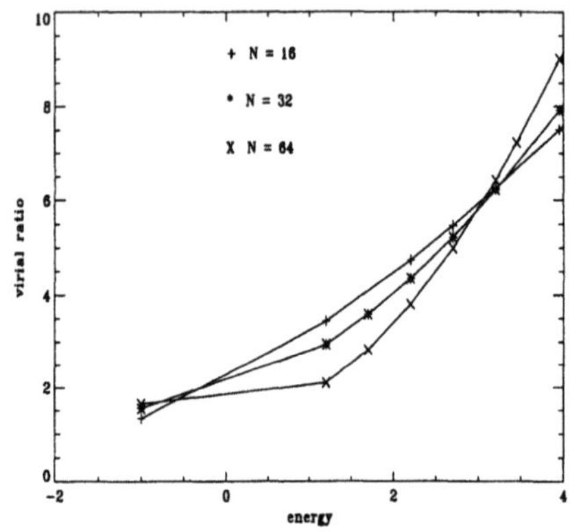

Fig. 12. Observed virial ratio from dynamical simulation for three different populations.

for long times in local regions of phase space in which the averages of dynamical quantities differ strongly from that of the predicted, microcanonical, mean. Consequently dynamical simulations require millions of crossing times to reproduce the mean values of physical observables predicted by theory. Using a diffusion model in the space of particle accelerations (Miller 1996), we recently predicted that the shortest time for the sheet system to relax is proportional to N, the system population. Future work will focus on testing this theory in both the single and double species system.

Simulations of the spherical shell system verify Keissling's theoretical predictions that a phase transition occurs in a spherically symmetric gravitating system. Although his analysis concerned a three dimensional system, we Performed the mean field calculations for the specific case of the one dimensional shell system (the only coordinate is the shell radius) and showed that a transition is predicted here as well. We determined the predicted phase equilibrium curve in the parameter space and, using real time simulations, demonstrated that the transition has dynamical as well as thermodynamic validity. Future work on this system will investigate the decay of fluctuations near the critical point, and the thermodynamics of the phase transition in different ensembles.

References

Binder, K. (1987): Finite Size Effects on Phase Transitions. Ferroelectrics **73**, 43–67

Froeschle, G., Scheidecker, J.P. (1975): Stochasticity of Dynamical Systems with Increasing Degrees of Freedom. Phys. Rev. A **12**, 2137–2143

Hohl, F. (1967): *Collective Effects in Stellar Dynamics and Plasma Physics* (Ph D dissertation, College of William and Mary, Virginia)

Hohl, F., Campbell, J. (1968): Statistical Mechanics of a Collisionless Self-Gravitating System. Astron. J. **73**, 611–615

Kiessling, M. (1989): On the Equilibrium Statistical Mechanics of Isothermal Classical Self-Gravitating Matter. J. Stat. Phys. **55**, 203–257

Miller, B.N. (1996): Source of Relaxation in a One Dimensional Gravitating System. Phys. Rev. E **53**, R4279–4282

Miller, B.N., Reidl, C.J., Jr. (1990): Gravity in One Dimension: Persistence of Correlation. Ap. J. **348**, 203–211

Miller, B.N., Youngkins, V.P. (1997): Dynamics of a Pair of Spherical Gravitating Shells. Chaos **7**, 187–197

Reidl, C.J., Jr., Miller, B.N. (1992): Gravity in One Dimension: Stability of Periodic Orbits Phys. Rev. E **46**, 837–843

Reidl, C.J., Jr., Miller, B.N. (1993): Gravity in One Dimension: The Critical Population. Phys. Rev. E **48**, 4250–4256

Reidl, C.J., Jr., Miller, B.N. (1987): Gravity in One Dimension: Selective Relaxation. Ap. J. **318**, 248–260

Rybicki, G. (1971): Exact Statistical Mechanics of a One-Dimensional Self-Gravitating System. Astrophys. Space Sci. **14**, 56–72

Severne, G., Luwel, M., Rousseeuw, P.J. (1984): Equipartition and Mass Segregation in 1-Dimensional Gravitational Systems. Astron. Astrophys. **138**, 365–370

Stahl, B., Kiessling, M., Schindler, K. (1965): Phase transition in gravitating systems and the formation of condensed objects. Planetary Space Sci. **43**, 271–282

Tsuchiya, T., Konishi, T., Gouda,N. (1994): Quasiequilibria in One-Dimensional Self-Gravitating Many-Body Systems. Phys. Rev. E **50**, 2607–2615

Tsuchiya, T., Konishi, T., Gouda, N. (1996): Relaxation Processes in One-Dimensional Self-Gravitating Many-Body Systems. Phys. Rev. E **53**, 2210–2216

Wright, H.L., Miller, B.N., Stein, W.E. (1982): The Relaxation Time of a One Dimensional Self-Gravitating System. Astrophys. Space Sci. **84**, 421–429

Yawn, K.R., Miller, B.N. (1997): Ergodic Properties and Equilibrium of One Dimensional Self-Gravitating Systems. Phys. Rev. E **56**, 2429–2436

Fluctuations of the Kinetic Energy and Stability of Coherent Structures in a One-Dimensional Gravitating Gas: Numerical Simulation

Jean-Louis Rouet and Marc R. Feix

Laboratoire de Mathématique, Applications et Physique Mathématique - UMR 6628, Université d'Orléans, UFR des Sciences, BP 6759, F-45067 Orléans Cedex 2, France

Abstract. The coupling between collective and individual modes for a meta-equilibrium one-dimensional gravitating gas is studied. The one-dimensional water-bag model allows the calculation of the theoretical values of these modes and they are compared with those obtained by simulations realised with an N Body code. Finally the influence on the time evolution and stability of holes localised in the water-bag equilibrium is numerically investigated.

1 Introduction

The evolution of systems such as plasma or gravitating systems exhibits usually two time scales. During the first, short one, collective effects take place and drive the system toward a kind of meta-equilibrium, not necessarily a stationary state. Then slowly the system relaxes toward the Maxwellian distribution function because of fluctuations due to the "grainy" nature of matter. Galaxies are a gravitational example of this kind of system and their evolution is driven by collective effects because the universe is too young to let individual effects take place (Lynden-Bell 1967).

In the case of one-dimensional weakly correlated plasma, such as fusion plasma or space plasma, this scenario has been tested with the help of numerical simulations and it works pretty well (Rouet and Feix 1991). For gravitating systems the situation is not so clear: Vlasov simulations show that the system relaxes toward a distribution function of the energy alone (which is the meta-equilibrium state in the Vlasov limit) only in average and in addition displays structures (holes in phase space) which remain for the duration of the simulation. The simulation of the same system, but described with particles, shows the same behaviour. This suggests that collective effects are strong enough to keep these collective structures (Mineau et al. 1990; Yamashiro 1992).

To avoid the formation of these holes we start from a meta-equilibrium. Nevertheless the strange evolution of a population of labelled particles suggests that collective and individual effects are coupled (Rouet 1989).

To have some insight into this idea we study the behaviour of a macroscopic quantity: the potential energy. In the plasma case it looks like noise. For weakly correlated plasma it is well-known that the different modes obtained after a Fourier transform on space are independent and that short wavelengths (with $kD \sim 1$, where D is the Debye length) look like noise while long wavelengths ($kD \ll 1$) display long time correlations which reveal the collective behaviour for distances much larger than the Debye length.

The gravitational case is very different. Firstly it does not possess a translation invariance on x space. Then there is no decomposition of the potential energy on Fourier modes. Secondly the time evolution of the potential energy is no longer noise but a kind of modulated oscillation.

In the plasma case a first order theory permits the computation of the spectrum of the potential energy. This computation is based on the solution of the linearised Vlasov equation around a homogeneous equilibrium. The homogeneity allows the independence of the space Fourier modes. The gravitational case is much more difficult. A model greatly simplifying the computations is the water-bag model (Rouet and Feix 1996) and its generalisation to the multiple water-bag (abbreviated MWB below). These models are described hereafter. We adopt this model in the gravitational case.

After this introduction Sect. 2 gives the calculus of the collective modes of the field in the case of the multiple water-bag model. Comparison between simulations and theoretical results is made in Sect. 3. A study of the behavior of holes and bumps created *ab initio* in phase space is given in Sect. 4. A conclusion follows.

2 The Multiple Water-Bag Model

In order to obtain analytical results we limit this work to a one-dimensional system and a very simple distribution function.

The simplest distribution function is one which has a single value A in a delimited area. This model, called water-bag, was introduced by De Pack (1962). He, and after him many authors, observed that, with this simple model, analytical treatment can be performed (Hohl and Feix 1967). An extension of this model is the multiple water-bag which presents several areas of constant value delimited by a "bag". The MWB model can be obtained by the discretisation of a continuous distribution function. Nevertheless it introduces discontinuities and the connection between the physical properties of the continuous distribution function and the disretised one deserves a careful treatment (Bertrand 1968).

Here we take an MWB, a function of the energy alone, which corresponds to a meta-equilibrium in the fluid limit (see Fig. 1). Then each bag is delimited by a contour which has a given energy ε_i. The velocity of a particle moving on the border of bag number i is $a_i(x)$ with

$$\varepsilon_i = \frac{1}{2} m a_i^2(x) + m\phi(x), \tag{1}$$

where $\phi(x)$ is the potential created by all the bags at point x. Equation (1) depends upon the square of the velocity a_i. Consequently this MWB is symmetric with respect to x. Let us call x_{si} the value for which $a_i(x_{si}) = 0$, that is:

$$\varepsilon_i = m\phi(x_{si}). \tag{2}$$

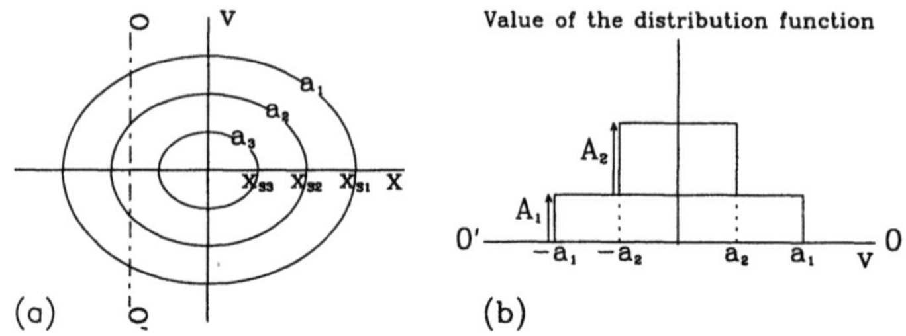

Fig. 1. (a) MWB model, (b) Cut off of Fig. 1(a) along the O′O axis

Let us perturb the MWB equilibrium. If we keep in mind the picture of the particle following the borders of the ith bag, its two velocities $V_i(x,t)$ are defined by:

$$\frac{\partial V_{i\pm}}{\partial t} + V_{i\pm} \frac{\partial V_{i\pm}}{\partial x} = E, \tag{3}$$

where

$V_{i+} = a + v_{i+}$ for the upper part of the phase space plane and
$V_{i-} = -a + v_{i-}$ for the lower part.

The gravitational field E is

$$E = E_0 + E_1, \tag{4}$$

where E_0 is the field created by the unperturbed MWB equilibrium and E_1 is the correction to first order for the perturbed one. E_0 is given by Poisson's law:

$$\frac{\partial E_0(x)}{\partial x} = -4\pi G \sum_i A_i 2a_i, \tag{5}$$

where $A_i(x)$ is the height of bag i at point x with $A_i(\varepsilon) = A_i$ for $\varepsilon \leq \varepsilon_i$ and $A_i(\varepsilon) = 0$ elsewhere.

The linearisation of (3) gives for the two perturbed velocities of bag i:

$$\frac{\partial v_{i+}}{\partial t} + \frac{\partial}{\partial x}(a_i v_{i+}) = E_1 \tag{6}$$

$$\frac{\partial v_{i-}}{\partial t} - \frac{\partial}{\partial x}(a_i v_{i-}) = E_1, \tag{7}$$

where $E_1(x,t)$ is the sum of the partial fields $E_{1i}(x,t)$ created by the particles of bag i:

$$E_1 = \sum_i E_{1i} \tag{8}$$

and

$$\frac{\partial E_{1i}}{\partial x} = -4\pi G A_i (v_{i+} - v_{i-}). \tag{9}$$

In addition we constrain the perturbed field E_{1i} to be equal to zero at $x = \pm x_{si}$. This means that the point x_{si} is held fixed. Obviously $E_{1i} = 0$ for $|x| > x_{si}$.

Calculating the difference between equations (6) and (7), with the help of (9), we find, after integrating on x:

$$\frac{\partial E_{1i}}{\partial t} = -4\pi G A_i a_i (v_{i+} + v_{i-}). \tag{10}$$

Substituting for $v_{i+} \pm v_{i-}$ by (9) and (10) into the result of (6)+(7) we obtain, after Fourier transform on t, for each bag i:

$$a_i \frac{\partial}{\partial x} a_i \frac{\partial}{\partial x} E_{1i}(\omega, x) + \omega^2 E_{1i}(\omega, x) = -\omega_{J_i}^2 E_1(\omega, x), \tag{11}$$

where $\omega_{J_i}^2$ is the Jean's frequency associated to the bag i. $\omega_{J_i}^2$ is

$$\omega_{J_i}^2(x) = 8\pi G A_i a_i(x) = 4\pi G n_i(x). \tag{12}$$

Equation (11) provides for each bag an equation connecting E_{1i} to E_1, while (8) will give the dispersion relation. For example, in the usual homogeneous plasma case, $a_i(x) = a_i$, and we can now take the Fourier transform on x of (11) to obtain:

$$E_{1i} = \frac{\omega_{pi}^2}{\omega^2 - k^2 a_i^2} E_1. \tag{13}$$

The plasma frequency ω_{pi}^2 replaces the Jean frequency, with formally $\omega_{pi}^2 = -\omega_{Ji}^2$ because of the change of sign in interaction. Indeed (8) gives the dispersion relation:

$$\sum_i \frac{\omega_{pi}^2}{\omega^2 - k^2 a_i^2} = 1, \tag{14}$$

which can be deduced from the general formulæ.

The eigenvalues ω of (11) are found with the constraint $E_{1i}(\pm x_{si}) = 0$. It can be observed that, if the velocity perturbation of the bags is taken to be independent of the time t, then $\omega = 0$ is a solution of (11). This marginal mode corresponds to a translation at constant velocity of the bags.

Now let us suppose that the system has just one bag. Equation (11) is

$$\frac{d^2}{d\theta^2}E_1(\omega,\theta) + (\omega^2 + \omega_J^2(\theta))E_1(\omega,\theta) = 0, \tag{15}$$

where we have introduced the new variable θ defined as

$$d\theta = \frac{dx}{a(x)}, \tag{16}$$

or

$$\theta = \int_0^x \frac{dx}{a(x)}. \tag{17}$$

θ is the time taken by a particle moving on the border of the bag to go from 0 to x.

Equation (15) has been numerically integrated by using a shooting method and a Runge-Kutta scheme of fourth order (Press et al. 1992). If we forget the marginal mode ($\omega = 0$ which is even), the first mode (which is odd) is obtained for $\omega = 0.71$, the second one (which is even) for $\omega = 1.14$. For values of ω large enough the Jean's frequency ω_J does not play any role and the solutions are those of a string with fixed end points.

The quarter of the period of rotation of a particle moving on the border of the bag is recovered by integrating (17) from 0 to xs. By varying the energy of the particle from 0 to ε, we recover all the other individual modes. When the energy of the particle increases, its rotation frequency decreases from 0.433 to 0.396.

3 Numerical Code and Results

As already mentioned, the restriction to a one-dimensional system allows us to obtain analytical results. Another interesting point is, always for a one dimensional system, that an exact code can be performed. As one dimensional particles are infinite plane sheets, they create a field which is a constant. Consequently the total field is piecewise constant and only depends on the relative position of the particles. Each one experiences an uniform acceleration as long as it does not cross its neighbour. Then it experiences a new field and a new accelerated motion. The program calculates the time at which crossings between two particles occur and keeps the position order relation between them. For more detail refer to Feix (1969). The crucial property of

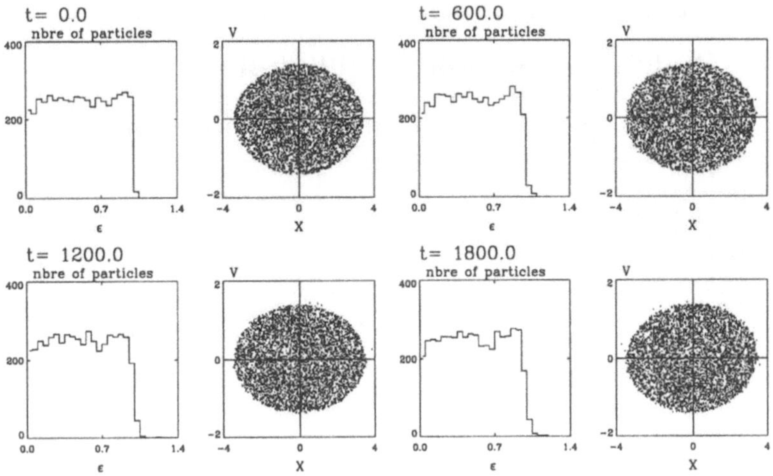

Fig. 2. Energy distribution function and time evolution in phase space of a water-bag function of the energy for almost 120 rotations of the system. The pictures show that the overall distribution does not change for the duration of the simulation.

this code is that it is exact and does not introduces error, except for the round-off errors due to the finite number of the bits treated by the computer.

Consequently this code will take into account all the effects, individual as well as collective ones. Moreover it will give precisely the individual modes.

In our particle code it is convenient to adopt the folowing normalisation: $4\pi G = 1$, the particle mass $m = 1/N$ and the maximun energy $\varepsilon = 1/N$. With these units, the maximum velocity $a(x = 0) = \sqrt{2}$, the value of the distribution function $A = 3/(32\sqrt{2})\, N$ and the Jeans' frequency ω_{J0} defined at $x = 0$ is equal to $\sqrt{3}/4$. Moreover all the systems simulated consist of $N = 5000$ particles.

Figure 2 gives the evolution of the meta-equilibrium water-bag which is just pertubated by the graininess of the system. It keeps its shape for the duration of the simulation. Figure 3(a) gives the total kinetic energy as function of time and Fig. 3(b) gives the time Fourier transform of Fig. 3(a). We observe that, indeed, the collective mode at $\omega = 0.7$ is excited while the individual modes are also present but at a much less intense level. These individual modes appear at a double frequency compared to the calculated one because, for a single rotation of the particles, the kinetic energy experiences two oscillations. This result indicates that collective modes are highly excited by the particular description of the system.

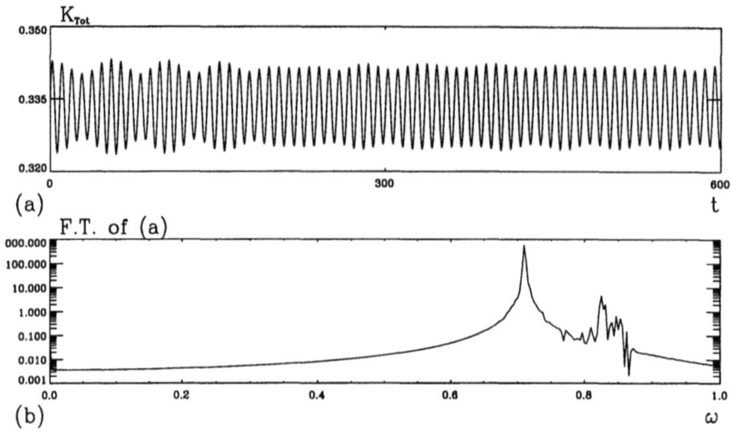

Fig. 3. (a) Total kinetic energy K_{Tot} function of time for the system represented in Fig. 2, (b) Time Fourier transform of K_{Tot} without its straight value.

4 Stability of Holes in Phase-Space

Numerical simulations of systems outside equilibrium show that they do not relax toward an equilibrium: they develop arms which carry the excess of kinetic energy and go around an empty zone of phase space. This process, which seems to be systematic, creates holes in phase space which remains for the duration of the simulation and prevent the system from reaching an equilibrium (Mineau et al. 1990).

In order to study the behavior of this structure we create *ab initio* a hole in the water bag equilibrium (Luwel and Severne 1967). To have a better insight into the behaviour of the hole, we fill it with "test" particles. Those particles experience the field of the other particles, but do not contribute to the field.

Figures 4 show the evolution of the holes for nearly 50 rotations of the system. The hole is still present at the end of the simulation and it resists the differential rotation of the particles which are localised at its border. Figures 5(a), (b), (c) and (d) give respectively, the kinetic energy of the particles of the system, its Fourier transform, the kinetic energy of the test particles and its Fourier transform. The modes excited (see Fig. 5(d)) are individual modes, indicating that the holes rotates at the same velocity as the particles. Moreover this hole triggers the collective mode as can be seen in Fig. 5(b).

Finally in the case of the complete water-bag equilibrium (with no holes), we follow the particles which are initially localised in the area of the previous hole. These particles are just "labeled" particles and they have the same physical properties as the others. Figure 6 show that the differential rotation

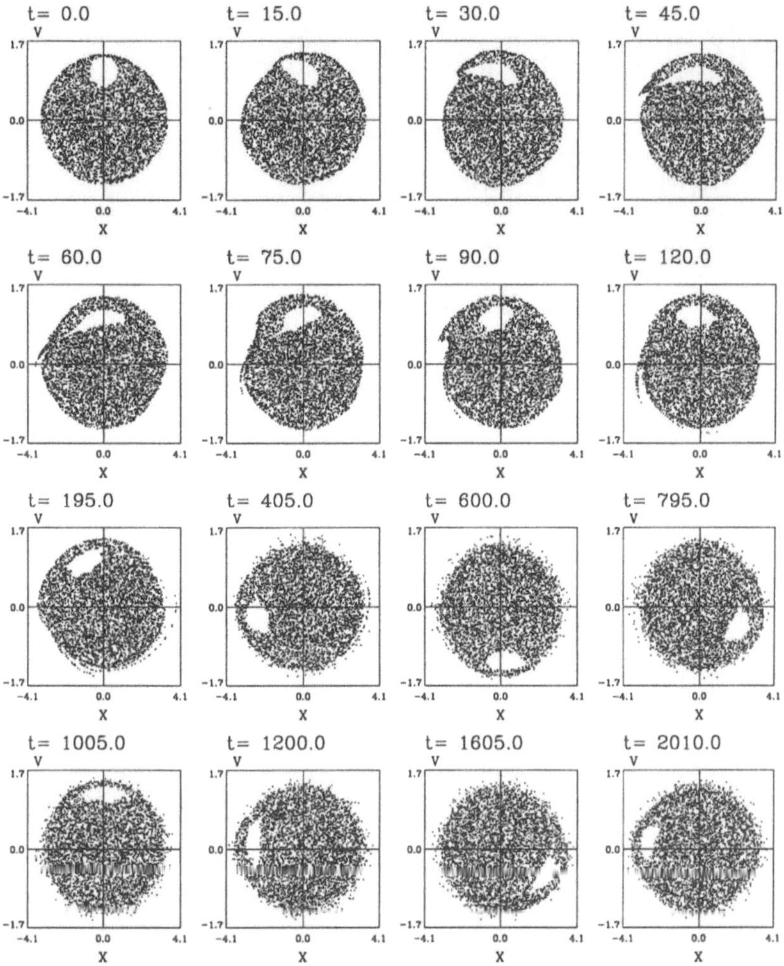

Fig. 4. Time evolution in phase space of a water-bag in which a hole is created *ab initio* for almost 130 rotations of the system

between particles of low and high energy stretches the area occupied by these labeled particles. Nevertheless, the presence of another effect can be detected because the stretching is not complete.

5 Conclusion

The value of the collective mode of a meta-equilibrium water-bag gravitating gas, which was obtained theoretically, has been numerically recovered. The initial distribution function is such that the collective effects are zero and the

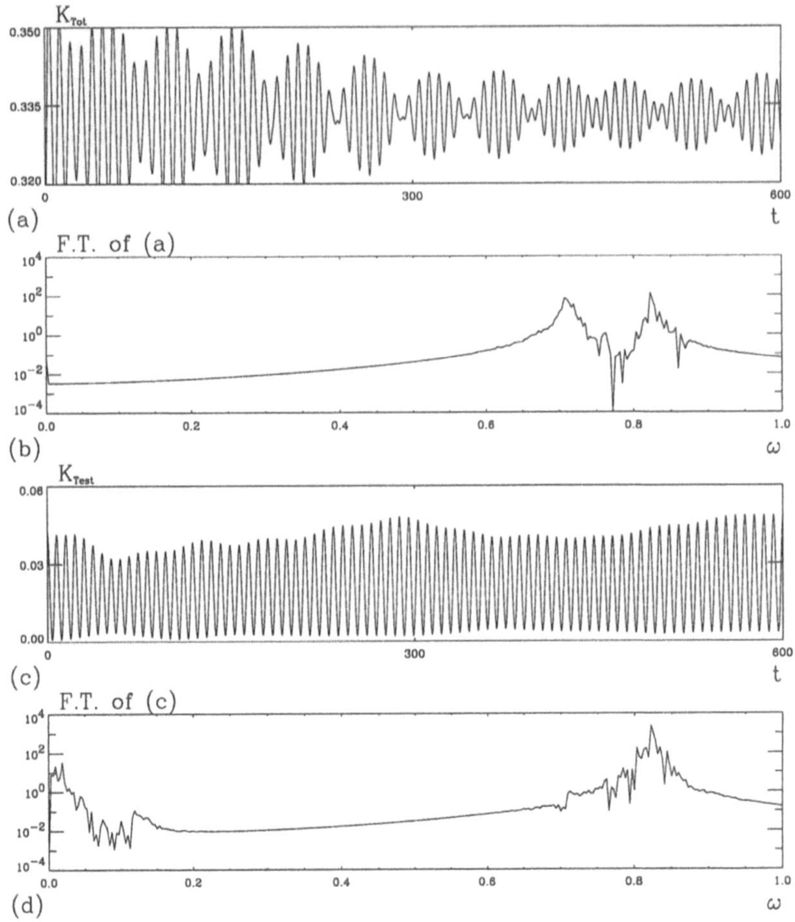

Fig. 5. (a) Total kinetic energy, K_{Tot}, as a function of time for the system represented in Fig. 4; (b) Time Fourier transform of K_{Tot}; (c) Kinetic energy K_{Test} as a function of time for the localised test particles initially filling the hole; (d) Time Fourier transform of K_{Test}

evolution of the system is driven by individual effects. Numerical simulations have shown that indeed the grainy effects trigger the collective modes, which explains the strange behaviour of a labeled population.

Numerical simulations show that holes are structures which seem to play an important role in one dimensional systems, observed for the first time in the plasma case. With an initial hole, the water-bag almost keeps its shape for a time large compared to the time necessary to destroy the same area filled with labeled particles.

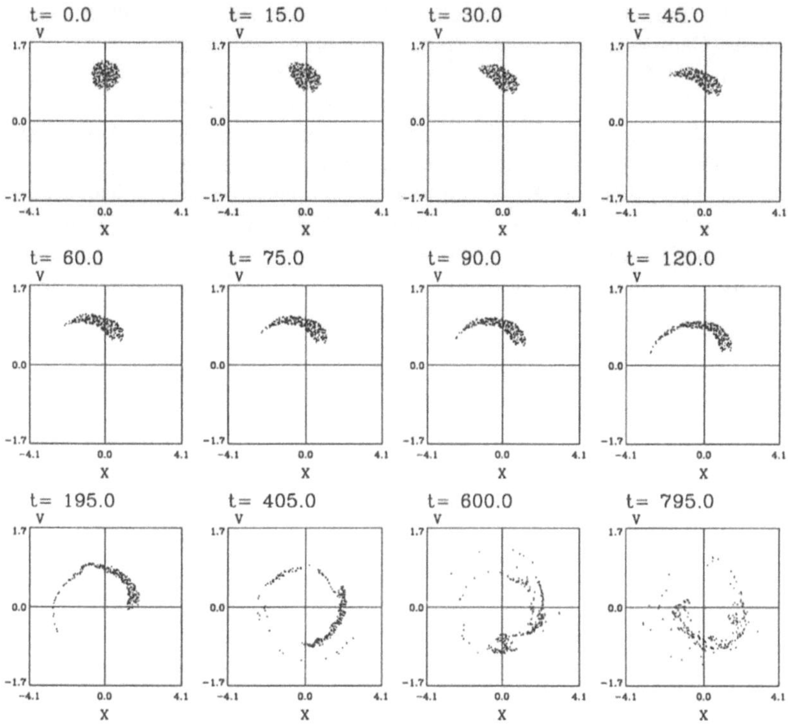

Fig. 6. Time evolution in phase space of a population of particles which belong to the water-bag represented in Fig. 1 and initially localised in the same area as the hole of Fig. 4

References

Bertrand, P., Feix, M.R. (1968): Non-Linear Electron Plasma Oscillation, the Water-Bag Model. Phys. Letters **28 A**, 68

De Pack, D.C. (1962): The Water-Bag Model of a Sheet Electron Beam. J. Electron. Control **13**, 417–424

Feix, M.R. (1969): *Non-Linear Effects in Plasmas*, Kalman, G., Feix, M.R. eds. (Gordon and Breach, New-York), 515–157

Hohl, F., Feix, M.R. (1967): Numerical Experiments with a One-Dimensional Model for a Self-Gravitating Star System. ApJ **147**, 1164–1180

Luwel, M., Severne, G. (1985): Collisionless Mixing in 1-Dimensional Gravitational Systems Initially in a Stationary Waterbag Configuration. A&A **152**, 305–314

Lynden-Bell, D. (1967): Statistical Mechanics of Violent Relaxation in Stellar Systems. MNRAS **136** (1), 101–121

Mineau, P., Feix, M.R., Rouet, J.L (1990): Numerical Simulations of Violent Relaxations and Formation of Phase Space Holes in Gravitational Systems. A&A **228**, 344–349

Press W.H., Teukolsky, S.A., Vetterling, W.T., Flannery, B.P., (1992): *Numerical Recipes* (Cambridge University Press, Melbourne), 745–753

Rouet, J.L (1989): *Simulations Numériques : Problèmes à N Corps pour les Systèmes Auto-gravitants et les Plasmas*, Thèse de l'Université d'Orléans

Rouet, J.L., Feix, M.R. (1991): Relaxation for One-Dimensional Plasma: Test Particles Versus Global distribution behavior. Phys. Fluids B **3**, 1830–1834

Rouet, J.L, Feix, M.R. (1996): Computer Experiments on Dynamical Cloud and Space-Time Fluctuations in One-Dimensional Meta-Equilibrium Plasmas. Phys. Plasmas **3** (7), 2538–2545

Yamashiro, T., Gouda, N., Sakagami, M. (1992): One-Dimensional Self-Gravitating System. Progress Theor. Phys. **88**, 269–282

Structure Formation in the One-Dimensional Gravitational Gas

Amador Muriel[1], Agnes Miciano-Cariño[1], and Ricolindo L. Cariño[2]

[1] World Laboratory Research Center for Fluid Dynamics, University of the Philippines Los Baños, 4031 College, Laguna, Philippines
[2] Institute of Computer Science, University of The Philippines Los Baños, 4031 College, Laguna, Philippines

Abstract. We show analytically that it is natural to develop a hierarchy of structures in a one dimensional self-gravitating gas exploded from an initial Gaussian space distribution and a smooth symmetric momentum distribution. We also conjecture the occurrence of an instability for some combination of time and gravitational coupling constant.

1 Introduction

One of the outstanding questions in cosmology is the formation of structures in the universe after the suspected "big bang" (Peebles 1980). The uncoupling of the radiation field from the material distributions is well understood from the point of view of elementary particle theory (Linde 1990), but the development of hierarchy of structures, from clusters of galaxies to galaxies and sub-galactic systems, is not yet completely understood. One would expect that in an expanding self-gravitating system, there is some room for the formation of substructure, or even microstructure, but it has been difficult to demonstrate this analytically. To understand the formation of substructure, it is quite useful to start with a toy system, the one dimensional gravitational gas (Camm 1950), to see if substructure formation is natural or not. There are numerical (Hohl and Feix 1966), and even analytic results (Mineau et al. 1990) that hint at the formation of substructure even for systems that possess smooth initial data. Numerical results show the development of gross features (Hohl and Feix 1966); and analytical results (Mineau et al. 1990) show the development of substructures, verified in numerical experiments, only if the smooth background is filtered. In some numerical analysis, the microstructure is only 10^{-4} of the gross density distribution, an explanation for why such substructures has been seen only after some decades of using the same numerical codes (Hohl and Feix 1966).

In this paper, we restrict ourselves to analytic methods, which interestingly enough, show gross structure and at least one substructure layer of description. We could thus conclude from our investigation that: *(1) the formation of substructure in Hohl and Feix (1966) and Mineau et al. (1990) are real, and not an artifact of perturbation calculations; (2) at least one layer of*

structure emerges, corresponding to (a) gross structure and possibly (b) substructure. Thus, it would seem that structure formation is a natural result of the dynamics of one dimensional self-gravitating systems, and perhaps for three dimensional systems as well. These observations render the problem of structure formation less severe. In addition, we explicitly show ***(3) some secular divergence which gives a hint of instability, and the suggestive inadequacy of the single particle description of a gravitational system.***

What is the relevance of this one dimensional investigation to cosmology? Our work is limited to one dimension, although it can be generalized to three dimensions. We may characterize this work as one dimensional Newtonian cosmology. With this limitation, we expect that conclusions obtained here are muted versions of real three dimensional systems. The weakening or absence of features in one dimension found in three dimensions is long recognized from statistical mechanics. Thus the new conclusions that we make in this paper, from items (1) to (3) in the preceding paragraph, are useful hints for the more difficult case in three dimensions.

2 Formulation of the Problem

Suppose that we concentrate N-bodies in the origin of a one dimensional space, and represent the initial spatial distribution by the function $\Pi_{i=1}^{N}\psi(x_i)$ where $\psi(x_i)$ is the Gaussian distribution for $i = 1, \cdots, N$. Let us explode this distributed assembly of particles by imparting to the system at time zero a statistical momentum distribution $\Pi_{i=1}^{N}\varphi(p_i)$. The initial probability is completely factored, without initial correlations. What is the ensuing mass distribution of this system? This system has been studied earlier for a one dimensional gravitational gas, using a numeric (Hohl and Feix 1966) and an approximate, analytic perturbation approach (Mineau et al. 1990). The results are consistent with a general diffusion phenomenon, with a hint of of some substructure formation. It is not clear if the substructure found is an artifact of the numeric work — the delta function initial spatial distribution used in Mineau et al. (1990) cannot be exactly simulated — or if the same substructure found analytically is a by-product of the approximate perturbation method. We provide a confirmation of the earlier results in Hohl and Feix (1966) and Mineau et al. (1990), and clarify the issue of substructure formation.

To treat this problem, we use a recent result in statistical many–body theory. According to Muriel and Esguerra (1996) and Muriel and Dresden (1997), the time evolution of the density of an arbitrary one dimensional classical system is given exactly by the six terms

$$n(x,t) = n_0 + n_1 + n_2 + n_3 + n_4 + n_5 \qquad (1)$$

where

$$n_0 = \int dp f_1^1(x - pt/m, p, 0) \tag{2}$$

$$n_1 = \int dp \int_0^t ds_1 \exp(-\frac{-ps_1}{m}\frac{\partial}{\partial x}) \int dx' \frac{\partial V(x-x')}{\partial x} \frac{\partial}{\partial p} f_2^1(x, x', p, 0) \tag{3}$$

$$n_2 = -\gamma \int dp \int_0^t ds_1 \int_0^{s_1} ds_2 \exp(-\frac{ps_2}{m}\frac{\partial}{\partial x}) \int dx' \frac{\partial V(x-x')}{\partial x} \times \\ \frac{\partial}{\partial p}\left(\frac{p}{m}\frac{\partial}{\partial x} f_2^1(x, x', p, 0)\right) \tag{4}$$

$$n_3 = -\gamma \int dp \int_0^t ds_1 \int_0^{s_1} ds_2 \exp(-\frac{ps_2}{m}\frac{\partial}{\partial x}) \int dp' \int dx' \frac{\partial V(x-x')}{\partial x} \times \\ \frac{\partial}{\partial p}\left(\frac{p'}{m}\frac{\partial}{\partial x} f_2^2(x, x', p, p', 0)\right) \tag{5}$$

$$n_4 = \frac{\gamma^2}{2} \int dp \int_0^t ds_1 \int_0^{s_1} ds_2 \exp(-\frac{ps_2}{m}\frac{\partial}{\partial x}) \int dx' \int dx'' \frac{\partial V(x-x')}{\partial x} \times \\ \frac{\partial V(x-x'')}{\partial x} \frac{\partial^2}{\partial p^2} f_3^1(x, x', x'', p, 0) \tag{6}$$

$$n_5 = \frac{\gamma^2}{2} \int dp \int_0^t ds_1 \int_0^{s_1} ds_2 \exp(-\frac{ps_2}{m}\frac{\partial}{\partial x}) \int dx' \left(\frac{\partial V(x-x')}{\partial x}\right)^2 \times \\ \frac{\partial^2}{\partial p^2} f_2^1(x, x', p, 0) \tag{7}$$

whose notation and convention we now explain.

The normalization used in Equations 2 – 7 is such that the density is expressed as inverse length. γ is the pair potential coupling parameter. We use the mixed distribution function $f_i^j(x_1, x_2, \cdots, x_i, p_1, p_2, \cdots, p_j, 0)$ to represent the initial combined probability that particles 1 to i will be found in their locations $x_1, x_2, \cdots, x_i$, such that the respective momenta are $p_1, p_2, \cdots, p_j$. The superscripts refer to the momentum labels while the subscripts refer to coordinate labels. $V(x - x')$ represents the pair potential of two particles in x, x'.

The derivation of the above exact general results for the more general three dimensional system is shown in Muriel and Esguerra (1996) and Muriel and Dresden (1997). In these two references, the authors break the BBGKY closure problem of the Liouville equation as described by Bogoliubov (1962), reducing the closure problem to a specification of initial data to include no more than two and three particle initial correlations. This result is true for any system with an exchange symmetry of particles. These authors also show the reduction of their formulation to a hydrodynamic regime, when conventional continuum approximations in fluid dynamics are used. We refer the reader to the above mentioned references for details of the derivation. Our main interest now is to apply this new technique from non-equilibrium statistical mechanics to stellar dynamics. We start with the favorite toy system of the one dimensional gas (Camm 1950). For the one dimensional gravitational gas, we use the following.

$$\frac{\partial V(x-x')}{\partial x} = \varepsilon(x-x') \tag{8}$$

$$\begin{aligned} f_1^1(0) &= \psi(x)\varphi(p) \\ f_2^1(0) &= \psi(x)\psi(x')\varphi(p) \\ f_2^2(0) &= \psi(x)\psi(x')\varphi(p)\varphi(p') \\ f_3^1(0) &= \psi(x)\psi(x')\psi(x'')\varphi(p) \end{aligned} \tag{9}$$

Equations 9 are consistent with the assumption of factored initial conditions. We use the definition

$$\varepsilon(x) = \begin{cases} 1 & \text{if } x > 0, \\ 0 & \text{if } x = 0, \\ -1 & \text{if } x < 0. \end{cases} \tag{10}$$

We now perform a term by term analysis of Equation 1. First, using the shift operator, n_0 becomes

$$n_0 = \sqrt{\frac{\alpha}{\pi}} \int dp e^{-\alpha(x-\frac{pt}{m})^2} \varphi(p,0). \tag{11}$$

Second,

$$\begin{aligned} n_1 &= \gamma \int dp \int_0^t ds_1 exp(-\frac{ps_1}{m}\frac{\partial}{\partial x})\psi(x) \int dx' \varepsilon(x-x')\psi(x') \frac{\partial \varphi(p,0)}{\partial p} \\ &= \frac{\gamma m}{4} \int dp \left[(erf[\sqrt{\alpha}x])^2 - (erf[\sqrt{\alpha}(x-\frac{pt}{m})])^2 \right] \frac{1}{p}\frac{\partial \varphi(p,0)}{\partial p}. \end{aligned} \tag{12}$$

Third,

$$\begin{aligned} n_2 &= \gamma \int dp \int_0^t ds_1 \int_0^{s_1} ds_2 exp(-\frac{ps_1}{m}\frac{\partial}{\partial x}) \int dx' \varepsilon(x-x')\psi(x') \\ &\quad \frac{\partial}{\partial p}\left(\frac{p}{m}\frac{\partial \psi(x)}{\partial x}\varphi(p,0) \right) \\ &= \gamma \int dp \left[\frac{m^2}{4p^2} \left\{ (erf[\sqrt{\alpha}(x-\frac{pt}{m})])^2 - (erf[\sqrt{\alpha}x])^2 \right] + \right. \\ &\quad \frac{m^2}{2p^2}\sqrt{\frac{2\alpha}{\pi}} \left\{ (x-\frac{pt}{m})(erf[\sqrt{2\alpha}x] - erf[\sqrt{2\alpha}(x-\frac{pt}{m})]) \right\} + \\ &\quad \left. \frac{m^2}{2\pi p^2}(e^{-\alpha x^2} - e^{-\alpha(x-\frac{pt}{m})^2}) + \frac{mt}{p}\sqrt{\frac{\alpha}{\pi}} e^{-\alpha x^2} erf[\sqrt{\alpha}x] \right] \times \\ &\quad \frac{\partial}{\partial p}\left(\frac{p}{m}\varphi(p) \right) \end{aligned} \tag{13}$$

of which the last term inside the bracket will not contribute because of the symmetric momentum distribution. Fourth, using the symmetric momentum distribution,

$$n_3 = 0. \tag{14}$$

Fifth,

$$n_4 = \frac{\gamma^2}{2}\int dp \int_0^t ds_1 \int_0^{s_1} ds_2 exp(-\frac{ps_2}{m}\frac{\partial}{\partial x})\int dx'\varepsilon(x-x')\psi(x')\times$$
$$\int dx''\varepsilon(x-x'')\psi(x'')\frac{\partial^2\varphi(p,0)}{\partial p^2}$$
$$= \frac{\gamma^2 m}{12}\int\frac{dp}{p}\int_0^t ds\left[(erf[\sqrt{\alpha}x])^3-(erf[\sqrt{\alpha}(x-\frac{ps}{m})])^3\right]\frac{\partial^2\varphi(p,0)}{\partial p^2} \tag{15}$$

Finally,

$$n_5 = \frac{\gamma^2}{2}\int dp \int_0^t ds_1 \int_0^{s_1} ds_2 exp(-\frac{ps_2}{m}\frac{\partial}{\partial x})\int dx'\varepsilon(x-x')^2\psi(x')\psi(x)\frac{\partial^2\varphi(p,0)}{\partial p^2}$$
$$= \frac{\gamma^2 m^2}{4}\int\frac{dp}{p^2}\left[\frac{1}{\sqrt{\pi\alpha}}\{e^{-\alpha(x-\frac{pt}{m})^2}-e^{-\alpha x^2}\}+\right.$$
$$\left.(x-\frac{pt}{m})\{erf[\sqrt{\alpha}(x-\frac{pt}{m})]-erf[\sqrt{\alpha}x]\}\right]\frac{\partial^2\varphi(p,0)}{\partial p^2}. \tag{16}$$

Combining all terms, we obtain

$$n = \sqrt{\frac{\alpha}{\pi}}\int dpe^{-\alpha(x-\frac{pt}{m})^2}\varphi(p,0)$$
$$+\frac{\gamma m}{4}\int dp\left[(erf[\sqrt{\alpha}x])^2-(erf[\sqrt{\alpha}(x-\frac{pt}{m})])^2\right]\frac{1}{p}\frac{\partial\varphi(p,0)}{\partial p}$$
$$+\gamma\int dp\left[\frac{m^2}{4p^2}\left\{(erf[\sqrt{\alpha}(x-\frac{pt}{m})])^2-(erf[\sqrt{\alpha}x])^2\right]+\right.$$
$$\frac{m^2}{2p^2}\sqrt{\frac{2\alpha}{\pi}}\left\{(x-\frac{pt}{m})(erf[\sqrt{2\alpha}x]-erf[\sqrt{2\alpha}(x-\frac{pt}{m})])\right\}+$$
$$\left.\frac{m^2}{2\pi p^2}(e^{-\alpha x^2}-e^{-\alpha(x-\frac{pt}{m})^2})+\frac{mt}{p}\sqrt{\frac{\alpha}{\pi}}e^{-\alpha x^2}erf[\sqrt{\alpha}x]\right]\times$$
$$\frac{\partial}{\partial p}\left(\frac{p}{m}\varphi(p)\right)$$
$$+\frac{\gamma^2 m}{12}\int\frac{dp}{p}\int_0^t ds\left[(erf[\sqrt{\alpha}x])^3-(erf[\sqrt{\alpha}(x-\frac{ps}{m})])^3\right]\frac{\partial^2\varphi(p,0)}{\partial p^2}$$
$$+\frac{\gamma^2 m^2}{4}\int\frac{dp}{p^2}\left[\frac{1}{\sqrt{\pi\alpha}}\{e^{-\alpha(x-\frac{pt}{m})^2}-e^{-\alpha x^2}\}+\right.$$
$$\left.(x-\frac{pt}{m})\{erf[\sqrt{\alpha}(x-\frac{pt}{m})]-erf[\sqrt{\alpha}x]\}\right]\frac{\partial^2\varphi(p,0)}{\partial p^2}. \tag{17}$$

which is applicable for (1) all initially even and physically reasonable momentum distribution. We summarize the other conditions for this solution to be valid for (2) completely factored many-body distributions, and (3) a spatial Gaussian distribution at the origin.

Note that n_0 and n_1 are well-behaved and bounded in x, p, and t. The remaining terms have integrable singularities at $p = 0$. Furthermore, the each of the terms n_2, n_4, and n_5, becomes zero as x approaches infinity; however, these terms diverge in t.

3 Graphical Results

In contrast to brute force difference methods, it is possible to use Equation 17 using substitution, and numerical integrations of analytic expressions. Of course, the initial data, in the form of initial distributions, must be specified. For the purpose of this work, we will assume, in addition to the Gaussian spatial distribution $\psi(x) = \sqrt{\frac{\alpha}{\pi}} \exp(-\alpha x^2)$, the "explosion" momentum distribution

$$\varphi(p) = \sqrt{\frac{\beta}{\pi}} \exp(-\beta p^2). \tag{18}$$

We have picked the above form for two reasons, that is, (1) to illustrate that a relatively simple structure in momentum space creates interesting density strucures, and (2) to avoid some singularities which will be treated elsewhere.

Substituting Equation 18 into Equation 17 with the parameters $\alpha = 1$, $\beta = 1$, $m = 1$, $\gamma = \frac{1}{100}$, for times $t = 0$ to $t = 10$, and using symbolic calculation routines, as in Maple, we produce Figs. 1, 2 and 3 which show the time evolution of n_0, $n_0 + n_1 + n_2$, and $n_0 + n_1 + n_2 + n_4 + n_5$, respectively. We observe the following:

- The most dominant behavior is given by n_0, the ballistic term, which shows smooth diffusion in phase space (see Fig. 1).
- The terms n_1 and n_2 account for the substructure, proportional to the parameter γ, the gravitational coupling parameter (see Fig. 2).
- The remaining terms constitute a possible microstructure, thus, the possibility that there is a hierarchy of substructures (see Fig. 3). These will be examined extensively in future studies.
- From an analytic-numeric standpoint, n_4 requires the main computational effort; we use the default numerical integrator (adaptive Clenshaw-Curtis quadrature with singularity checking) in Maple.

Figure 4 is a check for the conservation of mass. Observe that for time $t > 4$ using the above parameters, the total mass becomes less than one; this is because some particles have travelled outside the interval $-10 \leq x \leq 10$.

Although not visible in Figs. 2 and 3, negative probabilities of the order 10^{-3} occur in underlying single-particle probability distribution function (the

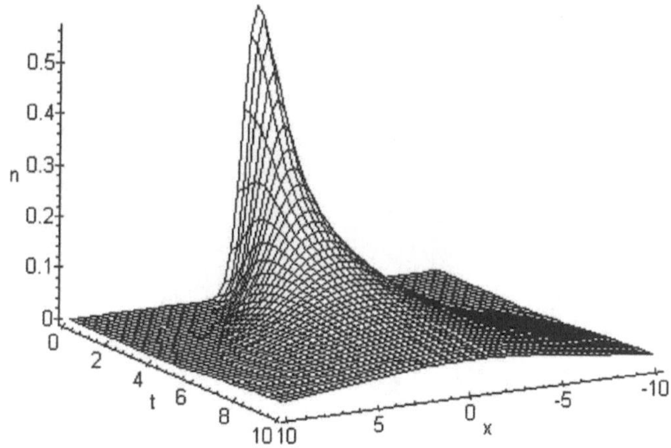

Fig. 1. n_0

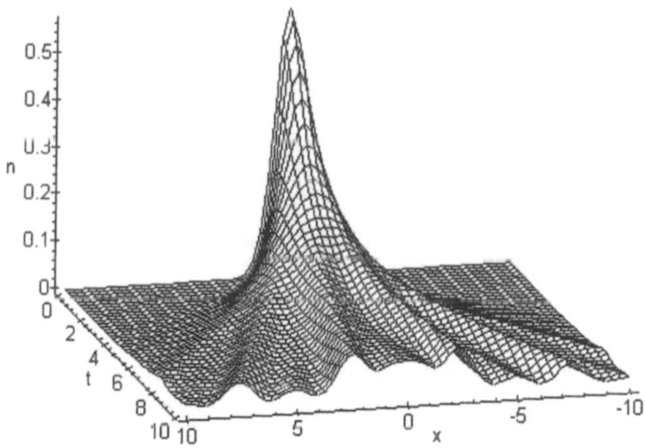

Fig. 2. $n_0 + n_1 + n_2$

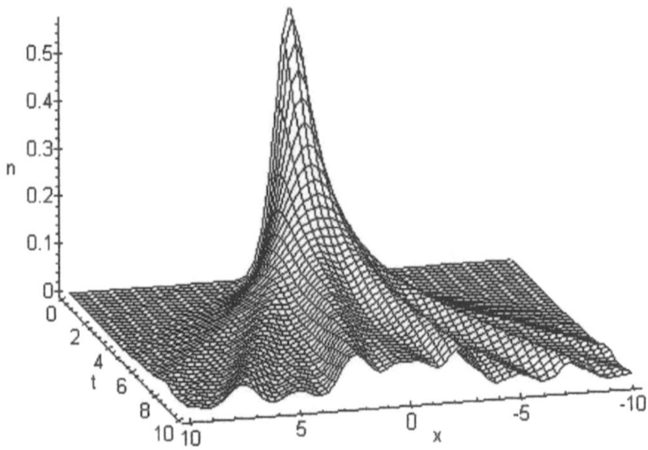

Fig. 3. $n_0 + n_1 + n_2 + n_4 + n_5$

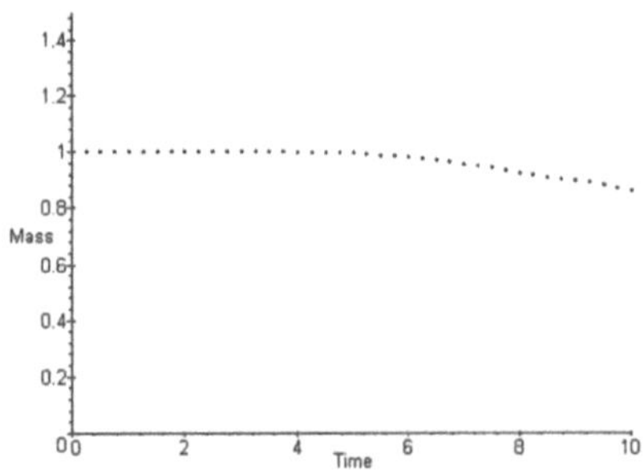

Fig. 4. Total Mass

integrand f in $n(x,t) = \int dpf(x,p,t))$ when $t > 4$ for the chosen parameters. We see that one of two things are possible: (1) the theory which produced Equation 17 is no longer valid for all times and all coupling parameters, or, (2) the negative probabilities indicate a form of instability where the single-particle distribution picture is no longer valid. This observation demands an intensive numeric and analytic study for a more general class of initial data to find out the critical combination of coupling strength and time for which negative probabilities appear. In the absence of any final conclusion regarding this, we offer the following remark: In a gravitational gas, one cannot exclude the formation of gravitational molecules, or binaries. Such binaries will occur for large coupling parameters, and long times. When binaries become dominant, the single-particle distribution function no longer describes the system adequately. The two-particle distribution function is needed, for which we have as yet no formulation. Over long times and large coupling strengths, correlations will develop, even for systems without initial correlations, especially for long range forces. Thus, the secular behavior is an indication of the limited applicability of the single-particle picture. The two-particle analogue of Equation 17 must be derived.

4 Remarks and Conclusions

The main conclusion of this work is that in a one dimensional self-gravitating system, a hierarchy of structures may emerge quite naturally, reducing the severity of the problem of structure formation. We suggest that this result may be applicable to three dimensions. In this work, the gross structure and substructures are of markedly different orders of magnitude, explaining why the latter may have not have been seen before in numerical experiments.

In the course of our investigation, which demonstrates the irreversible nature of our exploding system, we have observed secular divergence of the density. Such behavior, we suspect, is interpretable as an instability in the one-particle distribution function, to be physically resolved, we suspect, when we develop the analogue of Equation 17 for the time evolution of the pair correlation function, even for systems which is initially uncorrelated in momentum and space. Physically, we recognize that strong coupling, long range of the pair potential, and large times are all needed to develop correlations. Such correlations, we suggest, manifest themselves as binaries or gravitational molecules. These correlations are even more microscopic than the hierarchy of structures that we observe for the evaluation of Equation 17.

Acknowledgements

We thank P. Esguerra, L. Jirkovsky, M. Sarmiento, and E. Gutierrez for preliminary work. A. M. acknowledges the hospitality of the Institute for Advanced Study, Princeton, where most of this work was done. We acknowledge the assistance of the National Research Council of the Philippines, Department of Science and Technology of the Philippines and the ICSC World Laboratory, Lausanne.

References

Bogoliubov, N.N. (1962): *problems of a dynamical theory in statistical physics, Studies in Statistical Mechanics*, Vol. 1, De Boer, J. and Uhlenbeck, G.E. eds. (North Holland Publishing Company, Amsterdam)

Camm, G.L. (1950): Self gravitating star systems. MNRAS **110**, 305–324
Hohl, F., Feix, M.R (1966): A one-dimensional model to study collective effects in stellar dynamics. Phys. Lett. **56A**, 432–434
Linde, A.D. (1990): *Particle Physics and Inflationary Cosmology* (Harwood, New York)
Mineau, P., Feix, M.R, Muriel, A. (1990): A perturbation method for high-virial gravitational systems in the collisionless regime. Astron. Astrophys. **233**, 422–426
Muriel, A., Dresden, M. (1997): An integral formulation of hydrodynamics. Physica D, 299–316
Muriel, A., Esguerra, P. (1996): Exact time evolution of the density of a classical many-body system: the open one-dimensional gravitational gas. Phys. Rev. E. **54**, 1433–1441
Muriel, A., Esguerra, P., Dresden, M. (to be published)
Peebles, P.J. (1980) : *The Large-Scale Structure of the Universe* (The University Press, Princeton)

Interaction Forces and Space Symmetry

Alain Munier

Centre d'Études de Limeil-Valenton, BP 27, 94195, Villeneuve-St-Georges Cedex, France

Abstract. We determine the general expression for the interaction force between point elements or vector elements, based on symmetry group properties in space. In a particular case this expression is then shown to derive from a scalar form of energy.

1 Introduction

Point mass interactions like gravitational or coulombic interactions assume that the mutual force operates along the line defined by the position of the two masses (or charges). For point element interactions, this is well justified, since two points cannot define any direction other than the line connecting them. When the masses are of a vector nature, however, like for example interacting currents or spin-spin interactions, there is a priori no justification, other than experiment, for transferring this characteristic to a field in which the elements are arranged in definite directions. The problem is even more complicated when dealing with tensors which destroy the symmetry properties of space.

In this paper we analyse the assumptions underlying the definition of various interactions, and show that a general force may be derived that satisfies the symmetry group properties of space.

2 Basic Symmetries in Space

Call $\mathbf{F}(\mathbf{i},\mathbf{j};A,B)$ the force produced by the vector $\mathbf{j}$, located at point B, and acting upon the vector $\mathbf{i}$, located at point A. The two vectors $\mathbf{i}$ and $\mathbf{j}$ represent **masses** or **currents**, and the order of parameters is essential.

We wish to define several hypotheses and properties related to the symmetry of space. Firstly space may be **homogeneous**, allowing for parallel transport of vectors (Fig. 1). Secondly space may be **isotropic**, allowing for the invariance of vectors under any arbitrary rotation (Fig. 2).

These definitions, however, require that the vectors themselves do not modify space. A mass (or a spinning particle) modifies the space structure (the point where the mass is located is different from the others and the spin defines a privileged direction). This is why we require only **weak homogeneity** and **weak isotropy** as properties **(H)** and **(I)** of space that hold when the disturbing elements are removed.

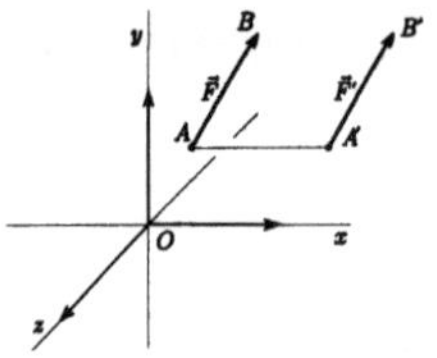

Fig 1. Parallel transport of a vector along Ox.

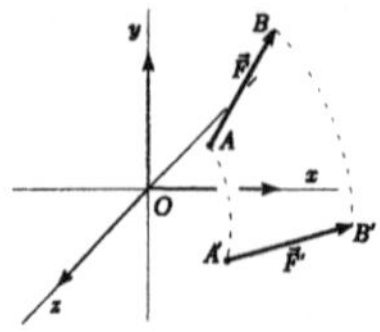

Fig 2. Arbitrary rotation of a vector.

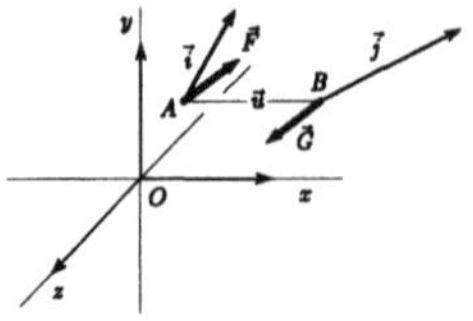

Fig 3a. Basic transformation T: Initial position.

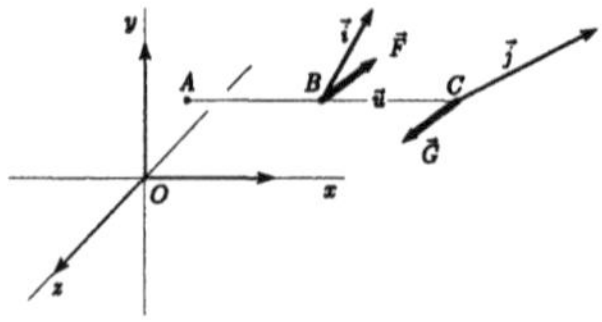

Fig 3b. First part of transformation T: Parallel transport.

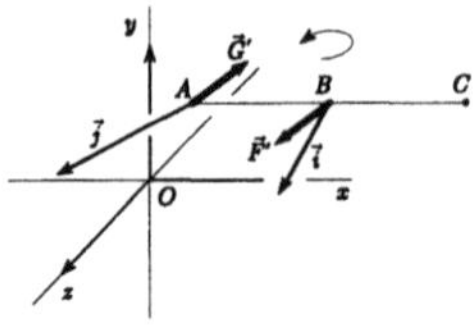

Fig 3c. Second part of transformation T: Rotation about B.

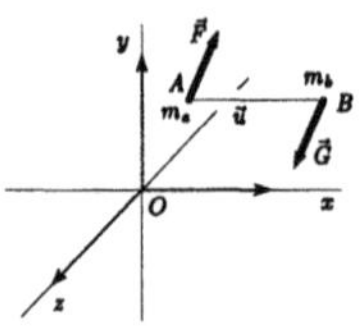

Fig 4a. Scalar interaction.

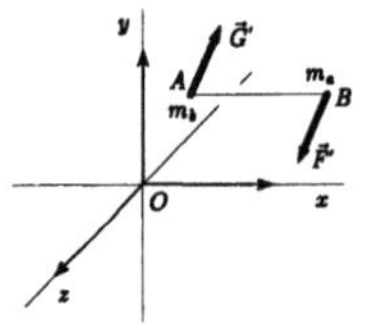

Fig 4b. Interchanging scalar masses and forces.

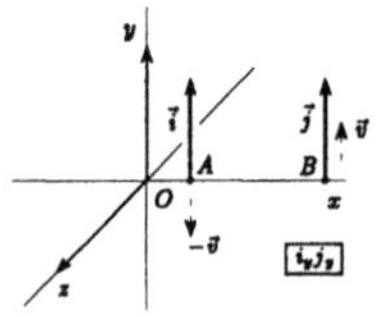

Fig 5. Parallel currents

Consider then the following basic transformation. We firstly define the unit vector $\mathbf{u} = \mathbf{AB}/|AB|$ and then choose a basis such that $\mathbf{e}_x$ is parallel to $\mathbf{u}$, the other two vectors being defined as in Fig. 1. The first part of the transformation consists in a translation defined by the vector $\mathbf{AB}$. The second part of the transformation consists in a rotation of 180°, centered on B, the axis of which is taken as we please (see Fig. 3b). We call T the transformation representing the combination of the translation and the rotation.

Examine now a scalar interaction between two masses, as in Fig. 4a. Translation and then rotation of the system allows for the interchange of the two masses, leading to Fig. 4b. It is clear that the system of forces acting on the two masses of Fig. 4b is the same as in Fig. 4a provided the two forces $\mathbf{F}$ and $\mathbf{G}$ are in the same plane. Hence

$$T[\mathbf{F}(m_a, m_b; A, B)] = \mathbf{F}'(m_a, m_b; B, A) = -\mathbf{F}(m_a, m_b; A, B), \qquad (1)$$

$$T[\mathbf{G}(m_b, m_a; B, A)] = \mathbf{G}'(m_b, m_a; A, B) = -\mathbf{G}(m_b, m_a; B, A). \qquad (2)$$

If the two masses are identical, we have that

$$\mathbf{G}'(m_b, m_a; A, B) = \mathbf{F}(m_a, m_b; A, B), \tag{3}$$

$$\mathbf{F}'(m_a, m_b; B, A) = \mathbf{G}(m_b, m_a; B, A). \tag{4}$$

Therefore

$$\mathbf{F}(m_a, m_b; A, B) = -G(m_b, m_a; B, A),. \tag{5}$$

which is **Newton's third principle** in a weak form (property **(N)**), where only the cancellation of forces is required, but nothing is said about the forces being aligned with **u**. If the two masses are different, the last relation may not be satisfied, because of the modification of space due to the existence of two spatial points with different properties, and also of a privileged direction (namely **u**).

For vector masses as in Fig. 3a, the situation is somewhat different. There is a priori no reason, based on symmetry properties, why the forces **F** and **G** should be aligned. Hence the axis of the transformation T may be different for the two forces **F** and **G**.

Making provision for the treatment of electrostatic and magnetostatic forces, we should also like the forces to have opposite signs whenever the sign of one of the masses or currents is changed. This may be expressed as an **odd parity** condition **(O)**

$$\mathbf{F}(-\mathbf{i}, \mathbf{j}; A, B) = -\mathbf{F}(\mathbf{i}, \mathbf{j}; A, B), \tag{6}$$

$$\mathbf{F}(\mathbf{i}, -\mathbf{j}; A, B) = -\mathbf{F}(\mathbf{i}, \mathbf{j}; A, B), \tag{7}$$

$$\mathbf{F}(\mathbf{i}, -\mathbf{j}; A, B) = -\mathbf{F}(\mathbf{i}, \mathbf{j}; A, B), \tag{8}$$

$$\mathbf{F}(-\mathbf{i}, -\mathbf{j}; A, B) = \mathbf{F}(\mathbf{i}, \mathbf{j}; A, B), \tag{9}$$

which means that the forces must be odd functions of the currents or masses.

We also want the forces to be expandable over the basis vectors, allowing any force of interaction to be resolved into components. This is simply the **parallelogram** law **(P)** for the addition of forces. Another reasonable requirement is that the force depends only on the **distance** between points A and B (property **(D)**, beside its dependance on the current vectors or the masses). Since the basis vector $\mathbf{e}_y$ is arbitrary, the system of forces must also be invariant with respect to any **rotation** about $\mathbf{e}_x$ (property **(R)**.

Another reasonable requirement, outside the scope of symmetry properties, is that the expression of the force be independent of our choice of the unit length. As a consequence the ratio of forces produced by identical currents at different distances must depend solely on the ratio of these distances. We call this independance with respect to the choice of the unit length **(S)** for **scaling**. We define now the most general system of forces compatible with these properties excluding, however, time dependency of the forces and all effects such as retardation or imbalance of forces at relativistic speeds.

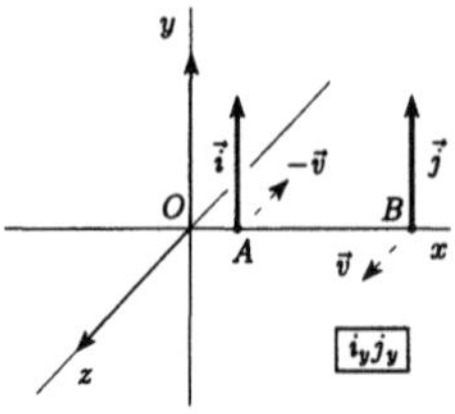

Fig 6a. Parallel currents

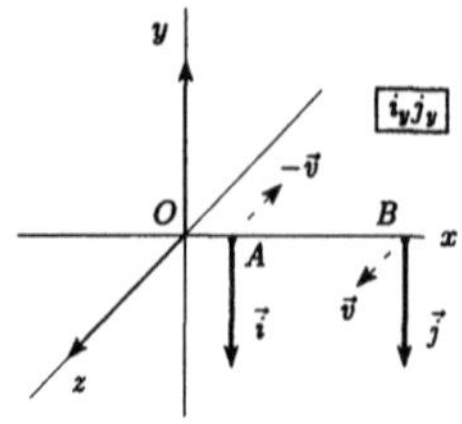

Fig 6b. Inversion of the two currents.

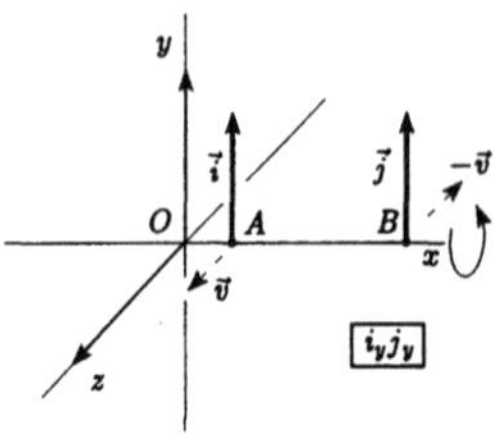

Fig 6c. Rotation about Ox.

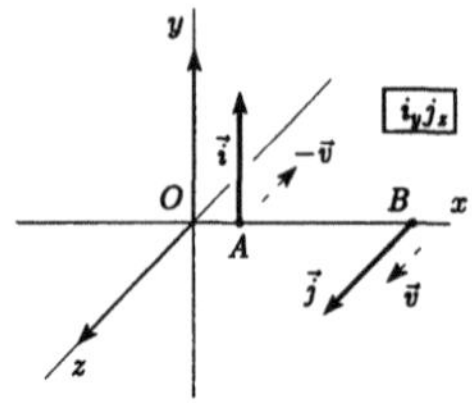

Fig 7a. Orthogonal currents

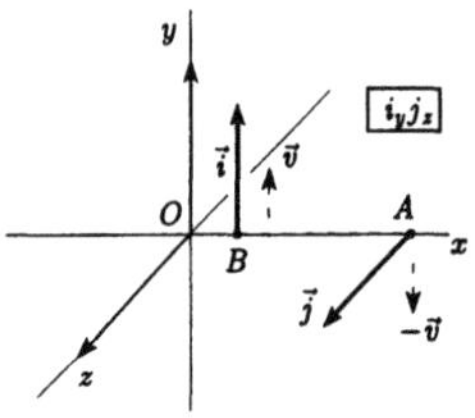

Fig 7b. Rotations about Oz and Ox.

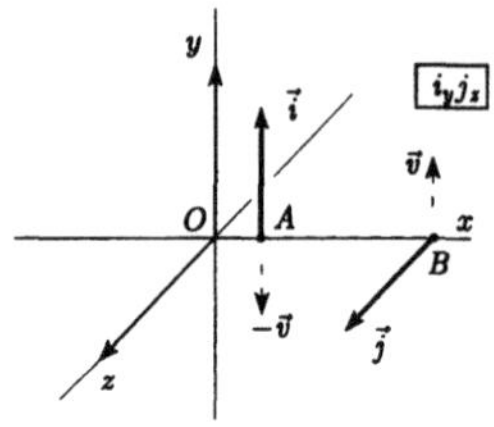

Fig 7c. Orthogonal currents

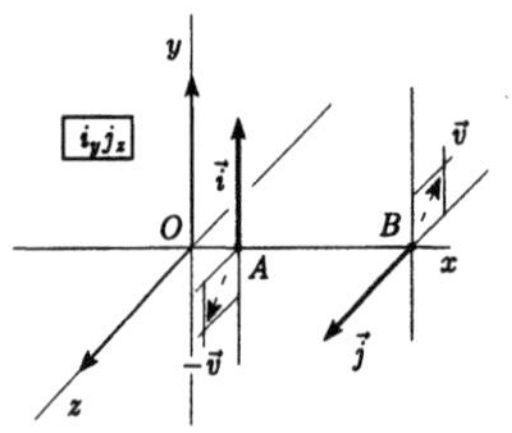

Fig 7d. Oblique motion as a result of $i_y j_z$

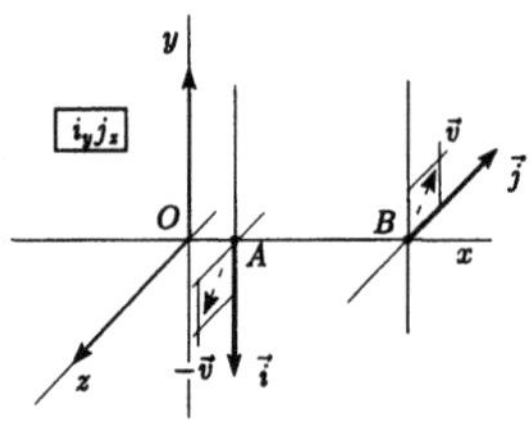

Fig 7e. Inversion of the currents.

3 Interaction Matrix

According to **P** the force of interaction $\mathbf{F}(\mathbf{i},\mathbf{j};A,B)$ between vector elements may be resolved into components. Making provision for the change of sign, we write the various components of $\mathbf{i}$ as i_x, i_y, i_z, $-i_x$, $-i_y$ and $-i_z$, and similar terms for $\mathbf{j}$. Hence, the interaction between $\mathbf{i}$ and $\mathbf{j}$ may be written as a sum of 36 combinations, which we write for short $x|x$, $x|y$,... $-x|x$,... $-x|-y$, etc... However, according to **R**, we have that

$$y|y = z|z = -y|-y = -z|-z, \qquad y|-y = z|-z = -y|y = -z|z, \tag{10}$$

$$z|y = y|-z = -z|-y = -y|z, \qquad z|-y = y|z = -z|y = -y|-z, \tag{11}$$

$$z|x = y|x = -z|x = -y|x, \qquad z|-x = y|-x = -z|-x = -y|-x, \tag{12}$$

$$x|z = x|y = x|-z = x|-y, \qquad -x|z = -x|y = -x|-z = -x|-y, \tag{13}$$

and we are left with only the 12 interactions:

$$x|x \quad x|y \quad x|-x \quad -x|x \quad -x|y \quad -x|-x, \tag{14}$$

$$y|x \quad y|y \quad y|-x \quad y|z \quad z|y \quad z|-z. \tag{15}$$

From **O** a change of the sign of one vector must change the sign of the interaction. Therefore

$$x|-x = -(x|x) \qquad -x|y = -(x|y), \qquad -x|-x = (x|x), \tag{16}$$

$$y|-x = -(y|x),\ z|y = y|-z = -(y|z),\ z|-z = -(y|y) = -(z|z). \tag{17}$$

We see that only five components remain, namely

$$x|x, \qquad x|y, \qquad y|x, \qquad y|y, \qquad y|z. \tag{18}$$

Using **P** again, we may write the force as a linear combination of these components

$$\begin{aligned}\mathbf{F}(\mathbf{i},\mathbf{j};A,B) = \phi_a(r)\, i_x j_x\, \mathbf{n}_a + \phi_b(r)\, i_x j_y\, \mathbf{n}_b + \phi_c(r)\, i_y j_x\, \mathbf{n}_c \\ +\phi_d(r)\, i_y j_y\, \mathbf{n}_d + \phi_e(r)\, i_y j_z\, \mathbf{n}_e,\end{aligned} \tag{19}$$

where $\phi_a(r)$, $\phi_b(r)$, $\phi_c(r)$, $\phi_d(r)$ and $\phi_e(r)$, are arbitrary functions of the distance r. Up to now the generic vectors $\mathbf{n}_\alpha$ are also arbitrary, but have to satisfy the symmetry properties that were defined earlier.

4 Generic Vectors

The main difficulty is now to determine the vectors $\mathbf{n}_\alpha$. Notice that the first component of the force, proportional to $i_x j_x$, defines one privileged direction in space only, namely the direction $\mathbf{e}_x$. This component leads to a motion that must be along $\mathbf{e}_x$. On the other hand the last component, $(i_y j_z)$, spans the full 3-D space since $\mathbf{u} \equiv \mathbf{e}_x$ is along the line AB joining the currents, and i_y and j_z define the two other diections $\mathbf{e}_y$ and $\mathbf{e}_z$. Hence, the basis vectors associated with the various components are

$$i_x j_x \to (\mathbf{e}_x \equiv \mathbf{u}), \qquad i_y j_y \to (\mathbf{e}_x \equiv \mathbf{u},\ \mathbf{e}_y,\ \mathbf{e}_z \equiv \mathbf{u} \times \mathbf{e}_y), \tag{20}$$

$$i_x j_y \to (\mathbf{e}_x \equiv \mathbf{u},\ \mathbf{e}_y,\ \mathbf{e}_z \equiv \mathbf{u} \times \mathbf{e}_y), \tag{21}$$

$$i_y j_x \to (\mathbf{e}_x \equiv \mathbf{u},\ \mathbf{e}_y,\ \mathbf{e}_z \equiv \mathbf{u} \times \mathbf{e}_y), \qquad i_y j_z \to (\mathbf{e}_x \equiv \mathbf{u},\ \mathbf{e}_y,\ \mathbf{e}_z). \tag{22}$$

We examine these components in turn.

— $i_x j_x$ — Since there is only one possible direction , we have that $\mathbf{n}_a \equiv \mathbf{u}$.

— $i_y j_y$ — Consider Fig. 5 and assume that the force has a component along $\mathbf{e}_y$ (and that consequently point B is submitted to an acceleration along this direction). After a while point B, starting from rest, has acquired a velocity $\mathbf{v}$ along this direction too. The relative motion of A is then $-\mathbf{v}$. Using however transformation T (translation plus rotation of 180° about $\mathbf{e}_y$), we obtain a system equivalent in principle to that of Fig. 5 (the currents are reversed and directed downwards), but with opposite velocities. Since these two configurations are in contradiction, we conclude that motion along $\mathbf{e}_y$ is impossible.

Consider next Fig. 6a, in which the velocity of B is along $\mathbf{e}_z$. According to **(O)** inversion of the two currents should not change the sign of the force, leading to Fig. 6b. Performing then a rotation about $\mathbf{e}_z$, we obtain Fig. 6c which clearly contradicts (6a). Motion along $\mathbf{e}_z$ must therefore also be excluded. Again, the only possibility is that $\mathbf{n}_b \equiv \mathbf{u}$.

— $i_y j_z$ — Examine now the last term as in Fig. 7a and assume first that the velocity impinged on B is along $\mathbf{e}_z$ (relative velocity $-\mathbf{v}$ impinged on A). After two successive rotations (first 180° about $\mathbf{e}_z$, then $-90°$ about $\mathbf{e}_x$) we obtain Fig. 7b which shows that the velocity at B is now along $\mathbf{e}_y$ (while that at A is along $-\mathbf{e}_y$). Clearly, the only solution is that the motion at B be a combination of these two vectors and that point A move along $\mathbf{e}_z - \mathbf{e}_y$.

Assuming then that the motion of B is along $\mathbf{e}_y$ (see Fig. 7c), by the same transformations, we arrive at the conclusion that the motion of B must be along $\mathbf{e}_y - \mathbf{e}_z$ (see Fig. 7d). Inverting then the two currents (Fig. 7e) and applying two successive rotations of 90 degrees about Ox, (Fig. 7f and 7g), we obtain a system which contradicts our previous result, and therefore must conclude that no motion is possible except along $\mathbf{e}_x$, consequently $\mathbf{n}_c \equiv \mathbf{u}$.

— $i_x j_y$ and $i_y j_x$ — These two cases must be treated together.

Consider first $i_x j_y$ in Fig. 8a in which point B moves along $\mathbf{e}_x$. A rotation of 180° about $\mathbf{e}_x$ leads to Fig. 8b, where the motion of the point is unchanged. This contradicts **(O)** since changing the sign of the current must change the sign of the force. Therefore, motion along $\mathbf{e}_x$ is impossible.

By the same token, in Fig. 9a and 9b, the motion of B as submitted to system $i_y j_x$ is impossible along $\mathbf{e}_x$.

Go now to Fig. 10a. Under a rotation of 180° about $\mathbf{e}_y$, the system $i_x j_y$ goes over to Fig. 10b. This new system however is identical to $-i_y j_x$. We may then conclude that the combination $i_x j_y - i_y j_x$ is to be treated as an entity.

By the same line of reasoning, starting from Fig. 10a, after a rotation of axis Oy, we obtain Fig. 10b. However, changing the signs of the currents directly on Fig (10a), we end up with Fig. 10c, which after a rotation about $\mathbf{e}_z$ goes over to Fig. 10d. This last configuration being identical to $i_y j_x$ (Fig. 9a) we again conclude that $i_x j_y + i_y j_x$ is to be treated as an entity.

Summing up all these results we have that the force is necessarily of the form

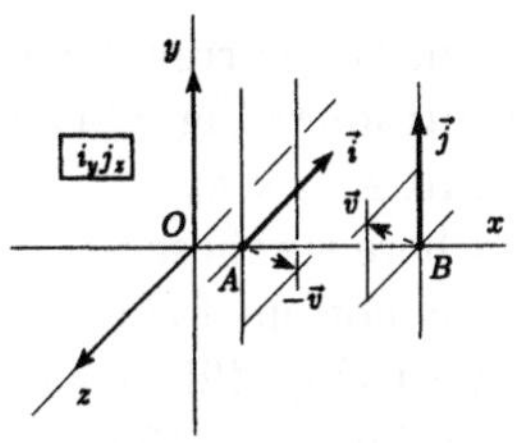

Fig 7f. Rotation about Ox.

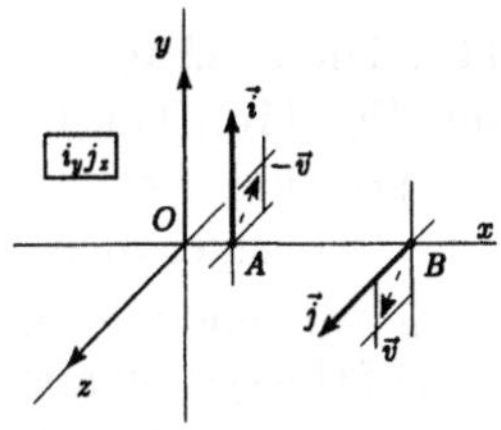

Fig 7g. Rotation about Ox.

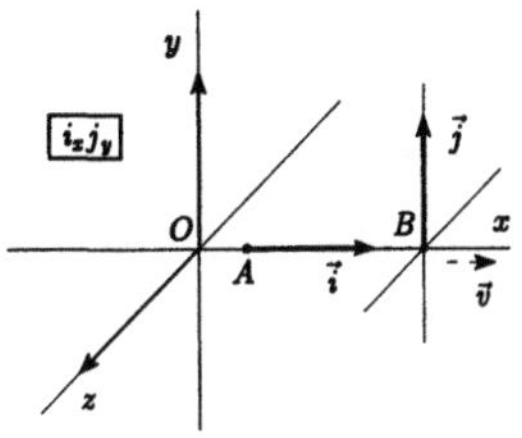

Fig 8a. Orthogonal currents

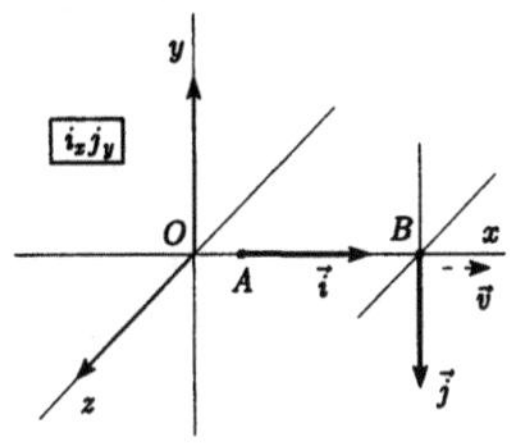

Fig 8b. Rotation about Oy.

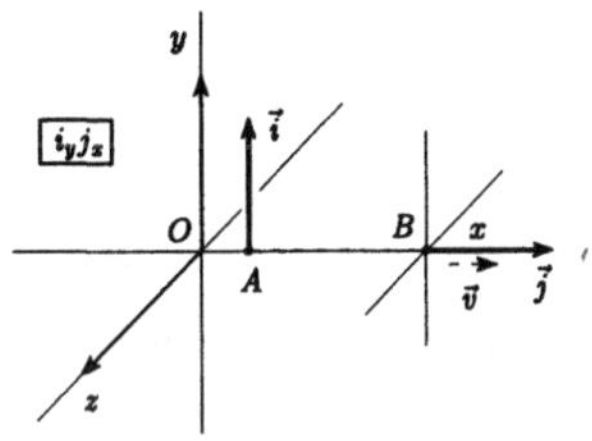

Fig 9a. Orthogonal currents

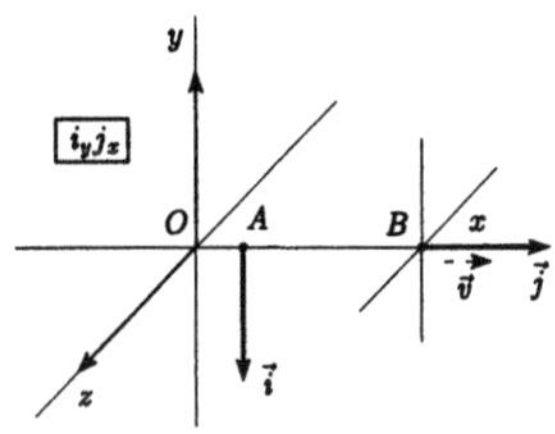

Fig 9b. Rotation about Ox.

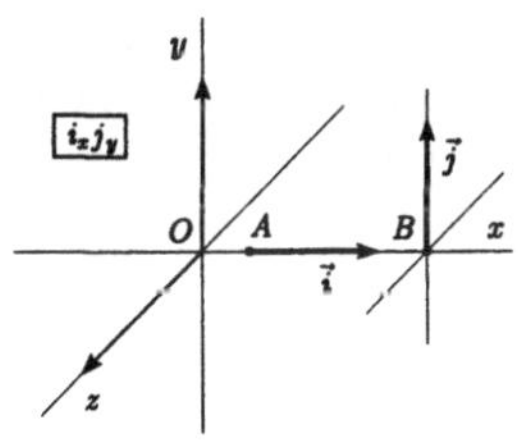

Fig 10a. Orthogonal currents

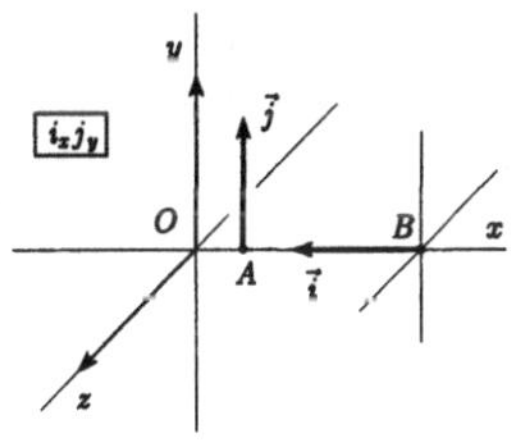

Fig 10b. Rotation about Oy.

$$\begin{aligned}\mathbf{F}(\mathbf{i},\mathbf{j};A,B) &= \phi_a(r)\,i_xj_x\,\mathbf{u} + \phi_b(r)\,(i_xj_y - i_yj_x)\,\mathbf{n}_b \\ &+\phi_c(r)\,(i_xj_y + i_yj_x)\,\mathbf{n}_c + \phi_d(r)\,i_yj_y\,\mathbf{u} + \phi_e(r)\,i_yj_z\,\mathbf{u}.\end{aligned} \tag{23}$$

We must now determine the last two generic vectors.

— $(i_xj_y - i_yj_x)$ — Having eliminated motion along $\mathbf{e}_x$, we want to see whether motions along the other directions are possible.

Consider Fig. 11a where we have chosen particular currents such that $i_x = i_y$, $j_x = -j_y = i_x$ and $i_z = j_z = 0$ (vectors $\mathbf{i}$ and $\mathbf{j}$ are perpendicular). The motion of B is assumed to be along $\mathbf{e}_y$. A rotation of the system about

$\mathbf{e}_y$ leads to Fig. 11b. Then a change of sign of the two currents must leave the velocities invariant (Fig. 11c). This however contradicts Fig. 11a and therefore such a motion is impossible. As a conclusion, $\mathbf{n}_b \equiv \mathbf{u} \times \mathbf{e}_y$.

— $(i_x j_y + i_y j_x)$ — Consider then Fig. 12a where we chose parallel currents such that $i_x = i_y$, $j_x = j_y$ and $i_z = j_z = 0$. The motion of B is assumed to be along $\mathbf{e}_z$. A rotation of the system about $\mathbf{e}_y$ leads to Fig. 12b. Then a change of sign of the two currents leaves the velocities invariant (Fig. 12c), contradicting Fig. 12a and rendering such a motion impossible. As a conclusion, $\mathbf{n}_c \equiv \mathbf{e}_y$.

Go back now to the general expression of the force that we obtained so far

$$\begin{aligned}\mathbf{F}(\mathbf{i},\mathbf{j};A,B) = \phi_a(r)\, i_x j_x\, \mathbf{u} + \phi_b(r)\,(i_x j_y - i_y j_x)\,\mathbf{u}\times\mathbf{e}_y \\ +\phi_c(r)\,(i_x j_y + i_y j_x)\,\mathbf{e}_y + \phi_d(r)\, i_y j_y\,\mathbf{u} + \phi_e(r)\, i_y j_z\,\mathbf{u}.\end{aligned} \tag{24}$$

We must not forget that many elements were eliminated on the grounds that they are equivalent to those already present in the expression of the force. For example, $i_z j_z = i_y j_y$ appears implicitly in the fourth term of $\mathbf{F}$. This means that another term of the form $\phi_d(r)\, i_z j_z\, \mathbf{u}$ may be added to the force. Combining these two elements we may replace the fourth term of $\mathbf{F}$ by $\phi_d(r)\,(i_y j_y + i_z j_z)\,\mathbf{u}$. Since the first term may also be written $(\phi_a(r) + \phi_d(r) - \phi_d(r))\, i_x j_x\, \mathbf{u}$, we conclude that the fourth term of the force may in fact be $\phi_d(r)\,(i_x j_x + i_y j_y + i_z j_z)\,\mathbf{u}$. Writing then $\phi_{\bar{a}}(r) = \phi_a(r) - \phi_d(r)$, we see that the force is

$$\begin{aligned}\mathbf{F}(\mathbf{i},\mathbf{j};A,B) = \phi_{\bar{a}}(r)\,(\mathbf{i}\cdot\mathbf{u})(\mathbf{j}\cdot\mathbf{u})\,\mathbf{u} + \phi_b(r)\,(i_x j_y - i_y j_x)\,\mathbf{u}\times\mathbf{e}_y \\ +\phi_c(r)\,(i_x j_y + i_y j_x)\,\mathbf{e}_y + \phi_d(r)\,(\mathbf{i}\cdot\mathbf{j})\,\mathbf{u} + \phi_e(r)\, i_y j_z\,\mathbf{u}.\end{aligned} \tag{25}$$

Examine now the second term of $\mathbf{F}$. A rotation about $\mathbf{e}_x$ must leave the system invariant. Therefore we may add to the previous relation

the term $\phi_b(r)\,(i_x j_z - i_z j_x)\,\mathbf{u}\times\mathbf{e}_z$ obtained by the substitution $y \to z$. Finally, since $i_y j_z$ is equivalent to $-i_z j_y$, we may modify the last term of the force into $\phi_e(r)\,(i_y j_z - i_z j_y)\,\mathbf{u}$. This last expression may now be written $(\phi_e(r) + \phi_b(r) - \phi_b(r))\,(i_y j_z - i_z j_y)\,\mathbf{u}$. It is easy to see that

$$\mathbf{i}\times\mathbf{j} = (i_x j_y - i_y j_x)\,\mathbf{u}\times\mathbf{e}_y + (i_x j_z - i_z j_x)\,\mathbf{u}\times\mathbf{e}_z + (i_y j_z - i_z j_y)\,\mathbf{u}. \tag{26}$$

Consequently on setting $\phi_{\bar{e}}(r) = \phi_e(r) - \phi_b(r)$ and using the fact that

$$(\mathbf{u}\times\mathbf{i})\times(\mathbf{u}\times\mathbf{j}) \equiv [\mathbf{u},\mathbf{i},\mathbf{j}]\;\mathbf{u} \equiv (i_y j_z - i_z j_y)\,\mathbf{u}, \tag{27}$$

(where $[\mathbf{u},\mathbf{i},\mathbf{j}]$ represents the scalar triple product) we find that the force becomes

$$\begin{aligned}\mathbf{F}(\mathbf{i},\mathbf{j};A,B) = \phi_{\bar{a}}(r)\,(\mathbf{i}\cdot\mathbf{u})(\mathbf{j}\cdot\mathbf{u})\,\mathbf{u} + \phi_b(r)\,\mathbf{i}\times\mathbf{j} \\ +\phi_c(r)\,(i_x j_y + i_y j_x)\,\mathbf{e}_y + \phi_d(r)\,(\mathbf{i}\cdot\mathbf{j})\,\mathbf{u} - \phi_{\bar{e}}(r)\,[\mathbf{i},\mathbf{u},\mathbf{j}]\,\mathbf{u}.\end{aligned} \tag{28}$$

For the same reasons as before the third term of $\mathbf{F}$ must be modified to include an equivalent expression obtained from it by the substitution $y \to z$. We may therefore add the contribution $\phi_c(r)\,(i_x j_z + i_z j_x)\,\mathbf{e}_z$. Writing the first term of the force as $\phi_{\bar{a}}(r) + \phi_c(r) - \phi_c(r)$ we have that

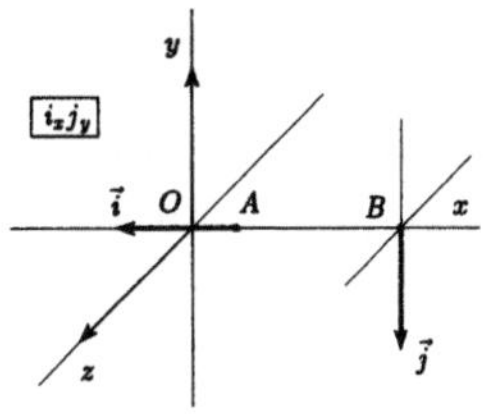

Fig 10c. Reversing the currents.

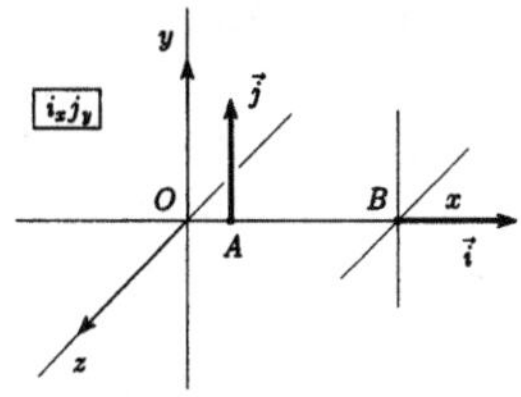

Fig 10d. Rotation about Oz.

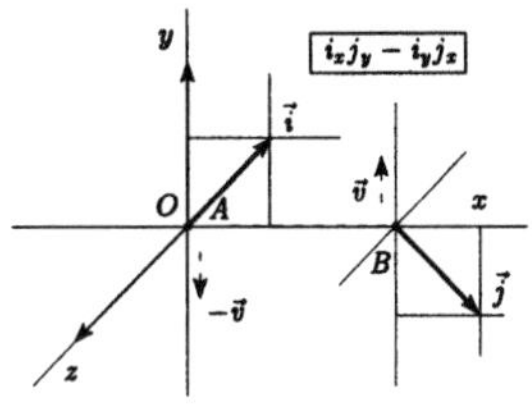

Fig 11a. Interaction $i_x j_y - i_y j_x$

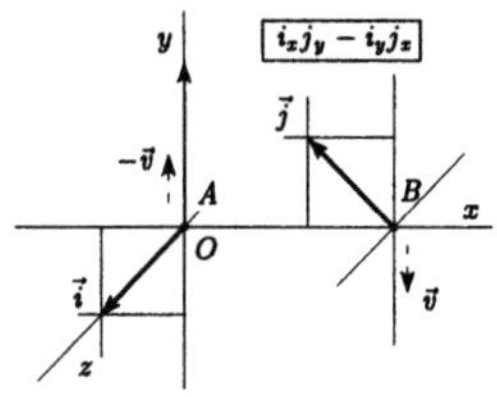

Fig 11b. Rotation about Oy.

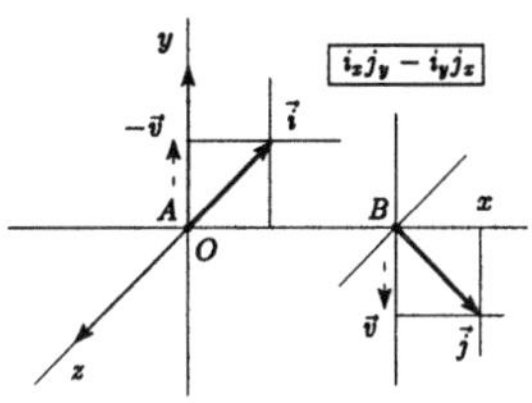

Fig 11c. Changing the signs of the currents.

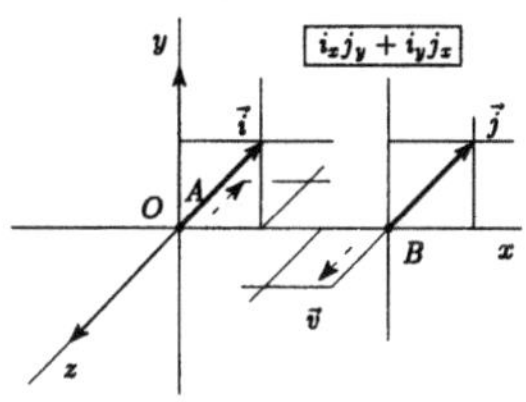

Fig 12a. System $i_x j_y + i_y j_x$

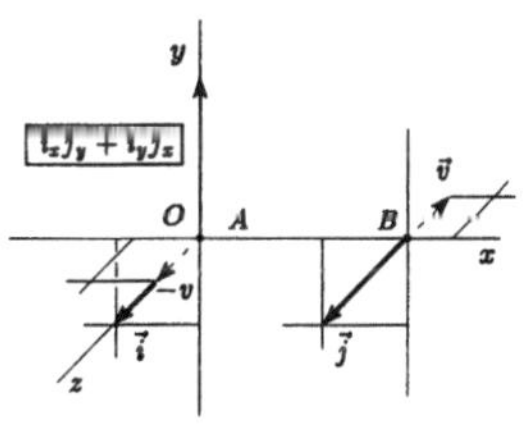

Fig 12b. Rotation about Oy.

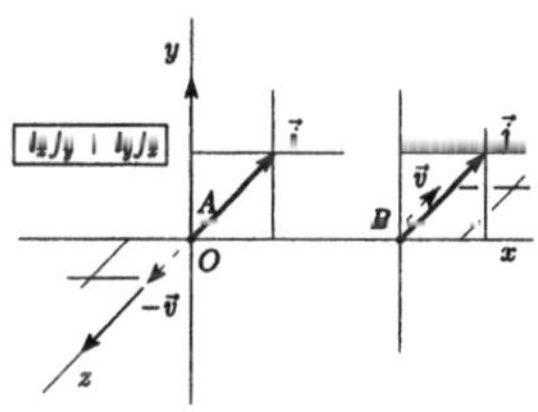

Fig 12c. Inversion of the currents.

$$(\mathbf{i}\cdot\mathbf{u})(\mathbf{j}\cdot\mathbf{u})\,\mathbf{u} + (i_x j_y + i_y j_x)\,\mathbf{e}_y + (i_x j_z + i_z j_x)\,\mathbf{e}_z = (\mathbf{i}\cdot\mathbf{u})\,\mathbf{j} + (\mathbf{j}\cdot\mathbf{u})\,\mathbf{i}. \quad (29)$$

Hence the general expression for the interaction between vector elements is

$$\begin{aligned}\mathbf{F}(\mathbf{i},\mathbf{j};A,B) &= \phi_a(r)\,(\mathbf{i}\cdot\mathbf{u})(\mathbf{j}\cdot\mathbf{u})\,\mathbf{u} + \phi_b(r)\,\mathbf{i}\times\mathbf{j} \\ &+\phi_c(r)\,\left((\mathbf{i}\cdot\mathbf{u})\,\mathbf{j} + (\mathbf{j}\cdot\mathbf{u})\,\mathbf{i}\right) + \phi_d(r)\,(\mathbf{i}\cdot\mathbf{j})\,\mathbf{u} + \phi_e(r)\,[\mathbf{i},\mathbf{u},\mathbf{j}]\,\mathbf{u}\end{aligned} \quad (30)$$

after renaming the arbitrary scaling functions $\phi_\alpha(r)$.

5 Scaling Functions

Our next step is to determine the functions $\phi_\alpha(r)$. In order to do that we assume that the force system satisfies the scaling postulate (**S**). This may be cast as (see Figs. 13a and 13b)

$$\frac{|\mathbf{F}(\mathbf{i},\mathbf{j};A,B_1)|}{|\mathbf{F}(\mathbf{i},\mathbf{j};A,B_2)|}=g\left(\frac{r_1}{r_2}\right), \tag{31}$$

where function g is arbitrary. Squaring this relation and defining $g\equiv f^2$ we conclude, after a lengthy calculation, that

$$\phi_\alpha(r_1)=f(r_1/r_2)\;\phi_\alpha(r_2). \tag{32}$$

Setting $r_1=r_2$ in this expression we obtain $f(1)=1$. Setting $r_2=1$ we have

$$\phi_\alpha(r_1)=f(r_1)\;\phi_\alpha(1). \tag{33}$$

Substituting r_2 for r_1 we get

$$\phi_\alpha(r_2)=f(r_2)\;\phi_\alpha(1). \tag{34}$$

Therefore, after elimination of $\phi_a(1)$, we obtain the relations

$$f(r_1)=f(r_2)\;f\left(\frac{r_1}{r_2}\right),\qquad f(1)=1. \tag{35}$$

All the scaling functions $\phi_a(r)$ are therefore proportional to that single $f(r)$ which we must now determine. The coefficients of proportionality will be written $\phi_a(1)=a$, $\phi_b(1)=b$, etc...

Substituting then $r_1=\lambda r$ and $r_2=r$ into (35), we obtain

$$f(kr)=f(r)\,f(\lambda). \tag{36}$$

Differentiation with respect to λ yields

$$r\,f'(\lambda r)=f(r)\,f'(\lambda). \tag{37}$$

Eliminating of $f(r)$ between these two relations, yields

$$\frac{r\,f'(\lambda r)}{f(\lambda r)}=\frac{f'(\lambda)}{f(\lambda)}. \tag{38}$$

As the right hand side of this expression is dependent upon λ only, we may write

$$\frac{f'(\lambda r)}{f(\lambda r)}=\frac{k}{\lambda r}. \tag{39}$$

Hence, in a very general way

$$f(r)=\frac{\alpha}{r^k}. \tag{40}$$

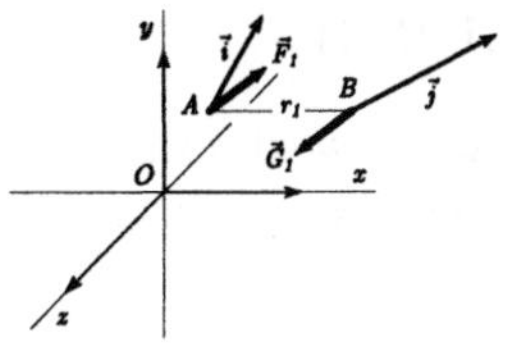

Fig 13a. Scaling functions.

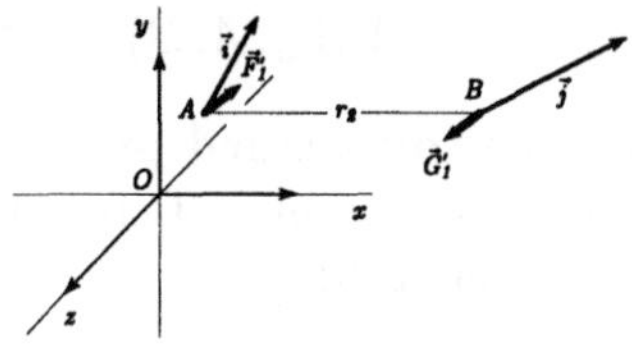

Fig 13b. Ratio of forces depends only on the ratio of distances.

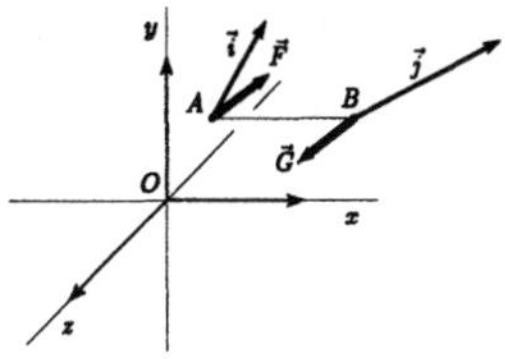

Fig 14a. Reflexivity property.

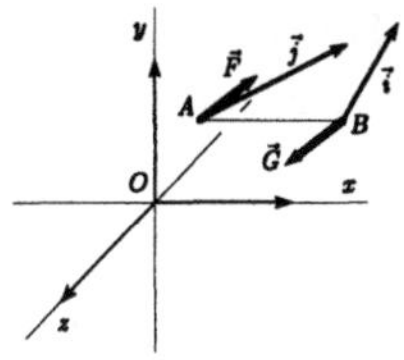

Fig 14b. Exchanging currents under parallel transport.

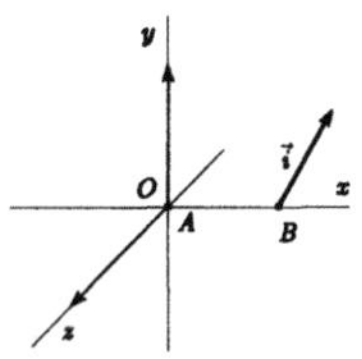

Fig 15. Interaction between a point and a vector.

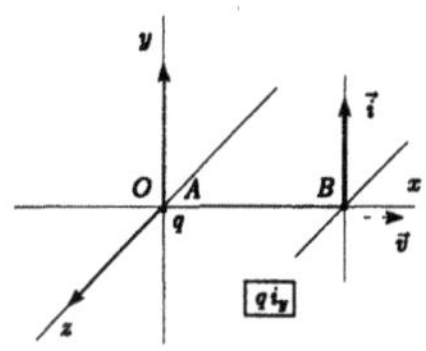

Fig 16a. Interaction $q\,i_y$

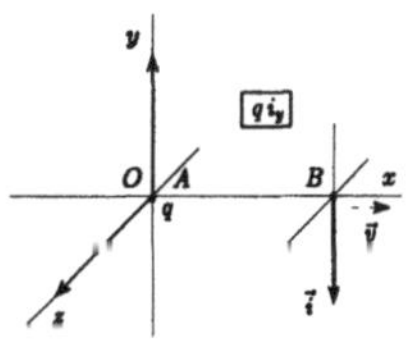

Fig 16b. Rotation about Ox.

The constant α plays no role since it can be absorbed into the arbitrary parameters a, b, c, d, e.

The general expression for the force is now (we have modified the order of appearance of the various terms)

$$\mathbf{F}(\mathbf{i},\mathbf{j};A,B) = \frac{1}{r^k}\ \{a(\mathbf{i}\cdot\mathbf{u})(\mathbf{j}\cdot\mathbf{u})\,\mathbf{u} + b\,(\mathbf{i}\cdot\mathbf{j})\,\mathbf{u} + c\,((\mathbf{i}\cdot\mathbf{u})\,\mathbf{j} + (\mathbf{j}\cdot\mathbf{u})\,\mathbf{i})\} + \frac{1}{r^k}\ \{d\,\mathbf{i}\times\mathbf{j} + e[\mathbf{i},\mathbf{u},\mathbf{j}]\,\mathbf{u}\}\,. \tag{41}$$

A short comment is in order. The first three terms of $\mathbf{F}$ are symmetric with respect to $\mathbf{i}$ and $\mathbf{j}$, and antisymmetric with respect to $\mathbf{u}$. The last two terms on the contrary are antisymmetric with respect to the currents and symmetric with respect to $\mathbf{u}$. As a consequence the force satisfies

$$\mathbf{F}(\mathbf{i},\mathbf{j};A,B) = -\mathbf{F}(\mathbf{j},\mathbf{i};B,A), \tag{42}$$

thus satisfying **Newton's third principle** in its weak form.

Only the first three terms satisfy a **reflexivity** property which we may write as (Fig. 14a and 14b)

$$\mathbf{F}(\mathbf{i},\mathbf{j};A,B) = \mathbf{F}(\mathbf{j},\mathbf{i};A,B), \tag{5.13}$$

that is, permuting the currents under a parallel transport will not change the value of the force. By taking the full form of $\mathbf{F}$ it is therefore possible to imagine a system that is not reflexive, although still in agreement with Newton's third principle.

6 Interaction Between a Point and a Vector

Consider now Fig. 15 representing an interaction between a mass q and a current $\mathbf{i}$. As before there are 12 combinations of q and $-q$ with i_x, $-i_x$, i_y, i_z and $-i_z$.

Using again ($\mathbf{R}$), i.e. the fact that the force is invariant under any rotation about $\mathbf{e}_x$, we have

$$q|y = q|z = q| - y = q| - z, \qquad -q|y = -q|z = -q| - y = -q| - z. \tag{43}$$

Since the odd property $\mathbf{O}$ of the force is also required, we may write

$$-q|\alpha = q| - \alpha = -(q|\alpha), \tag{44}$$

whatever the value of $\alpha = x, y, z$. Summing up these contributions we end up with only two terms

$$q|x, \qquad q|y, \tag{45}$$

and the force is

$$\mathbf{F}(q,\mathbf{i};A,B) = \phi_a\, qi_x\, \mathbf{n}_a + \phi_b(r)\, qi_y\, \mathbf{n}_b. \tag{46}$$

For the same reasons as before, the first component qi_x may induce a motion only along $\mathbf{e}_x$ and therefore $\mathbf{n}_a \equiv \mathbf{u}$. On the other hand (see Fig. 16a), if qi_y induces a motion along $\mathbf{u}$, then a rotation of the system about $\mathbf{e}_x$ leads to Fig. 16b, that is equivalent to $q| - y$. Since this would change the sign of the force according to $\mathbf{O}$, such a motion is ruled out. We are then left with the two possibilities that the motion be along $\mathbf{e}_y$ or $\mathbf{e}_z \equiv \mathbf{u} \times \mathbf{e}_y$. Hence

$$\mathbf{F}(q,\mathbf{i};A,B) = \phi_a\, qi_x\, \mathbf{u} + \phi_b(r)\, qi_y\, \mathbf{e}_y + \phi_{b'}(r)\, qi_y\, \mathbf{u} \times \mathbf{e}_y. \tag{47}$$

Since a rotation about $\mathbf{e}_x$ is always possible, we may add the terms obtained by the substitution $y \to z$.

We have therefore

$$\mathbf{F}(q,\mathbf{i};A,B) = \phi_a\, qi_x\, \mathbf{u} + \phi_b(r)\, (qi_y\, \mathbf{e}_y + qi_z\, \mathbf{e}_z) + \phi_{b'}(r)\, (qi_y\, \mathbf{e}_z - qi_z\, \mathbf{e}_y). \tag{48}$$

After a few manipulations the force becomes

$$\mathbf{F}(q,\mathbf{i};A,B)=\frac{1}{r^k}\,\{a'(\mathbf{i}\cdot\mathbf{u})\,\mathbf{u}+b'\,q\,\mathbf{i}+c'\,q\;\mathbf{i}\times\mathbf{u}\}\,. \tag{49}$$

It is clear that the last expression may be obtained from the interaction-force between vectors, by setting $\mathbf{j}=q\mathbf{u}$. By so doing we simply eliminate a possible degree of freedom. In order to be compatible with the vector interaction we may therefore write the general interaction between a point element and a vector as

$$\mathbf{F}(q,\mathbf{i};A,B)=\frac{1}{r^k}\,\{(a+b+c)\,(q\mathbf{i}\cdot\mathbf{u})\,\mathbf{u}+c\,q\mathbf{i}+d\;q\mathbf{i}\times\mathbf{u}\}\,. \tag{50}$$

By the same token the interaction between two points becomes

$$\mathbf{F}(q,q';A,B)=\frac{qq'}{r^k}\,(a+b+2c)\,\mathbf{u}. \tag{51}$$

7 Derivation of the Force from an Energy Principle

It should in principle be possible to derive an expression for the energy of interaction between the two vectors, **i** and **j**, by calculating the work done by the force while moving one of the vectors from infinity towards the other along an arbitrary curve. We always have

$$d\mathbf{r}=r\,d\mathbf{u}+\mathbf{u}\,dr, \tag{52}$$

and, since $\mathbf{u}\cdot d\mathbf{u}=0$, obtain

$$\begin{aligned}&\oint\mathbf{F}(\mathbf{i},\mathbf{j},A,B)\cdot d\mathbf{r}=\\ &\oint\frac{1}{r^k}\,\{a(\mathbf{i}\cdot\mathbf{u})(\mathbf{j}\cdot\mathbf{u})\,dr+b\,(\mathbf{i}\cdot\mathbf{j})\,dr+2c\,((\mathbf{i}\cdot\mathbf{u})(\mathbf{j}\cdot\mathbf{u})\;dr)+(e-d)\;[\mathbf{i},\mathbf{u},\mathbf{j}]\,dr\}\\ &+\oint\frac{c}{r^{k-1}}\,((\mathbf{i}\cdot\mathbf{u})(\mathbf{j}\cdot d\mathbf{u})+(\mathbf{j}\cdot\mathbf{u})(\mathbf{i}\cdot d\mathbf{u}))+\oint\frac{d}{r^{k-1}}(\mathbf{i}\times\mathbf{j})\cdot d\mathbf{u}.\end{aligned} \tag{53}$$

The first two integrals are easy to calculate since the integration over dr is independent of the other variables. We have

$$\begin{aligned}&\oint\mathbf{F}(\mathbf{i},\mathbf{j};A,B)\cdot d\mathbf{r}=\\ &\frac{1}{r^{k-1}}\,\{a(\mathbf{i}\cdot\mathbf{u})(\mathbf{j}\cdot\mathbf{u})+b\,(\mathbf{i}\cdot\mathbf{j})+2c\,((\mathbf{i}\cdot\mathbf{u})(\mathbf{j}\cdot\mathbf{u}))+(e-d)\;[\mathbf{i},\mathbf{u},\mathbf{j}]\}\\ &+\oint\frac{c}{r^{k-1}}\,((\mathbf{i}\cdot\mathbf{u})(\mathbf{j}\cdot d\mathbf{u})+(\mathbf{j}\cdot\mathbf{u})(\mathbf{i}\cdot d\mathbf{u}))+\oint\frac{d}{r^{k-1}}(\mathbf{i}\times\mathbf{j})\cdot d\mathbf{u}.\end{aligned} \tag{54}$$

Whenever the path of integration is arbitrary, the last two integrals seem intractable. However, if $c=d=0$, we can proceed to an expression for the energy. In this case only

$$E=\frac{1}{r^{k-1}}\,\{a(\mathbf{i}\cdot\mathbf{u})(\mathbf{j}\cdot\mathbf{u})+b\,(\mathbf{i}\cdot\mathbf{j})+e\,[\mathbf{i},\mathbf{u},\mathbf{j}]\}\,. \tag{55}$$

By calculating $\mathbf{F} = -\nabla E$ we recover that part of the interaction force between vectors which corresponds to this choice of constants. Note that the first two terms of this last relation are the usual terms of choice for describing the mutual interaction between spinning particles in quantum mechanics (see Messiah 1960).

References

Aspden, H. (1967): *Physics Unified.* (Sabberton Publications, Southampton).
Maxwell, J.C., (1954): *A Treatise on Electricity and Magnetism* (Dover, New York)
Messiah, A. (1960): *Mécanique Quantique* (Dunod, Paris), **vol 2**

Cellular Automaton Explosion Models

Jean Perdang

Institute of Astronomy, Madingley Road, Cambridge CB3 OHA, UK, and Institut d'Astrophysique, 5, Avenue de Cointe, B-4000 Liège, Belgium

Abstract. We translate a standard approach of Field Theory into the framework of Cellular Automata (CA) to construct a parametrised class of stochastic 2D CA models simulating general explosion phenomena. The models exhibit morphological instabilities and phase transitions from bounded to unbounded growth.

1 Introduction

The formal method of construction of a field theory proceeds as follows. (1) A configuration space is selected. (2) A field format (scalar field, vector field, tensor field,...) is chosen. (3) Mathematical constraints are introduced to specify the nature of the field equations (PDEs) : (3.a) differential order of the PDEs; (3.b) algebraic degree of the PDEs; and (3.c) symmetries. Depending on the interpretation of the field, the final equations subsume as a rule several physical situations.

As an illustration take a scalar field Φ over the 3D Euclidean space and over time. For order 2 and degree 1, and for homogeneity and isotropy of space and time, the final field equation then becomes, in standard notations

$$a\,\partial_t{}^2\Phi \;+\; b\,\partial_r{}^2\Phi \;+\; c\,\Phi \;=\; d. \tag{1.1}$$

Depending upon the choice of the free parameters a, b, c, d, this equation describes the Newtonian gravitational field equation, the scalar wave equation in a homogeneous medium, the Klein-Gordon equation etc.

The present paper presents a formal approach to stochastic CA modelling which replicates the above steps of construction. (1′) A lattice space of specific dimension and geometry is chosen. (2′) The state of each lattice cell is represented by a z-state CA field variable (values 0,1,..., or $z-1$). (3′) The CA evolution rules are further constrained by: (3′.a) a neighbourhood of specific size and geometry; (3′.b)conservation laws etc.; and (3′.c) symmetries. The construction consists in listing all individual evolution rules consistent with (3′) and by attaching transition probability parameters with each rule. In this way an f-parameter family is obtained. As in a field theory construction the procedure of generating the CA rules is formal. A specific model of the family may then mimic a variety of distinct physical phenomena, depending upon the interpretation of the states.

Formal systematic constructions of CA models have been investigated in Wolfram (1983) for 2-state 1D CA and in Packard and Wolfram (1985),

Packard (1986) for 2D CA. The approach of these authors does not deal with parameter-dependent classes of models, but rather with the construction of an exhaustive list of a basis of distinct *deterministic* models consistent with constraints of type (3′). The CA models we construct are *stochastic families.* Stochasticity introduces as a new element the non-reproducibility of an individual evolutionary sequence and, therefore, it can be viewed as providing an approximate description of chaotic evolution (Perdang 1995).
Although formal, our construction relies on a specific physical picture. We are dealing with a structural pattern – the growth cluster – in some medium. Growth at a given point depends on the nature of the neighbourhood of the point. We restrict our specific investigation to 2D models of 3-state automata in which a cell is either 'empty' (state value 0), 'active' (1) or 'inactive' (2) (cf the models describing the propagation of infectious diseases; Pandey 1991; Stauffer and Pandey 1992; Boccara at al. 1994; or star formation; Seiden and Schulman 1990). Growth is intrinsic, due to an instability (cf Langer 1980), in contrast with growth of the DLA models and their extensions (Stanley and Ostrowski 1986; Meakin 1995; Erzan at al. 1995).
We believe that the model family investigated here may be relevant for the simulation of explosive fluid dynamic phenomena when information on a single macroscopic magnitude is required. In fact it happens not infrequently that the long time behaviour of a fluid system, defined by the solution of the equations of hydrodynamics, is not accessible numerically, either because it involves too much computer time or because the solution is intrinsically chaotic and hence predictability is lost beyond the Lyapunov horizon. As an instance suppose we are just interested in the development of an overall density pattern in a given flow problem. When formulated in conventional fluid dynamics, the analysis involves all of the other properties as well. It is the evaluation of these – unwanted – fields that takes up a prohibitive amount of computer time. What we need in such a case is a computationally simple scheme that deals only with the observables of direct interested to us.
The fluid dynamical description of Type II Supernova explosions eventually leading to complex filamentary or spongy structures (cf the Crab nebula) is an instance of this situation. While an incipient development of irregular structure can be followed by hydrodynamic models (Fryxell at al. 1991; Todo 1991; Uchida 1991), the long time evolution remains inaccessible.

2 Basic 2D Families B(p;t_a)

Select a 2D lattice L of cells (requirement 1′). Except possibly for a finite number of 'singular' cells, the cells have same topological structure (same number of vertices, edges,...); we refer to the latter as the generic cells of the lattice. We require that there exist a collection of lattice transformations T such that any generic cell c can be transformed into any other generic cell c' by an element T of T, $c' = Tc$. The transformations T consist in shifts of the cell vertices and shifts combined with deformations of the cell edges; they preserve the topological cell features. For 'singular' cells the transformation requirement may be violated (case of central cell in a circular lattice). The cellstate at (integer) time, $t \in \mathrm{Z}$, of each cell c, $F(t;c)$, is a nonnegative integer ζ=0,1,..., or $z-1$ (2′); state 0 is called as empty. For the present purposes we understand by neighbourhood of a cell c, N(c), an ordered collection of γ+1 cells containing c itself, $(c,c_1,c_2,...,c_\gamma)$, such that any cell c_k in the neighbourhood is connected to the central cell by a chain of direct neighbours (at least one common vertex), (c,c_a), (c_a,c_b),...,(c_r,c_k). A pair of cells having at least a finite fraction of a cellside in common is dubbed a contiguous pair. A neighbourhood obeys the following condition: If c' is the cell c transformed by T, then the neighbourhood of cell c', $\mathrm{N}(c') = (c',c'_1,c'_2,..,c'_\gamma)$, is obtained from N($c$) by the same T, $\mathrm{N}(c') = \mathrm{N}(Tc) = T\,\mathrm{N}(c)$. We refer to the neighbourhood from which the central cell is discarded as the environment of cell c, N*(c).

Formal list of evolution rules.

At the present formal level we take account of a single special constraint (3′): (α)*An empty cell c at time t remains empty at time* $t+1$ *if the environment of c at time t is in the trivial empty configuration* K_0 (empty environment); equivalently the probability per timestep of the 0–0 transition is then 1:

$$F(t;c)=0 \wedge \forall c' \in \mathrm{N}{*}(c):\ F(t;c')=0 \rightarrow F(t{+}1;c)=0;\ p_{oo}(\mathrm{K}_0)=1. \quad (2.1)$$

(β) *If cell c at time t is in state* ζ *and the environment configuration of c is* K, *then at time* $t+1$ *the cellstate transforms into* ζ' *with probability* $p_{\zeta\zeta'}(\mathrm{K})$

$$F(t;c)=\zeta \wedge \mathrm{conf}_t\mathrm{N}{*}(c)=\mathrm{K} \ \rightarrow\ F(t{+}1;c)=\zeta' \text{ with prob } p_{\zeta\zeta'}(\mathrm{K}). \quad (2.2)$$

The environment configuration of cell c at time t conf_t N*(c), is the ordered collection of cellstates of the environment

$$\mathrm{conf}_t\mathrm{N}{*}(c) \ = \ (F(t;c_1),F(t;c_2),...,F(t;c_\gamma)). \quad (2.3)$$

($F(t;c_k) = \zeta_k$ = 0,1,..., or $z-1$, k=1,2,...,γ). We have $\mu = z^\gamma$ distinct environment configurations K_0, K′, K″, ..., $\mathrm{K}^{(\mu-1)}$ (including the empty configuration); the total number of probabilities $p_{\zeta\zeta'}(\mathrm{K})$ is $\mu \times z^2$. Among the latter we have the obvious constraints

$$\sum_{\zeta'} p_{\zeta\zeta'}(\mathrm{K}) \ = \ 1,\ \zeta=0,1,...,z-1, \mathrm{K}=\mathrm{K}_0,\mathrm{K}',\mathrm{K}'',...,\mathrm{K}^{(\mu-1)}. \quad (2.4)$$

Hence $\mu \times z$ probabilities, which we choose as the diagonal elements $p_{\zeta\zeta}(\mathrm{K})$, are not independent. Rule (α) further determines the z probabilities $p_{0\zeta}(\mathrm{K}_0)$, supplying $z-1$ extra constraints

$$p_{00}(\mathrm{K}_0) \ = \ 1 \text{ and } p_{0\zeta}(\mathrm{K}_0) \ = \ 0 \text{ for } \zeta=1,2,...,z-1. \quad (2.5)$$

Hence the general result: The maximum of the number $f(z;\gamma)$ of free probability parameters of a z-state CA with environment size γ, in the absence

of physical requirements (besides rule α), and in the absence of symmetry constraints, is max $f(z;\mu) = \mu \times z^2 - \mu \times z - (z-1)$ (not directly dependent on the dimensionality of the lattice).
Let S be the symmetry group, of elements S, representative of the physics to be simulated by our family. Two environment configurations K′, K″ are equivalent under S, $K' \simeq_S K''$, and belong into the same equivalence class [K] if there is an element S of S transforming K′ into K″. We then naturally have

$$\text{for } K', K'', ... \in [K] \; : \; p_{\zeta\zeta'}(K') \;=\; p_{\zeta\zeta'}(K'') \;=\; ... \;\equiv\; p_{\zeta\zeta'}([K]) \qquad (2.6)$$

Let μ_S be the number of equivalence classes $[K_0]$, $[K_1]$, ...,$[K_{\mu_S}]$ under S ($[K_0]$, equivalence class of empty configuration). Then the number $f(z;\mu_S)$ of free probability parameters reduces to $f(z;\mu_S) = \mu_S \times z^2 - \mu_S \times z - (z-1)$.

Particular physical requirements. A survey of the effect of the rules in the general z-state CA remains prohibitive, even for z small. Thus the square lattice with Moore neighbourhood and $z = 3$ involves 39 364 free probabilities (in the absence of symmetries). Therefore, in the remainder we limit ourselves to $z = 3$ with extra constraints such that the third state plays a passive role only.
(γ) *An active cell (state 1) at time t remains active at step t+1 if it has been active for less than t_a timesteps*

$$F(t;c) = 1 \;\wedge\; F(t-t_a;c) = 0 \;\rightarrow\; F(t+1;c) = 1. \qquad (2.7)$$

(δ) *An active cell at time t becomes passive (state 2) at step t + 1 if it has been active during t_a timesteps*

$$F(t;c) = 1 \;\wedge\; F(t-t_a;c) = 1 \;\rightarrow\; F(t+1;c) = 2. \qquad (2.8)$$

(ϵ) *A passive cell at timestep t remains passive at step t + 1*

$$F(t;c) = 2 \;\rightarrow\; F(t+1;c) = 2. \qquad (2.9)$$

(ζ) *An empty cell cannot transform into a passive cell.*
A passive state remains passive forever. If we lump together active and passive states into 'excited states', then excited states remain excited forever. These rules then determine the following 6μ ($6\mu_S$ respectively) probabilities (and $c_{1,2} = 4\mu$, or $4\mu_S$, new constraints) :

$$\begin{aligned} p_{10}(K) \;&=\; 0 \text{ and } p_{11}(K) \;=\; 1 \text{ and } p_{12}(K) \;=\; 0 \text{ (alternative } \beta), \\ p_{10}(K) \;&=\; 0 \text{ and } p_{11}(K) \;=\; 0 \text{ and } p_{12}(K) \;=\; 1 \text{ (alternative } \gamma), \\ p_{20}(K) \;&=\; 0 \text{ and } p_{21}(K) \;=\; 0 \text{ and } p_{22}(K) \;=\; 1 \text{ (rule } \epsilon). \end{aligned} \qquad (2.10)$$

Rule (ζ) is a rule of interdiction expressing $c_0 = \mu$-1 new constraints

$$p_{02}(K) \;=\; 0, \text{ K any configuration (rule } \zeta). \qquad (2.11)$$

We are then left with $f_c(3;\mu) = f(3;\mu) - c_0 - c_{1,2} \equiv \mu - 1$ free probabilities $p(\mathrm{K}) \equiv p_{01}(\mathrm{K})$, K any configuration $\neq \mathrm{K}_0$. The collection of free probabilities is symbolised by an array ordered according to a conventional order of the equivalence classes

$$\mathbf{p} = (p([\mathrm{K}'], p([\mathrm{K}''], ..., p([\mathrm{K}^{(\mu-1)}]) \equiv (p_1, p_2, ..., p_f), \qquad (2.12)$$

the equivalence classes of the environment configurations being ordered according to some rule. For a Moore neighbourhood $f_c(3;\mu) = 6560$. To reduce this number further, an exchange symmetry among the 'inert states' 0 and 2 in the environment is imposed.

(η) *For the purposes of evolution the 'inert states' are indistinguishable.*

The number of environment configurations under exchange is $\mu* = 2^\gamma$. If we choose as physical symmetry the permutation group P of the cells of the environment, then we have the coarse classes [0],[1],...,[γ] containing 0 (trivial class), 1, 2,..., and γ active cells respectively, or $\mu*_P = \gamma + 1$; the number of free parameters is equal to the size of the environment, γ.

A family of models conforming to rules (γ–η) defines a growth process. The probability array **p** and the lifetime of an active cell, t_a, determine the geometric nature of the pattern of excited cells.

Basic families on square lattice with Moore neighbourhood.

The planar regular square lattice possesses a local C_{4v} symmetry inherited by the Moore neighbourhood (central cell c plus 'ring' of $\gamma = 8$ cells surrounding c). We label the latter (1,2,...,8), starting with the right middle ring cell and turning counterclockwise; odd and even labels refer to 'edge' (e) and 'vertex' (v) cells respectively.

In Table 1 we show three classifications, referred to as coarse, intermediate and fine. The total number of active cells of the environment configuration, n = 0,1,... or γ, defines a coarse class [n] (first column) of multiplicity $\mu(n)$ (number of geometrically distinct environment configurations). The third column (π_ν) gives the intermediate classification (inequivalence of any two environment configurations of same coarse class with different groupings of active cells). The classes correspond to the partitions of n into ν groups; the notation $1^{\nu_1}2^{\nu_2}3^{\nu_3}$..., with $\nu_1+\nu_2+\nu_3+...=\nu$, designates a class containing ν_1 isolated active cells, ν_2 pairs of active cells separated from other active cells by inert cells, ν_3 chains containing 3 active cells etc.; $\mu(n;\pi_\nu)$ is the multiplicity of the intermediate class (number of geometrically distinct configurations under exchange). The fifth column (π_s) gives the fine classification with equivalence of two environment configurations under the local lattice symmetry C_{v4}; the class symbol is s1..10..01...10..1...1 (string of substrings of 1s alternating with substrings of 0s; first substring: chain of l_1 active cells starting at edge cell c_1, if $s = e$; or at vertex cell c_2, if s=v; followed by l_2 inert cells, l_3 active cells,...; prefixed v or e omitted if configurations v... and e... equivalent; $\mu(n;\pi_\nu;\pi_s)$, multiplicity).

Coarse		Intermediate		Fine	
$[n]$	$\mu(n)$	π_ν	$\mu(n;\pi_\nu)$	π_s	$\mu(n;\pi_\nu;\pi_s)$
0	1	0	1	0	1
1	8	1^1	8	$v1$	4
				$e1$	4
2	28	1^2	20	$v101$	4
				$e101$	4
				1001	8
				$v10001$	2
				$e10001$	2
		2^1	8	11	2
3	56	1^3	16	$v10101$	4
				$e10101$	4
				$e101001$	4
				$v101001$	4
		$1^1 2^1$	32	$v1011$	8
				$e1011$	8
				$v10011$	8
				$e10011$	8
		3^1	8	$v111$	4
				$e111$	4
4	70	1^4	2	$v1010101$	1
				$e1010101$	1
		$1^2 2^1$	24	$v1100101$	8
				$e1100101$	8
				1101001	8
		$1^1 3^1$	24	$v11101$	8
				$e11101$	8
				$v111001$	4
				$e111001$	4
		2^2	12	$v11011$	4
				$e11011$	4
				110011	4
		4^1	8	1111	8

Table 1. Classification of environment configurations on a square lattice (Moore neighbourhood), $n = 0,1,2,3$ and 4; alternatives $n = 5,6,7$ and 8 follow from duality with $n = 3,2,1$ and 0 respectively

For the square lattice with Moore neighbourhood we have $f_c(3;\mu*) = 255$ free probabilities $\mathbf{p}$ in the absence of extra symmetries besides exchange. Under

the C_{v4} symmetry the number of free probabilities drops to 50. The partition symmetry keeps 18 and the coarse classification eight free probabilities.
Basic families on triangular lattice with Moore neighbourhood.
By Moore neighbourhood of cell c we understand the central cell c, the three contiguous cells, and the three cells which have just a vertex in common with c. The local symmetry is C_{v3}. We have six free probabilities for the coarse, ten for the intermediate and 19 for the fine classification. In the absence of extra symmetries besides exchange we have 63 free probabilities.
There is a more basic difference between both lattices. We can paint the cells of the square lattice in black and white to obtain a checkerboard pattern. By a translation the black sublattice and the white sublattice can then be superposed. We can likewise paint the cells of the triangular lattice in black and white such that any black cell has as contiguous cells only white cells and vice versa; then all black cells (triangles) 'point up' while all white cells 'point down' so that the two sublattices cannot be superposed by a mere translation. On the triangular lattice all cells of the environment of a black cell are white and vice versa. On the square lattice the environment of a black cell contains four white cells and four black cells; there are evolution laws plus initial conditions for which both sublattices are independent. This is not the case on the triangular lattice.

3 Phase Transitions and Morphological Instabilities

We analyse several general properties of the growth cluster at time t, $C(t)$, t=0,1,2,...(collection of excited cells evolving from an initial cluster $C(0)$ for families obeying $\alpha - \eta$). These evolution rules are such that

$$\eta\ C(t) \supseteq C(t+1) \supseteq C(t), \tag{3.1}$$

$$\eta C(t) = \cup_{c\in C(t)} N(c); \tag{3.2}$$

(η, enlargement operation, such that $\eta C(t)$ represents the union of the neighbourhoods of all (excited) cells of $C(t)$). Iteration of (3.1) yields

$$\eta^t\ C(0) \supseteq C(t) \supseteq C(0). \tag{3.3}$$

For any neighbourhood (of size, or cardinality, ≥ 2) the cardinality of the iterated enlargement obeys

$$\lim_{t\to\infty} \mathrm{card}[\eta^t C(0)] = \infty. \tag{3.4}$$

For the basic families $B(\mathbf{p};t_a)$ listed above, for any initial $C(0)$, $\lim_{t\to\infty}\ \eta^t$ $C(0)$ covers the whole lattice.
Let $B(\mathbf{p};t_a)$ be a representative of activity time t_a and f-array of probabilities $\mathbf{p} = (p_1,p_2,...,p_f)$ of a basic family $B(\mathbf{p};t_a)$ (for given lattice, neighbourhood, and symmetries). We symbolise the cluster growth by

$$C(t+1) = B(\mathbf{p};t_a)C(t); \quad C(t) = B(\mathbf{p};t_a)^t C(0). \tag{3.5}$$

The probability array is a position vector in an f-dimensional closed unit cube $[0,1]^f$; the 2^f vertices define the subset of deterministic models; any other point of the cube defines a stochastic model. For the latter any two runs a and b starting with the same initial condition C(0) generate two distinct evolutionary sequences, say

$$a : \{C(0), C(1), ..., C(T), ...\} \ ; \quad b : \{C(0), C'(1), ..., C'(T), ...\}. \tag{3.6}$$

If $\mathbf{p} = \mathbf{0}$, then we have the empty model, $B(\mathbf{0};t_a)$, with the property

$$\forall C(0): \ B(\mathbf{0};t_a)C(0) = C(0) \ \text{ i.e. } \ B(\mathbf{0};t_a) \equiv \mathit{1}. \tag{3.7}$$

If $\mathbf{p} = \mathbf{1}$, then we have the covering or universal model of the family, $B(\mathbf{1};t_a)$,

$$\forall C(0): \ B(\mathbf{1};t_a)C(0) = \eta C(0) \ \text{ i.e. } \ B(\mathbf{1};t_a) \equiv \eta. \tag{3.8}$$

For any finite activity time t_a, these models exemplify two classes of growth, namely bounded and unbounded growth. For a representative $B(\mathbf{p};t_a)$ of family $\mathrm{B}(\mathbf{p};t_a)$ and an initial cluster C(0) the evolutionary sequence generated in a first run (Eq. 3.6, a) either keeps growing indefinitely or there exists a halting time, t_H, beyond which the clusters stop growing : $C(t_H+k) \equiv C(t_H)$, k any positive integer ($> t_a$). In a second run (Eq. 3.6, b) we may observe unbounded growth or bounded growth with halting time, $t_{H'} \neq t_H$, if the first run was bounded.
To define a reproducible characterisation of the sequences generated by the probabilistic models of the family $\mathrm{B}(\mathbf{p};t_a)$ we adopt a statistical average to distinguish bounded and unbounded growth. For a (nontrivial) initial configuration C(0) and model $B(\mathbf{p};t_a)$ compute a large number M of sequences over a large enough time T. Let m be the number of bounded sequences. Then the (ensemble averaged) growth of the initial cluster C(0) and model $B(\mathbf{p};t_a)$ is said to be bounded [unbounded] if

$$P_b(\mathbf{p};t_a;C(0);T) = \lim_{M\to\infty} \frac{m}{M} > \frac{1}{2} \ [\leq \frac{1}{2}\ ]. \tag{3.9}$$

For $T \to \infty$ $P_b(\mathbf{p};t_a;C(0);T)$ represents the probability of bounded growth. This probability can be evaluated analytically by enumerating all alternatives of growth at steps t = 1,2,....; for T finite this function is a polynomial in $\mathbf{p}$ and hence continuous in each component p_j, j = 1,2,...,f.
Given a fixed (finite) activity time, t_a, and a nontrivial initial cluster, C(0), the probability f-cube carries two complementary regions, $R_B(C(0))$ and $R_U(C(0))$, separated by an $(f-1)$-dimensional (algebraic) surface defined by

$$P_b(\mathbf{p};t_a;C(0);T) = 1/2, \tag{3.10}$$

with the origin $\mathbf{p} = \mathbf{0}$ in $R_B(C(0))$ (zone of bounded growth) and the vertex $\mathbf{p} = \mathbf{1}$ in $R_U(C(0))$ (zone of unbounded growth).

Bounded-Unbounded Growth Phase Transitions.
For a given family $B(\mathbf{p}; t_a)$ of fixed activity time, t_a, and for an initial cluster C(0), consider two models, $B(\mathbf{p}_B; t_a)$ and $B(\mathbf{p}_U; t_a)$, of probability vectors lying in the bounded and the unbounded zone of the probability cube respectively. Consider further an arbitrary smooth curve parametrised by p joining these two points

$$\mathbf{p} = \mathbf{h}(p), \; p \in [0,1], \quad \text{with} \; \mathbf{h}(0) = \mathbf{p}_B \; \text{and} \; \mathbf{h}(1) = \mathbf{p}_U. \tag{3.11}$$

Then the evolutionary sequences generated from the initial condition C(0), as we let p increase from 0 to 1, undergo a transition from bounded to unbounded growth at a critical parameter value $p = p_c$. Smoothness implies that the curve cuts (3.10) a first time at a critical point, $p = p_c$, as p increases from 0 to 1; at this point we have a transition (cf a percolation phase transition, Stauffer 1985; or self-organised criticality, Bak et al. 1987). Near criticality (p close to p_c) the size $S(p)$ of the asymptotic cluster $C(t)$, $t \to \infty$, obeys

$$S(p) \propto \frac{1}{(p_c - p)^{\alpha}}, \quad p < p_c. \tag{3.12}$$

Morphological Instabilities.
We understand by a subfamily $F(\underline{p}; t_a)$ of $B(\mathbf{p}; t_a)$ a k-parameter family $B(\mathbf{p}_O; t_a)$ of the full f-parameter family $B(\mathbf{p}; t_a)$, defined by an ordered subset $\underline{p} = (p_a, p_b, ..., p_s)$ of k among the probabilities of the full f-array, the remaining $f - k$ components being set zero:

$$\mathbf{p}_O = (0, ...0, p_a, 0, ...0, p_b, 0, ..., 0, p_s, 0, ...0) = (p_{O1}, p_{O2}, ..., p_{Of}); \tag{3.13}$$

the array $\underline{p}$ is such that the representative $F(\underline{1}; t_a)$ secures unbounded growth of an initial single cell $C(0) \equiv c$. This model is again referred to as the covering model of the subfamily. In general the operation $F(\underline{1}; t_a)$ is not equivalent with the enlargement η.
In fact suppose the lattice L is such that (i) the cells can be painted with q colours, a, b,..., any contiguous pair of cells having different colours; and (ii) there exists a proper sublattice L* of the lattice whose cells are coloured with $r < q$ distinct colours m,n,... among the colours a,b, ... such that $C(t)$, $t = 1,2,...$ generated from a single seed c on L*, remains on L*. The r-colour sublattice L* of L is then an invariant sublattice under the subfamily $F(\underline{p}; t_a)$, since for any individual model $F(\underline{p}; t_a)L* = L*$.
Define next the global density of growth cluster, $C(t)$ at time t, by $\rho(t) = S(t)/N(t)$ ($S(t)$, number of excited cells of cluster; $N(t)$, number of excited plus empty cells in zone of lattice L enclosed by outer connected border of $C(t)$). Define further a local density, $\rho(t, c) = \sigma(t, c)/(\gamma + 1)$ ($\sigma(t, c)$, total number of excited cells in $N(c)$ at time t). For growth with an invariant sublattice L*, $\rho(t, c) \leq \sigma_M/(\nu + 1)$, ($\sigma_M$, number of cells in $N(c) \cap L*$). In the presence of an invariant sublattice L* the global density of the asymptotic

cluster C(t), $t \to \infty$, generated by model $F(\underline{p}; t_a)$ from any initial cluster C(0) on L* cannot exceed the value $\rho_{\max} = r/q$.

On the square lattice with C_{4v} symmetry consider the subfamily $F^v(p_v; t_a)$, with p_v the growth probability of the vertex configuration $v1$ (Table 1). Label (colour) the cells of the lattice L alternatively 1 and 2. Then each sublattice L_1* and L_2* of cells 1 and 2 respectively is invariant under any model $F^v(p_v; t_a)$ of this family; the maximum global density is 1/2; the maximum local density at c is 5/9 or 4/9, depending on whether c is on the invariant lattice, say L_1*, or not.

On our triangular lattice under C_{3v} symmetry take the subfamily $F^v(p_v; t_a)$, with p_v the growth probability of a vertex configuration $v1$ (cell having with central cell a vertex only in common). Label a first row of cells of the triangular lattice L 1,2,3,4,1,2,3,4,...; a second row 4,3,2,1,4,3,2,1,... with 3 being contiguous to 1 in the first row; and again a third row 1,2,3,4,1,2,4,... with 4 contiguous to 2 in the second row; etc. With this labelling each 'coloured' sublattice L_1*, L_2*, L_3*, L_4*, is invariant under *any* model $F^v(p_v; t_a)$; maximum global density: 1/4; maximum local density: 4/7 or 1/7 if c on invariant sublattice of growth, or on one of the other three lattices; average maximum local density:$1/4 \equiv 1/4(4/7 + 3 \times 1/7)$.

Let F($\underline{p}; t_a$) be a subfamily of the basic family B($\mathbf{p}; t_a$), with invariant sublattice $\overline{\text{L}}$*. Let C(t), t = 1,2, ... be a growth pattern generated from an initial cluster C(0) on L*, with representative $F(\underline{p}; t_a)$ (= $B(\mathbf{p}_0; t_a)$ of B($\mathbf{p}; t_a$)); any C(t) then remains on L* and has a low density (cf $\rho_{\max}$).

Perturb model $F(\underline{p}; t_a)$, i.e., substitute for $F(\underline{p}; t_a)$ a generic representative $B(\mathbf{p}; t_a)$ of B($\mathbf{p}; t_a$), such that the probability components p_{oj} become $p_j = p_{oj} + \epsilon_j$, $|\epsilon_j| \ll 1$. The perturbed model $B(\mathbf{p}; t_a)$ then remains close to the original $F(\underline{p}; t_a)$, or the probability vectors $\mathbf{p}$ and $\mathbf{p}_0$ in the f-dimensional unit cube differ only slightly. By construction the basic family has no invariant sublattice; by genericity this property is inherited by $B(\mathbf{p}; t_a)$. A sequence C$'$(t), t=1,2, ... generated from the initial cluster C(0) on the invariant sublattice L* of F($\underline{p}; t_a$), with the perturbed model $B(\mathbf{p}; t_a)$ is not restricted to the sublattice $\overline{\text{L}}$*, and hence not subject to the low density constraint.

A growth sequence generated by $F(\underline{p}; t_a)$ of a subfamily is said to be morphologically unstable if it is on an invariant sublattice L* of L. Any generic arbitrarily close model $B(\mathbf{p}; t_a)$ of the basic family then generates a sequence of an essentially different aspect. The corrections ϵ_j are bound to enter at some later stages in the evolution to produce high local densities (ρ_{max} = 1), thereby inducing characteristic inhomogeneities. Such structures are reminiscent of cloudy and veiled patterns as encountered in the natural world. At the critical surface we have large pattern fluctuations since for half of the sequences the growth is bounded and for the other half it is unbounded. These large fluctuations survive close enough to the critical surface ($0.1 \leq P_b(\mathbf{p}; t_a; \text{C}(0); T) \leq 0.9$). Under morphological stability the growth pattern does not change significantly in structure as we let one or several parameters change continuously.

4 Results on Single Seed Growth

We analyse more closely the growth from a single seed c_0

$$\text{for } t = 0: \quad F(t;c) \; = 1 \; [0], \quad \text{if } c \; = \; c_0 \; = \; (0,0) \; [c \neq c_0], \tag{4.1}$$

on our square lattice, for the simplest subfamilies under C_{4v} symmetry. There are two 1-parameter subfamilies, namely one with free probability p_v associated with $v1$, $F^v(p_v; t_a)$, (vertex subfamily), and one with probability p_e associated with $e1$, $F^e(p_e; t_a)$, (edge subfamily). These subfamilies are embedded in the 2-parameter subfamily $F^{ve}(\underline{p}; t_a)$, $\underline{p} = (p_v, p_e)$. The patterns generated by these subfamilies, and their covering models, $F^v(1; t_a)$, $F^e(1; t_a)$, $F^{ve}((1,1); t_a)$, are discussed in Perdang (1996). For these subfamilies the evolution laws obey the C_{v4} symmetry probabilistically, and deterministically in the case of the covering models. Accordingly the growth patterns of the latter preserve the C_{v4} symmetry exactly. The growth patterns of individual realisations of the stochastic models violate global C_{v4}. Symmetry is restored statistically if we substitute to individual clusters at a given time t the ensemble-averaged fuzzy cluster $< C(t) >$.
Since $F^v(p_v; t_a)$ has two invariant sublattices $L_1{}^*$ and $L_2{}^*$, if c_0 is on $L_1{}^*$, then the evolution proceeds on $L_1{}^*$; this growth phenomenon is morphologically unstable. No invariant sublattice exists for $F^e(p_e; t_a)$ or, *a fortiori*, for the embedding subfamily $F^{ve}((p_v, p_e); t_a)$.

Analytical Estimates of the Halting Probability.
Denote by $P_b(T)$ the halting probability over T timesteps at least, for a growth from a single seed, under the vertex subfamily $F^v(p; t_a)$, for t_a=1 ($p \equiv p_v$). At timestep 0 we have a single active cell, c_0, surrounded by four possible growth sites; with t_a=1 the seed cell becomes passive at step 1; the probability of no growth beyond step 1 is then the probability of no growth on the first vertex, (1-p), *and* no growth on the second, ...,*and* no growth on the fourth vertex

$$P_b(1) \; = \; (1-p)^4. \tag{4.2.a}$$

A similar reasoning leads to the the following expressions of $P_b(2)$ and $P_b(3)$:

$$P_b(2) = (1-p)^4[1 + 4p - 2p^2 - 8p^3 + 9p^4 - 4p^5 + 2p^6], \tag{4.2.b}$$

$$P_b(3) = (1-p)^4 \; [1 + 4p + 10p^2 - 16p^3 - 37p^4 - 8p^5 + 288p^6 - 656p^7$$
$$+1007p^8 - 1260p^9 + 1178p^{10} - 740p^{11} + 278p^{12} - 44p^{13} - 4p^{14} + p^{16}]. \tag{4.2.c}$$

Setting $P_b(T) = 1/2$, we obtain the following estimates for $p(T)$

$$p(1) \; = \; 0.1591...; \;\; p(2) \; = \; 0.2772...; \;\; p(3) \; = \; 0.3358... \tag{4.3}$$

Write the halting probabilities in the form

$$P_b(T) = (1-p)^4 P_{n(T)}(p),$$

$$P_{n(T+1)}(p) = P_{n(T)}(p) + O(p^T); \tag{4.4}$$

$T = 1,2,\ldots$. The expression of $P_b(T)$ is then a correct approximation to $P_b(\infty)$, to order 0, 1, 2,...in the multiplicative polynomial for $T = 1,2,3,\ldots$. Keeping the correct terms only we have the estimates for the critical $p(T)$

$$p(1) = 0.1591...;\ p(2) = 0.3138...;\ p(3) = 0.4214\ldots. \tag{4.5}$$

To obtain a simple extrapolation scheme we assume that the ratio

$$\frac{p(T+2)-p(T+1)}{p(T+1)-p(T)} = \lambda(T) \tag{4.6}$$

is independent of T. From (4.3) and (4.5) we then find $p_c \equiv p(\infty) = 0.40$ and 0.63 respectively.

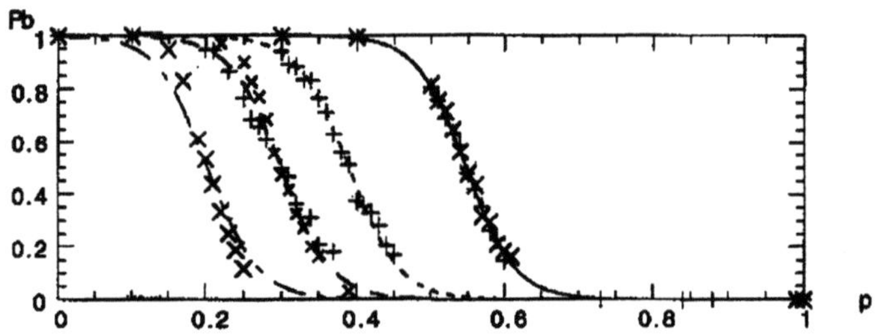

Fig. 1. Experimental estimates of the halting probabilities for the vertex (+) and edge (x) subfamilies, t_a=1,2,3, compared with the analytic profile (4.7)

Numerical Estimates of the Halting Probabilities.
The halting probability $P_b(p)$ was estimated from direct numerical experiments based on 500 realisations for each p value. The results of Fig. 1 (rightmost data points marked (+) for vertex growth, t_a=1) indicate that the following analytical representation holds

$$P_b(p) = \frac{1}{1+\exp\left[\dfrac{(p-p_c)}{\Delta p}\right]} \tag{4.7}$$

(p_c, critical probability; Δp, width of probability interval over which $P_b(p)$ differs significantly from 0 or 1), with a 'range of fuzziness' $p_m = p_c - 3\Delta p < p < p_c + 3\Delta p = p_M$. The numerical estimates of the parameters are : $p_c = 0.549$; $\Delta p = 0.033$; the range of fuzziness (0.45, 0.65) is closely approximated by the analytical estimates 0.40 and 0.63 given above. The unbounded growth patterns ($p > p_M$) display evolving structures reminiscent of a deformation of the dendritic pattern of the covering model $F^v(1; t_a)$. The bounded configurations ($p < p_m$) show irregular cloudy patterns. Representation (4.7) remains adequate for the vertex subfamilies $t_a = 2$ and 3, with $p_c = 0.291$ and 0.204; $\Delta p = 0.029$ and 0.02 (cf Fig. 1).
Estimates of the halting probabilities for the edge subfamily, $\mathrm{F}^e(p_e; t_a)$, were also made for $t_a = 1,2,3$. For $t_a = 1$ the halting probability curve (rightmost data points x of Fig. 1) is found to coincide with the halting probability curve of the vertex subfamily. The reason for this is due to an elementary geometrical equivalence of the subfamilies $\mathrm{F}^v(p; 1)$, and $\mathrm{F}^e(p; 1)$ which essentially implies that any statement about the statistics of one subfamily holds for the other as well. The difference between both growth processes, when viewed on the same lattice L, is that the edge subfamily has a denser aspect than the vertex subfamily. Moreover, vertex growth with $t_a = 1$ is morphologically unstable, while edge growth is stable. Note that the geometric equivalence between vertex and edge growth fails for activity times $t_a > 1$.
Representation (4.7) of the halting probabilities continues to hold for the edge subfamily $\mathrm{F}^e(p; t_a)$ (Fig. 1, data points marked x; $p_c(2) = 0.39$ and 0.30 for $t_a = 2$ and 3).

5 Numerical Simulation of an Observed Growth

The general properties (bounded-unbounded phase transition, morphological instabilities) enable us to anticipate to some extent the parameter values required to model a given type of explosive growth. As an illustration we simulate unbounded growth from a point source, developing in the long run a cloudy inhomogeneous pattern of statistically spherical shape with a locally irregular boundary. This problem can be formulated on both lattices.

Square Lattice.
With highest symmetry of the evolution rules (C_{4v}) the simplest unbounded growth with deformed dendritic border is the sequence $F^v(p; t_a)^t\ c_0$ with $p > p_c(t_a)$. Choose further p as small as possible to avoid patterns reflecting the lattice symmetry. By trial and error t_a, and p, are adjusted to provide a first match with the observed growth. Since this model is morphologically unstable, density inhomogeneities are produced by stabilisation. The subfamily $\mathrm{F}^{ve}((p_v, p_e); t_a)$ is the simplest embedding family for $\mathrm{F}^v(p; t_a)$ with no invariant sublattice. Hence we substitute for $F^v(p; t_a)$ a representative, $F^{ve}((p, \epsilon); t_a)$, as a second approximation. The small edge growth probability ϵ is adjusted to fit the pattern we wish to simulate. An eventual improvement

may be achieved by embedding $F^{ve}((p,\epsilon);t_a)$ in the full basic family B($\mathbf{p}$;t_a) (48 extra components, η_j; previous p and ϵ are kept; cf Perdang 1996 for illustrations).

Triangular Lattice.
With maximal symmetry (C_{3v}) in the evolution rules the sequence $F^v(p;t_a)^t\, c_0$ ($p > p_c(t_a)$) produces unbounded growth with an irregular border. Second and third approximations: $F^{ve}((p,\epsilon);t_a)^t\, c_0$; and $B(\mathbf{p});t_a)^t\, c_0$ (17 new components). These growth patterns are displayed in Fig. 2.
We notice in the first place that the underlying lattice symmetry is not observable; the square and triangular lattices produce globally very similar clusters. Secondly the $F^{ve}(p_v,p_e);t_a)$ subfamily gives already a good approximation to a realistic growth. On a finer scale, as expected from our discussion, the triangular lattice leads to enhanced density variations.

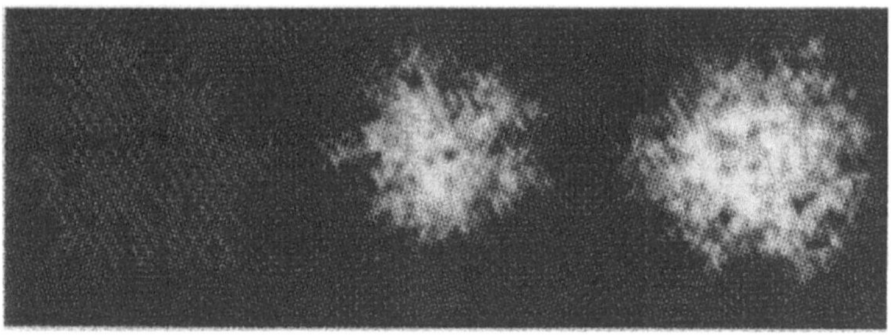

Fig. 2. Cloudy explosive growth from a single seed (triangular lattice): (a) vertex growth $p = 0.55$; (b) second approximation $\epsilon = 0.025$; (c) third approximation η_j random < 0.001; activity time 2; time $= 30$

The growth clusters are broadly reminiscent of the spongy structure of the remnants of supernova explosions (cf the Crab nebula). Detailed fluid dynamic simulations of the incipient supernova explosion reveal a statistical scale invariance in the density pattern: The cluster size, $S(t)$, the perimeter (number of cells of the outer border), $B(t)$, as well as the cluster diameter, $D(t)$, scale with time t as

$$S(t) \propto t^{\alpha},\ \alpha \le 2;\ B(t) \propto t^{\beta},\ \beta \le 2;\ D(t) \propto t^{\delta},\ \delta \le 1; \qquad (5.1)$$

or $S \propto D^{\frac{\alpha}{\delta}}$, $B \propto D^{\frac{\beta}{\delta}}$, so that α/δ and β/δ are fractal (mass) dimensions of the cluster and of the border respectively ($p \to 1$: $\delta \to 1$).
The close similarity of the simulated growth with explosive phenomena as encountered in fluid dynamics leads to the conjecture that the asymptotic

fluid dynamic equations representative of explosion processes – in particular of supernova explosions – and some representative of the basic family B($\mathbf{p}$;t_a) belong into the same universality class.

6 Concluding Remarks

We have analysed in this paper a multi-parameter family of growth models adaptable to various lattices, neighbourhoods and symmetries. Two general properties are observed: (a) typical one-parameter subfamilies exhibit a phase-transition from bounded to unbounded growth; and (b) a fraction of the subfamilies are morphologically unstable. Both properties are universal features of a variety of natural growth processes. The initial aim we pursued when devising the basic model family B($\mathbf{p}$;t_a) was to exhibit a category of flexible, algorithmically simple models capable of simulating complex fluid dynamic growth hard to reproduce with the standard equations of fluid mechanics.

We wish to mention that the systematic experimental survey we have carried out actually indicates that the patterns our models can generate are broader and far more general than those observed in fluid dynamics. In particular the structures are often reminiscent of biological growth, so that our model family might be relevant for the purposes of investigating biological growth.

The bulk of this work was carried out in the Summers of 1995 and 1996 at the Institute of Astronomy under a European Exchange Fellowship operated through the Royal Society and the Belgian FNRS. The author thanks the FNRS for financial support.

References

Bak, P., Tang, C., Wiesenfeld, K. (1987): Self organized criticality : an expansion of $1/f$ noise. Phys. Rev. Lett. **59**, 381–384

Boccara, N., Cheong, K., Oram, M. (1994): A probabilistic automata network epidemic model with births and deaths exhibiting cyclic behavior. J. Phys. **A27**, 1585–1597

Erzan, A., Pietronero, L., Vespignani, A. (1995): The fixed-scale transformation of fractal growth. Rev. Mod. Phys. **67**, 545–604

Fryxell, B., Arnett, D., Müller, E. (1991): Instabilities and clumping in SN 1987A. Ap. J. **367**, 619–634

Langer, J.S. (1980): Instabilities and pattern formation in crystal growth. Rev. Mod. Phys. **52**, 1–28

Meakin, M. (1995): Progress in DLA research. Physica **D86**, 104–112

Packard, N.H., Wolfram, S. (1985): Two dimensional cellular automata. J. Stat. Phys. **38**, 901–946

Packard, N.H. (1986): *Theory and Applications of Cellular Automata*, Wolfram, S. ed. (World Scientific, Singapore)

Pandey, R.B. (1991): Cellular automata approach to interacting cellular network models for the dynamics of cell population in an early HIV infection. Physica **A179**, 442–470

Perdang, J. (1993): The cellular automaton framework: A new tool for astrophysical modelling. *Cellular Automata: Prospects in Astrophysical Applications* (World Scientific, Singapore), 3–83

Perdang, J. (1995): Chaos and stellar evolution. *Stellar Evolution: What Should Be Done, Proceedings of the 32nd Liège International Astrophysical Colloquium*, Noels, A. ed. (Durouaux-Ordina, Liège), 147–167

Perdang, J. (1996): Cloudy, filamentary, and knotty structures : Simulating diffusive growth dynamics. *Proceedings on Knowledge Transfer*, Behrooz, A. ed. (Pace, London), 8–15

Seiden, P.E., Schulman, L.S. (1990): Percolation model of galactic structure. Adv. Phys. **39** 1, 1–54

Stanley, H.S., Ostrowski, N. (1986): *On Growth and Form*, (Nijhoff, Dordrecht)

Stauffer, D. (1985): *Introduction to Percolation Theory*, (Taylor & Francis, London)

Stauffer, D., Pandey, R.B. (1992): Immunology motivated simulations of cellular automata. Comp. Phys. **6**, 404–410

Todo, Y. (1991): Num. Astrophys. (Japan) **2**, 193

Uchida, Y. (1991): Num. Astrophys. (Japan) **2**, 219

Wolfram, S. (1983): Statistical mechanics of cellular automata. Rev. Mod. Phys. **55**, 601–644

Acoustic Earth Imaging by Seismic Migration

Ernesto Bonomi, Leesa Brieger, Carlo Nardone, and Enrico Pieroni

Centro di Ricerca, Sviluppo e Studi Superiori in Sardegna - Area Geofisica, Via Nazario Sauro, 10 I-09123 Cagliari, Italy

Abstract. Geophysical methods of prospecting for oil and gas resources are based on observations of elastic wave fields, treated to a first approximation as acoustic waves. Emitted pressure impulses penetrate the crust and then are backscattered to the surface where they are detected by an array of receivers. The goal of seismic processing is to recover from the recorded wave field as much information as possible about the earth subsurface.

Migration is an acoustic image reconstruction technique based on the inversion of the scalar wave equation. In the migration process the recorded pressure waves are used as initial conditions for a wave field governed by the scalar wave equation which propagates downward and in reverse time.

1 Introduction

In geophysical exploration the techniques for the echoreconstruction of subsurface structures are based on non-intrusive experiments in which emitted short impulses propagating downward are backscattered and measured at the surface (Boyd 1997). Collected seismic traces contain information on the spatial dependency of subsurface parameters of considerable consequence for prospecting for oil (Özdoğ 1997), gas, mineral and groundwater resources. These parameters include longitudinal and transversal propagation velocity, mass density etc.

To first order the earth's crust is a layered medium composed of unconsolidated formations (sand), consolidated and permeable formations (sandstone) and hard tight formations (rock), each one with specific continuum mechanical properties. A physical model that describes the wave propagation is required for the accurate echoreconstruction of the subsurface. A given model, although never complete, represents an idealisation which captures the essential features. For acoustic wave sources the model can be scalar, elastic or inelastic. In the context of seismic exploration, in a first approximation, waves are considered purely compressional so that their evolution is described by solutions of the scalar equation. In addition, ignoring spatial density variations, reflections and diffraction are only caused by the discontinuities of the compressional velocity field (see Fig. 1) typically a piecewise function taking values from 0.4 km/s (sand) up to 7.0 km/s (granite), normally increasing with depth.

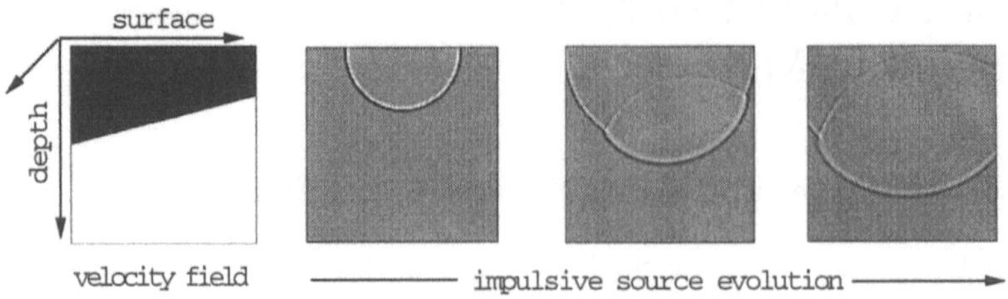

Fig. 1. Wave propagation, reflexion caused by discontinuities of the compressional velocity field.

Deterministic seismic analysis usually falls into one of the two following classes: the forward problem and the inverse problem. In the forward problem, an estimate of the subsurface model parameters, for instance the compressional velocity field in the case of the scalar wave equation, is used to produce synthetic data. Then the comparison between synthetic and real data allows the validation of the subsurface model. However, from the forward problem the updating of the model parameters and the minimisation of the discrepancy between synthetic and measured data is difficult. The inverse problem procedures are designed to infer these model parameters directly from recorded seismic data. Data migration is the way in which inversion was first formulated in the context of geophysical exploration (Scales 1997).

In the migration process the recorded pressure waves are used as initial conditions for a wave field governed by the "one-way" scalar wave equation, which backpropagates seismic data from the surface to the reflecting interfaces in reverse time. Those interfaces are then interpreted as the discontinuities of the velocity field model.

The entire migration process is thus based on the a priori estimation of the velocity field, obtained initially from well logs and from an empirical analysis on seismic traces (Sguazzero 1986). It is only by interpreting migrated data, namely by comparing the imaged interfaces with the discontinuities of the velocity model, that the latter can be iteratively improved. This turnaround process, illustrated by the flowchart of Fig. 2, can last several computing days and is particularly crucial for imaging complex geological structures including those of potential interest for hydrocarbon prospecting.

Another approach, recently discovered, to the direct determination of the subsurface characteristics, wave equation inversion, consists in formulating seismic inversion as an optimisation problem. Its solution is the velocity field whose seismic response, governed by the wave equation, gives the best fit to surface measurements and thus most accurately locates reflecting surfaces (Kolb 1986). However, the computational cost of this method is today out of reach for most real applications. For this reason a pragmatic approach such as seismic migration is still used in the oil industry for imaging large

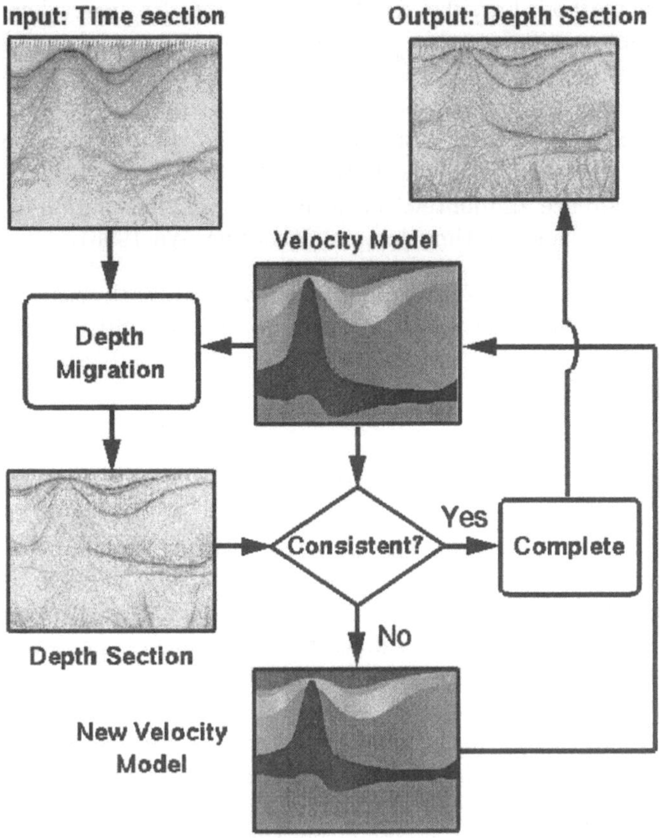

Fig. 2. Wave propagation, reflexion caused by discontinuities of the compressional velocity field.

3D data sets. In any case the computational cost of migration demands high performance computers.

Standard acoustic seismic techniques for imaging inhomogeneous media fall into five classes:

- Approximation of the Kirchhoff integral formula defined in the space-time domain (Schneider 1978).
- Finite-difference and finite-element methods for the solution of paraxial approximatations of the wave equation in the space-frequency domain (Collino 1995).

- Spectral methods formulated in the wavenumber-frequency domain based on the phase shift formula, solution of the exact paraxial equation (Stolt 1978).
- Explicit depth extrapolation in the space-frequency domain, obtained by convoluting seismic data with a designed, finite-length phase shift operator (Bagaini 1995; Hale 1991).
- Finite-difference and finite-element methods for the solution of the wave equation in the space-time domain (Wen-Jing Wu 1996).

The aim of this paper is to introduce the reader to depth seismic migration and to describe the Phase Shift Plus Interpolation (PSPI) method which generalises the phase shift formula to media with laterally variable velocity (Gazdag 1984).

2 Representation of Seismic Data

For the sake of simplicity we limit our discussion to idealised seismic data acquisition (coincident source-receiver experiment), which accomodates common midpoint (CMP) stacked data. CMP stacking is a pre-processing technique in which traces relative to different source-receiver pairs with a common midpoint (see Fig. 3) are gathered, transformed into zero-offset data by a hyperbolic time correction and then added to improve the signal-to-noise ratio (Sguazzero 1986). This is done to condense the information into a single seismic trace, simulating the response of a coincident source-receiver experiment at the common midpoint location $(X, Y, Z = 0)$. By repetition of CMP stacking for various common midpoint positions the resultant CMP dataset, denoted also as the zero-offset seismic section, is considered for all subsequent processing phases as a collection of traces, each recorded independently.

In a coincident source-receiver experiment, once the source has been initiated at time zero, the downward path and traveltime $t/2$ to a reflector point are identical with the upward path and traveltime $t/2$ to the receiver (see Fig. 3). As a consequence, an equivalent trace will result from a source initiated at the reflector with the appropriate strength R, provided the velocity of the medium c is divided by two. This is the so called exploding reflector imaging model (Stolt 1978). Hence, using the zero-offset seismic section $P(X, Y, Z = 0, t)$ as a boundary condition and solving the scalar wave equation

$$\frac{\partial^2 P}{\partial X^2} + \frac{\partial^2 P}{\partial Y^2} + \frac{\partial^2 P}{\partial Z^2} - \frac{1}{v(X, Z)^2} \frac{\partial^2 P}{\partial t^2} = 0, \quad v(X, Y, Z) = \frac{c(X, Y, Z)}{2} \tag{1}$$

in reverse time with zero-initial conditions, at time $t = 0$ we should obtain an acoustic picture $R(X, Y, Z) = P(X, Y, Z, t = 0)$ of the reflectors. The migrated section $R(X, Y, Z)$ represents the map of the local reflectivity.

Considering that the size of a typical industrial 3D dataset, including zero-offset seismic section and velocity field, is about 1GByte, even with modern

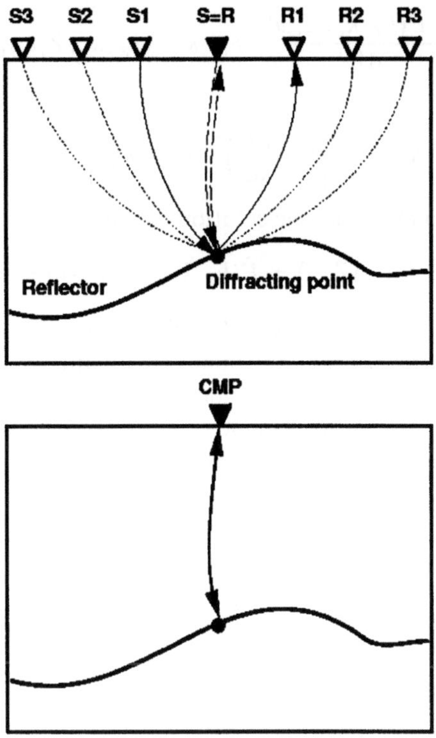

Fig. 3. Wave propagation, reflexion caused by discontinuities of the compressional velocity field.

supercomputers, the resulting numerical process is a heavy computational task demanding much in terms of data storage, data movement and computational load. For this reason an enormous effort has historically gone into altering the migration model based on the reverse time solution of equation (1) to reduce the cost of the operation while keeping the essential features of the wave propagation. The basic traditional strategy is described in the following section.

3 Downward Continuation

Assume for the moment that v, the halved velocity, is constant. A way of solving equation (1) is to use the depth, Z, as the advancing variable along which the zero-offset seismic section $P(X, Y, 0, t)$ must be extrapolated. Our target is thus to evaluate the migrated section $P(X, Y, Z, 0)$, representing for $Z > 0$ the picture of the exploding reflector points.

Writing equation (1) in the wavenumber-frequency domain (k_x, k_y, ω) we obtain the following second-order ordinary differential equation

$$\frac{\mathrm{d}^2 \hat{P}(k_x, k_y, Z, \omega)}{\mathrm{d}Z^2} = - \; k_z^2 \hat{P}(k_x, k_y, Z, \omega), \tag{2}$$

in which

$$k_z = \frac{\omega}{v}\left[1 - \left(\frac{v}{\omega}\right)^2 (k_x^2 + k_y^2)\right]^{1/2}. \tag{3}$$

Equation (2) has two characterisic solutions of the form

$$\hat{P}(k_x, k_y, Z + \Delta Z, \omega) = \hat{P}(k_x, k_y, Z, \omega) \exp(\pm i k_z \Delta Z), \tag{4}$$

relating the field computed at level Z and the new one at level $Z + \Delta Z$, both represented in the (k_x, k_y, ω)-domain, by a shift of phase. Since in the downward continuation process (equation (4)) ΔZ is positive, the sign agreement between phase and ω corresponds to a wave moving in reverse time, starting from the the initial condition $P(k_x, k_y, Z = 0, \omega)$. The other solution, with phase and ω of opposite sign, represents a wave moving forward in time, starting from the same initial condition. Because migration of seismic data is an inverse process, we are only interested in

$$\hat{P}(k_x, k_y, Z + \Delta Z, \omega) = \hat{P}(k_x, k_y, Z, \omega) \exp(i k_z \Delta Z). \tag{5}$$

Transforming equation (5) backward into the space-frequency domain, the imaging of migrated data at level $Z + \Delta Z$ is obtained by

$$P(X, Y, Z + \Delta Z, t = 0) = \sum_{\omega} \hat{P}(X, Y, Z + \Delta Z, \omega). \tag{6}$$

Equations (3), (5) and (6) are the basis for the Phase Shift migration algorithm (Gazdag 1984). They allow the exact inverse extrapolation of seismic data inside a homogeneous layer $]Z; Z + \Delta Z]$ with constant velocity $v = v_z$.

It is worth mentioning that, to avoid data aliasing and thus improve the accuracy of the extrapolated wave, it is often preferable to substitute the ideal exponential operator with a "designed" one, to provide a highest attenuation between the passband region and the spatial Nyquist frequencies.

To construct the migrated section, (6), seismic traces are first represented in the space-frequency domain (X, Y, ω). Since their power spectrum is band limited with a cutoff frequency far below the temporal Nyquist, mapping data into the space-frequency domain allows significant data compression.

Phase shift formulation of wavenumber domain migration leads to an elegant parallel implementation. $P(X, Y, Z = 0, \omega)$ is loaded from disk and distributed across processors by blocks of time frequencies ω. All blocks are then concurrently transformed by 2-D FFTs. During the downward continuation three steps are repeated for increasing values of Z. First, for each complex array, a parallel operator masks all spurious frequencies and shifts the phase of each entry. Second, partial sums over ω are concurrently performed, followed by the concurrent execution of 2-D FFTs, one per processor. Finally, the downward continuation ends by imaging the reflectors, namely summing up the transformed fields and then discharging on disk the resulting migrated section $P(X,Y,Z+\Delta Z,t=0)$.

To conclude this section we remark that expression (4) is also a solution of

$$\frac{d\hat{P}}{dZ}(k_x,k_y,Z,\omega) = ik_z\hat{P}(k_x,,k_y,Z,\omega), \tag{7}$$

the paraxial or one-way wave equation. Equation (7) is the starting point of a hierarchy of approximations defined in the space-frequency domain (X,Y,ω). Their numerical solution is based on finite-difference or finite element methods (Collino 1995).

4 Migration in Inhomogeneous Media: the PSPI Method

As previously seen, seismic imaging in a stratified medium is implemented using both (5) and (6). The case with lateral velocity variations requires more attention: in this context the wavenumber domain representation (2) of the scalar wave equation is meaningless and then no straightforward solution such as the phase shift formula is possible. To overcome this difficulty and keep the computational complexity of the migration process to a reasonable level, the wave propagation model is altered a step further in order to construct a pure spectral method for the data downward continuation also in an inhomogeneous medium.

The starting point is the phase shift formula, (5), written in a split form and then modified to handle the wave propagation inside the layer $]Z;Z+\Delta Z]$ with a laterally variable velocity field. The resulting first term governs vertically-travelling waves through the layer:

$$\hat{P}_0(X,Y,Z,\omega) = \hat{P}(X,Y,Z,\omega)\exp\left(i\,\frac{\omega}{v}\Delta Z\right), \qquad v = v_z(X,Y). \tag{8}$$

The resulting second term governs the horizontal correction for a reference velocity in a set $v_z^{(1)} < v_z^{(2)} < \cdots < v_z^{(n_z)}$:

$$\hat{P}^{(n)}(k_x,k_y,Z+\Delta Z,\omega) = \hat{P}_0(k_x,k_y,Z,\omega)\exp\left[i\left(k_z^{(n)} - \frac{\omega}{v_z^{(n)}}\right)\Delta Z\right], \tag{9}$$

with $k_z^{(n)}$, $n=1,2,\cdots,n_z$, given by (3) where $v=v_z^{(n)}$.

The transformed fields $\hat{P}^{(n)}(X,Y,Z+\Delta Z,\omega)$ serve as reference data from which the final result will be obtained by interpolation. In the case of linear interpolation, the depth continued wave field is given by

$$\hat{P}(X,Y,Z+\Delta Z,\omega) = \hat{P}^{(n)}(X,Y,Z+\Delta Z,\omega)\frac{v_z^{(n+1)} - v_z(X,Y)}{v_z^{(n+1)} - v_z^{(n)}} + \hat{P}^{(n+1)}(X,Y,Z+\Delta Z,\omega)\frac{v_z(X,Y) - v_z^{(n)}}{v_z^{(n+1)} - v_z^{(n)}} \tag{10}$$

for all points (X, Y) such that $v_z^{(n)} \leq v_z(X,Y) \leq v_z^{(n+1)}$. Finally, imaging of the reflectors at $Z + \Delta Z$ is obtained by the usual condition, (6). This is the essence of the Phase Shift Plus Interpolation Method (Gazdag 1984).

The PSPI algorithm, defined by (8), (9) and (10), allows the correct depth continuation of vertically-travelling plane waves and maintains high accuracy for small dips, characterised by $kv_z/\omega << 1$. Although the introduction of laterally-variable velocities and the use of interpolating reference solutions do not have a well established mathematical basis, computer experiments show that this approach is in fact very reliable. PSPI is a practical alternative to other forms of migration with the obvious advantage that each reference solution comes from a constant velocity extrapolation the implementation of which inherits the parallel structure of the phase shift algorithm.

It is clear from the above description that reference velocities play a crucial role in the migration process of seismic data (Bagaini 1995). The difficulty is the development of a strategy based on the selection of a few representative velocities to obtain a simplified but reliable description of the actual velocity field. This is the subject of the next section.

5 Rational Choice of Reference Velocities

We attempt now to give an answer to the problem of optimally choosing representative velocities using statistical arguments. The task is to highlight, in a certain statistical sense, a subset of velocity values that contribute predominantly to the propagation process. Any particular choice can erroneously make explicit certain information on the subsurface structure and cause the loss of other useful information, which is then hard to recover.

We consider a layer $[Z; Z + \Delta Z]$ and the associated velocity field $v = v_z(X,Y)$ that we discretise in L+1 equidistant values: $w_z^{(1)} = v_m < w_z^{(2)} < \cdots < w_z^{(L+1)} = v_M$ with v_m and v_M representing, respectively, the smallest and the largest values of the entire velocity field. Let us define on the collection $w_z^{(k)}, k = 1, \ldots, L+1$, the probability density distribution

$$P_z^{(k)} = \text{prob}\left\{w_z^{(k)} \leq v < w_z^{(k+1)}\right\}, \qquad \sum_{k=1}^{L} P_z^{(k)} = 1, \tag{11}$$

relative to each layer tagged by Z. The resulting distribution represents a form of information on the structure of the velocity field. From these data we would like, firstly, to characterise at each layer the minimal number of velocities, possibly smaller than $L+1$, necessary to represent this information. Secondly, we would like to determine which velocities are most representative.

Let us measure the statistical dispersion of the distribution (11) as follows :

$$S_z[P] = -\sum_{k=1}^{L} P_z^{(k)} \log P_z^{(k)}, \qquad 0 \leq S_z[P] \leq \log L. \tag{12}$$

This last expression is nothing but the definition of the "statistical entropy" of the probability density distribution. When there is no dispersion, namely when there is only one velocity interval, we find $S_z[P] = 0$; when the dispersion is maximal, namely all velocity intervals have the same statistical weight, we find $S_z[P] = \log L$.

It is possible to condense the L initial velocity intervals into new intervals, each one with the same statistical weight, and keep the statistical dispersion equal to $S_z[P]$. Using the definition of the statistical dispersion, (12), the minimal number N_z of such intervals, rounded to the closest integer, is

$$N_z = \lfloor \exp(S_z[P]) + \frac{1}{2} \rfloor, \qquad 1 \leq N_z \leq L. \tag{13}$$

Now, in order to preserve $S_z[P]$, the reference velocities $v_z^{(i)}, i = 1, \ldots, N_z + 1$, which define these N_z intervals of equal statistical weight must be such that

$$\frac{i}{N_z} = \int_{v_m}^{v_z^{(i+1)}} P_z(v) \mathrm{d}v, \qquad i = 1, \ldots, N_z, \tag{14}$$

where $P_z(v) = P_z^{(k)}$, for $w_z^{(k)} \leq v < w_z^{(k+1)}$, $k = 1, \ldots, L$, is the piecewise constant interpolation of the probability density distribution (11). By defining

$$y_z^{(k)} = \int_{v_m}^{w_z^{(k)}} P_z(v) \mathrm{d}v, \qquad k = 2, \ldots, L, \qquad y_z^{(1)} = 0, \quad y_z^{(L+1)} = 1, \tag{15}$$

the computation of each reference velocity, can be written as

$$v_z^{(i+1)} = w_z^{(k)} + \left[\frac{i}{N_z} - y_z^{(k)} \right] \frac{w_z^{(k+1)} - w_z^{(k)}}{y_z^{(k+1)} - y_z^{(k)}}, \qquad v_z^{(1)} = v_m, \quad i = 1, \ldots, N_z \tag{16}$$

where index k is such that $y_z^{(k)} < i/N_z \leq y_z^{(k+1)}$.

We recognise in the above approach, (13)–(16), a form of adaptive algorithm. The induced velocity grid $v_z^{(1)} < v_z^{(2)} < \cdots < v_z^{(n_z)}$, $n_z = N_z + 1$, has the intrinsic property of being fine around the maxima of $P_z(v)$ and coarse around the minima that we have identified. In geological situations, where $v_z(X, Y)$ is almost constant, only two or maybe three reference velocities should be necessary; in the presence of strong lateral fluctuations, the number of velocities will certainly be larger. With this adaptive mechanism, accuracy must increase on average because of the statistical importance confered to those velocities that contribute massively to the downward propagation of the wave field. Note that the constant velocity case, $v = v_z$, must be solved by hand, setting $n_z = 1$ and $v_z^{(1)} = v_z$.

6 Reconstruction of a complex subsurface model

Figure 4 illustrates the velocity field for a vertical slice of a subsurface model: $\Delta X = \Delta Y = \Delta Z = 11.25$ m. The first layer represents the ocean, $v = 1500$ m/s, and the subsequent stratification represents a typical complex structure of the earth's crust; inside this region the velocity varies from 2500 to 5650 m/s. The nested structure is characterized by a strong lateral variation of the velocity field.

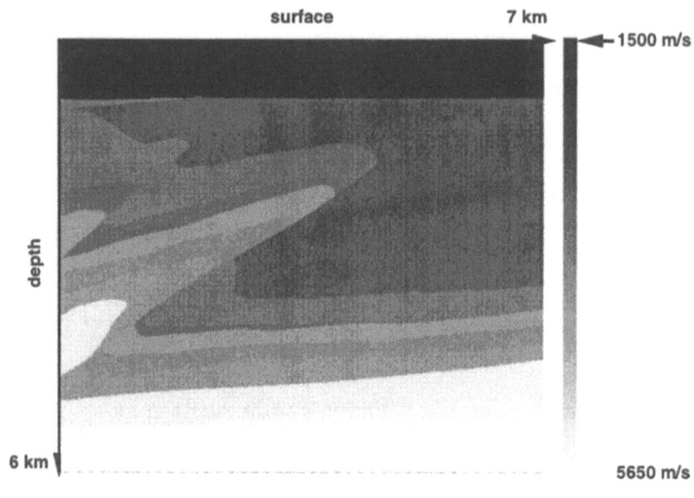

Fig. 4. Wave propagation, reflexion caused by discontinuities of the compressional velocity field.

Figure 5 shows the resulting synthetic zero-offset seismic section, $\Delta t = 4$ ms, obtained in two computational phases. In the first, the seismic measurements are simulated many times, repositioning a pressure source to produce different illuminations of the subsurface model shown in Fig. 4. Backward propagating waves are detected at $Z=0$. In the second phase all simulated data are finally collected and stacked in the zero-offset representation.

For the seismic data extrapolation the velocity field of Fig. 4 was discretised into 30 levels for each value of Z. Figure 6 displays the number of reference velocities as a function of Z and the corresponding relative empiric frequency. We see that the number of reference velocities required for this

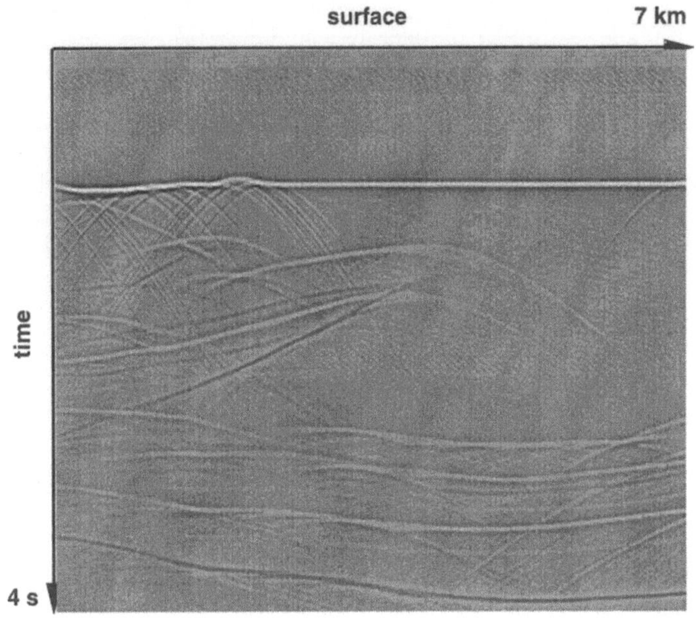

Fig. 5. Wave propagation, reflexion caused by discontinuities of the compressional velocity field.

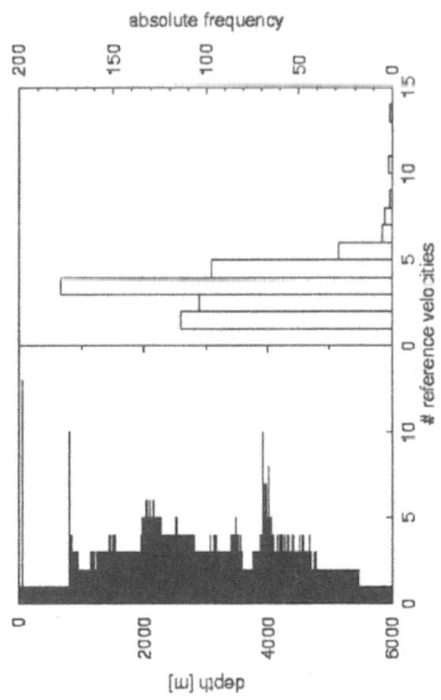

Fig. 6. Wave propagation, reflexion caused by discontinuities of the compressional velocity field.

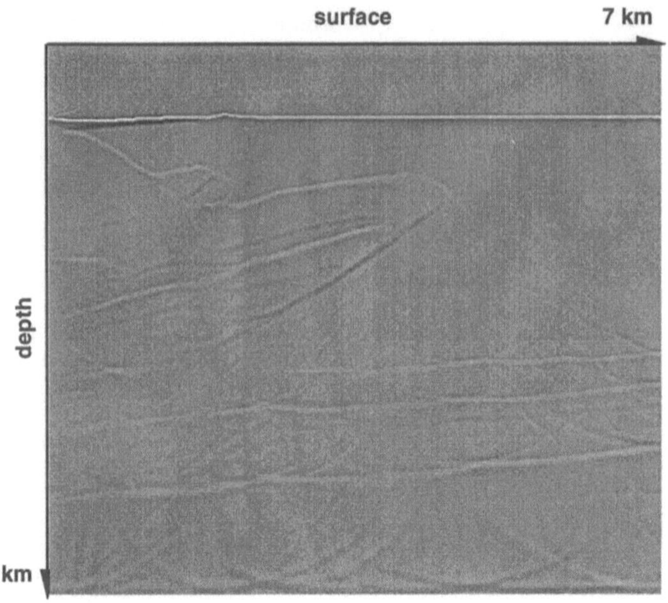

Fig. 7. Wave propagation, reflexion caused by discontinuities of the compressional velocity field.

example fluctuates from 1 to 13. On average, however, only three velocities are actually used in the extrapolation process.

The resulting migrated section, Fig. 7, was computed using the PSPI algorithm. The complex structure of the subsurface model represented, is largely reconstructed; the missing information in the migrated section is also absent in the zero-offset seismic section. The arcs of hyperbola appearing at the bottom of Fig. 7 are mostly due to scattering points present in the zero-offset section. Globally, the agreement between reflecting interfaces and discontinuities of the velocity field is very good.

Acknowledgments

The financial support of this work was partially provided by the Sardinian Regional Authorities.

References

Bagaini, C., Bonomi, E., Pieroni, E. (1995) : Split convolutional approach to 3D depth extrapolation, 65th Ann. Internat. Mtg., Soc. Expl. Geophys, Expanded Abstracts, Houston

Boyd, Thomas M. (1997): *Introduction to Geophysical Exploration*, Colorado School of Mines (http://magma.Mines.EDU/fs_home/tboyd/GP311/).

Collino, F., Joly, P. (1995): Splitting of operators, alternate directions, and paraxial approximations for the three-dimensional wave equation. SIAM J. Sci. Comput. **16**, 1019–1048

Gazdag, J., Sguazzero, P. (1984): Migration of seismic data. Geophysics **49**, 124–131

Hale, D. (1991): Stable explicit depth extrapolation of seismic wave-field. Geophysics **56**, 1770–1777

Kolb, P., Collino, F., Lailly, P. (1986): Pre-stack inversion of a 1D medium. Proceedings of the IEEE, **74**, 498–508

Özdoğan, Yilmaz (1997): Seismic data processing, Series: Investigation in Geophysics **2**, published by Soc. Expl. Geophys.

Scales, J.A. (1997): *Theory of Seismic Imaging* (notes for a graduate course) Colorado School of Mines (http://landau.Mines.EDU/~samizdat/)

Schneider, W. (1978): Integral formulation for migration in two and three dimensions. Geophysics **43**, 49–76

Sguazzero, P., Vesnaver, A., Proudian, V. (1986): Seismic velocity analysis on the IBM 3090 vector facility. IBM Tech. Rep., 513–5023

Stolt, R.H. (1978): Migration by Fourier transform. Geophysics **43**, 23–48

Wen-Jing Wu, Lines, L.R., Han-Xing Lu (1996): Analysis of higher-order, finite-difference schemes in 3-D reverse-time migration. Geophysics **61**, 845–856

Part III

Dynamical Systems

The purpose of modelling a physical or another system is to be able to understand the workings of that system in terms of general theories and to make quantitative predictions about the evolution of the system, usually in time. Quantitative predictions require equations which may come in a variety of forms: differential equations – ordinary and/or partial – for continuous deterministic systems, difference equations for discrete systems; probabilistic equations for statistical systems. To make predictions these equations must be integrated, if this be possible, or at least analysed for qualitative behaviour. Integrability may be very precise and give a solution in closed form. This is rare. More often we seek the existence of a sufficient number of first integrals which will guarantee integrability. The techniques of dynamical systems lead to first integrals by the methods of Darboux and Carleman, as discussed in the papers on the integrability of first order polynomial systems. Systems of even order with a Hamiltonian structure benefit from special techniques applicable to that structure. The central role of symmetry in the discussion of integrability transcends particular structures and shows a route to integrability which is fundamentally intrinsic to the internal dynamics of the differential equations.

Without specifically seeking a closed-form solution for a system, we may discuss its global properties in terms of its Poincaré map or its phase space diagram. The very nature of critical points often tells us enough about the behaviour of the system for our purposes. Then the nondeterministic systems can be studied by the methods of statistics and sometimes there are connections which transcend subject barriers in a way that reveals the remarkable generality of mathematics.

Integrability and Alegbraic Solutions for the 2-D Lotka–Volterra System

Laurent Cairó[1] and Jaume Llibre[2]

[1] Laboratoire de Mathématique, Applications et Physique Mathématique - UMR 6628, Université d'Orléans, UFR des Sciences, BP 6759, F-45067 Orléans Cedex 2, France

[2] Departament de Matemàtiques, Universitat Autònoma de Barcelona, 08193 – Bellaterra, Barcelona, Spain

Abstract. We apply the method of Darboux to study the integrability of the 2d Lotka-Volterra system. The method of Darboux, which consists of finding first integrals by using algebraic solutions, is reduced in the case of the Lotka-Volterra system to obtaining only one algebraic solution different from the two axes, which can be a polynomial of arbitrary degree. Here we are concerned with polynomials of degree less than five. We find that all the essential parameters of the Lotka-Volterra system must be fixed for the existence of a polynomial solution of degree greater than or equal to three whereas, for degrees one and two respectively, only one and two conditions on the parameters are required.

1 Introduction

The Lotka-Volterra system in two dimensions (2-d) is a particular quadratic system of the form

$$\dot{x} = x(ax + by + c), \qquad \dot{y} = y(Ax + By + C). \tag{1}$$

This system and its generalisation to n-d are called *Lotka-Volterra systems* (LVS) because they were initially studied by Lotka (1920) and Volterra (1931). Later Kolmogorov (1936) also studied these systems and for this reason some authors denote them by *Kolmogorov systems.* There are many natural phenomena that can be modeled by the LVS such as the time evolution of conflicting species in biology (May 1974), chemical reactions, plasma physics (Laval and Pellat 1975), hydrodynamics (Busse 1978), economics etc. Starting with Volterra (1931) mathematicians have been concerned particularly with the problem of integrability, which is only possible for a very restricted set of parameters. Cairó and Feix (1992) used a generalised Carleman method (Carleman 1932) to obtain some integrable cases, generalising the result to the n-d version of (1). This was facilitated by the fact that here $x = 0$ and $y = 0$ are invariant, thereby linearising the problem. However, for a general quadratic system

$$\dot{x} = P(x,y), \qquad \dot{y} = Q(x,y), \tag{2}$$

where P and Q are polynomials with coefficients in $\mathbf{F}$, denoting by this either the real field $\mathbf{R}$ or the complex $\mathbf{C}$, the Carleman method is less efficient and one is led to the use of methods leading to nonlinear systems such as the Hamiltonian method among others (Cairó et al. 1993). Another possibility is to return to the original Darboux theory (Darboux 1878) which splits the problem by starting with the search of algebraic solutions. Using the Darboux method it has been possible to study the integrability of systems (2) having an invariant conic (Cairó et al. 1997). Note here that the system (1) is one of these systems where the invariant conic degenerates into two straight lines. In the present work we complete the preceding work in Theorems 5 and 6 which concern the algebraic solutions of equations of third and fourth degree. After a short survey of the Darboux method in Sect. 2 we exhibit the different cases of integrability of Lotka-Volterra systems in Sect. 3. Our results about the integrability are summarised in Theorems 3, 4, 5 and 6. The first two were proved in Cairó et al. (1997), but we include them here for the sake of completeness.

2 The Darboux Method

Let us establish the link between the integrability of a polynomial system and its algebraic solutions. First we need some preliminary definitions. System (2) is *integrable* on an open subset U of $\mathbf{F}^2$ if there exists a nonconstant analytic function $H : U \to \mathbf{F}$, called a *first integral* of the system on U, which is constant on all solution curves $(x(t), y(t))$ of (2) on U, i.e. $H(x(t), y(t)) =$ constant for all values of t for which the solution $(x(t), y(t))$ is defined on U. Clearly H is a first integral of (2) on U if and only if $DH \equiv 0$ on U, where $D = P\partial/\partial x + Q\partial/\partial y$. An *invariant algebraic curve*, also called an *algebraic solution*, of (2) is an algebraic curve $f(x, y) = 0$ with $f \in \mathbf{F}[x, y]$ such that for some polynomial $K \in \mathbf{F}[x, y]$ we have

$$Df = Kf. \tag{3}$$

K is called the *cofactor* of the invariant algebraic curve $f = 0$. The function R is an *integrating factor* of system (2) on U if we have

$$\frac{\partial(RP)}{\partial x} = -\frac{\partial(RQ)}{\partial y}$$

or equivalently $\operatorname{div}(RP, RQ) = 0$, or $DR = -R\operatorname{div}(P, Q)$. The *first integral H associated to the integrating factor R* is given by $H(x, y) = \int R(x, y) P(x, y)\mathrm{d}y + f(x)$, satisfying $\partial H/\partial x = -RQ$. The following proposition is easy to prove.

Proposition 1. *Suppose that $f \in \mathbf{F}[x, y]$ and let $f = f_1^{n_1} \cdots f_r^{n_r}$ be the factorisation of f in irreducible factors over $\mathbf{F}[x, y]$. If $Df = Kf$, then f_i is a divisor of Df_i in $\mathbf{F}[x, y]$ for all $i = 1, \ldots, r$.*

The problem of integrating a polynomial system by using its algebraic solutions was first considered by Darboux (Darboux 1878). We state here only the result of Darboux that we need. All of its results can be found, for instance, in Cairó et al. (1997).

Darboux Theorem. *Suppose that a polynomial system (2) of degree m admits q invariant algebraic curves $f_i = 0$ with cofactors K_i for $i = 1, \ldots, q$. If $q = m(m+1)/2$, then the function $f_1^{\lambda_1} \cdots f_q^{\lambda_q}$ for suitable $\lambda_i \in \mathbf{F}$ not all zero is either a first integral and $\sum_{i=1}^{q} \lambda_i K_i = 0$, or an integrating factor and $\sum_{i=1}^{q} \lambda_i K_i = -\mathrm{div}(P, Q)$.*

Darboux invariants. A nonconstant analytic function I in the variables x, y and t such that $I(x(t), y(t), t)$ is constant for all solutions $(x(t), y(t))$ of system (2) in an open subset U of $\mathbf{F}^2$ is called an *invariant* of the system on U.

The existence of a first integral $H(x, y)$ for a planar differential system implies that, by drawing the curves $H(x, y) =$ constant, we can completely describe the phase portrait of such a system. The existence of an invariant gives us information on the $\alpha-$ and $\omega-$ limits of the orbits of the differential system.

The next proposition shows us how to obtain an invariant of (2) by knowing sufficient algebraic solutions of the system.

Proposition 2. *Suppose that a polynomial system (2) of degree m admits q algebraic solutions $f_i = 0$ with cofactors K_i for $i = 1, \ldots, q$. If there exist $\mu_i \in \mathbf{F}$ not all zero such that $\sum_{i=1}^{q} \mu_i K_i = -s$ with $s \in \mathbf{F}$, then the function $f_i^{\mu_1} \cdots f_q^{\mu_q} e^{st}$ is an invariant of system (2).*

See the proof in Cairó et al. (1997).

Christopher Lemma. *Suppose that a polynomial system (2) of degree m has the invariant algebraic curve $f = 0$ of degree n. Let P_m, Q_m and f_n be the homogeneous components of P, Q and f of degree m and n respectively. Then the irreducible factors of f_n must be factors of $yP_m - xQ_m$.*

See the proof in Christopher (1994).

3 Darboux Integrability for Lotka–Volterra Systems

The next theorem characterises all LVS systems which are integrable having at least three algebraic solutions of degree 1. Of course LVS in the normal form always have two invariant straight lines $x = 0$ and $y = 0$. We introduce the notation $\Delta = aB - bA$, $d_{12} = b - B$ and $d_{21} = A - a$ which will appear frequently in the text. In statement (3c) $R_{12} = cBd_{21} + Cad_{12}$.

Theorem 3. *Let $f_1 = x$ and $f_2 = y$. If an LVS has a third algebraic solution $f_3 = 0$ of degree 1, then, modulo the symmetry (x, y, a, b, c, A, B, C)*

$\rightarrow (y, x, B, A, C, b, a, c)$, *a Darboux invariant* $f_1^{\mu_1} f_2^{\mu_2} f_3^{\mu_3} e^{st}$ *and an integrating factor or a first integral* $f_1^{\lambda_1} f_2^{\lambda_2} f_3^{\lambda_3}$ *can be found, having the exponents given in Table 1, provided that the LVS satisfies the conditions indicated in the same table.* f_3 *takes the following corresponding values*
(3a) $f_3 = ax + c = 0$, *if* $caB \neq 0$,
(3b) $f_3 = d_{21}x + d_{21}y = 0$, *if* $cd_{12}d_{21} \neq 0$,
(3c) $f_3 = Cax + cBy + cC = 0$, *if* $C^2a^2 + c^2B^2 \neq 0$.

Table 1. Invariants and first integrals for Theorem 3.

st.	cond.	μ_1/s	μ_2/s	μ_3/s	λ_1	λ_2	λ_3
3a	$b = 0$	$-c^{-1}$	0	c^{-1}	$(C-c)c^{-1}$	-2	$\dfrac{cd_{21} - Ca}{ca}$
3b	$c = C$	$Bc^{-1}d_{12}^{-1}$	$ac^{-1}d_{21}^{-1}$	$\dfrac{\Delta}{cd_{12}d_{21}}$	$-bd_{12}^{-1}$	$-Ad_{12}^{-1}$	$-\dfrac{ad_{12} + Bd_{21}}{d_{12}d_{21}}$
3c	$R_{12} = 0$				Bd_{21}	ad_{12}	Δ

Proof. System (1) has always $f_1 = x = 0$ and $f_2 = y = 0$ as algebraic solutions. An easy computation shows that $K_1 = ax + by + c$ and $K_2 = Ax + By + C$. So, in general, we have only to find a third algebraic solution in order to apply the Darboux theorem. Under the assumptions of statement (3a) the cofactor of f_3 is $K_3 = ax$. Since $caB \neq 0$, the system $\lambda_1 K_1 + \lambda_2 K_2 + \lambda_3 K_3 = (2a + A)x + 2By + c + C$ has a unique solution $\lambda_1 = (C - c)/c, \lambda_2 = -2, \lambda_3 = [cd_{21} - aC]/(ac)$. Therefore by Darboux' Theorem we obtain that $f_1^{\lambda_1} f_2^{\lambda_2} f_3^{\lambda_3}$ is an integrating factor of LVS. The system $\mu_1 K_1 + \mu_2 K_2 + \mu_3 K_3 = -s$ has a unique solution $\mu_1 = -s/c, \mu_2 = 0, \mu_3 = s/c$. Hence, by Proposition 2, $f_1^{\mu_1} f_2^{\mu_2} f_3^{\mu_3} e^{st}$ is an invariant of LVS and so statement (3a) is proved.

Now we assume the hypotheses of statement (3b). Then the cofactor of f_3 is $K_3 = ax + By + c$. Since $cd_{21}d_{21} \neq 0$, the system $\lambda_1 K_1 + \lambda_2 K_2 + \lambda_3 K_3 = (2a + A)x + (b + 2B)y + 2c$ has a unique solution $\lambda_1 = -b/d_{12}, \lambda_2 = -A/d_{21}, \lambda_3 = (ab - 2aB + AB)/(d_{12}d_{21})$. Then by Darboux' Theorem we obtain that $f_1^{\lambda_1} f_2^{\lambda_2} f_3^{\lambda_3}$ is an integrating factor of LVS. The system $\mu_1 K_1 + \mu_2 K_2 + \mu_3 K_3 = -s$ has a unique solution $\mu_1 = Bs/(cd_{12})$, $\mu_2 = as/(cd_{21})$, $\mu_3 = \Delta s/[cd_{12}d_{21}]$. Therefore, from Proposition 2 it follows that $f_1^{\mu_1} f_2^{\mu_2} f_3^{\mu_3} e^{st}$ is an invariant of LVS.

Similarly, under the assumptions of statement (3c), the cofactor of f_3 is $K_3 = ax + By$. Since $cBd_{21} + Ca(b - B) = 0$ and $a^2C^2 + c^2B^2 \neq 0$, the homogeneous system $\lambda_1 K_1 + \lambda_2 K_2 + \lambda_3 K_3 = 0$ has, in particular, the solution

$\lambda_1 = d_{21}B$, $\lambda_2 = ad_{12}$ and $\lambda_3 = aB - bA$. Therefore by Darboux' Theorem we obtain that $f_1^{\lambda_1} f_2^{\lambda_2} f_3^{\lambda_3}$ is a first integral. Hence a first integral of the LVS is

$$H = x^{d_{21}B} y^{ad_{12}} (aCx + cBy + cC)^{aB-bA}.$$

Finally in order to check that, modulo the symmetry $(x, y, a, b, c, A, B, C) \to (y, x, B, A, C, b, a, c)$, the statements of the Theorem give all of the algebraic solutions of the LVS of degree 1 we must find all polynomials f_3 of degree 1 satisfying equation (3). We note that the Christopher Lemma states that f_3 must be of the form $f_3 = x + \text{constant}$ (3a) or $f_3 = -d_{21}x + d_{12}y + \text{constant}$ (3b) or (3c), the possibility $f_3 = y + \text{constant}$ being redundant due to the mentioned symmetry. ∎

Figure 1 is a typical phase portrait of statement (3c) ($a = -1$, $b = -2.5$, $c = 1$, $A = 2.5$, $B = 1$, $C = -1$). In this figure are represented the singular points, the three algebraic solutions forming the triangle, the three cofactors and some other solutions. We note firstly that the four singular points are distributed by pairs in the algebraic solutions and in the cofactors and secondly that these algebraic solutions coincide with the separatrices in the case of statement (3c).

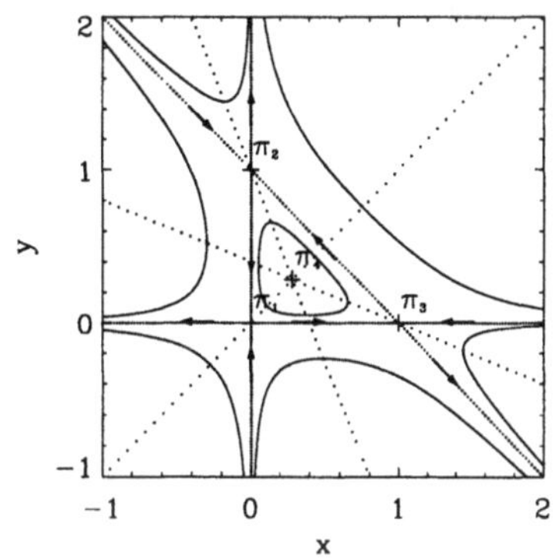

Fig. 1.

The next theorem characterises all Lotka-Volterra systems which are integrable having an invariant conic.

Table 2. Invariant conditions of Theorem 4.

statement	condition 1	condition 2
4a	$B - 2b = 0$	$(2c + C)a - cA = 0$
4b	$b + B = 0$	$(2c + C)a - cA = 0$
4c	$c = 2C$	$A(2B - b) = a(3B - 2b)$
4d	$(c + 2C)B = bC$	$(2c + C)a - cA = 0$

Theorem 4. *Let $f_1 = x$ and $f_2 = y$. If a LVS has a third algebraic solution $f_3 = 0$ of degree 2, then, modulo the symmetry $(x, y, a, b, c, A, B, C) \to (y, x, B, A, C, b, a, c)$, a Darboux invariant $f_1^{\mu_1} f_2^{\mu_2} f_3^{\mu_3} e^{st}$ and an integrating factor $f_1^{\lambda_1} f_2^{\lambda_2} f_3^{\lambda_3}$ can be found having the following values for f_3 and the exponents λ_i, μ_i under the conditions given in Table 2.*

(4a) $f_3 = C(ax+c)^2 + c^2By = 0$, $\mu_1 = -s/c$, $\mu_2 = 0$, $\mu_3 = s/(2c)$, $\lambda_1 = -2$, $\lambda_2 = (c-C)/C$, $\lambda_3 = -(2c+C)/(2C)$, if $cCa \neq 0$.

(4b) $f_3 = (c+C)(ax+c)^2 + 2caBxy = 0$, $\mu_1 = \mu_2 = -s/(c+C)$, $\mu_3 = (3c+C)s/[2c(c+C)]$, $\lambda_1 = -c/(c+C)$, $\lambda_2 = -(2c+C)/(c+C)$, $\lambda_3 = (c-C)/[2(c+C)]$ if $c(c+C)a \neq 0$.

(4c) $f_3 = cax + [ax + (2B-b)y]^2 = 0$, $\mu_1 = 2Bs/[c(b-2B)]$, $\mu_2 = 2s/c$, $\mu_3 = 2d_{21}s/[c(b-2B)]$ and $\lambda_1 = d_{21}/(b-2B)$, $\lambda_2 = -2$, $\lambda_3 = b/[2(b-2B)]$ if $ca(2B-b) \neq 0$.

(4d) $f_3 = (Cax + cBy + cC)^2 - 4cCaBxy = 0$, $\mu_1 = -s/(2c)$, $\mu_2 = -s/(2C)$, $\mu_3 = (c+C)s/(2cC)$ and $\lambda_1 = \lambda_2 = -1$, $\lambda_3 = -1/2$ if $cCa \neq 0$.

Proof. The cofactors of f_1 and f_2 are $K_1 = ax+by+c$ and $K_2 = Ax+By+C$ respectively. Under the suitable assumption for each statement it is easy to check that $f_3 = 0$ is an algebraic solution of system LVS, i.e. that f_3 satisfies equation (3) where K_3, the cofactor of f_3, is of the form $px + qy + r$. For statements (4a)–(4d) we must check that $\lambda_1 K_1 + \lambda_2 K_2 + \lambda_3 K_3$ is equal to minus the divergence of system LVS, i.e. equal to $-(2a+A)x - (b+2B)y - (c+C)$. Then, according to Darboux' Theorem, $f_1^{\lambda_1} f_2^{\lambda_2} f_3^{\lambda_3}$ is a Darboux integrating factor of the LVS. Again to see for statements (4a)–(4d) that $f_1^{\mu_1} f_2^{\mu_2} f_3^{\mu_3} e^{st}$ is an invariant of system LVS, from Proposition 2, it is sufficient to verify that $\mu_1 K_1 + \mu_2 K_2 + \mu_3 K_3 = -s$. Finally in order to check that, modulo the above mentioned symmetry, the statements of the theorem give all algebraic solutions of the LVS of degree two we must find all polynomials f_3 of degree two satisfying equation (3). This is a tedious task which we have performed with the help of the algebraic manipulation code, Maple, and the Christopher Lemma. We note that this lemma and the mentioned symmetry imply that the quadratic terms of f_3 must be x^2 (4a), xy (3a), $x(d_{21}x - d_{12}y)$ (4b).

Table 2 summarises the conditions that the coefficients of (1) must satisfy. Figure 2 is a sample phase portrait for statement (4c) ($a = 1, b = 6, c = -4, A = 0, B = 4, C = -2$) and Fig. 3 is a sample phase portrait for statement (4d) ($a = -1, b = 1.666, c = 0.5, A = 1, B = 1, C = -1.5$). In these figures only the singular points, the three algebraic solutions in solid lines and the three cofactors in dotted lines are represented. We must mention that statement (3c) of Theorem 3 and the invariants of statements (3a) and (3b) of Theorem 3, (4b), (4c) and (4d) of Theorem 4 were already found by Cairó and Feix (1992) using the Carleman method. The next theorem characterises all Lotka-Volterra systems which are integrable having an invariant cubic.

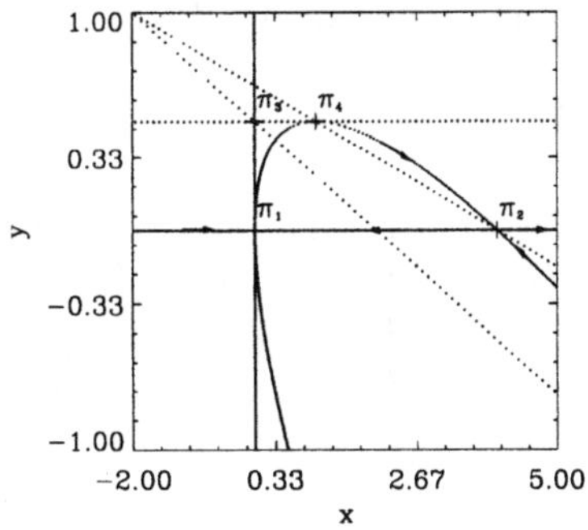

Fig. 2.

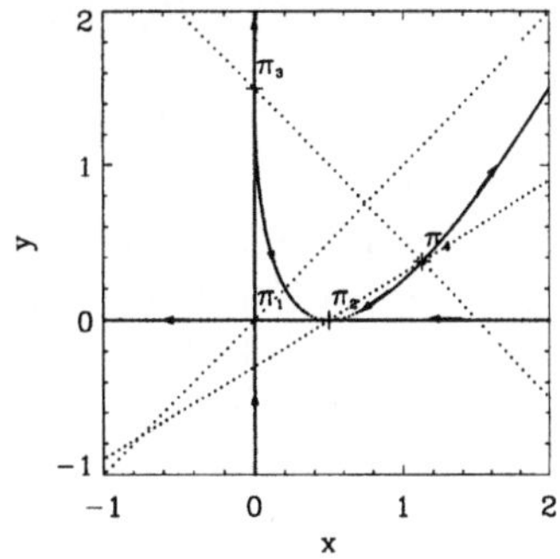

Fig. 3.

Theorem 5. *Let $f_1 = x$ and $f_2 = y$. If an LVS has a third algebraic solution $f_3 = 0$ of degree 3, then, modulo the symmetry $(x, y, a, b, c, A, B, C) \to (y, x, B, A, C, b, a, c)$, a Darboux invariant $f_1^{\mu_1} f_2^{\mu_2} f_3^{\mu_3} e^{st}$ and an integrating factor $f_1^{\lambda_1} f_2^{\lambda_2} f_3^{\lambda_3}$ can be found having the exponents given in Table 3 provided that the LVS satisfies the conditions indicated in the same table, with f_3 taking the following corresponding values, and provided that $CAB \neq 0$*

(5a) $f_3 = 5C(-25\,CBy + 3\,A^2x^2 - 30\,ABxy - 25\,B^2y^2) + 16\,A^3x^3 = 0$,

(5b) $f_3 = 5c(45\,cAx + 36A^2x^2 + 120ABxy + 50B^2y^2) + 36A^3x^3 = 0$,

(5c) $f_3 = 9\,c^3 + 27\,ac^2x - 24\,Bc^2y + 27\,a^2cx^2 + 72\,acBxy + 16\,cB^2y^2 + 9\,a^3x^3 = 0$,

(5d) $f_3 = -bc^2y - cA^2x^2 - 3\,bcAxy + b^2Axy^2 = 0$,

(5e₁) $f_3 = 4C(Ax - 3C)(Ax + 3by - 3C) + 3\,b^2Axy^2 = 0$,

(5e₂) $f_3 = 3\,c^3 + 12\,c^2Ax + 2\,bc^2y + 12\,cA^2x^2 + 36\,bcAxy + 18\,b^2Axy^2 = 0$,

(5f) $f_3 = -C^3 + 6\,AC^2x - 9\,A^2Cx^2 + 36ABCxy + 27\,AB^2xy^2 = 0$,

(5g) $f_3 = -3\,Cx(C + ax + 3\,by) + 9\,b^2Cy^2/(3\,a - 2\,A) - (8\,a - 3\,A)\eta(x, y)x^2 = 0$,

(5h) $f_3 = 25\,cby[3\,c + 2(2\,A - 3\,a)x + 2\,by]/A^2 - \eta(x, y)x^2 = 0$,

(5i₁) $f_3 = c^3 + 3\,ac^2x - 6\,bc^2y + 3\,a^2cx^2 - 12\,abcxy + 9\,cb^2y^2 + a^2(ax - 6\,by)x^2 = 0$,

(5i₂) $f_3 = 4c(c^2 + 3\,cax - 6\,cby + 3\,a^2x^2 + 15\,abxy + 9\,b^2y^2) + a^2(4ax + 3\,by)x^2 = 0$,

(5i₃) $f_3 = C^3/2 + 3\,AC^2x + 2\,bC^2y + 6\,A^2Cx^2 + 6\,bACxy + 2\,b^2Cy^2 + A^2(4\,Ax + 3\,by)x^2 = 0$,

(5j) $f_3 = -3\,(2\,c + C)[(c + C)(C + 2\,By + by)/(AB) - xy] - \eta(x, y)xy = 0$,

(5k) $f_3 = (c + C)(2\,c + C)[2C^2(c - Ax) + c(b + 2\,B)(2\,C + (b + B)y)y]/(bAC^2) - 2\,(2c + C)xy + \eta(x, y)xy = 0$,

(5l₁) $f_3 = 4\,B(c + C)[c/(3a - A) + x]y + d_{21}^2(3\,c - C)x^2/a + \eta(x, y)^2x = 0$,

(5l₂) $f_3 = 4\,B(c + C)[c/(3a - A) + x]y + d_{21}^2(3\,c - C)x^2/a + \eta(x, y)^2x = 0$,

(5m) $f_3 = cd_{21}^2(c^2 + 3\,cax + 3\,a^2x^2)/a^3 + d_{12}^2(c + 2\,C)xy/B + \eta(x, y)^2x = 0$,

(5n₁) $f_3 = cd_{21}^2(c^2C + 3\,cCax + c^2By + 3\,a^2Cx^2)/(a^3C) + 4\,B(c + 2\,C)xy + \eta(x, y)^2x = 0$,

(5n₂) $f_3 = -5\,C^2(C+By)/(16\,A) + 3\,cd_{21}^2x(c+ax)/a^2 + 4\,B(c+2\,C)xy + \eta(x,y)^2x = 0$,
(5n₃) $f_3 = 4\,c(c+2\,C)^2(C+By)/(9\,aC)+3\,d_{21}^2cx(c+ax)/a^2+4\,B(c+2\,C)xy+\eta(x,y)^2x = 0$,
(5o) $f_3 = Cd_{12}^3(C+2\,By)y/B^2 + 2\,Cd_{12}d_{21}^2xy/(2\,a-A) - \eta(x,y)^3 = 0$,
(5p₁) $f_3 = Cd_{12}^3y[C+2(3a-A)x+2\,By]/B^2 - Cd_{21}^3x^2/(2\,a) - \eta(x,y)^3 = 0$,
(5p₂) $f_3 = Cd_{12}^3y[C+2(3\,a-A)x+2\,By]/B^2 - Cd_{21}^3x^2/(2\,a) - \eta(x,y)^3 = 0$,
(5p₃) $f_3 = Cd_{12}^3y[C+2(3a-A)x+2\,By]/B^2 - Cd_{21}^3x^2/(2\,a) - \eta(x,y)^3 = 0$,
(5q₁) $f_3 = -4\,Cd_{21}^3x[C+ax-(b-3\,B)y]/a^2 + Cd_{12}^3y^2/B - \eta(x,y)^3 = 0$,
(5q₂) $f_3 = -4\,Cd_{21}^3x[C+ax-(b-3\,B)y]/a^2 + Cd_{12}^3y^2/B - \eta(x,y)^3 = 0$,
(5q₃) $f_3 = -4\,Cd_{21}^3x[C+ax-(b-3\,B)y]/a^2 + Cd_{12}^3y^2/B - \eta(x,y)^3 = 0$,
(5r) $f_3 = 3\,Cd_{21}^3x[3\,C/a+2x]/a - 6\,Cd_{21}d_{12}^2xy/(b-2\,B) + \eta(x,y)^3 = 0$,
where $\eta(x,y) = d_{21}x - d_{12}y$.

Proof. The proof follows the same lines of the preceding theorems. We note that the Christopher Lemma and the mentioned symmetry imply that the cubic terms of f_3 must be of the form x^3 (5a to 5c), xy^2 (5d to 5f), $\eta(x,y)x^2$ (5g to 5i), $\eta(x,y)xy$ (5j to 5k), $\eta(x,y)^2x$ (5l to 5n), $\eta(x,y)^3$ (5o to 5r). ∎

Figure 4 is a sample phase portrait for statement (5g) ($a = -0.4286$, $b = 1.5$, $c = 2$, $A = -0.5$, $B = 3$, $C = 1$). Here also only the singular points, the three algebraic solutions in solid lines and the three cofactors in dotted lines are represented. Table 3 summarises the conditions that the coefficients of (1) must satisfy.

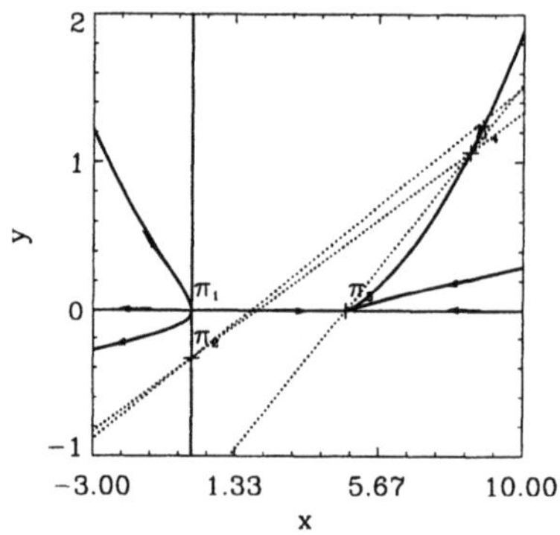

Fig. 4.

The next theorem characterises all Lotka-Volterra systems which are integrable having an invariant quartic.

Theorem 6. *Let $f_1 = x$ and $f_2 = y$. If an LVS has a third algebraic solution $f_3 = 0$ of degree four, then, modulo the symmetry $(x,y,a,b,c,A,B,C) \to (y,x,B,A,C,b,a,c)$, a Darboux invariant $f_1^{\mu_1}f_2^{\mu_2}f_3^{\mu_3}e^{st}$ and an integrating factor $f_1^{\lambda_1}f_2^{\lambda_2}f_3^{\lambda_3}$ can be found having the exponents given in Table 4 provided that the LVS satisfies the conditions indicated in the same table. f_3 takes the following corresponding values provided that $CAB \neq 0$.*
(6a) $f_3 = 20C(500\,C^2By+600\,CABxy+500\,CB^2y^2+27\,A^3x^3+45\,A^2Bx^2y) + 567\,A^4x^4 = 0$,
(6b) $f_3 = 8C(32C^2By+2\,CA^2x^2+32\,CABxy+32\,CB^2y^2+3\,A^3x^3+6\,A^2CBx^2y) + 9\,A^4x^4 = 0$,

Table 3. Invariants and integrating factors for Theorem 5.

st.	a	b	c	μ_1/s	μ_2/s	μ_3/s	λ_1	λ_2	λ_3
$5a$	$8A/15$	$2B/3$	$C/2$	$-6/C$	0	$2/C$	2	$-1/3$	$5/6$
$5b$	$2A/5$	$2B/3$	$2C$	$-3/(2c)$	0	$1/(2c)$	$1/4$	$1/2$	1
$5c$	$4A/9$	$2B/3$	$-4C/3$	$-1/c$	0	$1/(3c)$	$1/2$	$1/3$	1
$5d$	$2A$	$-B$	$C/2$	$-1/(3c)$	$-2/(3c)$	$1/(3c)$	$1/3$	$1/3$	1
$5e_1$	$2A$	$-B$	$-6C$	$1/(4C)$	$1/(2C)$	$-1/(4C)$	$7/6$	2	$1/6$
$5e_2$	$2A$	$-B$	$-2C/3$	$1/(2c)$	$1/c$	$-1/(2c)$	$1/2$	$2/3$	$5/6$
$5f$	$2A$	$-2B$	$-2C/3$	$-3/(4C)$	$-3/(2C)$	$3/(4C)$	$1/4$	$1/2$	1
$5g$	$6A/7$	$B/2$	$2C$	$-2/(3C)$	$-1/(2C)$	$5/(12C)$	$1/3$	2	$1/6$
$5h$	$7A/15$	$B/2$	$C/3$	$3/c$	$-7/(6c)$	$-1/(6c)$	2	$-1/6$	$5/6$
$5i_1$	$6A/5$	$B/2$	$-3C/2$	$-4/(3c)$	$-1/(2c)$	$7/(12c)$	$5/3$	2	$-1/6$
$5i_2$	$3A/7$	$B/2$	$-3C/2$	$-5/(6c)$	$1/(4c)$	$1/(12c)$	$2/3$	$1/2$	$5/6$
$5i_3$	$-A/3$	$B/2$	$-C/6$	$9/C$	$1/(2C)$	$-5/(2C)$	2	$7/6$	$1/6$
$5j$	$-A/2$	$-B$	$-7C/8$	$8/(3C)$	$4/(3C)$	$4/(3C)$	$-1/3$	$-1/6$	$5/6$
$5k$	$-A$	$-2B$	$-8C/7$	$-7/(6C)$	$-7/(3C)$	$-7/(6C)$	$-1/6$	$-1/3$	$5/6$
$5l_1$	$A/7$	$-B$	$C/2$	$-2/(3C)$	$2/(3C)$	$-4/(3C)$	$1/3$	$7/6$	$1/6$
$5l_2$	$3A/5$	$-B$	$C/3$	$-3/(4C)$	$-9/(4C)$	$3/(2C)$	$1/4$	$1/4$	1
$5m$	$4A/7$	$-2B$	$-4C/5$	$5/(6C)$	$-5/(3C)$	$5/(4C)$	$1/6$	$1/3$	1
$5n_1$	$3A/5$	$-B$	$6C$	$-1/(9C)$	$-1/(3C)$	$2/(9C)$	$5/6$	2	$-1/6$
$5n_2$	$3A/5$	$-B$	$-4/(9C)$	$-3C/4$	$-4/(3C)$	$8/(9C)$	$1/3$	$1/2$	$5/6$
$5n_3$	$A/7$	$-B$	$C/4$	$4/(3C)$	$-4/(3C)$	$8/(3C)$	$1/3$	$7/6$	$1/6$
$5o$	$5A/9$	$7B/3$	$C/3$	$-3/(2C)$	$-5/(2C)$	$2/C$	$1/2$	$1/6$	1
$5p_1$	$-A/5$	$5B/2$	$C/2$	$2/C$	$-1/(2C)$	$-3/(2C)$	$5/3$	$5/6$	$-1/6$
$5p_2$	$4A/7$	$5B/2$	$C/2$	$-1/C$	$-2/C$	$3/(2C)$	$2/3$	$1/3$	$5/6$
$5p_3$	$4A/3$	$7B/6$	$C/2$	$3/C$	$-2/C$	$-1/(2C)$	2	$1/3$	$1/6$
$5q_1$	$2A/5$	$-5B$	$2C$	$-1/(4C)$	$1/C$	$-3/(4C)$	$5/6$	$5/3$	$-1/6$
$5q_2$	$2A/5$	$7B/4$	$2C$	$-1/C$	$-1/(2C)$	$3/(4C)$	$1/3$	$2/3$	$5/6$
$5q_3$	$6A/7$	$3B/4$	$2C$	$-1/C$	$3/(2C)$	$-1/(4C)$	$1/3$	2	$1/6$
$5r$	$3A/7$	$9B/5$	$3C$	$-5/(6C)$	$-1/(2C)$	$2/(3C)$	$1/6$	$1/2$	1

(6c) $f_3 = 7C(392C^2Ax + 168\,CA^2x^2 + 588\,CABxy + 343\,CB^2y^2 + 24\,A^3x^3 + 28\,A^2Bx^2y) + 8\,A^4x^4 = 0$,

(6d) $f_3 = 7C(343\,C^3 - 392\,C^2Ax + 1029\,C^2By + 168\,CA^2x^2 - 588\,CAB\,xy + 1029\,CB^2y^2 - 32\,A^3x^3 + 84\,A^2Bx^2y + 588\,AB^2Cxy^2 + 343\,B^3y^3) + 16\,A^4x^4 = 0$,

(6e) $f_3 = 5C(125C^3 - 100\,C^2Ax + 250\,C^2By + 30\,CA^2x^2 + 200\,CABxy + 125\,CB^2y^2 - 4\,A^3x^3 - 5\,A^2Bx^2y) + A^4x^4 = 0$,

(6f) $f_3 = 2\,C^4 - 30\,AC^3x + 150\,A^2C^2x^2 - 50\,ABC^2xy - 250\,A^3Cx^3 + 750\,A^2BC\,x^2y + 625\,A^2B^2x^2y^2 = 0$,

(6g) $f_3 = C^4 + 3\,BC^3y + 2\,ABC^2xy + 3\,B^2C^2y^2 + 6\,AB^2Cxy^2 + CB^3y^3 + 2\,A^2B^2x^2y^2 = 0$,

(6h) $f_3 = 100CBy(100\,CBy + 9\,A^2x^2 + 120\,ABxy + 100\,B^2y^2) + 27A^3(21\,Ax + 20\,By)x^3 = 0,$

(6i) $f_3 = 11C(-5324\,C^3 + 5808\,C^2Ax - 15972\,C^2By - 2376\,CA^2x^2 + 8349\,CAB\,xy - 15972\,CB^2y^2 + 432\,A^3x^3 - 990\,A^2Bx^2y - 7260\,AB^2xy^2 - 5324\,B^3y^3) - 27A^3(12\,Ax + 11\,By)x^3 = 0,$

(6j) $f_3 = 4\,C(4\,C^3 - 24\,C^2Ax + 12\,C^2By + 54\,CA^2x^2 - 48\,CABxy + 12\,CB^2y^2 - 54\,A^3x^3 + 63\,A^2Bx^2y - 24\,AB^2xy^2 + 4\,B^3y^3) + 27A^3(3\,Ax - 4\,By)x^3 = 0,$

(6k) $f_3 = 48C(84C^3 + 78C^2Ax + 169CABxy + 2197A^2Bx^2y) + 28561A^2B(3\,Ax + 4\,By)x^2y = 0,$

(6l) $f_3 = 16CB(16\,CBy + 3A^2x^2 + 16ABxy + 16B^2y^2)y + A^2(3Ax + 4By)^2x^2 = 0,$

(6m) $f_3 = 5C(125\,C^3 - 700\,C^2Ax + 1470\,CA^2x^2 - 1225\,CABxy - 1372\,A^3x^3 + 1372\,A^2Bx^2y) + 2401A^2(\,Ax + 5By)^2x^2 = 0,$

(6n) $f_3 = 49\,C\,B\,(49B\,Cy + 4\,A^2x^2 + 84\,A\,B\,xy)y + 8A(Ax + 7By)^3x = 0,$

(6o) $f_3 = -cd_{21}^3(c^3C + 4\,c^2Cax + Bc^3y + 6\,a^2cCx^2 + 4\,a^3Cx^3)/(a^4C) - 27\,(c + 2\,C)B(c + 3\,C)xy/2 + 27\,BC(9\,a - 4\,A)x^2y/16 - 27\,B^2(c + 3\,C)xy^2 - \eta(x,y)^3x = 0,$

(6p) $f_3 = -5\,Cd_{21}^3(125\,C^3 + 100\,aC^2x + 125\,BC^2y + 30\,Ca^2x^2 + 4\,a^3x^3)/a^4 - 27(2\,C + \,c)B(c + 3\,C)xy/2 + 27\,BC(9\,a - 7\,A)\,x^2y - 27\,B^2(c + 3\,C)xy^2 - \eta(x,y)^3x = 0,$

(6q) $f_3 = cd_{21}^3(c + 2\,ax)(c^2 + 2\,cax + 2\,a^2x^2)/a^4 + 32\,(c + 2\,C)B(c + 3\,C)xy + 64\,AB(5\,c + 3\,C)x^2y/65 + 64\,B^2(c + 3\,C)xy^2 + \eta(x,y)^3x = 0,$

(6r₁) $f_3 = d_{21}^4[c^3x/a^3 + 3\,c^2x^2/a^2 - 3\,c^2(b - 4\,B)xy/a^3 + 3\,cx^3/a + 9\,cB(9\,a - 5\,A)x^2y/(16\,a^3) + 9\,c(b - 3\,B)(b - 4\,B)xy^2/(2\,a^3)] + Cd_{12}^4y^3/B + \eta(x,y)^4 = 0,$

(6r₂) $f_3 = d_{21}^4[c^3x/a^3 + 3\,c^2x^2/a^2 - 3\,c^2(b - 4\,B)xy/a^3 + 3\,cx^3/a + 9\,cB(9\,a - 2\,A)x^2y/a^3 + 9\,c(b - 3\,B)(b - 4\,B)xy^2/(2\,a^3)] + Cd_{12}^4y^3/B + \eta(x,y)^4 = 0,$

(6s) $f_3 = d_{21}^4[c^3x/a^3 + 3\,c^2x^2/a^2 - 4\,c^2(b - 4\,B)xy/a^3 + 3\,cx^3/a + 48\,bc(2\,a - A)x^2y/(35\,a^3) + 8\,c(b - 3\,B)(b - 4\,B)xy^2/a^3] + \eta(x,y)^4 = 0,$

(6t) $f_3 = d_{12}^4[C^3y/B^3 + 3\,(4\,a - A)C^2xy/B^3 + 3\,C^2y^2/B^2 + 9\,C(4\,a - A)(3\,a - A)x^2y/(2\,B^3) - 3\,C(5\,A - 8\,a)xy^2/(4\,B^2) + 3\,Cy^3/B] + Cd_{21}^4x^3/(3\,a) + \eta(x,y)^4 = 0,$

(6u) $f_3 = d_{12}^4[C^3y/B^3 + 3\,(4\,a - A)C^2xy/B^3 + 3\,C^2y^2/B^2 + 9\,C(4\,a - A)(3\,a - A)x^2y/(2\,B^3) - 9\,AC(9\,B - 2\,b)xy^2/(4\,B^3) + 3\,Cy^3/B] + Cd_{21}^4x^3/(3\,a) + \eta(x,y)^4 = 0,$

(6v) $f_3 = d_{21}^4[4\,C^2x^2/a^2 + 4\,Cx^3/a - 8\,(b - 2\,B)Cx^2y/a^2] + 12\,ACd_{12}^3xy^2/(5\,b - 10\,B) + \eta(x,y)^4 = 0,$

(6w) $f_3 = Cd_{12}^4[Cy^2 + 2\,By^3 + (2\,a - A)xy^2]/B^2 - 3\,BCd_{21}^3x^2y/(2\,a - A) + \eta(x,y)^4 = 0,$

Figure 5 is a sample phase portrait for statement (6p) ($a = -0.2273$, $b = -6$, $c = 5$, $A = -0.5$, $B = 3$, $C = 1$). Here also only the singular points, the three algebraic solutions in solid lines and the three cofactors in dotted lines are represented. Table 4 summarizes the conditions that the coefficients of (1) must satisfy.

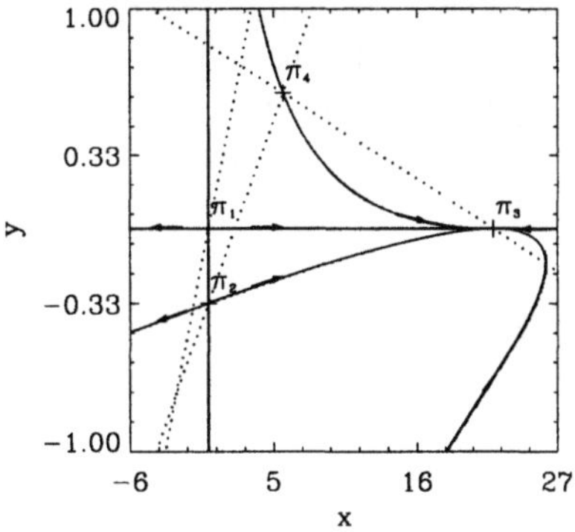

Fig. 5.

Table 4. Invariants and integrating factors for Theorem 6.

st.	a	b	c	μ_1/s	μ_2/s	μ_3/s	λ_1	λ_2	λ_3
6a	$7A/20$	$B/2$	$C/3$	$-12/C$	0	$3/C$	2	$-1/6$	$5/6$
6b	$3A/8$	$B/2$	$C/2$	$-4/C$	0	$1/C$	2	$-1/2$	1
6c	$2A/7$	$B/2$	$2C$	$-2/(3C)$	0	$1/(6C)$	3	$1/3$	1
6d	$9A/14$	$3B/4$	$-9C/4$	$4/(9C)$	0	$-1/(9C)$	$1/3$	$-1/2$	1
6e	$3A/10$	$B/2$	$-3C/2$	$2/(3C)$	0	$-1/(6C)$	$1/2$	$1/4$	1
6f	$2A$	$-B$	$-2C/5$	$-5/(3C)$	$-5/(3C)$	$5/(6C)$	$-2/3$	$1/3$	1
6g	$-A$	$B/2$	$-5C/2$	$2/(3C)$	$2/(3C)$	$-1/(3C)$	$1/3$	$-2/3$	1
6h	$13A/20$	$2B/3$	$C/2$	$8/C$	$-13/(3C)$	$-1/(3C)$	2	$-7/6$	$5/6$
6i	$7A/11$	$2B/3$	$-7C/3$	$5/(14C)$	$-1/(6C)$	$-1/(42C)$	$1/2$	$-1/6$	$5/6$
6j	$5A/4$	$2B/3$	$-5C/6$	$8/(5C)$	$1/(3C)$	$-7/(15C)$	2	$11/6$	$-1/6$
6k	$-A/2$	$-B$	$-7C/13$	$-13/(6C)$	$-13/(6C)$	$-13/(6C)$	$-7/6$	$-1/6$	$5/6$
6l	$5A/8$	$B/2$	$C/2$	$4/C$	$-5/C$	$1/C$	2	$-3/2$	1
6m	$3A/5$	$-B$	$-3C/7$	$-7/(4C)$	$-7/(4C)$	$7/(6C)$	$-3/4$	$1/4$	1
6n	$5A/7$	$-B$	$C/2$	$-2/(3C)$	$-10/(3C)$	$4/(3C)$	$1/3$	$-2/3$	1
6o	$7A/10$	$-2B$	$-7C/4$	$-4/(7C)$	$-2/C$	$6/(7C)$	$1/3$	$-1/6$	$5/6$
6p	$5A/11$	$-2B$	$5C$	$-1/(10C)$	$-1/(2C)$	$3/(10C)$	$5/6$	$11/6$	$-1/6$
6q	$9A/13$	$-3B$	$-9C/5$	$-5/(6C)$	$-5/(2C)$	$10/(9C)$	$1/6$	$-1/2$	1
$6r_1$	$3A/10$	$11B/4$	$3C$	$-8/(9C)$	$-2/(3C)$	$7/(9C)$	$1/3$	$1/2$	$5/6$
$6r_2$	$6A/11$	$-4B$	$3C$	$-2/(9C)$	$4/(3C)$	$-5/(9C)$	$5/6$	2	$-1/6$
6s	$4A/13$	$14B/5$	$4C$	$-5/(6C)$	$-2/(3C)$	$3/(4C)$	$1/6$	$1/3$	1
6t	$4A/11$	$10B/3$	$C/3$	$-2/C$	$-8/(3C)$	$7/(3C)$	$1/2$	$1/3$	$5/6$
6u	$-A/4$	$11B/6$	$C/3$	$4/C$	$-2/(3C)$	$-5/(3C)$	2	$5/6$	$-1/6$
6v	$2A/5$	$10B/7$	$2C$	$-7/(4C)$	$-1/(2C)$	$3/(4C)$	$-3/4$	$1/2$	1
6w	$7A/10$	$5B/2$	$C/2$	$-1/C$	$-7/(2C)$	$3/(2C)$	$1/2$	$-3/4$	1

Proof. The proof uses the same arguments of the preceding theorems. We note that the Christopher Lemma and the mentioned symmetry imply that the quartic terms of f_3 must be of the form x^4 (6a to 6e), x^2y^2 (6f to 6g), $\eta(x,y)x^3$ (6h to 6j), $\eta(x,y)x^2y$ (6k), $\eta(x,y)^2x^2$ (6l to 6m), $\eta(x,y)^3x$ (6n to 6q), $\eta(x,y)^4$ (6r to 6w). ∎

References

Busse, F.H. (1978): *Transition to turbulence via the statistical limit cycle route. Synergetics* (Springer Verlag, Berlin)

Cairó, L., Feix, M.R. (1992): Families of invariants of the motion for the Lotka-Volterra equations. The linear polynomial invariants. J. Math. Phys. **33**, 2440–2455

Cairó, L., Feix, M.R., Hua, D.D., Bouquet, S. (1993): Hamiltonian Method and invariant Search for 2D quadratic systems. J. Phys. A: Math. Gen. **26**, 4371–4386

Cairó, L., Feix, M.R., Llibre, J. (1997): Integrability and algebraic solutions for planar polynomial differential systems with emphasis on the quadratic systems. Université d'Orléans preprint 97-03

Carleman, T. (1932): Application de la théorie des équations intégrales linéaires aux systémes d'équations différentielles non linéaires. Acta Math. **59**, 63–87

Christopher, C.J. (1994): Invariant algebraic curves and conditions for a center. Proc. Roy. Soc. Edinburgh **124A**, 1209–1229

Darboux, G. (1878): Mémoire sur les équations différentielles algébriques du premier ordre et du premier degré (Mélanges), Bull. Sci. Math., 60–96; 123–144; 151–200

Kolmogorov, A. (1936): Sulla teoria di Volterra della lotta per l'esistenza. Giornale dell' Istituto Italiano degli Attuari **7**, 74–80

Kooij, R.J., Christopher, C.J. (1993): Algebraic invariant curves and the integrability of polynomial systems. Appl. Math. Lett. **6**, No. 4, 51–53

Laval, G., Pellat, R. (1975): *Plasma Physics. Proceedings of Summer School of Theoretical Physics* (Gordon and Breach, New York)

Lotka, A.J. (1920): Analytical note on certain rhythmic relations in organic systems. Proc. Natl. Acad. Sci. U.S. **6**, 410–415

May, R.M. (1974): *Stability and Complexity in Model Ecosystems* (Princeton University Press, Princeton, New Jersey)

Volterra, V. (1931): *Leçons sur la Théorie Mathématique de la Lutte pour la vie.* (Gauthier Villars, Paris)

Isochronous Centers of Cubic Reversible Systems

Javier Chavarriga and Isaac García

Departament de Matemàtica, Universitat de Lleida, Pl. Victor Siurana 1, 25003, Lleida, Spain

Abstract. In this paper we study isochronous centers of reversible two-dimensional autonomous system with linear part of center type and nonlinear part given by polynomials of third degree. Firstly we find necessary conditions for such isochronous centers in polar coordinates and finally we give a proof of the isochronicity of these systems using different methods.

1 Introduction

It is known that the problem of isochronicity appears only for *nondegenerate centers*, see Mardesic et al. (1995).

We consider cubic polynomial systems with a nondegenerate center at the origin. In an appropriate coordinate system and upon rescaling of the independent variable these systems take the form

$$\dot{x} = -y + X_2(x,y) + X_3(x,y), \quad \dot{y} = x + Y_2(x,y) + Y_3(x,y), \tag{1}$$

where $\dot{} = \mathrm{d}/\mathrm{d}\,t$ and $X_s(x,y)$ and $Y_s(x,y)$, $s = 2,3$, are *homogeneous polynomials* of degree s. We say that (1) is *reversible* if it is invariant under a rotation

$$\begin{pmatrix} \xi \\ \eta \end{pmatrix} = \begin{pmatrix} \cos\alpha & -\sin\alpha \\ \sin\alpha & \cos\alpha \end{pmatrix} \begin{pmatrix} x \\ y \end{pmatrix},$$

and a time inversion $t = -\tau$. Reversible systems are characterised by the existence of at least one straight line containing the origin, which is an axis of symmetry of the phase portrait. This line is given by the angle $\alpha/2$.

The integrable cases for homogeneous systems of the form

$$\dot{x} = -y + X_s(x,y), \quad \dot{y} = x + Y_s(x,y),$$

have been studied by several authors, namely *quadratic systems*, $s = 2$, and *cubic homogeneous systems*, $s = 3$, by Bautin (1952), Chavarriga (1994), Coppel (1996), Lloyd (1983), Lunkevich and Sibirskii (1982), Schlomiuk (1993) and Zoladek (1994). Some integrable cases of homogeneous systems when $s = 4,5$ have been determined by Chavarriga and Giné (1996), Chavarriga and Giné (1997).

A center is *isochronous* if the period of all integral curves in a neighbourhood of the origin is constant. The equations are most easily written when the arc length s is the variable: $\ddot{s} + k^2 s = 0$, where k is a real constant.

More recently several classes of isochronous systems have been studied: quadratic isochronous centers, see Loud (1964), Chicone and Jacobs (1989); homogeneous cubic systems, see Pleshkan (1969), Rousseau and Toni (1993a); the cubic system of Kukles, see Christopher and Devlin (1994), Rousseau and Toni (1993b); the class of "complex" systems $\dot{z} = iP(z)$ and the cubic reversible systems with $\dot{\theta} = 1$, see Mardesic et al. (1995); some isochronous cubic systems with four invariant lines, see Mardesic et al. (1995); and finally, isochronous centers of cubic systems with degenerate infinity, see Lloyd et al. (1996) and Chavarriga, Giné, García (1996a).

The aim of this paper is to find the reversible isochronous centers of (1).

2 Some Preliminary Results

In the study of this problem we have used polar coordinates. In Lemma 1 we give the expression for (1) in polar coordinates.

Lemma 1 *In polar coordinates* $x = r\cos\varphi$, $y = r\sin\varphi$ *we can write* (1) *as*

$$\begin{aligned} \dot{r} &= P_2(\varphi)r^2 + P_3(\varphi)r^3, \\ \dot{\varphi} &= 1 + Q_2(\varphi)r + Q_3(\varphi)r^2, \end{aligned} \tag{2}$$

where $P_s(\varphi)$ *and* $Q_s(\varphi)$, $s = 2, 3$ *are trigonometric polynomials.*

The following theorem classifies two integrable cases of (2).

Theorem 1 *The system* (2) *is integrable (in the sense of the vanishing of all Lyapunov constants) in the following two cases:*
(i) *if* $\sum_{s=2}^{3}(s+1)P_s(\varphi) + Q'_s(\varphi) \equiv 0$, *where* $' = \mathrm{d}/\mathrm{d}\,\varphi$, *and*
(ii) *if* $P_s(\varphi)$ *and* $Q_s(\varphi)$ *are of the form*

$$\begin{aligned} P_2(\varphi) &= R_3 \sin 3\omega + R_1 \sin\omega, \quad P_3(\varphi) = R_4 \sin 4\omega + R_2 \sin 2\omega, \\ Q_2(\varphi) &= R_3 \cos 3\omega + r_1 \cos\omega, \quad Q_3(\varphi) = R_4 \cos 4\omega + r_2 \cos 2\omega + r_0, \end{aligned} \tag{3}$$

where $\omega = \varphi + \varphi_0$, φ_0 *is arbitrary and* R_i, $r_j \in \mathbb{R}$, $(i = 1, 2, 3, 4,\ j = 0, 1, 2)$ *are arbitrary coefficients.*

Systems of class **(i)** are *hamiltonian* and systems of class **(ii)** correspond to *reversible systems*. In this work we study those reversible systems that have an isochronous center at the origin. Lemma 1 and Theorem 1 are proved by Chavarriga (1994).

Definition 1 *We ascribe the expression* reciprocal integrating factor *for the system* (1) *to a function* $V(x,y)$ *which is a particular solution of* (1) *and the divergence of the field*

$$C = \left(\frac{-y + X_2(x,y) + X_3(x,y)}{V(x,y)}, \frac{x + Y_2(x,y) + Y_3(x,y)}{V(x,y)} \right)$$

determined in $\{(x,y) \in \mathbb{R}^2 : V(x,y) \neq 0\}$ *is identically null.*

In particular $1/V(x,y)$ is an integrating factor which allows us to integrate (1) in the case that a reciprocal integrating factor is found.

The following theorem due to Poincaré and Liapunov characterises a non-degenerate center.

Theorem 2 *For a nondegenerate center of an analytic system there exists an analytic change of coordinates of the form* $u = x + o(|(x,y)|), v = y + o(|(x,y)|)$ *and an analytic function* Ψ *which transforms the system into the form*

$$\dot{u} = -v(1 + \Psi(u^2 + v^2)), \quad \dot{v} = u(1 + \Psi(u^2 + v^2)).$$

The following theorems are different methods to characterise an isochronous center.

Let $F(x,y) = x^2 + y^2 + o(|(x,y)|^2)$ be an analytic first integral of (1). Since all higher order terms are divisible either by x^2 or y^2, the first integral can be written in the form $F(x,y) = x^2(1 + f(x,y)) + y^2(1 + g(x,y))$. If we make the change of coordinates $X = x\sqrt{1 + f(x,y)}$, $Y = y\sqrt{1 + g(x,y)}$, (1) is transformed into

$$\dot{X} = -Y(1 + G(X,Y)), \ \dot{Y} = X(1 + G(X,Y)), \tag{4}$$

where $G(X,Y)$ is an analytic function.

Theorem 3 *The system* (4) *is isochronous if and only if*

$$T(r) = \int_0^{2\pi} \frac{d\Phi}{1 + G(r\cos\Phi, r\sin\Phi)} - 2\pi. \tag{5}$$

Theorem 4 *A center of an analytic system is isochronous if and only if there exists an analytic change of coordinates of the form* $u = x + o(|(x,y)|), v = y + o(|(x,y)|)$ *reducing the system to the linear isochronous system*

$$\dot{u} = -kv, \quad \dot{v} = ku, \tag{6}$$

where k *is a real constant.*

Theorem 5 *Let* (S) *and* (S_T) *be transversal plane differential systems of class* C^2. *Assume that the local flows defined by the solutions of* (S) *and* (S_T) *commute (in the sense of Lie's bracket). Then any center of* (S) *is isochronous.*

Theorems 3 and 4 are proved by Mardesic et al. (1995) and Theorem 5 is proved by Villarini (1992) and Sabatini (1995).

3 Computation of Isochronous Constants

Let the origin be an isochronous center. From Theorem 4 there exists an analytic change of coordinates $u = x + o(|(x,y)|), v = y + o(|(x,y)|)$ such that $\ddot{u} + u = 0$ and $\ddot{v} + v = 0$, where we have taken $k = 1$ by rescaling the time. Let $u = H(r, \varphi)$ be a power series of the form

$$H(r,\varphi) = H_1(\varphi)r + H_2(\varphi)r^2 + H_3(\varphi)r^3 + \ldots,$$

where $H_1(\varphi) = \cos\varphi$ and $H_i(\varphi)$ $(i = 1, 2, \ldots)$, are homogeneous trigonometric polynomials of degree i. Imposing $\ddot{H} + H = 0$ we obtain a system of recurent differential equations for $H_i(\varphi)$ (see Chavarriga, Giné, García (1996a)) and by means of analycity (imposing the above conditions) we obtain real numbers called *isochronous constants* or *period constants*.

The isochronous constants are polynomials in the parameters of system (2). By the Hilbert basis theorem there exists $m \in \mathbb{N}$ such that the ideal generated by the isochronous constants is finitely generated by these m isochronous constants. The vanishing of all isochronous constants is a necessary condition for the isochronicity of the origin. To prove that the origin is an isochronous center we must also use some of the theorems 3, 4 or 5.

In particular the vanishing of all isochronous constants implies the vanishing of all Lyapunov constants, see Gasull et al. (1995).

4 Cubic Reversible Systems with $R_3 = 0$ and $R_4 \neq 0$

In this section we study the isochronicity of (2) where $P_s(\varphi)$ and $Q_s(\varphi)$ are given by (3) with condition $R_3 = 0$. If $R_4 \neq 0$ (we consider the opposite case in the following section), we can take $R_4 = \pm 1$ without loss of generality because we can always make a rescaling of the form $R = kr$, where $k^2 = \pm R_4$ depending upon the sign of R_4. We only study the case $R_4 = 1$ because the other case is analogous.

The main result of this section is the following one.

Theorem 6 *The only cubic isochronous reversible systems* (2) *with* $R_3 = 0$ *and* $R_4 \neq 0$ *are the well known homogeneous systems,* i.e. $P_2 \equiv Q_2 \equiv 0$.

Proof of necessary condition. In order to simplify the massive expressions of the period constants we make the transformations $R_1 = \sqrt{k} - r_1$ and $r_1 = \sqrt{k}\, z$ with $k \geq 0$.

After the above changes the first period constant is $T_1 = 12(kz - 2r_0)$. From the vanishing of T_1 we obtain $r_0 = kz/2$ and the second period constant becomes $T_2 = 144(6 - 6kr_2 + 4r_2^2 + 4r_2R_2 - 3k^2z - 4kR_2z + 2k^2z^2)$.

If we suppose that the condition

$$kz - r_2 \neq 0 \tag{7}$$

is true, from the vanishing of T_2 we can express R_2 as a function of the remaining parameters as $4(kz - r_2)\, R_2 = 6 - 6kr_2 + 4r_2^2 - 3k^2z + 2k^2z^2$. The next period constants are

$$T_3 = \frac{3456}{(kz - r_2)^2}\,\alpha_3, \; T_4 = \frac{13063680}{(kz - r_2)^3}\,\alpha_4, \; T_5 = \frac{940584960}{(kz - r_2)^4}\,\alpha_5,$$

where $\alpha_3(r_2, k, z)$, $\alpha_4(r_2, k, z)$ and $\alpha_5(r_2, k, z)$ are polynomials.
Now we compute the following resultants relevant to r_2, viz.

$$\mathcal{R}(\alpha_3, \alpha_4, r_2) = 2123366400k^2(2 - 3k^2z + 2k^2z^2)^6(-3 - 6k + 2kz)\,\beta_{34}(k, z),$$
$$\mathcal{R}(\alpha_3, \alpha_5, r_2) = 95551488000000k^2(2 - 3k^2z + 2k^2z^2)^8\,\beta_{35}(k, z),$$

where $\beta_{34}(k, z)$ and $\beta_{35}(k, z)$ are polynomials.
With this information we begin the study of possible cases of isochronicity:

(1) Let $k = 0$. Then $r_0 = r_1 = R_1 = 0$ and $R_2 = -(3 + 2r_2^2)/(2r_2)$ with $r_2 \neq 0$ from relation (7). With these conditions the third and fourth period constants are

$$T_3 = \frac{207360}{r_2^2}(r_2 - 1)(r_2 + 1)(3 - 2r_2^2),$$

$$T_4 = \frac{1567641600}{r_2^3}(r_2 - 1)(r_2 + 1)(-81 - 20r_2 - 18r_2^2 + 20r_2^3 + 48r_2^4).$$

From the vanishing of T_3 and T_4 we obtain $r_2 = \pm 1$, which implies $R_2 = \mp\frac{5}{2}$, and the well known isochronous cubic homogeneous systems

$$\dot{r} = \left(\sin 4\varphi \mp \frac{5}{2}\sin 2\varphi\right)\, r^3, \quad \dot{\varphi} = 1 + (\cos 4\varphi \pm \cos 2\varphi)\, r^2. \tag{8}$$

(2) Let $2 - 3k^2z + 2k^2z^2 = 0$. If we make the change $z = u/k$, from the above condition we can express $k = 2(1+u^2)/(3u)$. In this case the only common real root of α_3 and α_4 is $r_2 = u$, which is in contradiction to assumption (7).

(3) Modulo a factorization alogorithm, the resultant of β_{34} and β_{35} related to k is $\mathcal{R}(\beta_{34}, \beta_{35}, k) = z^{35}(z - 2)(2z - 5)\,\beta_{345}(z)$, where $\beta_{345}(z)$ is a polynomial. Therefore the possible cases for the vanishing of β_{34} and β_{35} are given by $z = 0$, $z = 2$ or $z = 5/2$. However, in all these cases we find that β_{34} and β_{35} do not have common real roots.

(4) The last case consists of the simultaneous vanishing of $\beta_{35}(k, z)$ and $-3 - 6k + 2kz$. In this case we obtain $z = 3 + 3/(2k)$ and the only real root of β_{35} is $k = 1/4$. With these values of k and z we find that α_3 and α_4 do not have common real roots.

Finally, when the condition (7) does not hold, that is $r_2 = kz$, the second period constant is $T_2 = 432(2 - 3k^2 z + 2k^2 z^2)$ and from the vanishing of T_2 we obtain the same condition as for case **(2)**. □

5 Cubic Reversible Degenerated Systems

In this section we study the isochronicity of system (2) where $P_s(\varphi)$ and $Q_s(\varphi)$ are given by (3) when $R_3 = R_4 = 0$, i.e. *degenerated systems*. The main result of this section is the following theorem.

Theorem 7 *The only cubic reversible degenerated nontrivial isochronous systems,* i. e. $\dot\varphi \neq 1$, *are*

$$(i)\quad \begin{aligned} \dot r &= R_1 \sin\varphi\, r^2 + R_2 \sin 2\varphi\, r^3, \\ \dot\varphi &= 1 - R_1 \cos\varphi\, r - R_2 \cos 2\varphi\, r^2. \end{aligned}$$

$$(ii)\quad \begin{aligned} \dot r &= R_1 \sin\varphi\, r^2 + \tfrac{1}{2} r_1 (r_1 + R_1) \sin 2\varphi\, r^3, \\ \dot\varphi &= 1 + r_1 \cos\varphi\, r + \tfrac{1}{2} r_1 (r_1 + R_1)(1 - \cos 2\varphi)\, r^2. \end{aligned}$$

$$(iii)\quad \begin{aligned} \dot r &= R_1 \sin\varphi\, r^2, \\ \dot\varphi &= 1 - 2R_1 \cos\varphi\, r + R_1^2 (1 + \cos 2\varphi)\, r^2. \end{aligned}$$

$$(iv)\quad \begin{aligned} \dot r &= R_1 \sin\varphi\, r^2 + R_2 \sin 2\varphi\, r^3, \\ \dot\varphi &= 1 - R_1 \cos\varphi\, r. \end{aligned}$$

Proof of necessary condition. The first period constant is $T_1 = 12(r_1^2 + r_1 R_1 - 2r_0)$. From the vanishing of T_1 we obtain $2r_0 = r_1(r_1 + R_1)$. The second period constant is

$$\begin{aligned} T_2 = 144\, \big(&-r_1^4 - 5\, r_1^3\, R_1 - 7\, r_1^2\, R_1^2 - 3\, r_1\, R_1^3 - 6\, r_1^2\, r_2 - 12\, r_1\, R_1\, r_2 - \\ &6\, R_1^2\, r_2 + 4\, r_2^2 - 4\, r_1^2\, R_2 - 4\, r_1\, R_1\, R_2 + 4\, r_2\, R_2\big)\,. \end{aligned}$$

If we suppose

$$r_2 - r_1 R_1 - r_1^2 \neq 0, \tag{9}$$

from the vanishing of T_2 we can express R_2 as

$$R_2 = \frac{r_1^4 + 5 r_1^3 R_1 + 7 r_1^2 R_1^2 + 3 r_1 R_1^3 + 6 r_1^2 r_2 + 12 r_1 R_1 r_2 + 6 R_1^2 r_2 - 4 r_2^2}{4\left(-r_1^2 - r_1 R_1 + r_2\right)}.$$

With these conditions the next period constants are

$$\begin{aligned} T_3 &= (r_1 + R_1)(r_1^2 + r_1 R_1 + 2r_2)\; \alpha_3(R_1, r_1, r_2) \\ T_4 &= (r_1 + R_1)(r_1^2 + r_1 R_1 + 2r_2)\; \alpha_4(R_1, r_1, r_2) \\ T_5 &= (r_1 + R_1)(r_1^2 + r_1 R_1 + 2r_2)\; \alpha_5(R_1, r_1, r_2), \end{aligned}$$

where $\alpha_3(R_1, r_1, r_2)$, $\alpha_4(R_1, r_1, r_2)$ and $\alpha_5(R_1, r_1, r_2)$ are polynomials. On the other hand we compute the following resultants

$$\mathcal{R}(\alpha_3,\alpha_4,R_1) = r_1^{17}r_2^{15}(-2r_1^2+3r_2)^6(-r_1^2+4r_2)(r_1^2+27r_2)\ \beta_{34}(r_1,r_2)$$
$$\mathcal{R}(\alpha_3,\alpha_5,R_1) = r_1^{17}r_2^{18}(-2r_1^2+3r_2)^8(-r_1^2+4r_2)\ \beta_{35}(r_1,r_2),$$

where $\beta_{34}(r_1,r_2)$ and $\beta_{35}(r_1,r_2)$ are polynomials which do not have common real roots. So the possible cases of the vanishing of the period constants are given by:

(1) Let $r_1 = -R_1$. Then we have $r_0 = 0$ and $R_2 = -r_2$. This case corresponds to system **(i)**.

(2) Let $r_2 = -\frac{r_1}{2}(r_1+R_1)$. This implies $r_0 = R_2 = -r_2 = \frac{r_1}{2}(r_1+R_1)$, which corresponds to system **(ii)**.

(3) Suppose $r_1 + R_1 \neq 0$ and $r_2 + r_1(r_1+R_1)/2 \neq 0$. From the vanishing of $\mathcal{R}(\alpha_3,\alpha_4,R_1)$ and $\mathcal{R}(\alpha_3,\alpha_5,R_1)$ we obtain the following subcases:

(3.1) Let $r_1 = 0$. With this condition we obtain the polynomials

$$\alpha_3 = 4R_1r_2^2(23R_1^2-6r_2), \quad \alpha_4 = 48R_1r_2^3(-527R_1^4+57R_1^2r_2+24r_2^2).$$

Since the last factors of α_3 and α_4 do not have common real roots, from the vanishing of α_3 and α_4 we obtain:

(a) $r_1 = R_1 = 0$, but this is a particular case of system **(i)**.
(b) $r_1 = r_2 = 0$, but this is a contradiction of assumption (9).

(3.2) Let $r_2 = 0$. With this condition the polynomials α_3 and α_4 are

$$\alpha_3 = -r_1^2(r_1+R_1)^3(7r_1^2+32r_1R_1+37R_1^2),$$
$$\alpha_4 = r_1^3(r_1+R_1)^4(93r_1^4-160r_1^3R_1-4262r_1^2R_1^2-12640r_1R_1^3-11127R_1^4).$$

Since the last factors of α_3 and α_4 do not have common real roots, from the vanishing of α_3 and α_4 we obtain:

(a) $r_2 = r_1 = 0$, but this is impossible under hypothesis (9).
(b) $r_1 + R_1 = 0$, but this contradicts assumption $r_1 + R_1 \neq 0$.

(3.3) Let $r_2 = 2r_1^2/3$. With this condition we obtain the polynomials

$$\alpha_3 = -\tfrac{1}{27}r_1^2(r_1+3R_1)^2(137r_1^3+453r_1^2R_1+411r_1R_1^2+111R_1^3),$$
$$\alpha_4 = -\tfrac{1}{81}r_1^3(r_1+3R_1)^3(1055r_1^5+19197r_1^4R_1+90130r_1^3R_1^2+ 176334r_1^2R_1^3+133887r_1R_1^4+33381R_1^5).$$

Since the last factors of α_3 and α_4 do not have common real roots, from the vanishing of α_3 and α_4 we obtain:

(a) $r_1 = r_2 = 0$, but this is impossible under hypothesis (9).
(b) $r_1 = -3R_1$. Then $r_2 = 6R_1^2$, but this is impossible under (9).

(3.4) Let $r_2 = r_1^2/4$. With this condition we obtain the polynomials

$$\alpha_3 = r_1^2(r_1 + 2R_1)\,\gamma_3(r_1, R_1), \quad \alpha_4 = -r_1^3(r_1 + 2R_1)\,\gamma_4(r_1, R_1),$$

where the polynomials γ_3 and γ_4 do not have common real roots. From the vanishing of α_3 and α_4 we obtain

(a) $r_1 = r_2 = 0$, but this is impossible under hypothesis (9).
(b) $r_1 = -2R_1$. This condition implies $r_0 = r_2 = R_1^2$ and $R_2 = 0$ which corresponds to the system **(iii)**.

Finally we can suppose that assumption (9) does not hold, that is $r_2 = r_1(r_1 + R_1)$. In this case the second period constant is $T_2 = -432r_1(r_1 + R_1)^2(r_1 + 3R_1)$ and from the vanishing of T_2 we obtain the following cases:

(a) Let $r_1 = 0$. Then we have $r_0 = r_2 = 0$ which corresponds to a trivial isochronous system, i.e. $\dot{\varphi} = 1$.
(b) Let $r_1 = -R_1$. Then $r_0 = r_2 = 0$, which corresponds to the system **(iv)**.
(c) If $r_1 = -3R_1$ with $R_1 \neq 0$, the third and fourth period constants are

$$\begin{aligned} T_3 &= 1658880R_1^2(57R_1^4 + 25R_1^2R_2 + R_2^2), \\ T_4 &= 25082265600R_1^2(-1584R_1^6 - 711R_1^4R_2 + 53R_1^2R_2^2 + 7R_2^3), \end{aligned}$$

which do not have common real roots.

Proof of sufficient condition.

System (i) This system in polar coordinates admits a reciprocal integrating factor $V(r,\varphi)^{1/k}$, where $V(r,\varphi) = r^\alpha\left(1 + V_1(\varphi)\,r + V_2(\varphi)\,r^2\right)$ and $V_1(\varphi) = 2R_1(k-1)\cos\varphi$, $V_2(\varphi) = R_2(k-1)/k$, $\alpha = 6k - 2$ and $k = 1/2 \pm \sqrt{1 + 4R_2/R_1^2}$. Due to the arbitrariness of k we have two different reciprocal integrating factors. Therefore its quotient is a first integral of system (i).

In cartesian coordinates, $x = r\cos\varphi$, $y = r\sin\varphi$, the system (i) takes the form

$$\begin{aligned} \dot{x} &= y(-1 + 2R_1x + 3R_2x^2 - R_2y^2), \\ \dot{y} &= x - R_1x^2 + R_1y^2 - R_2x^3 + 3R_2xy^2. \end{aligned} \tag{10}$$

If $R_2 = 0$, the above system is a quadratic system and so we can assume $R_2 \neq 0$. If $R_1^2 + 4R_2 \geq 0$, an analytic first integral for system (10) is given by $H = V_+^{\lambda_+}\,V_-^{\lambda_-}$, where $\lambda_\pm = 2R_1/(\Delta \mp R_1)$ and

$$V_\pm(x,y) = (x^2 + y^2)^{-1}\left(1 - (R_1 \pm \Delta)x + \frac{R_2(\Delta \pm R_1)}{\Delta \mp R_1}(x^2 + y^2)\right),$$

where $\Delta = \sqrt{R_1^2 + 4R_2}$.
We note that, if $R_1^2 + 4R_2 < 0$ (that is $\Delta \in \mathbb{C}$), then $V_- = \bar{V}_+$ and $\lambda_- = -\bar{\lambda}_+$,

where the bar denotes complex conjugate. In this case the above first integral takes the form $H(x,y) = V_+^{\lambda_+}\,\bar V_+^{-\bar\lambda_+}$, that is

$$H(x,y) = \left[(\mathrm{Re}V_+)^2 + (\mathrm{Im}V_+)^2\right]^{\mathrm{i}\,\mathrm{Im}\lambda_+} \exp\left[-2\,\mathrm{i}\,\mathrm{Re}\lambda_+ \arctan\left(\frac{\mathrm{Im}V_+}{\mathrm{Re}V_+}\right)\right]$$

and $H(x,y)$ raised to the power of i is just a real first integral.

On the other hand the corresponding transversal commuting system is the orthogonal system,

$$\begin{aligned}\dot x &= x - R_1x^2 + R_1y^2 - R_2x^3 + 3R_2xy^2\\ \dot y &= y(1 - 2R_1x - 3R_2x^2 + R_2y^2),\end{aligned}$$

and by Theorem 5 (10) is isochronous at the origin.

System (ii) This system in cartesian coordinates takes the form

$$\begin{aligned}\dot x &= y\Big(-1 + (R_1 - r_1)x + r_1(r_1 + R_1)x^2 - r_1(r_1 + R_1)y^2\Big)\\ \dot y &= x + r_1x^2 + R_1y^2 + 2r_1(r_1 + R_1)xy^2.\end{aligned} \tag{11}$$

If we assume that $R_1 \neq r_1$, (11) admits the following transversal commuting system

$$\begin{aligned}\dot x &= x - R_1x^2 - r_1y^2 + 2r_1R_1\frac{r_1+R_1}{r_1-R_1}x^3 - \\ &\quad 2r_1^2\,\frac{r_1+R_1}{R_1-r_1}xy^2 - r_1^2\,\frac{(r_1+R_1)^2}{R_1-r_1}(x^4 - y^4),\\ \dot y &= y\Big(1 + (r_1 - R_1)x - r_1(r_1 - 3R_1)\frac{r_1+R_1}{r_1-R_1}x^2\\ &\quad + \frac{r_1(r_1+R_1)^2}{r_1-R_1}y^2 + \frac{2r_1^2(r_1+R_1)^2}{r_1-R_1}(x^3 + xy^2)\Big).\end{aligned}$$

By Theorem 5 (11) with $R_1 \neq r_1$ is isochronous at the origin.

Finally, when $R_1 = r_1$, system (11) in cartesian coordinates takes the form

$$\begin{aligned}\dot x &= y\Big(-1 + 2R_1^2(x^2 - y^2)\Big),\\ \dot y &= x + R_1x^2 + R_1y^2 + 4R_1^2xy^2.\end{aligned} \tag{12}$$

The above system admits the reciprocal integrating factor $V(x,y) = (x^2 + y^2)^2(1 + 2R_1x + R_1^2(x^2+y^2))$, which has the following associated C^∞ first integral

$$H(x,y) = \frac{x^2+y^2}{1 + 2R_1x + R_1^2(x^2+y^2)} \exp\left(\frac{2R_1x - 1}{R_1^2(x^2+y^2)}\right).$$

On the other hand (12) has the transversal commuting system

$$\dot{x} = (x + R_1x^2 - R_1y^2)(x^2 + y^2),$$
$$\dot{y} = (y + 2R_1xy)(x^2 + y^2).$$

By Theorem 5 (12) is isochronous at the origin.

Note. This transversal commuting system has been found using particular solutions of (12). The method consists of looking for a transversal commuting system of the form $S_T = fS'$, where f is a particular solution of the plane differential system S. So we have

$$[S, fS'] = f[S, S'] + (Sf)S' = f[S, S'] + \lambda fS' = f\big([S, S'] + \lambda S'\big),$$

where λ is the *cofactor* of the particular solution f, that is $\dot{f} = \lambda f$.

System (iii) We can suppose $R_1 = 1$ without loss of generality. After the change of variable $x = r\cos\varphi$, system (iii) becomes $r\ \mathrm{d}x = (-2x^2+3x-1)\ \mathrm{d}r$, which has the first integral $H(x,r) = (2x-1)^{-2}(x-1)^2r^2$.

We conclude that system (iii) in cartesian coordinates takes the form

$$\begin{aligned}\dot{x} &= y(1-x)(-1+2x)\\ \dot{y} &= x - 2x^2 + y^2 + 2x^3\end{aligned} \tag{13}$$

and an analytic first integral for system (13) is

$$H(x,y) = \frac{(x-1)^2(x^2+y^2)}{(2x-1)^2}.$$

The change of coordinates, $u = x(x-1)/(2x-1)$, $v = y(x-1)/(2x-1)$, transforms (13) into

$$\dot{u} = -v\big(1 + G(u,v)\big), \quad \dot{v} = u\big(1 + G(u,v)\big),$$

where $1 + G(u,v) = 1 - 2x + 2x^2$. By Theorem 3 the period function is given by the following integral

$$T(r) = \int_0^{2\pi} \frac{\mathrm{d}\theta}{1 + G(r\cos\theta, r\sin\theta)} = 2\pi - 2r\int_0^{2\pi} \frac{\cos\theta\ \mathrm{d}\theta}{\sqrt{1 + 4r^2\cos^2\theta}} = 2\pi.$$

We easily see that the last integral is zero by making the change $w = \theta + \pi$ and using the 2π-periodicity.

System (iv) In cartesian coordinates this system takes the form

$$\begin{aligned}\dot{x} &= y(-1 + 2R_1x + 2R_2x^2),\\ \dot{y} &= x - R_1x^2 + R_1y^2 + 2R_2xy^2.\end{aligned} \tag{14}$$

The change of variable, $z = y^2$, transforms (14) into the linear differential equation

$$\big(-1+2x(R_1+R_2x)\big)\frac{\mathrm{d}z}{\mathrm{d}x}=2x(1-R_1x)+2(R_1+2R_2x)z,$$

which by means of an immediate integration has the following first integral in the original variables

$$H(x,y)=\frac{x^2+y^2}{-1+2x(R_1+R_2x)}.$$

On the other hand (14) admits the following transversal commuting system,

$$\begin{aligned}\dot{x}&=x-R_1x^2+R_1y^2+2R_2xy^2\\ \dot{y}&=y(1-2R_1x+2R_2y^2),\end{aligned}$$

and by Theorem 5 (14) is isochronous at the origin. □

Note. We recall that the well-known quadratic isochronous centers are reversible. A first approximation to the problem of finding the cubic isochronous reversible centers would consist of the reversible perturbation of these systems. Making these computations we only obtain systems (i) and (iv) of Theorem 7.

6 Cubic Reversible Hamiltonian Systems

In this section we study the isochronicity of (2) where $P_s(\varphi)$ and $Q_s(\varphi)$ are given by (3) and with the aditional condition of null divergence, i.e.

$$\frac{1}{r}\frac{\partial\,(r\dot{r})}{\partial\,r}+\frac{\partial\,\dot{\varphi}}{\partial\,\varphi}\equiv 0,$$

which verifies that $r_1=3R_1$ and $r_2=2R_2$.

The main result of this section is the following theorem.

Theorem 8 *The only cubic reversible isochronous hamiltonian system is*

$$\dot{r}=\left(-\tfrac{1}{3}\sin\varphi+\sin 3\varphi\right)r^2+\tfrac{4}{9}\left(-2\sin 2\varphi+\sin 4\varphi\right)r^3$$

$$\dot{\varphi}=1+\left(-\cos\varphi+\cos 3\varphi\right)r+\tfrac{1}{9}\left(12-16\cos 2\varphi+4\cos 4\varphi\right)r^2.$$

Proof of necessary condition. We can take $R_3=1$ without loss of generality. Under this assumption the first period constant is $T_1=8(2+18R_1^2-3r_0)$. From the vanishing of T_1 we obtain $3\,r_0=2+18R_1^2$.

Computing the resultants of the period constants related to R_4 we obtain

$$\begin{aligned}\mathcal{R}(T_2,T_3,R_4)&=\Omega_1(R_1,R_2)\\ \mathcal{R}(T_2,T_4,R_4)&=\Omega_2(R_1,R_2)\\ \mathcal{R}(T_2,T_5,R_4)&=\Omega_3(R_1,R_2),\end{aligned}$$

where Ω_1, Ω_2 and Ω_3 are polynomials in the variables R_1, R_2 the resultants of which relevant to R_2 are

$$\mathcal{R}(\Omega_1, \Omega_2, R_2) = (1+3R_1)^4 \ \alpha(R_1), \ \ \mathcal{R}(\Omega_1, \Omega_3, R_2) = (1+3R_1)^4 \ \beta(R_1).$$

Since the polynomials α and β do not have common real roots, the only possible case of isochronicity is given by $R_1 = -1/3$.

The second and thirth period constants are

$$T_2 = \frac{32}{3}(-112 + 360R_2 + 324R_2^2 + 360R_4 + 81R_4^2),$$
$$T_3 = \frac{2560}{3}(-2304 - 12496R_2 - 1512R_2^2 + 972R_2^3 - 10928R_4 + 4320R_2R_4 - 6804R_2^2R_4 + 10152R_4^2 + 243R_2R_4^2 + 729R_4^3),$$

and its resultant related to R_4 is $\mathcal{R}(T_2, T_3, R_4) = 419904(8 + 9R_2)\ \gamma(R_2)$, where the only real root of the polynomial $\gamma(R_2)$ is not a root of the period constants. Consequently the only possible case of isochronicity is given by $R_2 = -8/9$.

With these values for the parameters of the system, the second period constant is $T_2 = 32(9R_4 - 4)(9R_4 + 44)/3$. From the vanishing of T_2 we now have two possibilities:

If $R_4 = -44/9$, we have the non-isochronicity condition $T_3 \neq 0$.
Let $R_4 = 4/9$. This case corresponds to the system of Theorem 8.

Proof of sufficient condition. The hamiltonian system of Theorem 8 in cartesian coordinates takes the form

$$\begin{aligned} \dot{x} &= y\left(-1 + \tfrac{8}{3}x - \tfrac{32}{9}y^2\right), \\ \dot{y} &= x - \tfrac{4}{3}y^2. \end{aligned} \tag{15}$$

An analytic first integral for (15) is $H(x,y) = 9y^2 + (3x - 4y^2)^2$. From the above first integral we obtain the linearising change of coordinates, $u(x,y) = 3y$, $v(x,y) = 3x - 4y^2$, bringing the (15) to the linear isochronous system $\dot{u} = v$, $\dot{v} = -u$.

By Theorem 4 the system (15) is isochronous at the origin. □

References

Bautin, N.N. (1952): On the number of limit cycles which appears with the variation of coefficients from an equilibrium position of focus or center type(R). Mat. Sb. **(30)** **(72)**, 181–196; Amer. Math. Soc. Transl. **100** (1954), 397–413

Chavarriga, J. (1994): Integrable systems in the plan with a center type linear part. Applicationes Mathematicae. **22**, 285–309

Chavarriga, J., Giné, J. (1996): Integrability of a linear center perturbed by fourth degree homogeneous polynomial. Publicacions Matemàtiques **40** (1), 21–39

Chavarriga, J., Giné, J. (1997): Integrability of a linear center perturbed by fifth degree homogeneous polynomial. To appear in Publicacions Matemàtiques **41** (2)

Chavarriga, J., Giné, J., García, I. (1996a): Isochronous centers of cubic systems with degenerate infinity. Submitted to Differential Equations and Dynamical Systems

Chavarriga, J., Giné, J., García, I. (1996b): Isochronous centers of a linear center perturbed by fourth degree homogeneous polynomial. Preprint. Universitat de Lleida, Spain. Submitted to Journal of Dynamics and Differential Equations

Chavarriga, J., Giné, J., García, I. (1996c): Isochronous centers of a linear center perturbed by fifth degree homogeneous polynomial. Preprint. Universitat de Lleida, Spain. Submitted to Journal of London Mathematical Society

Chicone, C., Jacobs, M. (1989): Bifurcation of critical periods for plane vector fields. Trans. Amer. Math. Soc. **312**, 433–486

Christopher, C.J., Devlin, J. (1994): Isochronous centres in planar polynomial systems. Preprint. University of Wales

Coppel, W.A. (1996): A survey of Quadratic Systems. J. Differential Equations. **122**, 293–304

Gasull, A., Guillamon, A., Mañosa, V. (1995): Centre and isochronicity conditions for systems with homogeneous nonlinearities. *Proceedings of the 2nd Catalan Days on Applied Mathematics*, Sofonea, M. and Corvellec, J.N. eds.(Presses Universitaires de Perpignan)

Giné, J. (1996): *Contribution to the integrability of systems of ordinary differential equations in the plane with linear part of center type* (Ph. D., Universitat Politècnica de Catalunya)

Lloyd, N.G. (1983): Small amplitude limit cycles of polynomial differential equations. *Lecture Notes in Mathematics* **1032** *Ordinary differential Equations and Operators* (Springer, Heidelberg, Berlin) 346–356

Lloyd, N.G., Christopher, J., Devlin, J., Pearson, J.M., Yasmin, N. (1996): Quadratic like cubic systems. Preprint. University of Wales

Loud, W.S. (1964): Behavior of the period of solutions of certain plane autonomous systems near centers. Contributions to Differential Equations **3**, 21–36

Lunkevich, V.A., Sibirskii, K.S. (1982): Integrals of a general quadratic differential system in cases of a center. Diff. Equations **18**, 563–568

Mardesic, P., Moser-Jauslin, L.,Rousseau, C. (1995): Darboux linearization and isochronous centers with a rational first integral. Preprint

Mardesic, P., Rousseau, C., Toni, B. (1995): Linearization of isochronous centers. J. Differential Equations **121** 67–108

Pleshkan, I. (1969): A new method of investigating the isochronicity of a system of two differential equations. Differential Equations **5**, 796–802

Rousseau, C., Toni, B. (1993): Local bifurcation in vector fields with homogeneous nonlinearities of the third degree. Canad. Math. Bull.

Rousseau, C., Toni, B. (1993): Local bifurcation of critical periods in the reduced Kukles system. Preprint

Sabatini, M. (1995): Quadratic isochronous centers commute (Preprint, U.T.M. **461** Università degli studi di Trento)

Schlomiuk, D. (1993): Algebraic and Geometric Aspects of the Theory of Polynomials Vector Fields. *Bifurcations and Periodics Orbits of Vectors Fields* (Kluwer Academic Publishers), 429–467
Villarini, M. (1992): Regularity properties of the period function near a centre of planar vector fields. Nonlinear Analysis, T.M.A. 19, **8**, 787–803
Zoladek, H. (1994): On certain generalization of the Bautin's Theorem. Nonlinearity **7**, 233–279

On the Polynomial First Integrals of the (a, b, c) Lotka–Volterra System

Simon Labrunie

L.P.T.M.C., Université Paris VII, 2 place Jussieu, 75251 Paris Cedex 05, France

Abstract. Using elementary differential algebraic techniques, we prove that the 3-D Lotka-Volterra dynamical system has no other nontrivial polynomial first integrals than the previously known ones.

1 Introduction

The (a, b, c) or 3-D Lotka-Volterra dynamical system (Lotka 1925; Volterra 1931)

$$\begin{cases} \dot{x} &= x\,(c\,y + z) \\ \dot{y} &= y\,(a\,z + x) \\ \dot{z} &= z\,(b\,x + y), \end{cases} \tag{1}$$

where (a, b, c) are three real (or complex) parameters, appears in various areas of science such as ecology (in which it models a three-species competition), plasma physics (in which it approximates the Vlasov-Poisson equation) etc.

Besides their biological and physical applications, the Lotka-Volterra systems with various numbers of "species" have been a matter of interest for mathematical study. The two species case has been exhaustively studied by Marc Feix and his collaborators (cf. the the paper by Cairó and Llibre in this volume). With three species or more, their properties rank them among some of the nicest dynamical systems not (generically) integrable by classical or special functions. In particular, the problem of existence of first integrals has been intensively studied by Cairó, Feix and others (Cairó et al. 1989; Cairó and Feix 1992) and also, for the (a, b, c) system, in Grammaticos et al. (1990). The techniques ranged from Carleman embedding to an extended Lie symmetry method and Painlevé analysis.

The objective of this work is somewhat different: it is to achieve a classification of the first integrals of the system (1). For the polynomial first integrals the classification is finished when the parameters are real (which is the case in biological and physical applications) and nearly finished when they are complex. This material was first published in Labrunie (1996), where the proofs that are sketched or even omitted here can be found. Then the study of complex cases was completed by Jean Moulin-Ollagnier (1997). As for the nonpolynomial first integrals, we give a theorem of generic nonexistence which may be valuable for both theory and applications.

This contribution is organised as follows: we recall the framework of Darboux elements in Sect. 2, study the Darboux elements of the (a, b, c) system in Sect. 3 and achieve the promised classification in Sect. 4. Finally in Sect. 5 we summarise our results.

2 Darboux Elements: a General Overview

A Darboux polynomial (Darboux 1878) of the autonomous polynomial dynamical system

$$\dot{x}_i = V_i(x_1, \ldots, x_n), \quad i = 1, \ldots, n \tag{2}$$

is a polynomial $f(x_1, \ldots, x_n)$ such that there exists a polynomial eigenvalue $p(x_1, \ldots, x_n)$ satisfying

$$\frac{\mathrm{d}f}{\mathrm{d}t} \equiv \sum_{i=1}^{n} \frac{\partial f}{\partial x_i} V_i = p\, f.$$

The couple (p, f) may also be called a *Darboux element.* The following properties of Darboux elements can be easily proven:

1. If m is the greatest of $\deg V_i$, $i = 1, \ldots, n$, then $\deg p \leq m - 1$.
2. The product of two Darboux polynomials is still Darboux, namely $(\mathrm{d}f_1/\mathrm{d}t = p_1\, f_1$ and $\mathrm{d}f_2/\mathrm{d}t = p_2\, f_2) \Longrightarrow \mathrm{d}(f_1\, f_2)/\mathrm{d}t = (p_1 + p_2)\,(f_1\, f_2)$. This can be generalised to any number of factors.
3. All irreducible factors of a Darboux polynomial are Darboux, i.e.

$$\left.\begin{array}{rcl} \mathrm{d}f/\mathrm{d}t & = & p\, f \\ f & = & \prod_{i=1}^{s} f_i^{\alpha_i} \end{array}\right\} \Longrightarrow \left\{\begin{array}{c} \forall i \in [[1, s]],\ \mathrm{d}f_i/\mathrm{d}t = p_i\, f_i \\ \text{with } p = \sum_{i=1}^{s} \alpha_i\, p_i. \end{array}\right.$$

   Thus the characterisation of the system's Darboux polynomials is equivalent to that of irreducible ones.
4. If there is a rational first integral equal to the irreducible fraction a/b, then a and b are Darboux polynomials of same eigenvalue.
5. If the dynamical system (2) is homogeneous of degree m, i.e. all V_i are homogeneous of degree m, then p is homogeneous of degree $m - 1$ and all homogeneous components of f are Darboux. Thus the search can be restricted to homogeneous Darboux polynomials.

   Less trivially, we have the following two theorems:
6. If a polynomial dynamical system admits an *algebraic* first integral, it also admits a *rational* one. This is a consequence of Bruns' theorem (Kummer et al. 1990).
7. If a *homogeneous* polynomial dynamical system admits a *meromorphic* first integral, it also admits a *rational* one (Ziglin 1982).

Darboux elements are the basis to first integral search techniques for polynomial dynamical systems, such as the Prelle-Singer procedure (Man 1994). Indeed a polynomial first integral f is a Darboux polynomial with $p = 0$. Only semidecisive procedures are known so far: there is no way in general to get an exhaustive characterisation of all Darboux elements of a given dynamical system. However, this becomes possible for some systems having distinctive features, such as the (a, b, c) system.

3 Darboux Elements of the Lotka–Volterra System

The following properties of the Lotka-Volterra system are essential and will play a key role in the characterisation of the polynomial first integrals:

1. The system (1) is homogeneous of degree two; hence all the Darboux polynomials can be chosen homogeneous and the eigenvalues are homogeneous of degree one. The equation defining the Darboux elements is:

$$x\,(c\,y + z)\,f_x + y\,(a\,z + x)\,f_y + z\,(b\,x + y)\,f_z = (\alpha x + \beta y + \gamma z)\,f, \qquad (3)$$

   where α, β, γ as well as the coefficients of f are *a priori* complex. For a Darboux polynomial f, the coefficients α, β, γ in Equation (3) may be denoted $\alpha(f)$, $\beta(f)$, $\gamma(f)$.
2. The system is invariant under simultaneous cyclical permutations of (x, y, z) and (a, b, c); and (3) is invariant under simultaneous cyclical permutations of (x, y, z), (a, b, c), and (α, β, γ).
3. The system is factorised, i.e. the variables (x, y, z) are Darboux polynomials. This means that the coordinate planes are invariant under the flow of (1) so that the dynamics can be projected onto these planes. Moreover, the Darboux elements survive through this projection, i.e. they become Darboux elements of the projected dynamics. This will allow one to obtain information on the Darboux elements of (1) because the reduced dynamics are easier to study.

Now we list some properties of the Darboux elements of the system (1).

Proposition 3.1 *Let f be a solution of (3). Then* $(\gamma \notin \mathrm{N} \Longrightarrow y \mid f)$ *and* $(\gamma \notin a\mathrm{N} \Longrightarrow x \mid f)$, *and, cyclically:* $(\beta \notin \mathrm{N} \Longrightarrow x \mid f)$, $(\beta \notin c\mathrm{N} \Longrightarrow z \mid f)$, $(\alpha \notin \mathrm{N} \Longrightarrow z \mid f)$, $(\alpha \notin b\mathrm{N} \Longrightarrow y \mid f)$.

These results are obtained by studying the reduced dynamics on suitable coordinate planes. For instance, if we examine the $y = 0$ plane, we find that the factorisation and multiplication properties of Darboux polynomials imply that, if $\gamma \notin \mathrm{N}$, $\tilde{f}(x, z) = f(x, 0, z)$ is divisible by any power of x. Similarly, if $\alpha \notin b\mathrm{N}$, it is divisible by any power of z. Hence in these two cases $\tilde{f} = 0$, i.e. $y \mid f$.

This leads to the following statement, a characterisation of the polynomial eigenvalues of (1), which will be a fundamental tool for proving our result.

Proposition 3.2 *Let f be a non-zero Darboux polynomial. Then there exist two positive integers γ_1 and γ_2 such that $\gamma(f) = \gamma_1 + \gamma_2\, a$.*

Suppose $\gamma \notin \mathbf{N} + \mathbf{N}a$. Then $\gamma \notin \mathbf{N}$ and $\gamma \notin \mathbf{N}a$. Hence by Proposition 3.1 f is divisible by x:

$$f = x\, f_1 \quad \text{and} \quad \mathrm{d}f_1/\mathrm{d}t = (\alpha\, x + (\beta - c)\, y + (\gamma - 1)\, z)\, f_1 \ .$$

However, $\gamma - 1$ cannot be in $\mathbf{N} + \mathbf{N}a$ unless γ be in it. This can be repeated indefinitely so that f is divisible by all powers of x and hence zero. Contradiction.

This result allows one to classify all the Darboux polynomials of (1) when $(a, b, c) \in (\mathbf{R} \setminus \mathbf{Q}^+)^3$ as is shown in the two following propositions.

Proposition 3.3 *Let f be a solution of (3). If $a \notin \mathbf{Q}^+$, then f is*

$$f = x^{\gamma_1}\, y^{\gamma_2}\, f_1, \tag{4}$$

where the positive integers γ_1 and γ_2 satisfy $\gamma(f) = \gamma_1 + a\,\gamma_2$ and f_1 is a solution to (3) such that $\gamma(f_1) = 0$.

Since $a \notin \mathbf{Q}^+$, $\mathbf{N} \cap \mathbf{N}a = \{0\}$. If γ is zero, setting $\gamma_1 = \gamma_2 = 0$ yields the result. If not, then by Proposition 3.1 f is divisible by x and/or by y. Let γ_1 and γ_2 be the highest powers of x and y contained in f. Setting $f = x^{\gamma_1}\, y^{\gamma_2}\, f_1$, we see that f_1 is a Darboux polynomial. If $\gamma(f_1)$ were non-zero, f_1 would be divisible either by x or by y, which would contradict the assumption on γ_1 and γ_2, hence the result.

Theorem 3.4 *Let f be a solution of (3). If none of the system parameters (a, b, c) belongs to $\mathbf{Q}^+$, then there exist three positive integers p, q, r and a polynomial first integral I of (1) – which may be trivial – such that*

$$f = x^p\, y^q\, z^r\, I \tag{5}$$

and

$$\left\{ \begin{array}{lcl} \alpha & = & q + b\, r \\ \beta & = & r + c\, p \\ \gamma & = & p + a\, q. \end{array} \right. \tag{6}$$

Using successively Proposition 3.3 and the corresponding statements on α and β as long as it is possible to factor f gives (5). I is a first integral since, if $\alpha(I)$, $\beta(I)$, $\gamma(I)$ were not all zero, it would be possible to factor f once more. The relation between (α, β, γ), (a, b, c), and (p, q, r) follows from the basic properties of Darboux elements.

The first consequence of this theorem is the absence of first integrals not reducible to polynomials for almost all (in the measure-theoretical sense) values of the parameters.

Theorem 3.5 *If none of the system parameters belongs to* $\mathbf{Q}^+$, *then the existence of an algebraic or meromorphic first integral implies that of a polynomial one and all rational first integrals are quotients of polynomial ones.*

By Bruns' and Ziglin's theorems, the existence of an algebraic or meromorphic first integral implies that of a rational one. The latter is the quotient of two Darboux polynomials, which in turn can be factored by Theorem 3.4, so that their quotient is simply that of two polynomial first integrals.

4 Polynomial First Integrals

Though it gives no information *directly* on the polynomial first integrals, the trivialisation result of Theorem 3.4 suggests that we make a distinction between those first integrals which are divisible by x, y or z and the others. We shall refer to the latter as the *strict* polynomial first integrals, and to the former as *nonstrict* ones.

The nonstrict first integrals are characterised by the following statement.

Proposition 4.1 *Let f be a nonstrict polynomial first integral. Then the system parameters (a,b,c) are negative rational numbers whose product is -1; and $f = x^p\, y^q\, z^r\, I$, where I is another polynomial first integral– which may be trivial – and (p,q,r) satisfy Equations (6) with $\alpha=\beta=\gamma=0$, i.e.*

$$\begin{cases} q+b\,r &= 0 \\ r+c\,p &= 0 \\ p+a\,q &= 0. \end{cases} \tag{7}$$

It is easy to prove (e.g. with the reduced dynamics) that a nonstrict first integral is indeed divisible by $x\,y\,z$. Setting $f = x\,y\,z\,Q$, $\mathrm{d}f/\mathrm{d}t = 0$ becomes

$$x\,y\,z\,\frac{\mathrm{d}Q}{\mathrm{d}t} + \Big((1+b)\,x + (1+c)\,y + (1+a)\,z\Big)\,x\,y\,z\,Q = 0\ .$$

Thus Q is a Darboux polynomial with $\alpha(Q) = -(1+b)$, $\beta(Q) = -(1+c)$, $\gamma(Q) = -(1+a)$. Proposition 3.2 implies that the parameters (a,b,c) are strictly negative. If e.g. a were ≥ 0, one would have $\gamma = \gamma_1 + \gamma_2\,a \geq 0$, but $\gamma = -1-a < 0$. Hence Theorem 3.4 holds for Q and $Q = x^{p_1}\,y^{q_1}\,z^{r_1}\,I$, or $f = x^p\,y^q\,z^r\,I$ with $p = p_1+1$ etc.

We check that (a,b,c) satisfy Equations (7). Those imply $a,b,c \in \mathbf{Q}^-$ since $p,q,r \geq 1$. Since this linear system admits a nonzero solution (p,q,r), its determinant $a\,b\,c+1$ is necessarily zero.

In order to finish the classification, we have to perform the following computation on strict first integrals.

Proposition 4.2 *Let f be a strict polynomial first integral of degree m. Then the system parameters satisfy $(-a\,b\,c)^m = 1$ and f can be written as*

$$f = \lambda\Big[(-a)^m\,\left((z-b\,x)^m - z^m - (-b\,x)^m\right) \\ + (y-a\,z)^m + (y-x/c)^m - y^m\Big] + x\,y\,z\,Q(x,y,z). \quad (8)$$

The analysis of the reduced dynamic in (x,z) leads to the following result (Labrunie 1996; Proposition 3.1):

$$f = \lambda_1\,\left((z-b\,x)^m + (z-y/a)^m - z^m\right) + x\,y\,k_1(x,y,z), \quad (9)$$

which, of course, comes with the cyclically permuted equalities

$$f = \lambda_2\,\left((y-a\,z)^m + (y-x/c)^m - y^m\right) + x\,z\,k_2(x,y,z), \quad (10)$$
$$f = \lambda_3\,\left((x-c\,y)^m + (x-z/b)^m - x^m\right) + z\,y\,k_3(x,y,z). \quad (11)$$

The compatibility condition of these three expressions is $(-a\,b\,c)^m = 1$, and in this case they coalesce to (8).

We consider only the two subcases $a\,b\,c = -1$ and $a\,b\,c = 1$ of the condition $(-a\,b\,c)^m = 1$. This, of course, will complete the classification for real parameters. We know (Grammaticos et al. 1990) that in those two cases there exist polynomial first integrals of degree respectively one and two. We shall compare them to (8), taking advantage of the fact that the polynomial Q which seems hard to determine – we have exhausted all the arithmetic arguments – is necessarily zero if $m = 1$ or 2.

It is also known (Grammaticos et al. 1990) that there exist other cases of polynomial first integrals when $-a\,b\,c$ is a primitive third, fourth or sixth root of unity which, of course, means that at least one parameter is complex. They cannot be found by the above theory because the first integrals are of degrees 3, 4 and 6 respectively and it is impossible to calculate them by (8). However, they have been characterised in Moulin-Ollagnier (1997) thanks to a cleverer representation of f and to smart combinatoric arguments about the linear system generated by $\mathrm{d}f/\mathrm{d}t = 0$ in this representation. The result is that they reduce to the cases given in Grammaticos et al. (1990).

Theorem 4.3 *If $a\,b\,c = -1$, the generators of the algebra of polynomial first integrals are, in all cases, the first integral $f_1 = a\,b\,x + y - a\,z$; when, moreover, $(a,b,c) \in (\mathbf{Q}^-)^3$, $x^p\,y^q\,z^r$, where (p,q,r) is the least positive integer solution to the system (7).*

We use the following algorithm:

1. If f is nonstrict, apply Proposition 4.1 as many times as possible and factor f as $(x^p\, y^q\, z^r)^i\, g$, where g is either a constant, or a strict first integral of strictly lower degree. In the latter case, apply Step 2 to g.
2. If f is strict, it satisfies Equation (8). With $c = -1/(a\,b)$ and $m = 1$, this formula is $f = \lambda\, f_1$ and, for $m = 2$, $f = \lambda(f_1)^2$. If $\deg f < 3$, the term $x\,y\,z\,Q$ must be zero. We check that f_1 is indeed a first integral.
   Suppose $m = \deg f \geq 3$. The terms in the expansion of $(f_1)^m$ which are not divisible by $x\,y\,z$ are those in factor of λ in Equation (8). Consequently $f = \lambda\,(f_1)^m + x\,y\,z\,Q_1$, and $x\,y\,z\,Q_1$ is a nonstrict polynomial first integral, which may be zero. If it is not, apply Step 1 to it.

As the first step produces an integral of strictly lower degree, this algorithm terminates in a finite number of steps and produces an expression of f in function of f_1 and $x^p\, y^q\, z^r$ only.

Theorem 4.4 *If $a\,b\,c = 1$, polynomial first integrals exist under the extra condition that $1+b+a\,b = 0$ and in this case the algebra they span is generated by $f_2 = a^2\,b^2\,x^2 - 2\,a\,b\,x\,y + y^2 - 2\,a\,y\,z - 2\,a^2\,b\,x\,z + a^2\,z^2$.*

If $a\,b\,c = 1$, the degree of the integral is necessarily even: $m = 2\,k$. When $k = 1$ and $c = 1/(a\,b)$, (8) is $f = \lambda\, f_2$ since then $x\,y\,z\,Q = 0$. The derivative of f_2 is

$$\mathrm{d}f_2/\mathrm{d}t = -4\,a\,x\,y\,z\,(1+b+a\,b).$$

Thus, as $a \neq 0$, f_2 cannot be a first integral unless $1 + b + a\,b = 0$.

For $k > 1$, (8) is $f - \lambda\,(f_2)^k = x\,y\,z\,Q_1$, or, since f is assumed to be a first integral,

$$\frac{\mathrm{d}}{\mathrm{d}t}\left(f - \lambda\,(f_2)^k\right) = 0 - \lambda,\frac{\mathrm{d}}{\mathrm{d}t}\left((f_2)^k\right) = \frac{\mathrm{d}}{\mathrm{d}t}(x\,y\,z\,Q_1)$$

which, once simplified by $x\,y\,z$, gives

$$4\,\lambda\;a\,k\,(1+b+a\,b)\,(f_2)^{k-1} = \frac{\mathrm{d}Q_1}{\mathrm{d}t} + \left[(1+b)\,x + (1+c)\,y + (1+a)\,z\right] Q_1. \qquad (12)$$

The linear differential operator in Q_1 in the right-hand side has been studied in the proof of Proposition 4.1: a necessary condition for its kernel to be non-trivial is $a\,b\,c = -1$. Hence this operator is injective and, if $1 + b + a\,b = 0$, we have $Q_1 = 0$ and $f = \lambda(f_2)^k$.

The rest of the proof consists in showing that the case $1 + b + a\,b \neq 0$ is excluded. To do this, one projects (12) on the coordinate planes, determines the corresponding projections of Q_1 and studies under what conditions they can be polynomial. These conditions are:

$$\begin{cases} 2 + b + 1/c = -n_1 \\ 2 + a + 1/b = -n_2 \\ 2 + c + 1/a = -n_3, \end{cases} \qquad (13)$$

where n_1, n_2, $n_3 \in \mathbf{N}$. Since $a\,b\,c = 1$, there are two possibilities: either the three parameters are positive, or one is positive, e.g. a and the other two are negative. The first possibility is excluded by (13).

On the other hand, if $a > 0$ and $b, c < 0$, the second equation in (13) yields $0 < a = -n_2 - 2 - 1/b$ or $-1/b > n_2 + 2 \geq 2$. Hence $|b| < 1/2$. Similarly $|1/c| < 1/2$. Then the first equation yields $-n_1 = 2 + b + 1/c > 2 - 1/2 - 1/2 = 1$, a contradiction.

5 Conclusion

We summarise the classification of the polynomial (and of some other) first integrals of the Lotka-Volterra system:

1. If $a\,b\,c = -1$, we have the first integrals $f_1 = a\,b\,x + y - a\,z$ and $x^p\,y^q\,z^r$, where (p, q, r) satisfy (7). The latter can be made a polynomial when, moreover, $(a, b, c) \in (\mathbf{Q}^-)^3$, but it always exists. The biological interpretation of f_1 is the total mass of the three species, whereas $x^p\,y^q\,z^r$ expresses the skew-symmetric character of the equations in this case.
2. If $a\,b\,c = 1$ and $1 + b + a\,b = 0$, we have the first integral $f_2 = a^2\,b^2\,x^2 - 2\,a\,b\,x\,y + y^2 - 2\,a\,y\,z - 2\,a^2\,b\,x\,z + a^2\,z^2$, the interpretation of which is less evident.
3. No other polynomial first integrals can exist when the parameters are real, as is the case in the applications. There exists a handful of triples (a, b, c), satisfying $(-a\,b\,c)^m = 1$ with $m = 3$, 4 or 6, such that the system (1) possesses a polynomial first integral of degree m (Moulin-Ollagnier 1997).
4. Algebraic or meromorphic first integrals which do not reduce to polynomials cannot exist unless at least one of the parameters is a positive rational number. This possibility, too, may be discarded for applications.

This classification result is one of the first of its kind for a nonlinear, not an explicitly integrable dynamical system. Darboux elements, from which the classification is obtained, are promising tools, but the efficiency of their use relies upon specific properties such as factorisability, homogeneity and discrete symmetries.

References

Cairó, L., Feix, M.R., Goedert, J. (1989): Invariants for models of interacting populations. Phys. Lett. A **140**, 421–427

Cairó, L., Feix, M.R. (1992): Families of invariants of motion for the Lotka-Volterra equations: the linear polynomials family. J. Math. Phys. **33**, 2440–2455

Darboux, G. (1878): Mémoire sur les équations différentielles algébriques du premier ordre et du premier degré. Bull. Sc. Math. 2ème série **2**, 60–96, 123–144, 151–200

Grammaticos, B., Moulin-Ollagnier, J., Ramani, A., Strelcyn, J.M., Wojciechowski, S. (1990); Integrals of quadratic ordinary differential equations in $\mathbf{R}^3$: the Lotka-Volterra system. Physica A **163**, 683–722

Kummer, M., Churchill, R.C., Rod, D.L. (1990): On a result of Bruns. Can. Math. Bull. **33** No. 2 175–180

Labrunie, S. (1996): On the polynomial first integrals of the (a, b, c) Lotka-Volterra system. J. Math. Phys. **37**, 5539–5550

Lotka, A.J. (1925): *Elements of Physical Biology* (Williams and Wilkins, Baltimore)

Man, Y.K. (1994): First integrals of autonomous systems of differential equations and the Prelle-Singer procedure. J. Phys. A **27**, L327–L332

Moulin-Ollagnier, J. (1997): Polynomial first integrals of the Lotka-Volterra vector field, to appear in Bull. Sc. Math. **121**

Volterra, V. (1931): *Leçons sur la théorie mathématique de la lutte pour la vie* (Gauthier-Villars, Paris)

Ziglin, S.L. (1982): Branching of solutions and nonexistence of first integrals in Hamiltonian mechanics. Funct. Appl. Anal. **16**, 181–189

Local Integrability for Nilpotent Critical Point

Javier Chavarriga and Jaume Giné

Departament de Matemàtica, Universitat de Lleida, Pl. Victor Siurana 1, 25003, Lleida, Spain

Abstract. We study the two-dimensional autonomous systems of differential equations of the form $\dot{x} = y + a_{20}x^2 + a_{11}xy + a_{02}y^2, \dot{y} = b_{20}x^2 + b_{11}xy + b_{02}y^2$, where a_{ij} and b_{ij} are arbitrary real constants. The origin is a degenerate critical point of this system. In this work we give a sufficient conditions for local integrability at the origin.

1 Introduction

We consider two-dimensional autonomous systems of differential equations of the form

$$\dot{x} = y + a_{20}x^2 + a_{11}xy + a_{02}y^2, \quad \dot{y} = b_{20}x^2 + b_{11}xy + b_{02}y^2, \tag{1}$$

where a_{ij} and b_{ij} are arbitrary real coefficients.

These systems are particular cases of systems of the form

$$\dot{x} = y + F(x, y), \quad \dot{y} = G(x, y), \tag{2}$$

where F and G are analytic functions in a neighborhood of the origin starting in quadratic terms in the variables x and y. Systems (2) have been studied by several authors (see for instance Andreev 1958; Moussu 1982; Takens 1974 and Żoladek 1996).

Takens (1974) proves that the system (2) can be formally reduced to

$$\dot{x} = y + a(x), \quad \dot{y} = b(x), \tag{3}$$

where $a(x) = a_r x^r + a_{r+1}x^{r+1} + \ldots$ and $b(x) = b_{s-1}x^{s-1} + b_s x^s + \ldots$ are formal power series.

Bogdanov and some years later Żoladek (1996) proved that the system (3) reduces to the form

$$\dot{x} = y + a(x), \quad \dot{y} = x^{s-1}. \tag{4}$$

In this last work Żoladek has shown that the normal forms (3) and (4) can be chosen analytic.

Moussu (1982) finds necessary and sufficient conditions to have a center at the origin for systems (3). His result is the following:

Theorem (Nilpotent Center Theorem). *System* (3) *has the following expression if and only if it has a center at the origin.*

$$\dot{X} = Y + a(X), \quad \dot{Y} = -X^{2s-1},$$

where a(X) is an even function, that is, $a(X) = a(-X)$.

The relation between the Bogdanov normal form and the Moussu normal form for the centers is the change of variable $X^{2s} + 2x^s = 1$ and a change of time.

In his work Moussu says that, contrary to what occurs in the case of a nondegenerate center, in the case of degenerate centers there does not exist, in general, a first integral at the origin. Poincaré's result on the center states that the necessary and sufficient condition for a weak focus of an analytic system in a neighbourhood of the origin to be a center is that the system must be analytically integrable in that neighbourhood. Schlomiuk (1993) asks the following question: Is there a relationship between a degenerate center, ie with zero eigenvalues, and local integrability at this point?

We give a partial answer to this question in this work for cases where the origin is a nilpotent critical point for systems of the form (1). In this paper firstly we find solutions $V(x,y) = 0$ of system (1), where $V(x,y)$ is a reciprocal integrating factor (this notion will be defined below), and we find conditions for the coefficients of system (1). Secondly we study the conditions to have local integrability, that is, systems that have an analytical first integral in a neighbourhood of the origin. Finally we show the relationship between these systems and systems which have a center at the origin.

Let (P, Q) be a C^1 vector field defined in a open subset $U \in \mathbb{R}^2$. We call a C^1 solution $V(x,y)$ of the equation

$$P\frac{\partial V}{\partial x} + Q\frac{\partial V}{\partial y} = \left(\frac{\partial P}{\partial x} + \frac{\partial Q}{\partial y}\right) V \tag{5}$$

a reciprocal integrating factor. In recent works it has been shown that this function plays a fundamental role in the problem of the nondegenerate center (see for instance Chavarriga 1994; Chavarriga 1995; Chavarriga and Giné 1996; Chavarriga and Giné 1997 and Giacomini and Ndiaye 1996), in the determination of limit cycles (see Chavarriga et al. 1997a and Giacomini et al. 1996) and in the construction or generation of systems with a known reciprocal integrating factor(see Chavarriga et al. 1997b).

In this paper we use the reciprocal integrating factor to determine systems of the form (1) which have local integrability at the origin. The reason for the use of the reciprocal integrating factor instead of the first integral is because in many cases in which the first integral is not analytic the reciprocal integrating factor is analytic and we can find this reciprocal integrating factor by proposing a real formal power series. This implies that the function $V(x,y)$ is a good tool to approach the problem of degenerate critical points.

In Sect. 2 we give some results which are necessary for proving Theorem 1. In Sect. 3 we prove Theorem 1 and Theorem 2. Theorem 1 characterises the systems of the form (1) which have an analytic reciprocal integrating factor. Consequently these systems can have a local integral at the origin. Theorem 2 characterises systems of the form (1) which have an analytic first integral.

2 Some Preliminary Results

In the study of this problem we use polar coordinates because they simplify the computations. In Lemma 1 we give the expression of system (1) in polar coordinates.

Lemma 1. *In polar coordinates $x = r\cos\varphi$, $y = r\sin\varphi$ we can write system* (1) *as*

$$
\begin{aligned}
\dot{r} &= \frac{\sin 2\varphi}{2} r + (A_1\cos 3\varphi + A_2\sin 3\varphi + A_3\cos\varphi + A_4\sin\varphi)r^2 \\
\dot{\varphi} &= -\frac{1}{2} + \frac{\cos 2\varphi}{2} + (A_2\cos 3\varphi - A_1\sin 3\varphi + B_3\cos\varphi + B_4\sin\varphi)r,
\end{aligned}
\tag{6}
$$

where

$$
\begin{aligned}
A_1 &= \frac{a_{20} - a_{02} - b_{11}}{4}, \quad A_2 = \frac{a_{11} + b_{20} - b_{02}}{4}, \quad A_3 = \frac{a_{02} + 3a_{20} + b_{11}}{4}, \\
A_4 &= \frac{a_{11} + 3b_{02} + b_{20}}{4}, \quad B_3 = \frac{b_{02} + 3b_{20} - a_{11}}{4} \text{ and } B_4 = \frac{b_{11} - 3a_{02} - a_{20}}{4}.
\end{aligned}
$$

The proof follows from direct computations.

We can assume $A_1 = 0$ without lost of generality because under the linear transformation $x = au + bv$ and $y = av$, where a and b are arbitrary real constants, the system (1) takes the form

$$
\dot{u} = v + \bar{a}_{20}u^2 + \bar{a}_{11}uv + \bar{a}_{02}v^2, \quad \dot{v} = \bar{b}_{20}u^2 + \bar{b}_{11}uv + \bar{b}_{02}v^2, \tag{7}
$$

where

$$
\begin{aligned}
\bar{a}_{20} &= aa_{20} - bb_{20}, \quad \bar{a}_{11} = \frac{a^2a_{11} + 2aba_{20} - abb_{11} - 2b^2b_{20}}{a}, \\
\bar{a}_{02} &= \frac{a^3a_{02} + a^2ba_{11} + ab^2a_{20} - a^2bb_{02} - ab^2b_{11} - b^3b_{20}}{a^2}, \\
\bar{b}_{20} &= ab_{20}, \quad \bar{b}_{11} = ab_{11} + 2bb_{20}, \quad \bar{b}_{02} = \frac{a^2b_{02} + abb_{11} + b^2b_{20}}{a}.
\end{aligned}
$$

After the transformation to polar coordinates the new coefficient $\overline{A}_1$ in system (7) becomes

$$\overline{A}_1 = 4A_1a^3 + (b_{02} - a_{11} - 3b_{20})ba^2 + (b_{11} - a_{20})ba + b_{20}b^3. \tag{8}$$

Therefore, if $A_1 \neq 0$, we always can find a real root such that $\overline{A}_1 = 0$.

Finally we can take cartesian coordinates and system (6) becomes

$$\begin{aligned} \dot{x} &= y + \alpha_1 x^2 + (3\alpha_4 + \alpha_5 - 5\alpha_3)xy - \alpha_2 y^2, \\ \dot{y} &= 2\alpha_3 x^2 + (\alpha_1 + \alpha_2)xy + (\alpha_3 - \alpha_4 + \alpha_5)y^2, \end{aligned} \tag{9}$$

where $A_2 = \alpha_4 - \alpha_3,\ A_3 = \alpha_1,\ A_4 = \alpha_5,\ B_3 = 3\alpha_3 - \alpha_4$ and $B_4 = \alpha_2$.

In polar coordinates (5) becomes

$$\dot{r}\frac{\partial W}{\partial r} + \dot{\varphi}\frac{\partial W}{\partial \varphi} = W\left[\frac{1}{r}\frac{\partial(r\dot{r})}{\partial r} + \frac{\partial \dot{\varphi}}{\partial \varphi}\right], \tag{10}$$

where $W = W(r,\varphi) = V(r\cos\varphi, r\sin\varphi)$. The method to establish the results of Theorem 1 works as follows. We propose a given form for the function $W(r,\varphi) = r^k \sum_{i=1}^{N} W_i(\varphi) r^i$ where $W_i(\varphi)$, $i = 1, 2, \dots N$, are arbitrary functions and $k \in N$. By replacing in (10) $\dot{r}$ and $\dot{\varphi}$ by the right hand sides of equations (6) and, since the functions $W_i(\varphi)$ must be homogeneous polynomials of degree i, we obtain the conditions of Theorem 1.

3 Main Results

Theorem 1. *System (9) has an analytic reciprocal integrating factor at the origin if one of the following conditions holds:*

(1) $\alpha_1 = \alpha_2 = 0$,
(2) $3\alpha_1 + \alpha_2 = 3\alpha_3 - \alpha_4 - 3\alpha_5 = 0$,
(3) $3\alpha_1 + \alpha_2 = \alpha_1^2 + 4\alpha_3\alpha_4 = 0$,
(4) $\alpha_3 = 2\alpha_1\alpha_4 + \alpha_2\alpha_4 + \alpha_2\alpha_5 = 0$,
(5) $\alpha_3 = \alpha_2 - 2\alpha_1 = 25\alpha_1^2 + (\alpha_5 - 3\alpha_4)(\alpha_5 + 7\alpha_4) = 0$.

Proof. Assume that system (9) satisfies condition (1). Then it is easy to check that the function $V(x,y) = (1 + (3\alpha_4 + \alpha_5 - 5\alpha_3)x)^{\frac{\alpha_4 + 3\alpha_5 - 3\alpha_3}{3\alpha_4 + \alpha_5 - 5\alpha_3}}$ is a reciprocal integrating factor of system (9) if $3\alpha_4 + \alpha_5 - 5\alpha_3 \neq 0$, and $V(x,y) = e^{2(6\alpha_3 - 4\alpha_4)x}$ is a reciprocal integrating factor of system (9) if $3\alpha_4 + \alpha_5 - 5\alpha_3 = 0$.

Suppose that system (9) satisfies condition (2). Then the system is Hamiltonian, i.e. $\dot{x} = -\partial H/\partial y$, $\dot{y} = \partial H/\partial x$ with

$$H(x,y) = \frac{1}{2}y^2 + \frac{1}{3}\left[-2\alpha_3 x^3 + 3\alpha_1 x^2 y + 6(\alpha_3 - 2\alpha_5)xy^2 + 3\alpha_1 y^3\right]$$

which is a first integral defined in a neighbourhood of the origin.

We assume that system (9) satisfies condition (3). In this case system (9) has two reciprocal integrating factors. The first one is

$$V(x,y) = (8\alpha_3 + (\alpha_1^2 - 20\alpha_3^3 + 4\alpha_3\alpha_5)(2\alpha_3 x - \alpha_1 y))^{\frac{12\alpha_3\alpha_5 - \alpha_1^2 - 12\alpha_3^2}{4\alpha_3\alpha_5 + \alpha_1^2 - 20\alpha_3^2}}$$

if $4\alpha_3\alpha_5 + \alpha_1^2 - 20\alpha_3^2 \neq 0$, and the second one is

$$V(x,y) = \left[\frac{1}{\alpha_1^2 - 4\alpha_3^2 - 4\alpha_3\alpha_5} - \frac{x}{2\alpha_3} + \frac{\alpha_1 y}{4\alpha_3^2} + \frac{3\alpha_1^2 - 28\alpha_3^2 - 4\alpha_3\alpha_5}{16\alpha_3^3}\right.$$
$$\left.(\alpha_3 x^2 - \alpha_1 xy + 3\alpha_3 y^2)\right]^{\frac{12\alpha_3\alpha_5 - \alpha_1^2 - 12\alpha_3^2}{2(4\alpha_3\alpha_5 - \alpha_1^2 + 4\alpha_3^2)}},$$

if $4\alpha_3\alpha_5 - \alpha_1^2 + 4\alpha_3^2 \neq 0$. When $4\alpha_3\alpha_5 + \alpha_1^2 - 20\alpha_3^2 = 0$ and $4\alpha_3\alpha_5 - \alpha_1^2 + 4\alpha_3^2 = 0$, we obtain $\alpha_1 = 2\sqrt{3} \mid \alpha_3 \mid$ and $\alpha_5 = 2\alpha_3$ respectively. Hence the system (9) is Hamiltonian and the function

$$H(x,y) = \frac{y^2}{2} + \frac{2\alpha_3}{3}(-x^3 \pm 3\sqrt{3}x^2 y - 9xy^2 \pm 3\sqrt{3}y^3)$$

is a first integral defined in a neighbourhood of the origin.

Now we suppose that system (9) satisfies condition (4). Then it is easy to check that the function

$$V(x,y) = \frac{(\alpha_1 + \alpha_2)y^{\frac{3\alpha_1 + \alpha_2}{\alpha_1 + \alpha_2}}}{[2\alpha_2(\alpha_2 - \alpha_1)]}$$

is a reciprocal integrating factor of system (9) if $\alpha_2 \neq 0$, $\alpha_2 - \alpha_1 \neq 0$ and $\alpha_1 + \alpha_2 \neq 0$. If $\alpha_2 = 0$, condition (4) implies $\alpha_4 = 0$ and in this case a reciprocal integrating factor is $V(x,y) = y^3$. If $\alpha_2 = \alpha_1$, condition (4) implies $\alpha_5 = -3\alpha_4$ and a reciprocal integrating factor is $V(x,y) = y^2$. Finally, if $\alpha_2 = -\alpha_1$, condition (4) implies $\alpha_4 = \alpha_5$ and a reciprocal integrating factor is $V(x,y) = y + 4\alpha_5 xy + \alpha_1(x^2 + y^2)$.

Suppose that system (9) satisfies condition (5). Then it is easy to see that the function

$$V(x,y) = (50\alpha_1(3\alpha_4 - \alpha_5)y + (7\alpha_4 + \alpha_5)(-25\alpha_1^2 + 12\alpha_4^2 + 2\alpha_4\alpha_5 - 2\alpha_5^2)x^2 +$$
$$10\alpha_1(3\alpha_4 - \alpha_5)(7\alpha_4 + \alpha_5)xy - 25\alpha_1^2(7\alpha_4 + \alpha_5)y^2)^{5/2}$$

is a reciprocal integrating factor of system (9).

Theorem 2. *System* (9) *has local integrability at the origin if one of the first three conditions of Theorem 1 holds.*

Proof. We are going to prove for the first three conditions that there always exists an analytic change of variables where system (9) takes the form

$$\dot{x} = y, \quad \dot{y} = b(x)$$

and therefore the system has a first integral of the form $H(x,y) = \frac{1}{2}y^2 - \int b(x)dx$.

Suppose that system (9) satisfies condition (1) of Theorem 1. The analytic change of variables

$$x = -\frac{u}{1+cu} \qquad y = v\frac{1-(1+cu)^{\frac{a}{c}}}{au(1+cu)^{\frac{a}{c}}}$$

tranforms system (9) into

$$\dot{u} = v \qquad \dot{v} = \frac{bu(1+cu)\left((1+cu)^{\frac{a}{c}}-1\right)}{a(1+cu)^{\frac{a}{c}}},$$

where $a = 3\alpha_4 + \alpha_5 - 5\alpha_3$, $b = 2\alpha_3$ and $c = \alpha_3 - \alpha_4 + \alpha_5$, if $\alpha_3 - \alpha_4 + \alpha_5 \neq 0$. If $\alpha_3 - \alpha_4 + \alpha_5 = 0$ and $3\alpha_4 + \alpha_5 - 5\alpha_3 \neq 0$, the analytic change of variables

$$x = \frac{e^{(4\alpha_4 - 6\alpha_3)u} - 1}{4\alpha_4 - 6\alpha_3}, \; y = v$$

transforms system (9) into

$$\dot{u} = v \qquad \dot{v} = 2\alpha_3\left[\frac{e^{(4\alpha_4 - 5\alpha_3)u} - 1}{4\alpha_4 - 6\alpha_3}\right]^2.$$

If $\alpha_3 - \alpha_4 + \alpha_5 \neq 0$ and $3\alpha_4 + \alpha_5 - 5\alpha_3 = 0$, the analytic change of variables

$$x = \frac{1}{d}\ln\left[\frac{1}{1+du}\right], \quad y = \frac{v}{1+du}$$

transforms system (9) into

$$\dot{u} = v, \quad \dot{v} = \frac{2\alpha_3}{d^2}(1+du)\ln\left[\frac{1}{1+du}\right],$$

where $d = 6\alpha_3 - 4\alpha_4$. If $\alpha_3 - \alpha_4 + \alpha_5 = 0$ and $3\alpha_4 + \alpha_5 - 5\alpha_3 = 0$, sysyem (9) takes the form $\dot{x} = y$, $\dot{y} = 2\alpha_3 x^2$ which is Hamiltonian and a first integral is given by $H(x,y) = y^2/2 - 2\alpha_3 x^3/3$.

Suppose that system (9) satisfies condition (2) of Theorem 1. This system is Hamiltonian and in the Takens normal form, $\dot{x} = y + a(x)$, $\dot{y} = b(x)$, it has $a(x) = 0$.

Suppose that system (9) satisfies condition (3) of Theorem 1. The analytic change of variables

$$x = \frac{4\alpha_3}{\alpha_1^2 - 20\alpha_3^2 + 4\alpha_3\alpha_5}\left[\frac{k_4 u}{c_1 - k_4 u}\right]^{\frac{k}{k_4}} + \frac{\alpha_1 v}{2\alpha_3(c_1 - k_4 u)}, \qquad y = \frac{v}{c_1 - k_4 u},$$

where

$$k = \frac{\alpha_1^2 - 20\alpha_3^2 + 4\alpha_3\alpha_5}{4\alpha_3} \qquad k_4 = \frac{-\alpha_1^2 + 4\alpha_3^2 + 4\alpha_3\alpha_5}{4\alpha_3}$$

and c_1 is an arbitrary real constant that takes positive or negative values depending on the values of the parameters, transforms system (9) into

$$\dot{u} = v, \quad \dot{v} = (c_1 - k_4 u)\left[k_1 + 8\alpha_3^2 k_2\left[\frac{c_1}{c_1 - k_4 u}\right]^{\frac{k}{k_4}} + 64\alpha_3^4 k_3\left[\frac{c_1}{c_1 - k_4 u}\right]^{\frac{2k}{k_4}}\right],$$

where

$$\begin{aligned} k_1 &= \frac{32\alpha_3^3}{(\alpha_1^2 - 20\alpha_3^2 + 4\alpha_3\alpha_5)^2} \\ k_2 &= \frac{-8\alpha_3}{(\alpha_1^2 - 20\alpha_3^2 + 4\alpha_3\alpha_5)^2} \\ k_3 &= \frac{1}{2\alpha_3(\alpha_1^2 - 20\alpha_3^2 + 4\alpha_3\alpha_5)^2}, \end{aligned} \qquad (11)$$

if $\alpha_1^2 - 20\alpha_3^2 + 4\alpha_3\alpha_5 \neq 0$ and $k_4 \neq 0$. If $\alpha_1^2 - 20\alpha_3^2 + 4\alpha_3\alpha_5 = 0$ and $k_4 \neq 0$, the analytic change of variables

$$x = -\frac{\alpha_1 v}{2\alpha_3(1 + k_5 u)} - \frac{1}{k_5}\ln(1 + k_5 u), \qquad y = -\frac{v}{1 + k_5 u},$$

where $k_5 = \frac{12\alpha_3^2 - \alpha_1^2}{2\alpha_3}$ transforms system (9) to

$$\dot{u} = v \qquad \dot{v} = -\frac{2\alpha_3}{k_5^2}(1 + k_5 u)\ln^2(1 + \mathrm{k_5 u}).$$

If $\alpha_1^2 - 20\alpha_3^2 + 4\alpha_3\alpha_5 \neq 0$ and $k_4 = 0$, then system (9) with the analytic change of variables

$$x = \frac{e^{k_6 u} - 1}{k_6} + \frac{\alpha_1 v}{2\alpha_3} \qquad y = v,$$

where $k_6 = \frac{\alpha_1^2 - 12\alpha_3^2}{2\alpha_3}$, becomes

$$\dot{u} = v \qquad \dot{v} = k_7 + 8\alpha_3^2 k_8 e^{ku} + 64\alpha_3^4 k_9 e^{2ku},$$

where

$$k_7 = \frac{32\alpha_3^3}{(2\alpha_1^2 - 24\alpha_3^2)^2}$$
$$k_8 = \frac{-8\alpha_3}{(2\alpha_1^2 - 24\alpha_3^2)^2}$$
$$k_9 = \frac{1}{2\alpha_3(2\alpha_1^2 - 24\alpha_3^2)^2}. \tag{12}$$

If $\alpha_1^2 - 20\alpha_3^2 + 4\alpha_3\alpha_5 = 0$ and $k_4 = 0$, $\alpha_1^2 - 12\alpha_3^2 = 0$ and $\alpha_5 = 2\alpha_3$ and in this case we have seen that system (9) is Hamiltonian.

Now we suppose that system (9) satisfies condition (4) of Theorem 1. In this case system (9) has a first integral of the form

$$H(x,y) = \frac{2\alpha_2 y + (\alpha_1 - \alpha_2)(\alpha_2 x^2 - \alpha_4 xy + \alpha_2 y^2)}{y^{\frac{2\alpha_1}{\alpha_1+\alpha_2}}},$$

if $\alpha_2 \neq 0$, $\alpha_2 - \alpha_1 \neq 0$ and $\alpha_2 + \alpha_1 \neq 0$. In this case, if $2\alpha_1/(\alpha_1 + \alpha_2) < 0$, then $H(x,y)$ is a C^1 function and, if $2\alpha_1/(\alpha_1 + \alpha_2)$ is a negative integer, then $H(x,y)$ is an analytic function. In the other cases the first integral of system (9) is $H(x,y) = (\alpha_1 x^2 + 2y + 2\alpha_5 xy)/2y^2$ if $\alpha_2 \neq 0$, $H(x,y) = 4\alpha_4 x - \alpha_1 x^2/y - \alpha_1 y + \ln y$ if $\alpha_2 = \alpha_1$ and $H(x,y) = y$ if $\alpha_2 = -\alpha_1$. The first two are not well-defined in a neighbourhood of the origin and the last one is polynomial.

Now we suppose that system (9) satisfies condition (5) of Theorem 1. With $\alpha_4 = (\beta_2 - \beta_1)/10$ and $\alpha_5 = (7\beta_1 + 3\beta_2)/10$ and in terms of the variable $z = \beta_1 x + 5\alpha_1 y$ the system takes the form

$$\dot{z} = \frac{1}{5\beta_1}(5\beta_1^2 y - 50\alpha_1^2 yz + 2\beta_1^2 yz + 5\alpha_1 z^2), \quad \dot{y} = \frac{y}{5\beta_1}(-100\alpha_1^2 y + 4\beta_1^2 y + 15\alpha_1 z),$$

where the reciprocal integrating factor is $V(x,y) = \frac{25\alpha_1}{\beta_1}(\alpha_1 z^2 - 2\beta_1^2 y)^{\frac{5}{2}}$ and a first integral is

$$H(x,y) = \frac{(2(25\alpha_1^2 - \beta_1^2)z(3\beta_1^2 y + \alpha_1 z^2) - 15\beta_1^4 y)^2}{(\alpha_1 z^2 - 2\beta_1^2 y)^3},$$

which is not well-defined in a neighbourhood of the origin.

4 Final Remarks

(a) From the nilpotent critical point theorem (see the appendix of Llibre et al. 1997) it is easy to prove that system (1) does not have a center at the origin. We prove in this work that there is a local integral in many cases (see Theorem 2), but these cases are not centers at the origin. All of them have a cusp at the origin. This is easy to prove by using also the nilpotent critical point theorem.

(b) In the cases of Theorem 2 we have analytic integrability in a neighbourhood of the origin and therefore local integrability. In Cases 4 and 5 of Theorem 1 it seems that the analytic integrability is not generic. We must mention that in the cases of Theorem 2 the C^1 function $V(x, y)$ takes a value different from zero at the origin. On the other hand for Cases 4 and 5 of Theorem 1 the C^1 function $V(x, y)$ is zero at the origin.

(c) It seems that for systems of degree greater than two having a nilpotent critical point at the origin we shall again find that the critical points with local integrability are centers, saddles or cusps.

References

Andreev, A. (1958): Investigation of the behaviour of the integral curves of a system of two differential equations in the neighborhood of a singular points. Translation of AMS **8**, 187–207

Bogdanov, R.I. (1979): Local orbital forms of vector fields on a plane. Trans. of the Petrovski Sem. **5**, 50–85

Chavarriga, J. (1994): Integrable systems in the plan with a center type linear part. Applicationes Mathematicae **22**, 285–309

Chavarriga, J. (1995): A class of integrable polynomial vector fields. Applicationes Mathematicae **23**, 339–350

Chavarriga, J., Giné, J. (1996): Integrability of a linear center perturbed by fourth degree homogeneous polynomial. Publicacions Matemàtiques **40** (1), 21–39

Chavarriga, J., Giné, J. (1997): Integrability of a linear center perturbed by fifth degree homogeneous polynomial. To appear in Publicacions Matemàtiques **41** (2)

Chavarriga, J., Giacomini, H., Giné, J. (1997a): On a new type of bifurcation of limit cycles for a planar cubic system. To appear in Nonlinear Analysis, Theory, Methods and Applications

Chavarriga, J., Giacomini, H., Giné, J. (1997b): The null divergence factor. To appear in Publicacions Matemàtiques **41** (1)

Giacomini, H., Llibre, J., Viano, M. (1996): On the nonexistence, existence, and uniquennes of limit cycles. Nonlinearity **9**, 501–516

Giacomini, H., Ndiaye, M. (1996): New sufficient conditions for a center and global phase portraits for polynomial systems. Publicacions Matemàtiques **40** (1), 351-372

Llibre, J., Mañosa, V., Mañosas, F. (1997): The center-focus problem for semihomogeneous systems. Preprint Universitat Autònoma de Barcelona.

Moussu, R. (1982): Symétrie et forme normale des centres et foyers dégénérés. Ergodic Theory Dynamical Systems **2**, 241–251

Schlomiuk, D. (1993): Algebraic and Geometric Aspects of the Theory of Polinomial Vector Fields, *Bifurcations and Periodics Orbits of Vector Fields* (Kluwer, Dordrecht), 429–467

Takens, F. (1974): Singularities of vector fields. Inst. Hautes Études Sci. Publ. Math. **43**, 47–100
Żoladek, H. (1996): The analytic normal form for the nilpotent singularity. Preprint Institute of Mathematics, University of Warsaw

Lie Symmetries of Generalized Ermakov Systems

Fernando Haas and Joao Goedert

Instituto de Física-UFRGS, Caixa Postal 15051, 91500-970 Porto Alegre, RS - Brazil

Abstract. The Lie point symmetry analysis of Ermakov systems is extended to the case where the frequency function in the equations of motion depends also on the dynamical variables and their derivatives. From this new standpoint, a quite general class of two dimensional oscillators with nonlinear coupling can be viewed as a Hamiltonian Ermakov system possessing a Lie point symmetry and is shown to be completely integrable.

1 Introduction

Ermakov systems, first studied around the end of the nineteenth century (Ermakov 1880), were rediscovered in the late sixties (Lewis 1967) and have since then deserved much attention as a paradigm for nonlinear integrable dynamical system. The scope of application of Ermakov systems covers several interesting fields, such as the exact quantisation (Lewis and Riesenfeld 1969; Ray and Hartley 1982) of the time-dependent oscillator, the search of coherent states for nonlinear, nonautonomous systems (Ray 1982), nonlinear elasticity (Shahinpoor and Nowinski 1971), shallow water waves theory (Rogers and Rangulam 1989), cosmological particle creation (Ray 1979), the optics of elliptic gaussian beams (Goncharenko et al. 1991) and the stability of orbits in accelerators (Courant and Snyder 1958). The interested reader may consult (Rogers et al. 1993; Schief et al. 1996) and references therein for a more detailed account of the literature.

The Ermakov equations (Schief et al. 1996) are currently written as,

$$\ddot{x} + \omega^2 x = \frac{1}{yx^2} f(y/x) \,, \qquad \ddot{y} + \omega^2 y = \frac{1}{xy^2} g(x/y) \,, \tag{1}$$

where ω is an arbitrary frequency that may depend on time, the coordinates and their derivatives. The functions f and g depend on the dynamical variables as indicated by their arguments. As is easy to verify (Ray and Reid 1979a), equations (1) admit the Ermakov invariant

$$I = \frac{1}{2}(x\dot{y} - y\dot{x})^2 + \int^{y/x} f(\tau)\mathrm{d}\tau + \int^{x/y} g(\tau)\mathrm{d}\tau \tag{2}$$

whereby a nonlinear superposition law (Reid and Ray 1980) can be constructed explicitly whenever one of the equations decouples, a situation that is more likely to occur when ω is a function of time only. The usual assumption that ω is a function of time only can be relaxed and, in general, ω may depend also on the dynamical variables and their derivatives (Reid and Ray 1980; Goedert 1989) with no impact on the existence of the Ermakov invariant and the corresponding superposition law. This extra freedom has not been exploited in any detail except for very few exceptions (Goedert 1989; Haas and Goedert 1996). For instance, the Lie point symmetry analysis of Ermakov systems for $\omega = \omega(t)$ has been thoroughly investigated (Leach 1991; Govinder et al. 1993; Govinder and Leach 1994a; Govinder and Leach 1994b) but not for ω dependent on the dynamical variables. The purpose of this work is to fill this gap in the literature and perform the Lie symmetry analysis of Ermakov systems in the general case. Theoretically ω could depend on the dynamical variables and their derivatives up to any arbitrary order, besides time. For physical reasons, however, we restrict ourselves to the cases where $\omega = \omega(t, x, y, \dot{x}, \dot{y})$ only.

The paper is organized as follows. In Sect. 2 it is shown how the Ermakov system may be recast in a more compact form by redefining appropriately the functions ω, f and g. This compact form is employed in the calculations of Sect. 3, where the most general Ermakov systems that admit Lie point symmetries are determined. Section 4 presents a coupled nonlinear oscillator which is an interesting example of generalized Ermakov systems endowed with geometric symmetries. The conclusions are presented in Sect. 5.

2 Compact Representation

Ermakov systems have been presented traditionally in the form (2) which involve three arbitrary functions ω, f and g. However, and in spite of the appearance of the equations (1), only two and not three independent arbitrary functions are necessary to specify the most general Ermakov system. As explained in Haas and Goedert (1996), let the relations

$$\Omega^2 = \omega^2 - \frac{1}{xy^3} g(x/y) \tag{3}$$

$$F(y/x) = f(y/x) - \frac{x^2}{y^2} g(x/y) \tag{4}$$

define new functions Ω and $F(y/x)$ in terms of the original ω, f and g. In terms of these functions, the pair of equations (1) is

$$\ddot{x} + \Omega^2 x = \frac{1}{yx^2} F(y/x), \tag{5}$$

$$\ddot{y} + \Omega^2 y = 0. \tag{6}$$

and the Ermakov invariant becomes

$$I = \frac{1}{2}(x\dot{y} - y\dot{x})^2 + \int^{y/x} F(\tau)\mathrm{d}\tau\,. \tag{7}$$

Except for very special cases, Ω will now depend on the dynamical variables (x, y) and possibly on their derivatives. Nevertheless, the term "frequency" will still be used to refer to Ω. The representation (5–6) does not clog the usual calculations with unnecessary functions, but still covers appropriately all previously discussed Ermakov systems.

The use of generalized frequencies considerably expands the reach of the Ermakov equations. In particular several dynamical systems can be cast in the Ermakov form (5–6), although not so in their first appearance. Two examples will be mentioned here, with no intend of exhausting the full span of such possibilities. The first example is the Kepler-Ermakov system (Athorne 1991a)

$$\ddot{x} = -x\,h(y/x)/r^3 + f(y/x)/(yx^2) \tag{8}$$

$$\ddot{y} = -y\,h(y/x)/r^3 + g(x/y)/(xy^2), \tag{9}$$

where $h(y/x)$ is arbitrary and $r^2 = x^2 + y^2$. The reduction to the form (5–6) is obtained by choosing

$$\Omega^2 = h(y/x)/r^3 - g(x/y)/(xy^3) \tag{10}$$

$$F(y/x) = f(y/x) - (x^2/y^2)g(x/y)\,. \tag{11}$$

Another interesting example of a nonstandard Ermakov system is Lutzky's integrable system (Lutzky 1980),

$$\ddot{x} + \omega^2(t)x = F_1(x, y) \tag{12}$$

$$\ddot{y} + \omega^2(t)y = F_2(x, y), \tag{13}$$

where F_1 and F_2 satisfy

$$xF_2 - yF_1 = \psi(x/y)/(x\,y) \tag{14}$$

for some given ψ. To cast these equations in Ermakov form, it is only necessary to define

$$\Omega^2 = \omega^2 - F_2/y \tag{15}$$

$$F(y/x) = -(x/y)\psi(x/y). \tag{16}$$

In fact F_1 and F_2 could depend on velocities and time while still preserving the Ermakov character of the dynamical equations. The first integral presented in Lutzky's paper is nothing but the Ermakov invariant (7).

3 Lie Point Symmetries

The outstanding examples presented in Sect. II attest the importance of specifying the Ermakov systems in terms of only two rather than three arbitrary functions, as our starting point for its Lie symmetry study. Following Govinder and Leach (Leach 1991; Govinder and Leach 1994b), we first consider the symmetries of the system equivalent to (5–6) given by

$$x\ddot{y} - y\ddot{x} + \frac{1}{x^2}F(y/x) = 0 \tag{17}$$

$$\ddot{y} + \Omega^2(\mathbf{x}, \dot{\mathbf{x}}, t)y = 0, \tag{18}$$

where we introduced the vectorial notation $\mathbf{x} = (x, y)$ to shorten the representation. The advantage in dealing with equation (17) relies in its independence on Ω, a feature that allows the straightforward determination of its symmetry generator. This has already been done when performing the symmetry analysis of usual Ermakov systems (Govinder and Leach 1994b). The symmetry is

$$G = \rho^2\frac{\partial}{\partial t} + \rho\dot{\rho}\mathbf{x}\cdot\frac{\partial}{\partial \mathbf{x}}, \tag{19}$$

where $\rho(t)$ is an arbitrary differentiable function. The first integral I of the Ermakov system remains invariant under the first extension of G as pointed out in Govinder and Leach (1994b). When $\rho = \{1, \sqrt{t}, t\}$, we obtain time translation, self-similarity and conformal transformations respectively, that is, the full $SL(2, R)$ group is recovered.

For the complete Ermakov system (equations (17–18)) to present the same symmetry obeyed by (17), that is, in order for equation (18) to also possess the symmetry generated by G as given by (19), some restriction must be imposed on the functional form of Ω. Applying the twice extended generator G on (18) and standard Lie techniques, one conclude that the most general admissible frequencies are of the form

$$\Omega^2 = -\frac{\ddot{\rho}}{\rho} + \frac{1}{\rho^4}\sigma(\mathbf{x}/\rho, \rho\dot{\mathbf{x}} - \dot{\rho}\mathbf{x}), \tag{20}$$

where σ is an arbitrary function of the indicated arguments. Note that velocity dependence is also possible, a fact that considerably expands the class of Ermakov systems that admit symmetry transformations.

Let us summarize the results so far: without any *a priori* assumption on the functional form of the frequency, we obtained the most general Ermakov system (5–6) invariant under a Lie point symmetry

$$\ddot{x} + \left(-\frac{\ddot{\rho}}{\rho} + \frac{\sigma}{\rho^4}\right)x = \frac{1}{yx^2}F(y/x), \tag{21}$$

$$\ddot{y} + \left(-\frac{\ddot{\rho}}{\rho} + \frac{\sigma}{\rho^4}\right)y = 0. \tag{22}$$

The symmetry group is generated by G as given by (19) and σ has the form

$$\sigma = \sigma(\mathbf{x}/\rho, \rho\dot{\mathbf{x}} - \dot{\rho}\mathbf{x}) . \tag{23}$$

It is apparent that not all generalised Ermakov systems are invariant under the Lie point symmetry G. The Kepler-Ermakov problems, for instance, do not belong to the class of invariant systems for any conceivable function ρ, σ or F.

Let us show now that all previous results on the group theoretical structure of Ermakov systems are included as special cases of our present treatment. For the particular choice

$$\ddot{\rho} + \omega^2(t)\rho = \frac{k}{\rho^3} \tag{24}$$

$$\sigma = k - \frac{\rho^4}{xy^3} g(x/y), \tag{25}$$

where k is a constant, we arrive at the Ermakov system (1) with time-dependent frequency $\omega(t)$. At least for such simple frequencies there exists a quasi-invariance transformation (Burgan et al. 1979; Athorne et al. 1990)

$$\bar{x} = x/C \qquad \bar{y} = y/C \qquad \bar{t} = \int \mathrm{d}t/C^2, \tag{26}$$

where $C(t)$ satisfies the time dependent harmonic oscillator equation

$$\ddot{C} + \omega^2(t)C = 0, \tag{27}$$

which removes ω from the equations of motion

$$\bar{x}'' = \frac{1}{\bar{y}\bar{x}^2} f(\bar{y}/\bar{x}) , \qquad \bar{y}'' = \frac{1}{\bar{x}\bar{y}^2} g(\bar{x}/\bar{y}) . \tag{28}$$

In this case the symmetry group of the transformed Ermakov system is indeed $SL(2, R)$.

The symmetry analysis performed here does not specify any particular form for the functions σ and ρ, nor depend on any previous coordinate transformation like (26). Despite its apparent simplicity, the transformation (26) would not be well defined when ω depends on the dynamical variables, because in such a case the equation (27) is meaningless. Even for $\omega = \omega(t)$, the time dependent harmonic oscillator (27) generally does not possess a closed form solution, a fact that can prevent the application of (26) in practice.

4 Two Dimensional Oscillator with Nonlinear Coupling

Unlike the simple $SL(2,R)$ invariant Ermakov systems in two spatial dimensions, the invariant Ermakov systems (21–22) do not always admit complete integrability. This is the price to pay for the extra generality. Some additional structure, perhaps a Hamiltonian formalism, must exist in order to solve exactly the equations of motion. To gain insight in this respect, let us consider the equations for a two dimensional oscillator with nonlinear coupling introduced by Ray and Reid (Ray and Reid 1979b)

$$\ddot{x} + \omega^2(t)x = x\rho^{-4}G(xy/\rho^2) \tag{29}$$
$$\ddot{y} + \omega^2(t)y = y\rho^{-4}G(xy/\rho^2), \tag{30}$$

where $\rho(t)$ is an auxiliary function constrained by

$$\ddot{\rho} + \omega^2(t)\rho = 1/\rho^3 \,. \tag{31}$$

Ray and Reid found the first integral of (29–30)

$$J = (\rho\dot{x} - \dot{\rho}x)(\rho\dot{y} - \dot{\rho}y) + \frac{xy}{\rho^2} - \int^{xy/\rho^2} G(\tau)\mathrm{d}\tau. \tag{32}$$

This system is in fact an Ermakov system with $F = 0$ and coordinate dependent frequency Ω given by

$$\Omega^2 = \omega^2(t) - \frac{1}{\rho^4}G\left(\frac{xy}{\rho^2}\right). \tag{33}$$

The corresponding Ermakov invariant is

$$I = \frac{1}{2}(x\dot{y} - y\dot{x})^2. \tag{34}$$

Moreover, it is a Hamiltonian Ermakov system (Cerveró and Lejarreta 1991; Athorne 1991b; Haas and Goedert 1996) with Hamiltonian

$$H = p_x p_y + \omega^2 xy - \frac{1}{\rho^2}\int^{xy/\rho^2} G(\tau)\mathrm{d}\tau. \tag{35}$$

H is not a constant of motion since the system is not conservative. However, J and the Ermakov invariant (34) are sufficient to assure complete integrability.

We now show that (33) is indeed of the form (20) which guaranties the invariance of (29–30). Using the auxiliary equation (31) to substitute $\omega(t)$ for ρ we find that the frequency becomes

$$\Omega^2 = -\frac{\ddot{\rho}}{\rho} + \frac{1}{\rho^4}\left(1 - G\left(\frac{xy}{\rho^2}\right)\right), \tag{36}$$

which in fact is of the general form (20).

The Hamiltonian structure and the symmetry can be now further exploited to reach complete integrability, that is to reduce the problem to quadrature. For this purpose canonical group coordinates (u, v) are the adequate transformations

$$u = x/\rho\,, \qquad v = y/\rho\,, \qquad T = \int \rho^{-2} \mathrm{d}t, \tag{37}$$

under which the symmetry can be viewed as a translation, $G = \partial/\partial T$. In the new variables the equations of motion reduce to the autonomous system

$$u'' + u = uG(uv) \tag{38}$$
$$v'' + v = vG(uv) \tag{39}$$

possessing the Hamiltonian

$$K = p_u p_v + uv - \int^{uv} G(\tau)\mathrm{d}\tau\,, \tag{40}$$

which is exactly the transformed first integral J. Since there are two constants of motion in involution for a Hamiltonian system, there is complete integrability. The Hamiltonian K of the Ermakov system in canonical group coordinates is quadratic in momentum. As explained in (Haas and Goedert 1996), the appropriated variables for quadrature of this Ermakov system are given by

$$q = uv\,, \quad s = v/u\,. \tag{41}$$

In these coordinates the constants of motion become

$$\sqrt{2I} = q\,s'/s, \tag{42}$$
$$J = (q'^2 - 2I)/4q + q - \int^q G(\tau)\mathrm{d}\tau\,, \tag{43}$$

where the prime denotes the derivative with respect to the new time. The quadrature of the last two equations gives successively $q(T)$ and $s(T)$. The map to the original variables is a purely algebric task performed by the equations

$$x^2 = \rho^2 q/s\,, \qquad y^2 = \rho^2 q\,s\,. \tag{44}$$

One final step is the determination of $T(t)$ which is obtained by integrating the last equation in (37). In the quadrature sense, this completes the integration process. Recall that all the required structure and the correct choice of variables for the problem was dictated by the Hamiltonian Ermakov character and by the presence of symmetry. In particular, the analytical form of the solution can be found explicitly in terms of elliptic functions for $G(\tau) = c_1/\tau^2 + c_2 + c_3\tau + c_4\tau^2$, where the c_i are arbitrary constants.

5 Conclusion

The generalised Ermakov systems (21–22) are very general in nature and the width of their application scope is still not completely determined. In this paper we determined their Lie point symmetry group and show that a wide set of nonlinearly coupled harmonic oscillators belonging to this class is in fact completely integrable. This gives a clear indication of the variety of possible applications into which generalised Ermakov systems may enter. In particular it can be shown that Lie symmetries of the type generated by (19) and some modifications of it occur in several problems in Physics. Subclasses of two and three dimensional charged particle motion, some perturbed magnetic monopole and time-dependent Kepler problems fall into the category of systems presenting such symmetries (Haas and Goedert 1997). Extensions of this work to include contact or nonlocal symmetries can be valuable. It is also expected that the results of this paper may serve to generalise the Ermakov system to more dimensions in the spirit of Leach's approach (Leach 1991).

Acknowledgments

This work has been partially supported by the Brazilian agencies Conselho Nacional de Desenvolvimento Científico e Tecnológico (CNPq), Fundação de Amparo a Pesquisa do Rio Grande do Sul (FAPERGS) and Financiadora de Estudos e Projetos (FINEP).

References

Athorne, C. (1991): Kepler-Ermakov problems. J. Phys. A: Math. Gen. **24**, L1385–L1389

Athorne, C. (1991): On generalized Ermakov systems. Phys. Lett. A **159**, 375–378

Athorne, C., Rogers, C., Ramgulam, U., Osbaldestin, A. (1990): On linearization of the Ermakov system. Phys. Lett. A **143**, 207–212

Burgan, J.R., Feix, M.R., Fijalkow, E., Munier, A. (1979): Solution of the multidimensional quantum harmonic oscillator with time-dependent frequencies. Phys. Lett. A **74**, 11–14

Cerveró, J.M., Lejarreta, J.D. (1991): Ermakov Hamiltonians. Phys. Lett. A **156**, 201–205

Courant, E.D., Snyder, H.D. (1958): Theory of the alternating-gradient synchroton. Ann. Phys. **3**, 1–48

Ermakov, V.P. (1880): Second order differential equations. Conditions of complete integrabilty. Univ. Izv. Kiev Ser III **9**, 1–25

Goedert, J. (1989): Second constants of motiom for generalized Ermakov systems. Phys. Lett. A **136**, 391–394

Goncharenko, A.M., Logvin, Y.A., Samson A.M., Shapovalov, P.S., Turovets, S.I. (1991): Ermakov Hamiltonian systems in nonlinear optics of elliptic Gaussian beams. Phys. Lett. A **160**, 138–142

Govinder, K.S., Athorne, C., Leach, P.G.L. (1993): The algebraic structure of generalized Ermakov systems in three dimensions. J. Phys. A: Math. Gen. **26**, 4035–4046

Govinder, K.S., Leach, P.G.L. (1994): Integrability of generalized Ermakov systems. J. Phys. A: Math. Gen. **27**, 4153–4156

Govinder, K.S., Leach, P.G.L. (1994): Ermakov systems: a group theoretical approach. Phys. Lett. A **186**, 391–395

Haas, F., Goedert, J. (1996): On the Hamiltonian structure of Ermakov systems. J. Phys. A: Math. Gen. **29**, 4083–4092

Haas, F., Goedert, J. (1997): On the Lie Symmetries of Ermakov Systems. Submitted for publication in Phys. Lett. A.

Leach, P.G.L. (1991): Generalized Ermakov systems. Phys. Lett. A **158**, 102–106

Lewis, H.R (1967): Classical and quantum systems with time-dependent harmonic-oscillator-type hamiltonians. Phys. Rev. Lett. **18**, 510–512

Lewis, H.R., Riesenfeld, W.B. (1969), An exact quantum theory of the time-dependent harmonic oscillator and of a charged particle in a time-dependent electromagnetic field. J. Math. Phys. **10**, 1458–1473

Lutzky, M. (1980): Generalized Ray-Reid systems. Phys. Lett. A **78**, 301–303

Ray, J.R. (1979): Cosmological particle creation. Phys. Rev. D **20**, 2632–2633

Ray, J.R. (1982): Minimum-uncertainty coherent states for certain time-dependent systems. Phys. Rev. D **25**, 3417–3419

Ray, J.R., Hartley, J.G. (1982): Solutions to N-dimensional time-dependent Schrödinger equations. Phys. Lett. A **88**, 125–127

Ray, J.R., Reid, J.L. (1979a): More exact invariants for the time-dependent harmonic oscillator. Phys. Lett. A **71**, 317–318

Ray, J.R., Reid, J.L. (1979): Exact time-dependent invariants for N-dimensional systems. Phys. Lett. A **74**, 23–25

Reid, J.L., Ray, J.R. (1980): Ermakov systems, nonlinear superposition, and solutions of nonlinear equations of motion. J. Math. Phys. **21**, 1583–1587

Rogers, C., Hoenselaers, C., Ray, J.R. (1993): On (2 + 1)-dimensional Ermakov systems. J. Phys. A: Math. Gen. **26**, 2625–2633

Rogers, C., Ramgulam, U. (1989): A non-linear superposition principle and Lie group invariance: application in rotating shallow water theory. Int. J. Non-Linear Mech. **24**, 229–236

Schief, W.K., Rogers, C., Bassom, A.P. (1996): Ermakov systems of arbitrary order and dimension: structure and linearization. J. Phys. A: Math. Gen. **29**, 903–911

Shahinpoor, M., Nowinski, J.L. (1971): Exact solution to the problem of forced large amplitude radial oscillatons of a thin hyper-elastic tube. Int. J. Non-Linear Mech. **6**, 193–207

Hamilton's Principal Function and Integration by Quadratures for an N-Degree-of-Freedom Non - autonomous System Given N Invertible Invariants in Involution

H. Ralph Lewis[1] and Serge Bouquet[2]

[1] Dartmouth College, Department of Physics and Astronomy, Hanover, New Hampshire 03755-3528, USA

[2] Commissariat à l'Energie Atomique, Centre de Bruyères-le-Châtel, DPTA, BP 12, 91680 Bruyères-le-Châtel, France

Abstract. Hamilton's principal function for an N-degree-of-freedom non autonomous Hamiltonian system is expressed in terms of quadratures involving N, possibly time-dependent, invariants in involution. This determines a set of $2N$ canonical coordinates and momenta, each of which is an invariant.

Keywords. Hamilton's principal function, integrability, invariant

1 Introduction

One of the long-standing interests of Prof. Marc R. Feix has been the search for exact invariants of time-dependent Hamiltonian systems (Feix and Lewis 1985; Feix et al. 1987; Lewis et al. 1992). It is appropriate on the occasion of Prof. Feix' retirement to present an account of our recent contribution to integration by quadratures and solution of the Hamilton-Jacobi equation. The collaboration that has led to our results began with an extended visit by one of us (HRL) to the group of Prof. Feix at the Université d'Orléans. A more complete account will be presented elsewhere (Lewis and Bouquet 1998).

Integrability of classical Hamiltonian systems is much studied and has a variety of meanings (Arnold et al. 1988; Tabor 1989; Gutzwiller 1990; Lichtenberg and Lieberman 1992; Ott 1993). In discussions about integrability, attention is frequently restricted to autonomous Hamiltonians and invariants, and it is common to consider the properties of solutions of the equations of motion as functions of a complex time variable. This has led to much information about the chaotic or non-chaotic nature of the solutions of the equations of motion of Hamiltonian systems. The present paper concerns the concept known as integration by quadratures (Arnold et al. 1988), whose applicability depends only weakly on the nature of the solutions of the equations of motion. The discussion applies equally to non autonomous and autonomous Hamiltonians, and solutions of the equations of motion are considered as functions of a real time variable.

A system with N degrees of freedom is said to have been integrated by quadratures if either the solutions of the equations of motion or $2N$ invariants have been expressed in terms of quadratures (*i.e.*, integrals of definite functions). The possibility of integration by quadratures for an autonomous system was first discussed by E. Bour

(1855). For an autonomous system, Bour proved that, given N invariants in involution, an additional N invariants can be constructed in terms of quadratures. Subsequently, Liouville showed for a general non autonomous Hamiltonian that, given N possibly time-dependent invariants in involution, an additional N invariants could be constructed in terms of quadratures (Liouville 1855a; Liouville 1855b; Whittaker 1944; Arnold et al. 1988) Liouville did not construct the additional N invariants as canonically conjugate to the N given invariants (Whittaker 1944).

In the present work, we consider solutions of a Hamiltonian system with N-degrees-of-freedom as functions of a real time variable, and we allow the system to be non autonomous. We postulate N possibly time-dependent invariants in involution that can be inverted locally in the phase space to obtain the momenta as functions of the coordinates, the invariants and time. Then, in terms of quadratures, we construct a Type Two generating function for a canonical transformation to new conjugate coordinates and momenta, each of which is an invariant, where the given N invariants are the new momenta. The construction is an explicit realization of integration by quadratures that yields N canonically conjugate pairs of invariants. It is also an explicit realization of the transformation to canonical variables that is considered by the theorem of Darboux (Arnold et al. 1988). When the value of the new Hamiltonian function is chosen to be zero, then the generating function is known as Hamilton's principal function, which solves the Hamilton-Jacobi equation. Our determination of N additional invariants is a generalization of the treatment for one-degree-of-freedom systems that was presented by Bouquet and Lewis (1996). When applied to autonomous systems, the results provide the angle variables (the $\alpha-$variables) for the zeroth-order Hamiltonian in the perturbation theory for constructing invariants of non autonomous Hamiltonian systems that was recently presented by Lewis et al. (1996).

It is always possible in principle to choose N invariants in involution for any Hamiltonian system. For example, the N initial values of the coordinates or momenta, expressed in terms of the coordinates and momenta at a later time, are a set of N invariants in involution. Our integration by quadratures applies in principle to any Hamiltonian system for which the chosen N invariants in involution are locally invertible to give the momenta. (Note that we do not require that the N invariants in involution be given analytically.) To carry out the construction of N additional invariants for any particular Hamiltonian requires that N invariants in involution be provided in some form at the outset.

In Section 2 we construct the Type Two generating function with which a Hamiltonian system can be integrated by quadratures given N invertible invariants in involution. The existence and canonical nature of the initial value functions are discussed in the Appendix.

2 The Generating Function

Denote N invariant functions in involution by $\tilde{I}^{(m)}(q,p,t)$, $1 \le m \le N$:

$$\frac{\partial}{\partial t}\tilde{I}^{(m)}(q,p,t)+\left[\tilde{I}^{(m)},H\right]=0\ , \tag{1}$$

$$\left[\tilde{I}^{(m)},\tilde{I}^{(n)}\right]=\sum_k\left(\tilde{I}^{(m)}_{q^{(k)}}\tilde{I}^{(n)}_{p^{(k)}}-\tilde{I}^{(m)}_{p^{(k)}}\tilde{I}^{(n)}_{q^{(k)}}\right)=0\ , \qquad 1\le m,n\le N\ , \tag{2}$$

where $[A,B]$ denotes the Poisson bracket of A with B. Denote by $G^{(n)}(q,I,t)$ the inverse of the invariant functions to give the momenta $p^{(n)}$:

$$I^{(m)}=\tilde{I}^{(m)}(q,p,t) \qquad \leftrightarrow \qquad p^{(n)}=G^{(n)}(q,I,t)\ . \tag{3}$$

This inverse transformation exists if the Jacobian of the transformation from the variables p to the variables I does not vanish. We let the given invariants in involution be the new momenta of a canonical transformation from (q,p). The new canonical coordinates,

$$J^{(n)}=\tilde{J}^{(n)}(q,p,t)\ , \qquad 1\le n\le N\ , \tag{4}$$

are also to be invariants. They can be determined by constructing a Type Two generating function, $S(q,I,t)$, for the canonical transformation $(q,p)\to(J,I)$ The defining equations for the generating function are (Goldstein 1980)

$$S_{q^{(m)}}(q,I,t)=p^{(m)}\ , \tag{5}$$

$$S_{I^{(m)}}(q,I,t)=J^{(m)}\ , \tag{6}$$

$$H(q,p,t)=K(t)-S_t\left\{q,\tilde{I}(q,p,t),t\right\}\ . \tag{7}$$

The new Hamiltonian, $K(t)$, is taken to be an arbitrary function of time only, as a result of which the new coordinates J as well as the new momenta I are invariants. If the arbitrary function $K(t)$ were be taken to be zero, then $S(q,I,t)$ would satisfly the Hamilton-Jacobi equation and be known as Hamilton's principal function. In order to construct $S(q,I,t)$, we first solve (5); then we satisfy (7); finally, the new coordinates can be calculated from (6).

In order to solve the system (5), we use (3) to express it as a system of first-order PDEs in the variables q:

$$S_{q^{(m)}}(q,I,t)=G^{(m)}(q,I,t)\ , \qquad 1\le m\le N\ . \tag{8}$$

The compatibility conditions for (8) are

$$G^{(m)}_{q^{(n)}}(q,I,t)=G^{(n)}_{q^{(m)}}(q,I,t)\ ; \tag{9}$$

they are necessary and sufficient conditions for the existence of a solution of (8). The conditions (8) can be expressed in terms of derivatives of $\tilde{I}^{(m)}(q,p,t)$ by using the identity

$$p^{(m)} \equiv G^{(m)}\left\{q, \tilde{I}(q,p,t), t\right\} . \tag{10}$$

Differentation of (10) with respect to $q^{(n)}$ and $p^{(n)}$ gives

$$0 = G^{(m)}_{q^{(n)}} + \sum_k G^{(m)}_{I^{(k)}} \tilde{I}^{(k)}_{q^{(n)}} , \tag{11}$$

$$\delta_{mn} = \sum_k G^{(m)}_{I^{(k)}} \tilde{I}^{(k)}_{p^{(n)}} . \tag{12}$$

The involution conditions, (2), combined with relations (11) and (12), imply satisfaction of the compatibility conditions (9), that is, the compatibility conditions for (8) are satisfied by virtue of the fact that the invariants $\tilde{I}$ are in involution. The demonstration can be carried out most easily by writing (2), (9), (11), and (12) in terms of matrices (Lewis and Bouquet 1998).

Since the compatibility conditions are satisfied, the system of equations (8) can be integrated one by one in any order. Although the detailed form of the solution for $S(q,I,t)$ will look different depending on the order of integration, the result will be independent of the order. As an illustration of the procedure, consider the case $N=2$. Integrating (8) with $m=1$ we get

$$S(q,I,t) = F_2^{(1)}(q^{(2)},I,t) + \int_{q_0^{(1)}}^{q^{(1)}} G^{(1)}\left(q', q^{(2)}, I, t\right) dq' . \tag{13}$$

Substitution this into (8) with $m=2$ and integration give

$$F_2^{(1)}(q^{(2)},I,t) = F_1^{(1)}(I,t) + \int_{q_0^{(2)}}^{q^{(2)}} G^{(2)}\left(q_0^{(1)}, q', I, t\right) dq' , \tag{14}$$

with the result

$$S(q,I,t) = F_1^{(1)}(I,t) + \int_{q_0^{(2)}}^{q^{(2)}} G^{(2)}\left(q_0^{(1)}, q', I, t\right) dq' + \int_{q_0^{(1)}}^{q^{(1)}} G^{(1)}\left(q', q^{(2)}, I, t\right) dq' . \tag{15}$$

The functions $F_1^{(1)}(I,t)$ and $F_2^{(1)}(q^{(2)},I,t)$ are arbitrary functions of their arguments. Integration of (8) in the reverse order gives an equivalent result that is similar in form (Lewis and Bouquet 1998).

The result of the integration of (8) in any order for any value of N will be of the form

$$S(q,I,t) = F(I,t) + R(q,I,t) , \tag{16}$$

where $R(q,I,t)$ is expressible in terms of integrals of definite functions and where $F(I,t)$ is arbitrary. To finish the construction of $S(q,I,t)$, we choose $F(I,t)$ to satisfy (7):

$$F(I,t)-F(I,t_0)$$

$$=-R(q,I,t)+R(q,I,t_0)+\int_{t_0}^{t}K(t')\,dt'-\int_{t_0}^{t}H\{q,G(q,I,t'),t'\}\,dt' \,. \tag{17}$$

This formula for $F(I,t)$ will yield an expression that is independent of q by virtue of the fact that the functions $\tilde{I}(q,p,t)$ are invariants of $H(q,p,t)$. Bouquet and Lewis (1996) verified this property directly for the case $N=1$. Now that the generating function $S(q,I,t)$ has been determined, the new invariant coordinates can be calculated directly from (6).

It is interesting to note that the canonical transformations constructed in this paper can be used in principle to define a canonical transformation that will carry any particular N-degree-of-freedom Hamiltonian with N invertible invariants in involution into any other N-degree-of-freedom Hamiltonian with N invertible invariants in involution. The canonical transformation that carries Hamiltonian $H_1(q,p,t)$ to any function of time $K(t)$ can be composed with the inverse of the transformation that carries Hamiltonian $H_2(q,p,t)$ to $K(t)$. The resulting transformation will carry $H_1(q,p,t)$ to $H_2(q,p,t)$.

Acknowledgment

Helpful discussions with B. Cano concerning the proof in the Appendix are gratefully acknowledged.

Appendix. The initial-value functions

The initial values of the coordinates or momenta, expressed in terms of the coordinates and momenta at a later time, are in involution because the mapping to the initial values from the coordinates and momenta at the later time is canonical. The canonical property can be demonstrated as follows. Hamilton's equations of motion are

$$\frac{\partial}{\partial t}q^{(i)}(q_0,p_0;t)=H_{p^{(i)}}(q,p,t)\,, \qquad \frac{\partial}{\partial t}p^{(i)}(q_0,p_0;t)=-H_{q^{(i)}}(q,p,t)\,, \tag{A.1}$$

where $1\leq i\leq N$, and the initial data are

$$q^{(i)}(q_0,p_0;t_0)=q_0^{(i)}\ ,\qquad p^{(i)}(q_0,p_0;t_0)=p_0^{(i)}\ . \tag{A.2}$$

The solutions of (A.1) exist in the space of differentiable functions of the real time variable t. Denote the values of the coordinates and momenta at a time t_1 by

$$q^{(i)}(q_0,p_0;t_1)=q_1^{(i)}\ ,\qquad p^{(i)}(q_0,p_0;t_1)=p_1^{(i)}\ . \tag{A.3}$$

The mapping (A.3) from (q_0,p_0) to (q_1,p_1) is canonical because it is generated as a solution of Hamilton's equations. The inverse mapping, from (q_1,p_1) to (q_0,p_0), can be expressed in terms of functions $Q^{(i)}(q_1,p_1;t)$ and $P^{(i)}(q_1,p_1;t)$, which are $q^{(i)}(q_0,p_0;t)$ and $p^{(i)}(q_0,p_0;t)$ evaluated with time running in the opposite sense:

$$Q^{(i)}(q_1,p_1;t)=q^{(i)}\big(q_0,p_0;(t_0+t_1)-t\big)\ , \tag{A.4a}$$

$$P^{(i)}(q_1,p_1;t)=p^{(i)}\big(q_0,p_0;(t_0+t_1)-t\big)\ . \tag{A.4b}$$

The inverse mapping is given by

$$Q^{(i)}(q_1,p_1;t_1)=q_0^{(i)}\ ,\qquad P^{(i)}(q_1,p_1;t_1)=p_0^{(i)}\ . \tag{A.5}$$

It is canonical because $Q^{(i)}(q_1,p_1;t)$ and $P^{(i)}(q_1,p_1;t)$ also solve a system of Hamilton's equations. The Hamiltonian for the inverse mapping is

$$\overline{H}(Q,P,t)=-H\big(Q,P,(t_0+t_1)-t\big)\ , \tag{A.6}$$

that is, (A.4a) and (A.4b) solve

$$\frac{\partial}{\partial t}Q^{(i)}(q_1,p_1;t)=\overline{H}_{P^{(i)}}(Q,P,t)\ ,\quad \frac{\partial}{\partial t}P^{(i)}(q_1,p_1;t)=-\overline{H}_{Q^{(i)}}(Q,P,t)\ , \tag{A.7}$$

which reduce to (A.1). It should be noted that the solutions of (A.7) also exist in the space of differentiable functions of the real time variable t, along with the solutions of (A.1).

References

Arnold, V.I., Kozlov, V.V, Neishtadt, A. I. (1988): Mathematical Aspects of Classical and Celestial Mechanics, in *Dynamical Systems III*, vol. 3, *Encyclopaedia of Mathematical Sciences*, V. I. Arnold, Ed. (Springer-Verlag, Berlin), 1-291

Bouquet, S., Lewis, H.R. (1996): A second invariant for one-degree-of-freedom time-dependent Hamiltonians given a first invariant. J. Math. Phys. **37**, 5509-5517

Bour, E. (1855): Sur l'integration des equations differentielles de la mécanique analytique. J. Math. Pures et Appl. **20**, 185-200

Feix, M.R., Bouquet, S., Lewis, H.R. (1987): Invariants for time-dependent potentials: use of self-similar techniques. Physica D **28**, 80-102

Feix, M. R., Lewis, H.R. (1985): Invariants for Dissipative Nonlinear Systems by Using Rescaling. J. Math. Phys. **26**, 68-73

Goldstein, H. (1980): *Classical Mechanics*, 2nd ed (Addison-Wesley, Reading)

Gutzwiller, M.C. (1990): *Chaos in Classical and Quantum Mechanics*, Vol. 1 (Springer-Verlag, New York)

Lewis, H. R., Bates, J. W., Finn, J. M. (1996): Time-dependent perturbation theory for the construction of invariants of Hamiltonian systems. Phys. Lett. A **215**, 160-166

Lewis,H. R., Bouquet, S. (1998): Integration by Quadratures for an *N*-Degree-of-Freedom Non Autonomous Hamiltonian System: Application to the Hénon-Heiles Hamiltonian, submitted for publication

Lewis, H. R., Leach, P. G. L., Bouquet, S., Feix, M.R. (1992): Representations of one-dimensional Hamiltonians in terms of their invariants. J. Math. Phys. **33**, 591-598

Lichtenberg, A.J., Lieberman, M.A. (1992): *Regular and Chaotic Dynamics*,Vol. 38, 2nd ed (Springer-Verlag, New York)

Liouville, J. (1855a): Note sur l'intégration des équations différentielles de la Dynamique. J. Math. Pures et Appl. **20**, 137-138

Liouville, J. (1855b): Note à l'occasion du mémoire précédent de M. Edmond Bour. J. Math. Pures et Appl. **20**, 201-202

Ott, E. (1993): *Chaos in Dynamical Systems* (Cambridge University Press, Cambridge)

Tabor, M. (1989): *Chaos and Integrability in Nonlinear Dynamics* (Wiley, New York)

Whittaker, E. T. (1944): *A Treatise on the Analytical Dynamics of Particles and Rigid Bodies*, 4th ed (Dover, New York) Sec Sec. 148.

Hierarchies of Similarity Symmetries and Singularity Analysis

Peter G.L. Leach

Department of Mathematics and Applied Mathematics, University of Natal, Durban 4041, South Africa

Abstract. The singularity analysis of ordinary differential equations is based on the concept of the balance of the dominant terms. The equivalence of these terms is reflected by the possession of the same similarity symmetry. Different possible singular behaviours are reflected by different similarity symmeties. Hierarchal structures are observed. The generalised Chazy equation provides a simple example of this structure. The singularity analysis is based on the recently introduced 'g function' method.

1 Introduction

The singularity analysis of ordinary differential equations is an important tool in the determination of their integrability. Since the work of Kowalevski (1889), Painlevé (1900, 1902) and others (Gambier 1909; Garnier 1914; Chazy 1911) some hundred years ago the existence of only movable singularities in the solution of a differential equation has been taken to indicate integrability. The development of the ARS algorithm (Ablowitz, Ramani and Segur 1978; Ablowitz, Ramani and Segur 1980a; Ablowitz, Ramani and Segur 1980b; Ramani, Grammaticos and Bountis 1989) as a quick method to establish integrability has been a practical reflection of the utility of the result. The more careful considerations of Conte (1994) underscore the necessity to place the analysis on a precise theoretical basis.

The essence of the ARS algorithm is to determine whether the solution of a given differential equation can be expressed as a Laurent expansion about a movable singularity in the form

$$x = \sum_{i=-n}^{\infty} a_i \tau^i \qquad \tau = t - t_0, \tag{1}$$

where n is an integer and t_0 is the location of the movable singularity, which contains arbitrary constants equal in number to the order of the equation. A subsequent development (Lemmer and Leach 1993; Fordy and Pickering 1991) has been the recognition of Laurent expansions of the form

$$x = \sum_{i=-\infty}^{-n} a_i t^i, \tag{2}$$

which are asymptotic solutions. Hence $\tau = t - t_0$ is replaced by t; the additional arbitrary constant comes from a_{-n-1} which is always arbitrary (Feix et al. 1997). The expansion (1) is called a Right Painlevé Series (RPS) and (2) a Left Painlevé Series (LPS).

Problems with the analysis of equations which do not fit into the mould required by the ARS algorithm led recently to the introduction of the 'g function' method (Feix et al. 1997a) which is based on *leading order* analysis and the variation of the function about the leading order behaviour so that the solution is written as

$$x = \alpha\tau^p + g(\tau), \tag{3}$$

where the leading order behaviour is given by $\alpha\tau^p$ and $\tau^{-p}g(\tau) \to 0$ as $\tau \to 0$ or $\tau^{-p}g(\tau) \to 0$ as $\tau \to \infty$ depending upon the region under investigation. In practice g is determined by successive approximation.

Central to the concept of integrability is the existence of a sufficient number of 'useful' Lie symmetries (Govinder and Leach 1995; Govinder and Leach 1997) to permit the reduction of the differential equation(s) to an algebraic equation. We recall that the differential equation

$$E(t, x, \dot{x}, \ddot{x}, \dddot{x}, \ldots) = 0 \tag{4}$$

possesses a Lie symmetry

$$G = \tau\frac{\partial}{\partial t} + \eta\frac{\partial}{\partial x} \tag{5}$$

if

$$G^{[n]}E_{|_{E=0}} = 0, \tag{6}$$

where $G^{[n]}$ is the nth extension of G given by (Mahomed and Leach 1990)

$$G^{[n]} = G + \sum_{j=1}^{n} \eta^{[j]}\frac{\partial}{\partial x^{(j)}} \tag{7}$$

and

$$\eta^{[j]} = \eta^{(j)} - \sum_{i=1}^{j} \binom{j}{i} x^{(j+1-i)}\tau^{(i)}. \tag{8}$$

If one considers the leading order behaviour

$$x = \alpha\tau^p, \tag{9}$$

there exists the symmetry

$$G = \frac{t}{p}\frac{\partial}{\partial t} + x\frac{\partial}{\partial x} \tag{10}$$

under which (9) is invariant. G is a similarity symmetry.

The terms in a differential equation which contribute to the leading order behaviour all possess the same similarity symmetry.

In the discussions of singularity analysis (equally the g function analysis) various opinions are expressed as to which leading order behaviours must be considered in instances where more than one exist. Some would take only those which contain the highest derivative. Others would consider any possible combinations of terms even if the highest derivative was excluded. In the latter case the full quota of arbitrary constants cannot be obtained. Must one then judge that the equation be nonintegrable even though another set of leading order behaviour gave acceptable results? This question suggests that our understanding of the implications of the results of this method of analysis is incomplete.

In this paper we consider equations invariant under the symmetries both of time translation and similarity. This includes all equations invariant under Lie's Type II algebra (Lie 1967). Other equations can be transformed to this structure by an appropriate increase of order. We cannot provide a full answer to the question posed, but hope that the considerations raised here will provoke a more careful examination of the implications for integrability of leading order analysis.

Given that the terms which contribute to a class of leading order behaviour must share the same similarity symmetry, the possible different classes of leading order behaviour must possess different similarity symmetries. To provide a concrete basis for our analysis we examine how these can arise in the case of the generalised Chazy equation.

2 The Generalised Chazy Equation

The general third order ordinary differential equation invariant under the two symmetries

$$\begin{aligned} G_1 &= \frac{\partial}{\partial t} \\ G_2 &= -qt\frac{\partial}{\partial t} + x\frac{\partial}{\partial x} \end{aligned} \tag{11}$$

has the form

$$\dddot{x} + x^{3q+1} f\left(\frac{\dot{x}}{x^{q+1}}, \frac{\ddot{x}}{x^{2q+1}}\right) = 0, \tag{12}$$

where f is an arbitrary function of its arguments each of which must necessarily be invariant under (11). For the purposes of our discussion we take $q = -1$ and use an explicit form for f so that we consider

$$\dddot{x} + x\ddot{x} + k\dot{x}^2 = 0. \tag{13}$$

Both (12) and (13) are called generalised Chazy equations (Feix et al. 1997b). When $k = -3/2$, (13) is *the* Chazy equation (Chazy 1911) and, when $k = 1$, (13) is known as Bureau's equation (Bureau 1964).

When $q = 1$, the similarity variable of (11) is xt so that the leading order behaviour of (13) with all terms dominant is

$$x = \alpha\tau^{-1}, \qquad \alpha = \frac{6}{k+2}. \tag{14}$$

If one sets

$$x = \alpha\tau^{-1} + g(\tau), \tag{15}$$

(13) becomes

$$\tau^3\dddot{g} + \alpha\tau^2\ddot{g} - 2k\alpha\tau\dot{g} + 2\alpha g + \tau^3\left(g\ddot{g} + \dot{g}^2\right) = 0 \tag{16}$$

which is the exact equation g must satisfy. If we impose the condition $\tau g(\tau) \to 0$ near $\tau = 0$, (16) can be linearised to the Euler equation

$$\tau^3\dddot{g} + \alpha\tau^2\ddot{g} - 2k\alpha\tau\dot{g} + 2\alpha g = 0 \tag{17}$$

with solution

$$g = K_1\tau^{-2} + K_2\tau^{\lambda_1} + K_3\tau^{\lambda_2}, \tag{18}$$

where

$$\lambda_{1,2} = \tfrac{1}{2}\left\{5 - \alpha \pm \sqrt{25 - 14\alpha + \alpha^2}\right\}, \tag{19}$$

provided $\alpha \neq 7 \pm \sqrt{6}$,

$$g = K_1\tau^{-2} + (K_2 + K_3\log\tau)\,\tau^{-1\pm\sqrt{6}} \tag{20}$$

if $\alpha = 7 \pm \sqrt{6}$ and

$$g = (K_1 + K_2\log\tau)\,\tau^{-2} + K_3\tau^{-7} \tag{21}$$

if $\alpha = 14$, where K_1, K_2 and K_3 are arbitrary constants. (There is no value of α which gives the possibility of a $\tau^{-1}\log\tau$ term in g.)

The introduction of logarithmic terms is not acceptable and cases (20) and (21) are excluded. For the behaviour of the solution about the singularity at $t = t_0$ terms with powers of less than -1 must be excluded. In (18) this immediately excludes K_1 so that for a general solution to be possible both K_2 and K_3 must be nonzero and so $\lambda_{1,2} > -1$, $\lambda_{1,2} \neq 0$ ($\lambda_{1,2} = 0$ implies $\alpha = 0$ which contradicts the requirement of leading order behaviour). For a Laurent expansion about t_0 in the form of an RPS to exist we have

$$\alpha = \frac{m(5-m)}{m+1}, \quad \lambda_1 = \frac{5-m}{m+1}, \quad \lambda_2 = m, \tag{22}$$

where m is a positive integer such that λ_1 is also integral. The only possible value of m is 1 which corresponds to the known integrable case of $k = 1$.

The asymptotic solution requires that all powers in g be less than -1 and, as τ is replaced by t, all of K_1, K_2 and K_3 must be included to give the correct number of arbitrary constants. An LPS exists only when $m = -3, -4$

and both correspond to the case $k = -3/2$ in which case (13) is the Chazy equation.

The only possibility for a different type of leading order behaviour in (13) is from the second and third terms since a combination of the first and second (third) must include the third (second) due to the symmetry G_2 in (11) (with $q = -1$). The second and third of (13) are homogenous in both x and t. A homogeneity symmetry $G_{2a} = x\partial/\partial x$ (equally $G_{2b} = t\partial/\partial t$) excludes the first term of (13). As there is no t in G_{2a} (or x in G_{2b}), the index in a posited leading order behaviour of

$$x = \alpha\tau^p \tag{23}$$

cannot be determined by a balance of indices and coefficients must be balanced. Thus α is arbitrary and

$$p = 0, \frac{1}{1+k}. \tag{24}$$

For the second and third terms to provide the leading order behaviour the solution they provide must dominate the first term. From the first of (24),

$$x = \alpha \tag{25}$$

and we can draw no conclusion. From the second

$$\begin{aligned} x &= \alpha\tau^{1/(1+k)} \\ \dddot{x} &\sim \tau^{-3+1/(1+k)} \end{aligned} \tag{26}$$

so that $-2 < k < -1$ for the two terms to dominate near $\tau = 0$ and $-\infty < k < -2 \cup -1 < k < \infty$ for the two terms to dominate as $t \to \infty$.

In the case $p = 0$ we find that setting

$$x = \alpha + g \qquad g \to 0 \quad \text{as} \quad \tau \to 0 \tag{27}$$

leads to the linearised third order ordinary differential equation

$$\dddot{g} + \alpha\ddot{g} = 0 \tag{28}$$

with solution

$$g = K_1 + K_2\tau + K_3 e^{-\alpha t} \tag{29}$$

in which both K_1 and K_3 must be set at zero. The next iteration is to set

$$x = \alpha + K_2\tau + g, \tag{30}$$

where now g is the solution of the nonhomogeneous equation

$$\dddot{g} + \alpha\ddot{g} + 2kK_2\dot{g} = -kK_2^2 \tag{31}$$

and the nonhomogeneous term introduces a logarithm into the solution.

When $p = 1/(1+k)$, we set

$$x = \alpha\tau^{\frac{1}{1+k}} + g \tag{32}$$

and find that

$$g = K_1\tau^{\frac{1}{1+k}} + K_2\tau^{\frac{-k}{1+k}} - \frac{k(1+2k)}{(1+k)(2+k)}\tau^{-1}, \tag{33}$$

where the coefficient K_1 must be set at zero. For behaviour near $t = t_0$ the term with coefficient K_2 must be discarded as it is more singular than the term in $\tau^{1/(1+k)}$. However, it is acceptable for the asymptotic régime. The term in τ^{-1} is satisfactory both near $t = t_0$ and as $t \to \infty$.

3 Discussion

We have considered some aspects of the leading order behaviour of an ordinary differential equation invariant under the two symmetries of time translation and similarity using the generalised Chazy equation, (13), as a vehicle for the discussion and the 'g function' method instead of the ARS algorithm. We admit that the ARS algorithm is faster and neater, but there are some disadvantages. The ARS algorithm is based on the presumption of a Laurent expansion starting at a negative power, indeed in its original conception an RPS. This already excludes solutions of equations, such as constant coefficient linear equations which do not have movable singularities. In the 'g function' method we impose nothing *à priori*. The solutions obtained are not required to be analytic. Where the equation contains a parameter we may look *à postiori* to those values of the parameter which will give an analytic solution. However, the attraction of the 'g function' method is that the introduction of logarithmic terms and considerations of compatrability come about as part of the solution process and need not be inferred. This feature is of particular importance when a parameter is present since then the determination of resonances and compatability, separate procedures in the ARS algorithm, is a single process (Hua et al. 1996).

The connection between the similarity symmetry and the leading order behaviour is both robust and delicate. It is robust in that the terms which contribute to the leading order behaviour share the same similarity symmetry. Different possible combinations of terms having leading order behaviour require that the similarity symmetries be distinct for each combination. In the case of the generalised Chazy equation we saw that the second type of leading order behaviour followed from the existence of a homogeneity symmetry in a subset of the three terms of the equation. This symmetry allowed for the possibility of different leading order behaviour in the neighbourhood of t_0 and infinity depending upon the values of the parameter, k.

Several questions remain unresolved. An equation such as

$$y'' = y^3 + 1 \tag{34}$$

may not have all terms dominant and yet be integrable because there exists a Laurent expansion about a movable pole. Since not all terms are dominant, (34) does not have a similarity symmetry. The only obvious symmetry is that of invariance under time translation. However, integrability in the sense of Lie is guaranteed by the presence of the nonlocal symmetry (Abraham-Shrauner 1993; Govinder and Leach 1995)

$$G = \int \frac{F\left(y'^2/2 - y^3/4 - y\right)}{y'^2} \mathrm{d}y \frac{\partial}{\partial x}. \tag{35}$$

Is it possible to look at an equation with symmetry breaking terms and identify whether it will be integrable? Certainly this will be the case if a sufficient number of 'useful' symmetries can be identified. In the case of Lie point symmetries this is not difficult. The art of the determination of 'useful' nonlocal symmetries is yet in its infancy.

The matter of different types of leading order behaviour poses some problems. In the case of the generalised Chazy equation, (13), we know that it is integrable for the two cases $k = 1$ and $k = -3/2$ and we identified the existence of an RPS for the former and an LPS for the latter. The behaviour near infinity (resp t_0) is given by the terms admitting the homogeneity symmetry. Yet the expansion (31) is not analytic due to the presence of the logarithmic term. This certainly suggests that an expansion not based on the highest derivative present in the equation provides information about behaviour near t_0 or infinity, but does not determine integrability. What happens when the highest derivative is in the analysis is another question the answer to which may be contained in the answer to the question above.

Acknowledgements

The Foundation for Research Development of South Africa and the University of Natal are thanked for their continuing support.

References

Ablowitz, M.J., Ramani, A., Segur, H. (1978): Nonlinear evolution equations and ordinary differential equations of Painlevé type. Lett. Nuovo Cimento **23**, 333–338

Ablowitz, M.J., Ramani, A., Segur, H. (1980): A connection between nonlinear evolution equations and ordinary differential equations of P-type. I. J. Math. Phys. **21**, 715–721

Ablowitz, M.J., Ramani, A., Segur, H. (1980): A connection between nonlinear evolution equations and ordinary differential equations of P-type. II. J. Math. Phys. **21**, 1006–1015

Abraham-Shrauner, B. (1993): Hidden symmetries and linearization of the modified Painlevé-Ince equation. J. Math. Phys. **34**, 4809–4816

Bureau, F.J. (1964): Differential equations with fixed critical points. Annali di Matematica pura ed applicata **LXIV**, 229–364

Chazy, Jean (1911): Sur les équations différentielles du troisiéme ordre et d'ordre supérieur dont l'intégrale générale a ses points critiques fixes. Acta Math. **34**, 317-385

Conte, R. (1994): Singularities of differential equations and integrability in *Introduction to methods of complex analysis and geometry for classical mechanics and nonlinear waves* Benest, D. and Frœschlé, C. eds. (Éditions Frontières, Gif-sur-Yvette), 49–143

Feix, M.R., Geronimi, C., Cairó, L., Leach, P.G.L., Lemmer, R.L., Bouquet, S.É. (1997): On the singularity analysis of ordinary differential equations invariant under time translation and rescaling. J. Phys. A (to appear)

Feix, M.R., Geronimo, C., Pillay, T., Leach, P.G.L. (1997): The behaviour of the solution of an ordinary differential equation about its singular solutions and the connection with the Painlevé series. (preprint MAPMO, Université d'Orléans, UFR Sciences, BP 6759, F-45067 Orléans Cedex 2, France)

Feix, M.R., Geronimi, C., Leach, P.G.L. (1997): Periodic solutions and associated limit cycle for the generalised Chazy equation. (preprint MAPMO, Université d'Orléans, UFR Sciences, BP 6759, F-45067 Orléans Cedex 2, France)

Fordy, A., Pickering, A. (1991): Analysing negative resonances in the Painlevé test. Phys. Lett. A **160**, 347–354

Gambier, B. (1909): Sur les équations différentielles du second ordre et du premier degré dont l'intégrale générale est a points critiques fixes. Acta Math. **33**, 1–55

Garnier, R. (1912): Sur des équations différentielles du troisième ordre dont l'intégrale générale est uniforme et sur une classe d'équations nouvelles d'ordre supérieur dont l'intégrale générale a ses points critiques fixes. Annales Scientifiques de l'École Normale Supérieur **XXIX**, 1–126

Govinder, K.S., Leach, P.G.L. (1995): On the determination of nonlocal symmetries. J. Phys. A **28**, 5349–5359

Govinder, K.S., Leach, P.G.L. (1997): A group theoretic approach to a class of second order ordinary differential equations not possessing Lie point symmetries. J. Phys. A **30**, 2055–2068

Hua, D.D., Cairo, L., Feix, M.R., Govinder, K.S., Leach, P.G.L. (1996): Connection between the existence of first integrals and the Painlevé property in two-dimensional Lotka-Volterra and Quadratic Systems. Proc. Roy. Soc. Lond. A **452**, 859–880

Kowalevski, Sophie (1889): Sur les problème de la rotation d'un corps solide autour d'un point fixe. Acta Math. **12**, 177–232

Kowalevski, Sophie (1889): Sur une propriété du système d'équations différentielles qui définit la rotation d'un corps solide autour d'un point fixe. Acta Math. **14**, 81–93

Lemmer, R.L., Leach, P.G.L. (1993): The Painlevé test, hidden symmetries and the equation $y'' + yy' + ky^3 = 0$. J. Phys. A **26**, 5017–5024

Lie, Sophus (1967): *Differentialgleichungen* (Chelsea, New York)

Mahomed, F.M., Leach, P.G.L. (1990): Symmetry Lie algebras of nth order ordinary differential equations. J. Math. Anal. Appln. **151**, 80–107

Painlevé, P. (1900): Mémoire sur les équations différentielles dont l'intégrale générale est uniforme. Bull. Math. Soc. France **28**, 201–261

Painlevé, P. (1902): Sur les équations différentielles du second ordre et d'ordre supérieur dont l'intégrale générale est uniforme. Acta Math. **25**, 1–85
Ramani, A., Grammaticos, B., Bountis, T. (1989): The Painlevé property and singularity analysis of integrable and non-integrable systems. Phys. Rep. **180**, 159–245

Area Preserving Flows on Compact Connected Surfaces

Anna Cima[1], Li Jibin[2], and Jaume Llibre[3]

[1] Departament de Matemàtica Aplicada II, E.T.S. d'Enginyers Industrials de Terrassa, Universitat Politècnica de Catalunya, Colom 11, 08222 Terrassa, Barcelona, Spain

[2] Department of Basic Science, Kunming University of Science and Technology, Kunming, P.N. 650093, P.R. China

[3] Departament de Matemàtiques, Universitat Autònoma de Barcelona, 08193 Bellaterra, Barcelona, Spain

Abstract. Let S be an orientable compact connected surface without boundary and let X be a C^2 vector field such that its flow is area preserving on S. Assume that X has finitely many singular points, satisfies a Lojasiewicz condition at all its singular points and has singular points if S is the 2-dimensional torus. Then every singular point is either a center or it has a neighbourhood which splits into a finite union of an even number of hyperbolic sectors and all orbits of X are closed except finitely many of them which are either singular points or separatrices joining singular points. The same result is true for measure preserving flows on S. In this case it is not necessary that S be orientable.

1 Introduction and Statement of the Main Results

Let S be a C^∞ compact connected surface without boundary endowed with a metric. Then S always has a C^1 isometric imbedding in $\mathbf{R}^3$. Moreover S always has a C^∞ isometric imbedding in some $\mathbf{R}^n$, see Nash (1956). Therefore without loss of generality we can assume that the surface S is imbedded in some $\mathbf{R}^n$ and that its metric is the induced metric by the Euclidean metric of $\mathbf{R}^n$.

A C^2 *flow* on S is a C^2 map $\varphi : \mathbf{R} \times S \to S$ such that

(1) $\varphi(0,p) = p$ for all $p \in S$;

(2) $\varphi(t,\varphi(s,p)) = \varphi(t+s,p)$ for all $p \in S$ and all $t, s \in \mathbf{R}$.

Sometimes we denote $\varphi(t,\cdot)$ by φ_t. The pair (S,φ_t) is called a *dynamical system.*

We denote by T_pS the *tangent plane* to S at p and let TS be the *tangent bundle* to S. A vector field $X : S \to TS$ is a C^2 map such that $X(p) \in T_pS$ for all $p \in S$.

Let X be a C^2 vector field on S. We say that X is *area preserving* if the flow defined by X preserves the area defined by the metric of S.

In order to determine all possible local phase portraits at isolated singular points of a C^2 area preserving vector field on the surface S we assume that

X satisfies a *Lojasiewicz condition* at each isolated singular point $p \in S$, i.e. $|X(q)|$ does not decrease as $q \to p$ more rapidly than some power of the distance to the singular point, more precisely $|X(q)| \geq C|q-p|^r$ for some positive constants C and r (see Dumortier 1977 or Anosov and Arnold 1988 (pages 82 and 86)). We note that an analytic vector field satisfies a Lojasiewicz condition at all of its isolated singular points.

Our main result on the dynamics of C^2 area preserving vector fields or equivalently on C^2 area preserving flows on orientable compact connected surfaces is the following.

Theorem 1 *Let S be an orientable boundaryless compact connected surface and let X be a C^2 area preserving vector field on S. Assume that*

(1) X has finitely many singular points,

(2) X satisfies a Lojasiewicz condition at all its singular points and

(3) X has singular points if S is the 2-dimensional torus.

Then the following statements hold.

(a) Every singular point is either a center or it has a neighbourhood which splits into a finite union of an even number of hyperbolic sectors.

(b) All orbits of X are closed except singular points and separatrices joining singular points.

A *transversal segment* on a local chart of S is a segment transversal to the flow. Theorem 1 without the Lojasiewicz condition but with the following two additional assumptions:

(1) all the singular points are either centers or saddles; and

(2) on every local chart, since the flow is area preserving, the vector field is Hamiltonian and can be written as $(H_{y_2}, -H_{y_1})$, then on every transversal segment to the flow parametrized by s the function dH/ds vanishes only on finitely many points;

was proved by Gavrilov (1974). We note that a *saddle* is a singular point which has a neighbourhood which splits into union of exactly four hyperbolic sectors.

The proof of Theorem 1 is based into two main results. The first one is the characterization of the local phase portraits at isolated singular points of vector fields satisfying the Lojasiewicz condition due to Dumortier (1977); and the second is the generalization of the Poincaré-Bendixson Theorem due to Schwartz (1963). We remark that the proof of Gravilov does not use the first result, but implicitly uses the second one.

By a *metric dynamical system* we mean a quadruple $(S, \mathcal{M}, \mu, \varphi_t)$, where $(S, \mathcal{M}, \mu)$ is a normalized measure space (S denoting the space, $\mathcal{M}$ a σ-algebra and μ a measure with $\mu(S) = 1$) and φ_t denotes a measure preserving flow of S onto S, i.e. $\mu(\varphi_t(A)) = \mu(A)$ for all $t \in \mathbf{R}$ and $A \in \mathcal{M}$.

Theorem 1' *Let S be a boundaryless compact connected surface and let X be a C^2 vector field on S. If φ_t is the flow defined by X, then we assume that*

(1) $(S, \mathcal{M}, \mu, \varphi_t)$ is a metric dynamical system,
(2) X has finitely many singular points,
(3) X satisfies a Lojasiewicz condition at all its singular points and
(4) X has singular points if S is the 2-dimensional torus.

Then the following statements hold.

(a) Every singular point is either a center or it has a neighbourhood which splits into a finite union of an even number of hyperbolic sectors.
(b) All orbits of X are closed except singular points and separatrices joining singular points.

The paper is organized as follows. In Sect. 2 we prove statement (a) of Theorem 1, in Sect. 3 we prove statement (b) of Theorem 1 and finally in Sect. 4 we recall the Poincaré-Hopf Theorem and the Kneser result on the flows over the Klein bottle.

2 Singular Points

In this section we study the local phase portrait of an area preserving vector field on a C^∞ orientable boundaryless compact connected surface S.

We recall that a singular point is a *center* if there is a neighbourhood of it where all orbits, except the singular point, are closed.

Lemma 2 *In the assumptions of Theorem 1, if p is an isolated singular point of Y, then either p is a center or p has a neighbourhood which splits into a finite union of an even number of hyperbolic sectors.*

Proof: From Dumortier (1977) (or Anosov and Arnold 1988 page 86) and since X satisfies a Lojasiewicz condition at its singular points, we have that every isolated singular point p of X is either a focus or a center, or p has a neighbourhood which splits into a finite union of hyperbolic, parabolic, and elliptic sectors. Hence, since X is area preserving, it follows that p is either a center or it has a neighbourhood which splits into a finite union of h hyperbolic sectors. In this last case the index of p is equal to $1 - h/2$ (see Andronov et al. (1973) for more details on index theory of singular points of planar vector fields). Since $1 - h/2$ must be an integer number, it follows that h is even. ∎

Clearly statement (a) of Theorem 1 follows from Lemma 2.

3 The Poincaré–Bendixson–Schwartz Theorem

Let S be a boundaryless compact connected surface and let X be a C^2 vector field on S. Let $\varphi(t) = \varphi(t,p)$ be the orbit of X passing through $p \in S$ when $t = 0$. Since S is compact, every orbit $\varphi(t,p)$ is defined for all $t \in \mathbf{R}$ (see Sotomayor (1979) page 17). We define the *ω-limit set* of p as

$$\omega(p) = \{q \in S : \text{ there exist } t_n \to \infty \text{ such that } \varphi(t_n) \to q \text{ when } n \to \infty\}.$$

In a similar way we define the *α-limit set* of p as

$$\alpha(p) = \{q \in S : \text{ there exist } t_n \to -\infty \text{ such that } \varphi(t_n) \to q \text{ when } n \to \infty\}.$$

The *ω-limit set* (respectively *α-limit set*) of an orbit γ is the set $\omega(p)$ (respectively $\alpha(p)$) for some $p \in \gamma$.

The generalization of the Poincaré-Bendixson Theorem to C^2 vector fields on boundaryless compact connected surfaces is due to Schwartz (1963) (see also Sotomayor (1979) page 320).

Schwartz Theorem *Let X be a C^2 vector field on a boundaryless compact connected surface S. Let γ be an orbit of X. If $\omega(\gamma)$ does not contain singular points, then $\omega(\gamma)$ is a closed orbit or $\omega(\gamma) = S$, where S is equal to the 2-dimensional torus $\mathbf{T}^2$, and X has no singular points.*

Of course the Schwartz Theorem can be stated with α-limit sets instead of ω-limit sets.

Let X be a C^1 vector field defined on a boundaryless compact connected surface S or on an open set of the real plane. A *graph* of X is a closed curve formed by finitely many singular points $p_1, \ldots, p_n$ and finitely many orbits $\gamma_1, \ldots, \gamma_n$ formed by regular points such that $\omega(\gamma_i) = \alpha(\gamma_{i+1}) = p_i$ for $i = 1, \ldots, n-1$, $\omega(\gamma_n) = \alpha(\gamma_1) = p_n$, and eventually some of the points p_is are the same.

From the Schwartz Theorem we can prove:

Corollary 3 *Let X be a C^2 vector field on a boundaryless compact connected surface S. Let γ be an orbit of X. Assume that X has finitely many singular points.*

(a) If $\omega(\gamma)$ has no singular points, then either $\omega(\gamma)$ is a periodic orbit or $\omega(\gamma) = S = \mathbf{T}^2$ and X has no singular points.

(b) If $\omega(\gamma)$ has regular and singular points, then $\omega(\gamma)$ is a graph.

(c) If $\omega(\gamma)$ has no regular points, then $\omega(\gamma)$ is a single singular point.

Proof: From the Schwartz Theorem statement (a) follows.

Let Σ be a local transversal section through the regular point $q \in \omega(\gamma)$ for the flow of X. Such a section exists because q is a regular point (use the flow box theorem, see Sotomayor (1979) page 222).

Let $p \in \gamma$. Since $q \in \omega(\gamma)$, there exists a sequence $\varphi_{t_n}(p) \to q$ as $t_n \to +\infty$ such that $\varphi_{t_n}(p) \in \Sigma$. Take an orientation on Σ and consider the interval $I = [\inf_{t_n} \varphi_{t_n}(p), \sup_{t_n} \varphi_{t_n}(p)] \subset \Sigma$.

Now let U be the set defined by the union of all positive semiorbits of the flow φ_t which pass through the set $I \setminus \{q\}$. Then, either the interior of U is homeomorphic to an annulus or the closure of U is homeomorphic to a Mobius strip. Of course this second possibility only can occur if S is nonorientable.

By the usual Poincaré-Bendixson Theorem, if the interior of U is an annulus (see Sotomayor (1979) page 248), statement (b) follows. If the closure of U is a Mobius strip, then we consider its double covering on an annulus and the unique flow on it which covers the flow on U. Hence, since statement (b) is true for the annulus, it follows for the Mobius strip.

If $\omega(\gamma)$ has no singular points, since $\omega(\gamma)$ is connected (see Sotomayor (1979) page 245) and X has finitely many singular points, it follows that $\omega(\gamma)$ is a single singular point. Therefore, (c) is proved. ∎

Of course a similar corollary can be stated for the α-limit sets of X instead of the ω-limit sets.

Statement (b) of Theorem 1 follows immediately from the next lemma.

Lemma 4 *In the assumptions of Theorem 1, let γ be an orbit of X different from a singular point.*

(a) If $\omega(\gamma)$ has no singular points, then γ is a periodic orbit.

(b) If $\omega(\gamma)$ has singular points, then γ is a separatrix of a hyperbolic sector of a singular point.

Proof: Assume that $\omega(\gamma)$ has no singular points. Then, from the Schwartz Theorem, it follows that $\omega(\gamma)$ is a closed orbit or $\omega(\gamma) = S = \mathbf{T}^2$ and X has no singular points. This last case is not possible because by assumption, if $S = \mathbf{T}^2$, then X has singular points. If $\omega(\gamma) \neq \gamma$, since $\omega(\gamma)$ is a closed orbit we obtain a contradiction to the fact that the flow is area preserving. Hence $\omega(\gamma) = \gamma$ and consequently γ is a periodic orbit. This proves (a).

Suppose that $\omega(\gamma)$ has a singular point. We consider two cases, either $\omega(\gamma)$ is formed by a single singular point or not. In the first case, since $\omega(\gamma) \neq \gamma$, from Lemma 2 it follows that γ is a separatrix. In the second case, from Corollary 3, $\omega(\gamma)$ is a graph. Since $\omega(\gamma) \neq \gamma$, we obtain a contradiction with the fact that the flow is area preserving. So the second case is not possible. This completes the proof of (b). ∎

Of course a similar lemma can be stated for the α-limit sets of X instead of the ω-limit sets.

The proof of Theorem 1′ uses the same kind of arguments as the proof of Theorem 1.

4 The Poincaré–Hopf Theorem

A boundaryless orientable compact connected surface S_g of genus g is diffeomorphic to the sphere if $g = 0$, to the torus if $g = 1$, or to the connected sum of g copies of the torus if $g \geq 2$. The Euler characteristic of S_g is $\chi(S_g) = 2-2g$. For more details see for instance Massey (1977).

A boundaryless nonorientable compact connected surface P_g of genus g is diffeomorphic to the real projective plane if $g = 1$, or to the connected sum of g copies of the real projective plane if $g > 1$. The Euler characteristic of P_g is $\chi(P_g) = 2 - g$.

Poincaré-Hopf Theorem *If X is a C^1 vector field on a boundaryless compact connected surface S having finitely many singular points, then the sum of their indices is equal to the Euler characteristic of S.*

For a proof of this theorem see, for instance, Hirsh (1976).

From the Poincaré-Hopf Theorem, it follows that, if S is neither the torus nor the Klein bottle, then $\chi(S) \neq 0$ and consequently every C^1 vector field X on S has singular points.

If S is the Klein bottle, then we know that either X has a singular point or X has a periodic orbit, see Kneser (1924).

Acknowledgements

We thank Kenneth Meyer, Eduardo Gallego and Agustí Reventós for their comments which helped us to improve the paper.

The first and the third authors are partially supported by a DGICYT grant, number PB93–0860. The second author is partially supported by a DGICYT grant, number SAB95–0453.

References

Andronov, A.A., Leontovich, E.A., Gordon, I.I., Maier, A.G. (1973): *Qualitative Theory of Second Order Dynamical Systems* (John Wiley & Sons, New York)

Anosov, D.V., Arnold, V.I. (1988): *Dynamical Systems* (Vol. 1 Encyclopedia of Mathematical Science, Springer-Verlag, New York)

Dumortier, F. (1977): Singularities of vector fields inthe plane. J. Diff. Eq. **23**, 53–106

Gavrilov, N.I. (1974): Dynamic systems with invariant Lebesgue measure on closed, connected, oriented surfaces. Diff. Eq. **12**, 139–144

Hirsh, M.W. (1976): *Differential Topology* (Springer-Verlag, New York)

Kneser, H. (1924): Reguläre Kurvensharen auf den Ringflähen. Math. Ann. **91**, 135–154

Massey, W.S. (1977): *Algebraic Topology: An introduction* (Springer-Verlag, New York)

Nash, J. (1956): The imbedding problem for Riemannian manifolds. Ann. of Math. **63**, 20–63

Schwartz, A.J. (1963): A generalization of a Poincaré-Bendixson Theorem to closed two-dimensional manifolds. Amer. J. of Math. **85**, 453–458

Sotomayor, J. (1979): *Lições de equações diferenciais ordinárias* (Instituto de Matemática Pura e Aplicada, Projecto Euclides, Rio de Janeiro)

A Lattice Model for Tropical Rain Forests

Jérôme Chave[1], Marc A. Dubois[1], and Bernard Riéra[2]

[1] Service de Physique de l'Etat Condensé, DRECAM/ DSM, CEA Saclay, l'Orme des Merisiers, F91191 Gif sur Yvette Cedex, France
[2] Laboratoire d'Ecologie Générale, Muséum National d'Histoire Naturelle, CNRS/ URA 1183, 4 avenue du Petit Château, F91800 Brunoy, France

Abstract. We propose a model of tropical rain forests over large time scales. Discrete rate equations are derived from the elementary physical phenomena, which drive the evolution of the ecosystems, using methods from quantum field theory. This leads to a set of equations essentially similar to the Lotka-Volterra system – though more complex because of the presence of transport terms. Considerations on the integrability of this system are also given.

1 Introduction

Rain forests represent a large part of the earth's vegetation and therefore are environmentally very important. As they exhibit highly competitive behaviour, they are also of great interest as an example of interacting populations dynamics. Anthropic actions upon climate, especially variations of CO_2 and greenhouse gases in the atmosphere directly act on the equilibrium of these tropical ecosystems. On the other hand there is strong evidence that 'catastrophs' have already occured in tropical ecosystems during the past 10 000 years. For example, it has been shown that, around 6 000 BP, the Amazonian forest has been strongly reduced (ECOFIT 1996). This is probably due to a global drought which led to extended forest fires, whereas the reduction of the African tropical forest around 2 800 BP seems to come rather from anthropic factors. Thus it is interesting to compare the present situation with the available data of the past.

Another important question concerns the understanding of the biodiversity persisting in a interspecific competitive ecosystem. Such a preservation of diversity is a fact experimentally verified a long time ago. Still a crucial issue is to determine whether this phenomenon is due to competition itself or whether it is a consequence of external climatic conditions which maintain the whole system far from equilibrium and thus more resistant to perturbations, especially of anthropic origin. To answer this question requires a deep understanding of the underlying dynamical processes of a tropical forest.

This can only be done through a global analysis as well as through well-defined numerical simulations incorporating both the knowledge of present tropical forest dynamics and paleoecological data. We do believe that the correctness of such a pluridisciplinary approach can only be insured if mathematical, physical and experimental abilities are properly associated.

2 Spatially Extended Mechanistic Model

We propose a model of rain forest areas of about 10×10 to 1000×1000 km^2. Our approach is discrete both in space, with a fixed spatial grid step of 100 metres, and in time: we take four time steps per year. The time scale considered may be over 10 000 years, but species mutation phenomena are assumed irrelevant. The variables of interest are biomasses for vegetals, and densities for animals the role of which is either to predate vegetals (e.g. herbivora) or to disseminate seeds (frugivora). Such phase space variables are kept real. Then we deal with a so-called *coupled map lattice* formalism (Chaté and Manneville 1992; Kaneko 1992). This is a quite different approach to those available until now because it is specific to rain forests – unlike some more formal attempts (McGlade 1995; Solé et al. 1992) – and it is *not* an individual based model (i.e. one spatial point represents one tree) which can give a valuable information only over a very small spatial scale.

A few *classes* of vegetal species are defined with respect to (w.r.t.) their dynamical behaviour. In a first approach we choose to restrict ourselves to three classes, namely grasses, heliophil trees and sciaphil trees. Some classes (especially trees) are not suitably described by only one real variable. More information about the dynamical state of the trees needs to be included. A recent account of available theories on this topic is found in McNair (1995). A three states description turns out to be useful: *young* state, *active* state (grouping fructiferous individuals) and *passive* state (grouping mature trees not in fructification condition). Such a three-states class can be dynamically described by a set of three equations. We call them *state structured* species[1]. Unstructured classes are those which do not require this three states description, e.g. grasses.

We now derive in a somewhat intuitive way the dynamical equations governing the evolution, disregarding the transport phenomena at first. Let us denote by $E_i(t)$ the biomass of species i at time t. Assuming that the only interaction is the competition for solar resources we propose the following set of Lotka-Volterra (LV) equations

$$\frac{\mathrm{d}E_i}{\mathrm{d}t} = \tau_i^b \left(1 - S_i^1 \Big(\sum_j \sigma_j E_j^{2/3} - S_i^0\Big)^2\right) E_i - \tau_i^d E_i. \tag{1}$$

τ_i^b is the time dependent birth rate embodying any climatic variations, τ_i^d is the death rate, S_i^1 is a sensitivity to the solar competition and $\sum_j \sigma_j E_j^{2/3}$ models the vegetal light screening effect. We assume that the screening surface due to an amount of biomass E is proportional to $E^{2/3}$ according to trivial dimensional considerations. Then S_i^0 is the optimal amount of light screening. For species needing light, this is 0, whereas for those growing only under a shadowing precanopy, $S_i^0 \neq 0$.

[1] Note that our states are not too dissimilar from the compartmental patches in population dynamics theory.

3 From a Realistic Forest to Diffusive Lotka–Volterra

We claim that, in a spatially extended model, intuition does not provide sufficient arguments to propose a set of dynamical equations for our model. Therefore we develop an individual-based approach to forests dynamics by taking into account only *elementary interactions* between a few individuals and over a given time scale. From such elementary processes, we are able to establish a analogue to many-particle quantum field theory from which we derive mean field (e.g. macroscopic) equations. Although quite abstract this theory has the huge advantage of relying only on mechanistic processes.

To be more specific we consider a set of interactions between the individuals of a population. For one species which does not interact with the others, we regroup the possible reactions in Table 1.

Table 1. Elementary processes for one species

Elementary process	Associated reaction	Local rate	Diffusive rate
Birth	$X \to X + X$	t_1^2	D
Death	$X \to \emptyset$	t_1^0	$\emptyset$
Saturation	$X + X \to X$	t_2^1	s_2^1
Anihilation	$X + X \to \emptyset$	t_2^0	s_2^0

Here we sketch the quantum mechanical analogy to this situation following Grassberger and Scheunert (1980), Peliti (1985) and Cardy and Täuber (1997). Let us denote by $n_i(t)$ the number of individuals at (discrete) point i and at time t. A *configuration* of the whole system is given by the sequence of integers $|\mathbf{n}\rangle = \{n_1, \ldots, n_N\}$, where N is the number of sites. Then we call a *state* of the system the vector

$$|\Psi(t)\rangle = \sum_{\{\mathbf{n}\}} P_\Psi(\mathbf{n};t)|\{\mathbf{n}\}\rangle, \tag{2}$$

where $P_\Psi(\mathbf{n};t)$ is the probability for the state $|\Psi\rangle$ to be in the configuration $|\{\mathbf{n}\}\rangle$ at time t. An Hilbert space is created from the set of all the states $|\Psi\rangle$ once we have defined a scalar product between the pure states (Grassberger and Scheunert 1980), viz.

$$\langle\{\mathbf{m}\}|\{\mathbf{n}\}\rangle = \prod_i m_i!\delta_{m_i}^{n_i}. \tag{3}$$

We assume the Markov property for the elementary processes under consideration and write the master equation describing the evolution of $P_\Psi(\mathbf{n};t)$:

$$\frac{\mathrm{d}P(\mathbf{n};t)}{\mathrm{d}t} = \sum_{\mathbf{m}\neq\mathbf{n}} \left[\Pi_t(\mathbf{n}|\mathbf{m})P(\mathbf{m};t) - \Pi_t(\mathbf{m}|\mathbf{n})P(\mathbf{n};t)\right], \tag{4}$$

where $\Pi_t(\mathbf{m}|\mathbf{n})$ is the transition rate from configuration $\mathbf{n}$ to configuration $\mathbf{m}$ at time t. This is straightforwardly rewritten in the form of an imaginary time Schrödinger equation for $|\Psi(t)\rangle$:

$$\frac{\mathrm{d}|\Psi(t)\rangle}{\mathrm{d}t} = H|\Psi(t)\rangle \tag{5}$$

with $\langle \mathbf{n}\,|H(t)|\,\mathbf{m}\rangle = \Pi_t(\mathbf{n}|\mathbf{m})$ if $\mathbf{n} \neq \mathbf{m}$, and $\langle \mathbf{n}\,|H(t)|\,\mathbf{n}\rangle = -\sum_{\mathbf{m}\neq\mathbf{n}} \Pi_t(\mathbf{m}|\mathbf{n})$ otherwise. H is an Hamiltonian – or Liouvillean – and it is a linear combination of the elementary processes.

We then turn to a Bargmann-Fock space representation by means of the definition of suitable creation-annihilation fields $\overline{u}(x;t)$ and $u(x;t)$ respectively in the continuous limit, thence yielding the possibility of a Lagrangian description. The normalisation of the density probability gives a condition on the creation field, which is $\overline{u}(x;t) = 1$. Thus the saddle point approximation reduces to only one equation

$$\frac{\delta L}{\delta \overline{u}}(1, u; x, t') = \frac{\delta\{\overline{u}\partial_t u - H\}}{\delta \overline{u}}(1, u; x, t') = 0, \tag{6}$$

where $\delta/\delta\overline{u}$ is the functional derivative w.r.t. $\overline{u}$ and L is the Lagrangian[2]. After a rather cumbersome calculation we obtain a PDE in the field u, which is nothing but the average density n, since $n = \langle \overline{u}u\rangle = \overline{u}u = u$. In our case this can be written quite generally as

$$\partial_t u = au - bu^2 + \nabla((c + du)\nabla u) + e(\nabla u)^2, \tag{7}$$

where the coefficients $\{a, \ldots, e\}$ are determined from the microscopic rates defined in Table 1. This is a second order diffusive generalisation of the well-known Verhulst equation. Such methods generalise straightforwardly to more complex systems with several species (leading to diffusive LV equations to lowest order), diffusive interactions, memory effects and nonlocal transport.

In two dimensions, which is typical of most realistic forests, the mean field theory does not actually apply because the critical dimension is precisely $d_c = 2$, as shown by Peliti (1986), but, as numerically we deal with spatially discrete variables, we can still consider our mean field approach as a good way to derive rate equations of an extended lattice ecosystem.

4 Integrability of the Local System

Having justified the form of the diffusive equations we discuss the integrability of our model. Restricting ourselves to the homogeneous limit $D \to \infty$ we now refer to equation (1). This is obviously a special form of the LV system extensively studied since the pionneering work of Volterra (1931) and, later, of Goel et al. (1971). Recent progress on this topic has been performed making use of modified Carleman embedding methods (Cairó and Feix 1992), hamiltonian analogies (Nutku 1990) or Painlevé analysis.

[2] One can check afterwards that $\overline{u} = 1$ is a solution of the second equation $\frac{\delta L}{\delta u}(\overline{u}, u; x, t') = 0$ for all u.

Our scope is much less ambitious than such approaches. Indeed we have constructed a LV system on mechanistic grounds and we want to know whether it is integrable, at least partially, so that we can deal with invariants of motion which can be destroyed by temporal perturbations of the parameters. Let us consider equation (1) for N unstructured classes. If we assume that all classes have the same light screening optimum S^0, then the system is almost fully integrable. Indeed, we find $N-1$ first integrals when all the parameters are time independent

$$E_i^{\alpha_i} \exp(-\beta_i t) = E_j^{\alpha_j} \exp(-\beta_j t) \tag{8}$$

for $j < i$, $\alpha_i = 1/(\tau_i^b S_i^1)$ and $\beta_i = (\tau_i^b - \tau_i^d)/(\tau_i^b S_i^1)$. The problem is reduced to a one class evolution problem, which obeys a nonpolynomial evolution equation but without singularities. Let E represent species i such that $\beta_i \geq \beta_j, \forall j$. Then

$$\frac{\mathrm{d}E}{\mathrm{d}t} = \tau^b \left(1 - S^1 \left(\sum_j \sigma_j E^{\frac{2\alpha}{3\alpha_j}} \exp\left[\frac{2(-\beta+\beta_j)}{3\alpha_j} t\right] - S^0\right)^2\right) E - \tau^d E. \tag{9}$$

If $\beta < \beta_j, \forall j$, the following asymptotic equation holds

$$\frac{\mathrm{d}E}{\mathrm{d}t} \simeq \tau^b \left(1 - S^1 \left(\sigma E^{2/3} - S^0\right)^2\right) E - \tau^d E$$

the solution of which is, if $u = E^{2/3}$, $u_0 = (S - \sqrt{\beta})/\sigma$ and $u_1 = (S + \sqrt{\beta})/\sigma$,

$$u^{\frac{2\sqrt{\beta}\sigma}{\beta - S^2}} \left[\frac{(u-u_0)^{\frac{1}{u_0}}}{(u_1-u)^{\frac{1}{u_1}}}\right] = u(0)^{\frac{2\sqrt{\beta}\sigma}{\beta - S^2}} \left[\frac{(u(0)-u_0)^{\frac{1}{u_0}}}{(u_1-u(0))^{\frac{1}{u_1}}}\right] \exp\left(\frac{4\sqrt{\beta}\sigma}{3\alpha} t\right). \tag{10}$$

There is no explicit solution here unlike for the Verhulst equation. Besides, if more than one species has an equal maximal β, even less trivial solutions appear. In the case of different light screening optimum we have only $N-2$ first integrals and then the problem reduces to a two interacting classes evolution problem, which could allow the emergence of singularities, even if we have not yet been able to check numerically this fact. As a conclusion to this section we remark that there is very little hope to achieve analytically this work because of the intricacy of the equations. There is also no great incentive because we are basically interested in time varying parameters. In this more general case we expect invariants to be destroyed even for adiabatic perturbations so that any integrability analysis becomes irrelevant. Thus invariants can just give us indications on the unperturbed behaviour, whereas the perturbed dynamics must be approached through simulation.

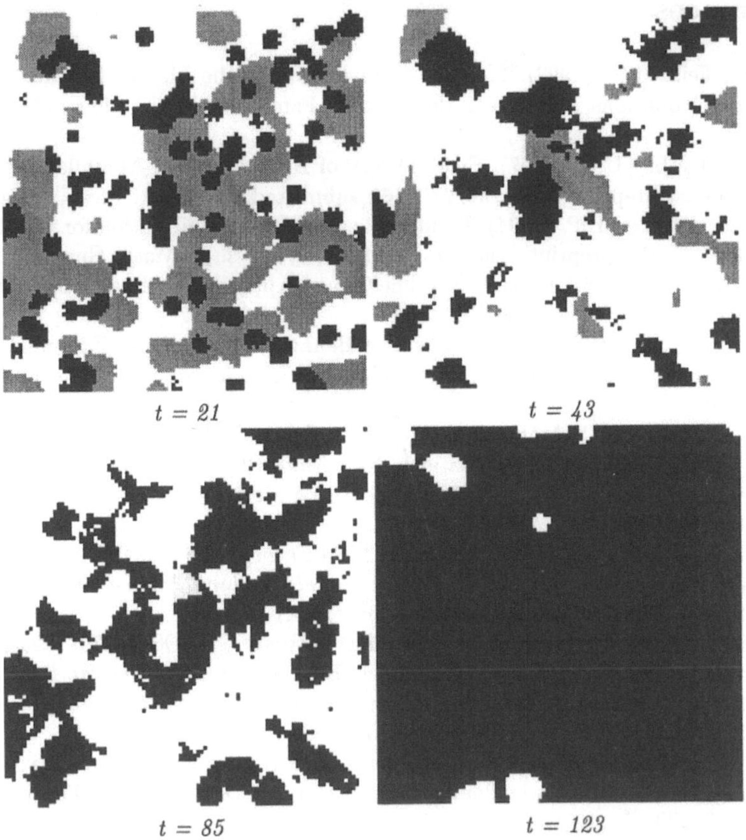

Fig. 1. Rain forest at different time steps (t in years)

5 Conclusion

We have developed a C++ code which embodies all the specifications of our model. The results shown in Fig. 1 were obtained with a three classes forest: one unstructured class – grasses (white) – and two state-structured classes – heliophil (grey) and sciaphil (black) trees. The dynamics are typical of a large scale perturbation due to climatic forcing. After a transition time when the heliophil trees disappear – only residuals remain – climatic variations favour the rain forest's recruitment over grasses. This is just one of different scenarii which can be studied. We now have a carefully established framework in which numerical simulations can be performed. We believe that our approach is rigorous and practical enough to be a useful tool in the understanding of spatially extended dynamical phenomena in rain forests incorporating climatic perturbations, aging effects and recruitment.

References

Cairó, L., Feix, M.R. (1992): Families of Invariants of the Motion for the Lotka-Volterra Equations: the Linear Polynomials Family. J. Math. Phys. **33**, 2440–2445

Cardy, J., Täuber, U.C. (1997): Field Theory of Branching and Annihilating Random Walks, preprint `cond-mat/9704160` submitted to J. Stat. Phys., OUTP97 14S. See also Lee, B.P. (1994): Renormalization Group Calculation for the Reaction $kA \to \emptyset$", preprint `cond-mat/9311064`. J. Phys. A: Math. Gen. **27**, 2633 and Cardy, J. (1996): Renormalisation Group Approach to Reaction-Diffusion Problems. preprint `cond-mat/9607163`

Chaté, H., Manneville, P. (1992): Collective Behaviors in Spatially Extended Systems with Local Interactions and Synchronous Updating. Prog. Theor. Phys. **87**, 1–60

Goel, N.S., Maitra, S.C., Montroll, E.W. (1971): On the Volterra and Other Nonlinear Models of Interacting Populations. Rev. Mod. Phys. **43**, 231–276

Grassberger, P., Scheunert, M. (1980): Fock-Space Methods for Identical Classical Objects. Fortschr. Phys. **28**, 547–583

Itzykson, C., Zuber, J.-B. (1980): *Quantum Field Theory* (Mc Graw-Hill, New York). See also Negele, J.W., Orland, H. (1980): *Quantum Many-Particle Systems*, coll. Frontiers in Physics (Addison-Wesley, New York)

Kaneko, K. (1992): Overview of Coupled Map Lattices. CHAOS **2**, 279–282

McGlade, J.M. (1995): Dynamics of Complex Ecologies in *Modelling the Dynamics of Biological Systems - Nonlinear Phenomena and Pattern Formation*, Moselkilde, E., Mouritsen, O.G. eds. (Springer, Heidelberg, Berlin), 251–268

McNair, J.N. (1995): Ontogenetic Patterns of Density-dependent Mortality: Contrasting Stability Effects in Populations with Adult Dominance. J. Theor. Biol. **175**, 207–230

Nutku, Y. (1990): Hamiltonian structure of the Lotka-Volterra equations. Phys. Lett. A, **145**, 27. See also Plank, M. (1995): Hamiltonian Structures for the n-dimensional Lotka-Volterra Equations. J. Math. Phys. **36**, 3520–3534 and Kerner, E.H. (1997): Comment on Hamiltonian Structures for the n-dimensional Lotka-Volterra Equations. J. Math. Phys., **38**, 1218–1223

Peliti, L. (1985): Path-Integral Approach to Birth-Death Processes on a Lattice. J. Physique **46**, 1469–1483

Peliti, L. (1986): Renormalisation Fluctuation Effects in the $A + A \to A$ Reaction. J. Phys. A: Math. Gen. **19**, L365–L367

Rapport d'activité (1996) ECOFIT (ECOsystèmes Forestiers Inter-Tropicaux), joint program ORSTOM/ CNRS/ CEA

Solé, R.V., Bascompte, J., Vallis, J. (1992): Nonequilibrium Dynamics in Lattice Ecosystems: Chaotic Stability and Dissipative Structures. CHAOS **2**, 387–395

Volterra, V. (1931): *Leçons sur la Théorie Mathématique de la Lutte pour la Vie* (Gauthier-Villars, Paris)

Periodic Solutions and Associated Limit Cycle for the Generalised Chazy Equation

Claude Géronimi[1], Marc R. Feix[1], and Peter G.L. Leach[1,2,3]

[1] Laboratoire de Mathématique, Applications et Physique Mathématique - UMR 6628, Université d'Orléans, UFR des Sciences, BP 6759, F-45067 Orléans Cedex 2, France

[2] Permanent address: Department of Mathematics and Applied Mathematics University of Natal, Durban 4041, Republic of South Africa

[3] Member of the Centre for Theoretical and Computational Chemistry, University of Natal, Durban

Abstract. We study the generalised Chazy equation, $\dddot{x}+x^q\ddot{x}+kx^{q-1}\dot{x}^2=0$, which is characterised by the symmetries of time translation and rescaling. For a large class of initial conditions numerical computations reveal the asymptotic appearence of periodic solutions for $k=q+1$. These solutions are identical after rescaling and, in this sense, exhibit the property of a limit cycle in the three dimensional phase space. The periodic solutions are related to a conventional limit cycle of a class of second order ordinary differential equations which are connected to the existence of a first integral of the generalised Chazy equation.

1 Introduction

The possibility of demonstrating the existence of periodic orbits and limit cycles and finding their properties in three-dimensional systems is not so frequent. If we consider systems with only the symmetry of time translation, we may obtain limit cycles in which the initial conditions still play a rôle. However, we are looking for limit cycles where these conditions can be absorbed. To this end we can turn to systems with two symmetries. An additional symmetry, common in physical models, is that of self-similarity or rescaling. The general third order ordinary differential equation invariant under the two symmetries (Feix et al. 1996)

$$\begin{aligned} G_1 &= \frac{\partial}{\partial t} && \text{(time translation) and} \\ G_2 &= -qt\frac{\partial}{\partial t}+x\frac{\partial}{\partial x} && \text{(self similarity/rescaling)} \end{aligned} \tag{1}$$

is

$$\dddot{x}+x^{3q+1}f(\xi,\eta)=0, \tag{2}$$

where

$$\xi=\frac{\dot{x}}{x^{q+1}}\quad,\quad\eta=\frac{\ddot{x}}{x^{2q+1}} \tag{3}$$

and f is an arbitrary function of its two arguments.

A well known representation is the Chazy equation (Chazy 1911) for which

$$f(\xi, \eta) = k\xi^2 + \eta \tag{4}$$

and q and k take the special values of one and $-3/2$ respectively. Numerical investigations of the actual Chazy equation do not reveal any periodic solutions. Indeed none are found for general k. However, in the case that $q = 2$ we detect oscillatory solutions which decay for $2 < k < 3$ and diverge for $k > 3$. For $k = 3$ there is initially some transient behaviour and, then, a periodic solution is observed for a large class of initial conditions (see Fig. 1).

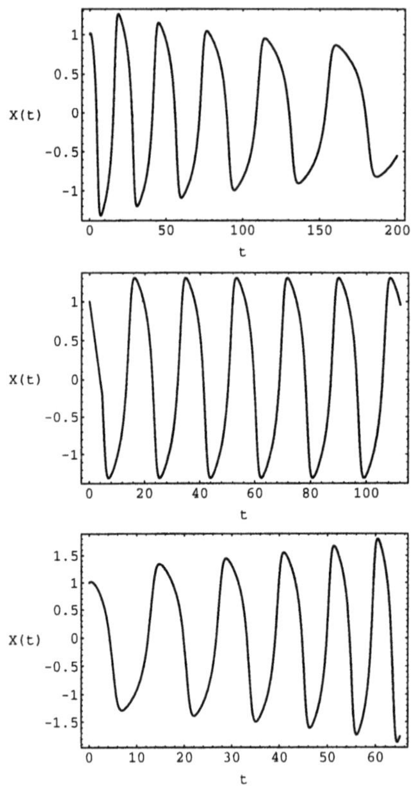

Fig. 1. Oscillatory solutions of the generalised Chazy equation with $q = 2$, $x(0) = 1$, $\dot{x}(0) = 0.1$, $\ddot{x}(0) = -0.3$, $k = 2.9$,3 and 3.1 respectively.

Similar periodic behaviour is observed for q even and $k = (q + 1)$. These observations lead us to consider the equation

$$\dddot{x} + |x|^q\ddot{x} + k|x|^q\frac{\dot{x}^2}{x} = 0. \tag{5}$$

Now the periodic behaviour is found for all real positive values of q and the same relation $k = (q+1)$. In this paper we report our investigations of the properties of (5). In §2 we discuss the boundaries of the space of initial conditions through the phase plane of the reduced equation. In §3 we present an heuristic argument for $k = (q+1)$ to be a critical value for the existence of these periodic orbits. We show that under appropriate rescaling the periodic orbits appear as a single universal orbit. Moreover we show that the rescaling factor is given by a first integral of motion which exists for $k = (q+1)$. In §4 we connect a class of second order ordinary differential equations with the above mentioned first integral. Although the form of this equation suggests that the Nemytskii and Stepanov (1989) theorem, a generalisation of Liénard's theorem (Davis 1960), could be applied to give the conditions for the existence of a limit cycle, a more precise study shows that some of the conditions are not fulfilled. Nevertheless the reduced equation has a limit cycle which is the projection of the periodic orbit in the three dimensional $(x, \dot{x}, \ddot{x})$ phase space onto the two dimensional $(x, \dot{x})$ phase plane. In §5 we present our conclusions.

As a practical point we remark that in the computations for the third order ordinary differential equation (5) we can, without loss of generality, take $x(0) = 1$. This is a consequence of the invariance of (5) under rescaling.

2 Phase Plane of the Reduced Equation

The space of initial conditions for which the periodic orbits are obtained is found from a study of the (ξ, η) phase plane of the reduced equation,

$$\frac{\mathrm{d}\eta}{\mathrm{d}\xi} = \frac{f(\xi,\eta) + (2q+1)\xi\eta}{(q+1)\xi^2 - \eta}, \tag{6}$$

obtained by the double reduction of order of (2) using the two symmetries G_1 and G_2 as given by (1), where ξ and η are defined in (3) and, for the generalised Chazy equation f is given by (4). The critical curves in the (ξ, η) phase plane are given by the parabola $\eta = (q+1)\xi^2$ upon which the denominator of (6) vanishes (curve D in Fig. 2).

On this curve the trajectory must have a vertical tangent. Another critical curve is

$$\eta = \frac{-k\xi^2}{1 + (2q+1)\xi},$$

upon which the numerator of (6) vanishes, thereby imposing an horizontal tangent when the trajectory crosses it (noted N in Fig. 2). The intersection of these curves gives the two equilibrium points marked A and B in Fig. 2. The point A corresponds to the selfsimilar solution

$$x(t) = a(t - t_0)^{-1/q} \tag{7}$$

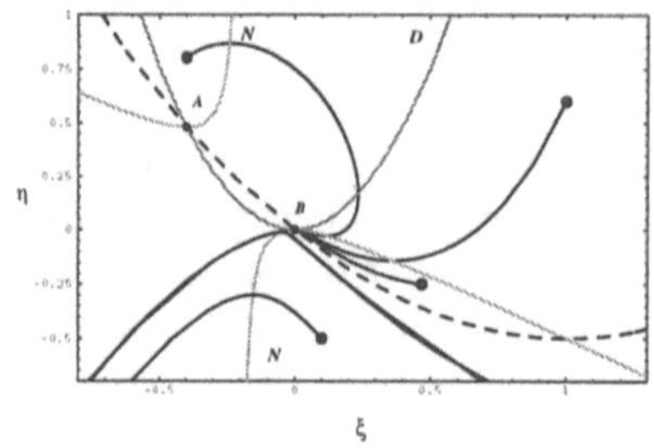

Fig. 2. Phase space portrait in the (ξ, η) plane for the reduced equation (6) in the case $q = 2$, $k = 3$, $x(0) = 1$ and selected values of $\dot{x}(0) = \xi(0)$ and $\ddot{x}(0) = \eta(0)$ to cover the different regions of the plane. The grey curves represent the numerator (N) and denominator (D), the dashed curve is the dividing line and the black curves are the trajectories. The black dots denote the starting points. The points A and B are the equilibrium points.

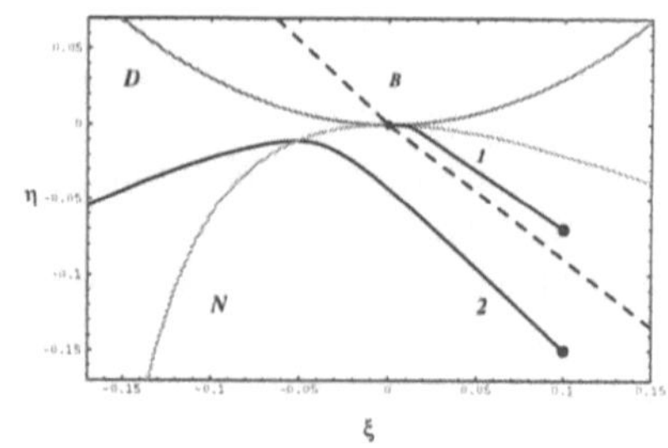

Fig. 3. Phase space portrait in the (ξ, η) plane for the reduced equation (6) around the point B for $q = 2$, $k = 3$, which shows the trajectory 1 with the initial conditions $\xi(0) = 0.1$, $\eta(0) = -0.07$ and the trajectory 2 with the initial conditions $\xi(0) = 0.1$, $\eta(0) = -0.15$.

which, according to the sign of t_0, corresponds either to a singularity taking place at a finite time (for $t_0 > 0$) or an asymptotically vanishing solution (for $t_0 < 0$). This singularity analysis provides the criteria for the Painlevé test. This is not the purpose of the present paper. Primarily we are interested in the structure of the solution closed to point B. Linearisation of (6) leads to $\mathrm{d}\eta/\mathrm{d}\xi = -1$ and, consequently, to trajectories of the form $\eta + \xi = C$ (where C is a constant). However, we cannot keep only the linear term because of the properties of the solution on the curves D and N defined above both of which are tangent to the ξ axis at the point $(0,0)$. Firstly we consider a trajectory in the neighbourhood of the point B for which $\eta + \xi = C$, where now C is a small positive number. Moreover we suppose a trajectory coming from an initial positive ξ and negative η (see Fig. 3 which is an enlargement of Fig. 2 around the equilibrium point B). The trajectory parallel to the second bisectrix cannot cross the N-curve from below. Consequently it converges to the origin along N (*ie* for small ξ and η with $\eta = -k\xi^2$). It can be proven that this corresponds to an asymptotic behaviour in $x(t) = at^{1/k+1}$ (trajectory 1 in Fig. 3). We consider now the same trajectory but with a negative and C still small. The trajectory cuts the N-curve in the $\xi < 0$ region and has an horizontal tangent with the origin acting as a repulsor (trajectory 2 on the Fig. 3). As a matter of fact we see that the "dividing line" (*ie* the boundary between the two above mentioned behaviours) will be the solution of (6) which goes through the point B with a slope equal to -1. In fact, for $k = (q + 1)$, it is easily verified that (6) has the solution

$\eta = \frac{1}{2}\xi^2 - \xi$, which is just the "dividing line" we seek. Another "dividing line" can be found for the case $k = q$ in which case its equation is $\eta = -\xi$. The "dividing line" is the dotted line on Figs. 2 and 3. The periodic behaviour for $k = (q+1)$ has been found for the initial values $x(0) = 1$ (with no loss of generality as pointed out in §1), $\dot{x}(0) = \xi(0)$ and $\ddot{x}(0) = \eta(0)$ with the points $(\xi(0), \eta(0))$ located below the "dividing line" with $\xi(0) > 0$ and $\eta(0) < 0$. Moreover we stayed in the region $0 < \xi(0) < 1$ and $0 > \eta(0) > -4$. In that zone we systematically recovered the asymptotic periodic solution for $k = (q+1)$.

3 Periodic Orbits for $k = q + 1$

To provide an heuristic description of the periodic motion we resort to the integrodifferential equation obtained by integrating (5). This is

$$\ddot{x} + |x|^q \dot{x} + (k-q) \int_0^t |x|^q \frac{\dot{x}^2}{x} \mathrm{d}\tau = F, \tag{8}$$

where F is a constant of integration. We assume that $x(t)$ is periodic with period T and, in a sense which we will make more precise below, of mean value zero. Over a period

$$\int_0^T F\dot{x}\mathrm{d}t = F(x(T) - x(0)) = 0. \tag{9}$$

We now compute the work done during the period, T, by the pseudoforces on the left side of (8). We need not consider the "acceleration" term as the "kinetic energy" is periodic. For the damping type term the contribution is

$$\int_0^T |x|^q \dot{x}^2 \mathrm{d}t. \tag{10}$$

For the third term we have, by integrating by parts,

$$\begin{aligned} (k-q) \int_0^T \dot{x}\mathrm{d}t \int_0^t |x|^q \frac{\dot{x}^2}{x}\mathrm{d}\tau &= (k-q)\left\{\left[x \int_0^t |x|^q \frac{\dot{x}^2}{x}\mathrm{d}\tau\right]_0^T - \int_0^T |x|^q \dot{x}^2 \mathrm{d}t\right\} \\ &= (k-q)\left\{x(T)\int_0^T |x|^q \frac{\dot{x}^2}{x}\mathrm{d}t - \int_0^T |x|^q \dot{x}^2 \mathrm{d}t\right\}. \end{aligned} \tag{11}$$

We now make precise the meaning of the average value of $x(t)$ being zero by requiring that the first integral in (11) be zero. We are then left with

$$-(k-q-1)\int_0^T |x|^q \dot{x}^2 \mathrm{d}t. \tag{12}$$

The integral is positive and so the forces balance over a period only if $k = (q+1)$. We note that (12) is the negative of the work done by the pseudoforces. For $k > (q+1)$ this work is consequently positive and increases the energy. For $k < (q+1)$ the work is negative and so decreases the energy in agreement with the numerical results shown on Fig. 1. We emphasise that the argument is heuristic, but the result has been clearly demonstrated numerically.

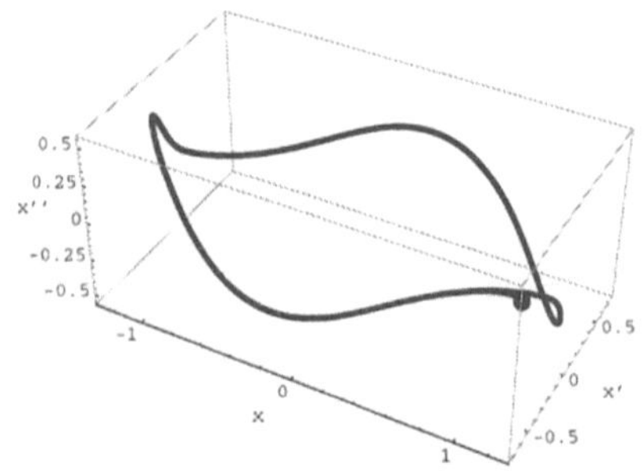

Fig. 4. Trajectory for (5) in the three dimensional $(x, \dot{x}, \ddot{x})$ phase space with $q = 2$ and $k = 3$ and for $x(0) = 1$, $\dot{x}(0) = 0.1$ and $\ddot{x}(0) = -0.26$.

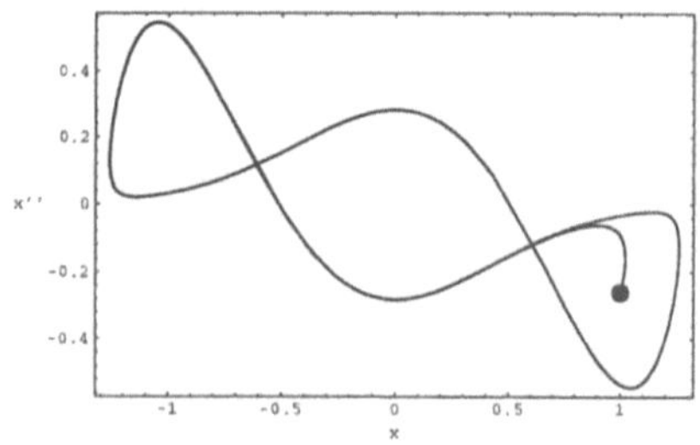

Fig. 5. Projection of the trajectory of the Fig. 4 onto the phase plane $(x, \ddot{x})$.

In Fig. 4 we show a view of the trajectory in the three dimensional space and in Fig. 5 we exhibit the projection of this three dimensional trajectory in the phase plane $(x, \ddot{x})$. In Fig. 6 we show the projection on the phase plane $(x, \dot{x})$ of trajectories with the same $x(0)$, $\dot{x}(0)$ and with different values of $\ddot{x}(0)$. In this figure, although the orbits are asymptotically periodic, it is evident that we do not have an universal limit cycle as the asymptotic orbits depend upon the initial conditions. However, we can recover an universal periodic orbit through rescaling x, $\dot{x}$ and $\ddot{x}$ by the factors a, a^{q+1} and a^{2q+1} respectively, where a is, for instance, the amplitude of the $x(t)$ curve after it has reached its periodic behaviour. This is due to the existence of both symmetries under which both ξ and η are invariant. This universal orbit is now independent of the initial conditions except through the rescaling factor and so plays the rôle

of an universal limit cycle (see Fig. 7). In fact we need not introduce an "experimentally measured quantity" as a to rescale the orbits. We just observe the existence of a first integral for $k = (q+1)$ which is

$$I = x\ddot{x} - \frac{1}{2}\dot{x}^2 + |x|^q x\dot{x}. \tag{13}$$

Consider the effect of rescaling on (13). Recall that the rescaling is represented by

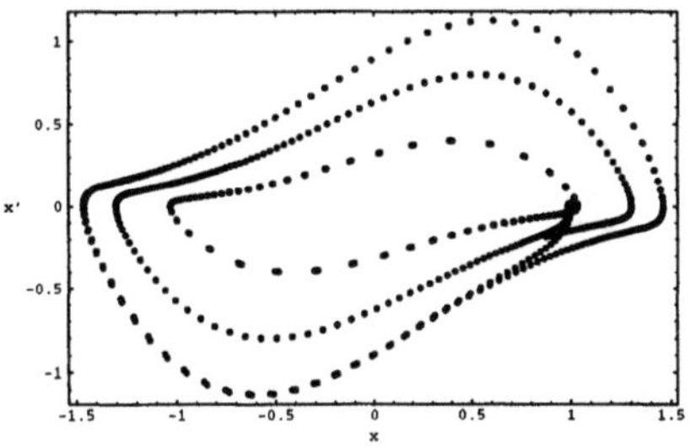

Fig. 6. Projection of the trajectories of the solution of (5) onto the $(x, \dot{x})$ plane with $x(0) = 1$, $\dot{x}(0) = 0$, taking for $\ddot{x}(0)$ the values -0.05, -0.2 and -0.4, the first being associated to the innermost orbit.

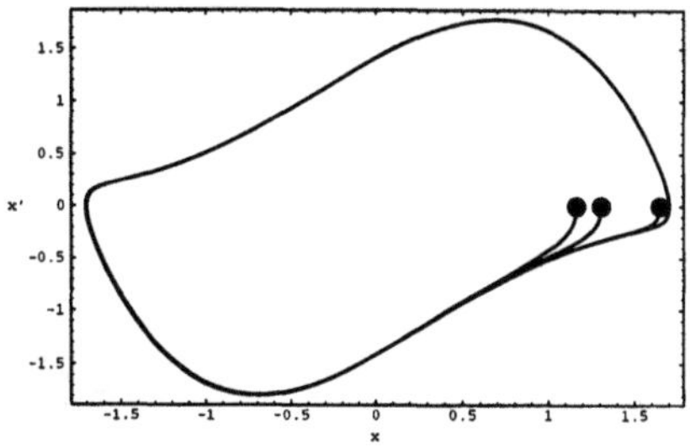

Fig. 7. Renormalisation of Fig. 6 using $a = (-I)^{1/6} = (-\ddot{x}(0))^{1/6}$, since $\dot{x}(0) = 0$. The values of $\ddot{x}(0)$ are successively -0.4, -0.2, -0.05.

$$\bar{x} = ax \qquad \dot{\bar{x}} = a^{q+1}\dot{x} \qquad \ddot{\bar{x}} = a^{2q+1}\ddot{x}, \tag{14}$$

where $a > 0$. Then

$$\begin{aligned} \bar{I} &= a^{2(q+1)}x\ddot{x} - \tfrac{1}{2}a^{2(q+1)}\dot{x}^2 + |ax|^q a^{q+2}x\dot{x} \\ &= a^{2(q+1)}I. \end{aligned} \tag{15}$$

Consequently we can take as rescaling factor $I^{1/2(q+1)}$ or, more precisely, $(-I)^{1/2(q+1)}$ (since I is negative below the "dividing line"). The parameter I is obtained from the initial conditions which are introduced in (13). Note that, starting from ξ_0 and η_0 on the "dividing line", we obtain $I = 0$ which indicates an orbit of vanishing dimension. We remark also that $I = 0$ gives $\eta - \frac{1}{2}\xi^2 + \xi = 0$, which is the equation for the "dividing line" previously obtained by solving the reduced equation (6).

4 The Reduced Equation and the Limit Cycle

The periodic orbit in the three dimensional phase space is on the surface specified by $I = I_0$, where I_0 is a constant. The particular surface is determined by the initial conditions. Now we change our point of view and consider (13) as a second order ordinary differential equation containing a parameter I. The phase space orbit is now two dimensional and, as the asymptotic orbit is the same, for a given value of I, for all initial conditions $(x(0), \dot{x}(0))$, this will be a limit cycle in the reduced phase space. In effect the periodic orbit on the surface $I = I_0$ is projected onto the $(x, \dot{x})$ plane. In the three dimensional phase space there is no limit cycle since the surface is a set of measure zero. This is no longer the case for the $(x, \dot{x})$ plane. One needs to distinguish carefully between the projections onto the $(x, \dot{x})$ plane of various periodic orbits

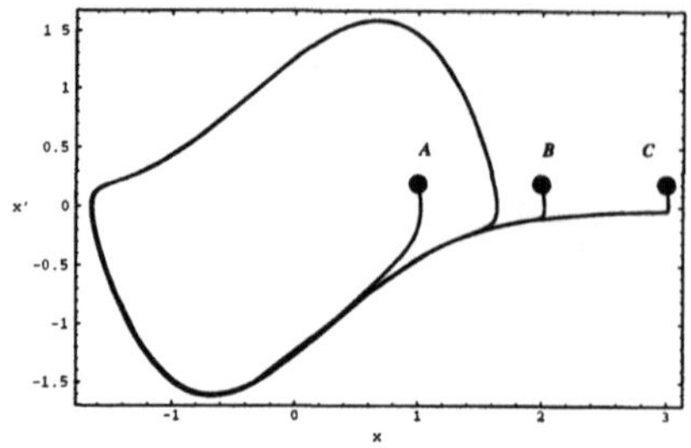

Fig. 8. Trajectories of the solution of (13) in the $(x, \dot{x})$ phase plane for $I = -0.82$ and $\dot{x}(0) = 0.2$ for $x(0) = 1$, 2 and 3 (labels A, B and C respectively).

in the three dimensional phase space (as in Fig. 6) and the limit cycle of (13) for a particular value of I (see Fig. 8).

The projections in Fig. 7 can be mapped to the projection of the universal curve by the rescaling factor a as explained above. The effect of the rescaling on (13) is to change the value of the parameter I and so change the second order ordinary differential equation. This is because I is not invariant under both symmetries. We note that in numerical computations we cannot set $x(0) = 1$ in (13) without loss of generality as we could in the case of the third order ordinary differential equation (5).

5 Conclusion

The existence of periodic behaviour and limit cycles absorbing all initial conditions for three dimensional systems is not common. We believe that this example, provided by the generalised Chazy equation (with $q = 2$ or, alternatively, the form (5) for any q), is sufficiently interesting to be discussed. Note that, although a second order differential equation was obtained, it was not possible to apply the theorem of Levinson and Smith to it. Indeed, strangely, it was on an integrodifferential form of the equation that some heuristic arguments (based on a balance of the work done by pseudoforces) highlighted the critical value $k = (q + 1)$. Finally it was shown that although these periodic orbits were dependent of the initial conditions a simple rescaling gives to these solutions universal properties as in the case of limit cycles.

Acknowledgements

PGLL thanks Professor M R Feix and MAPMO, Université d'Orléans, for their kind hospitality while this work was undertaken and the Foundation for Research Development of South Africa and the University of Natal for their continuing support.

References

Chazy, J. (1911): *Sur les équations différentielles du troisième ordre et d'ordre supérieur dont l'intégrale générale a ses points critiques fixes* (Thèse Paris, 1910). Acta Math. **34**, 317–385

Davis, Harold T. (1960): *Introduction to Nonlinear Differential and Integral Equations* (Dover, New York), p 304

Feix, M.R., Géronimi, C., Cairó, L., Leach, P.G.L., Lemmer, R.L., Bouquet, S. É. (1997): On the Singularity Analysis of Ordinary Differential Equations Invariant under Time Translation and Rescaling. J.Phys. A: Math. Gen. **30**, 7437–7461

Géronimi, C., Feix, M.R., Leach, P.G.L. (1997): Integrability of Third Order Equations Invariant under Time Translation and Rescaling. (Preprint: MAPMO-UMR 6628, Université d'Orléans, UFR Sciences, BP 6759, F-45067 Orléans Cedex 2, France)

Nemytskii, V.V., Stepanov, V.V. (1989): *Qualitative Theory of Differential Equations* (Dover, New York), p 140

A Branch and Bound Algorithm for the Flow-Shop Scheduling Problem

Khoan Vo Khac

Laboratoire de Mathématique, Applications et Physique Mathématique - UMR 6628, Université d'Orléans, UFR des Sciences, BP 6759, F-45067 Orléans Cedex 2, France

Abstract. A branch and bound algorithm for the flow-shop scheduling problem is proposed. Dominated nodes are removed from the tree.

1 Introduction

The job-shop scheduling problem has been studied by Carlier and Pinson (1989) using the results obtained in the case of one machine (Carlier 1982). However, that method is not good for the flow-shop scheduling problem. The present article proposes a special algorithm for the flow-shop scheduling problem. It is not much longer than the usual algorithms (Baker 1975) for the permutation flow-shop scheduling problem.

2 Formulation

We are given a set of M machines and a set $\mathcal{J}$ of J jobs. For every machine m and for every job j, the processing time $a(m,j)$ of the operation (m,j) is given. Every job j must pass through machine 1 first, then machine 2, $\cdots$ and machine M last, and has a preparation time $a(0,j)$ and a clean-up time $a(M+1,j)$. We seek to find for every machine m an order, i.e a permutation S_m of $\mathcal{J}$, such that the completion time is minimised; such a sequence $S = (S_1, \cdots, S_M)$ will be called an optimal sequence of permutations.

3 Preliminary Study

Let (m,j) be an operation. The fictive preparation time for this operation is $r(m,j) = a(0,j) + \cdots + a(m-1,j)$. Its fictive clean-up time is $q(m,j) = a(M+1,j) + \cdots + a(m+1,j)$. Let s be a partial sequence of permutations, i.e. a set of sequences of permutations such that the first values are known. The fictive preparation time $r(s,m,j)$ of (m,j), actualised by s, is the value of a longest path from the root to the node (m,j) in the network which represents all the constraints. If s is the empty sequence, then $r(s,m,j) = r(m,j)$. In the general case, this actualised preparation time can be calculated by the following result.

Let s be a partial sequence and s' be the partial sequence obtained by adding a new operation (m', j'). Then for every operation $(m, j) : r(s', m, j) = r(s, m, j) + r(s, m' + 1, j) - r(s, m', j)$, if $m \geq m'$, $s'(m', j)$ is unknown and $r(s, m', j) < r(s, m' + 1, j')$, $r(s', m, j) = r(s, m, j)$ in the other cases.

4 Algorithm

We use a branch and bound algorithm and we add a domination principle which removes a great number of nodes.

Each node of the tree is a partial sequence, i.e. a set of sequences. The root of the tree is the set of all sequences. It has J successors. The successor i is the set of sequences $S = (S_1, \cdots, S_M)$ such that $S_1(1) = I$. This node has $(J - 1)$ successors, ... When S_1 is known, we examine S_2, $\cdots$

A sequence S is called dominated by a sequence S' if the scheduling associated with S is greater (in the large sense) than the scheduling associated with S'. A sequence is called passive if it is dominated by another sequence. Now, let us consider a partial sequence s. Let s' be the partial sequence obtained by adding to s a new operation (m', j''). If $r(s, m'+1, j') \leq r(s, m', j'')$, then, for every sequence S' extending s', there exists a sequence S' extending s' such that S'' is dominated by S'. The node s' is called passive and may be removed from the tree.

5 Bounds

A good lower bound (of the completion time) at every node s of the tree can be obtained.

Let m be a machine. We consider the one-machine sequencing problem with preparation time $r(s, m, j)$, processing time $a(m, j)$ and clean-up time $q(m, j)$ for every job j. Then the completion-time D_m of an optimal scheduling is a lower bound at node s. On the other hand, for every subset $\mathcal{H}$ of $\mathcal{J}$, a lower bound for D_m is $\inf r(s, m, \mathcal{H}) + \sum a(m, \mathcal{H}) + \inf q(m, \mathcal{H})$. By setting $\mathcal{H} = \mathcal{J}$ we obtain for every machine m a lower bound $g(s, m)$; so, $g(s) = \sup \{g(s, 1), \cdots, g(s, M)\}$ is a lower bound.

On the other hand, for every job j and by setting $\mathcal{H} = \{j\}$ and $m = M$, we obtain another lower bound $f(s, j) = r(s, M, j) + a(M, j) + a(M + 1, j)$. Thus $f(s) = \sup f(s, \mathcal{J})$ is also a lower bound, which is an efficient lower bound, i.e. which verifies the coincidence property at terminal nodes. Then $\sup \{g(s), f(s)\}$ is also an efficient lower bound.

Another lower bound better than $g(s)$ can be found. Let m be a machine. Let $\mathcal{J} = \{j_1, \cdots, j_J\}$ be such that $r(s,m,j_1) \leq \cdots \leq r(s,m,j_J)$. By setting $\mathcal{H}_1 = \mathcal{J}$, $\mathcal{H}_2 = \{j_2, \cdots, j_J\}$, ..., $\mathcal{H}_J = \{j_J\}$, we obtain lower bounds $k(s,m,1)$, $k(s,m,2)$, ... , $k(s,m,J)$, which can be calculated by recurrence from $k(s,m,1) = g(s,m)$. By setting $k(s,m) = \sup k(s,m,J)$, $k(s) = \sup\{k(s,1),...,k(s,M)\}$, we obtain a lower bound $k(s)$ better than $g(s)$.

It can be proved that, if s is a partial sequence such that s_m is known, then $k(s,m) \leq f(s)$.

Note: Carlier (1982) gives a method to find a subset $\mathcal{H}$ which gives the greatest lower bound of D_m, but this method is not very simple.

6 A Numerical Example

We have four jobs a, b, c, d and three machines 1, 2, 3.

Table 1.

Jobs	a	b	c	d
Preparation time	0	9	4	5
Processing time 1	4	1	9	1
Processing time 2	2	1	2	5
Processing time 3	4	1	2	2
Clean-up time	6	4	0	1

Bound $\sup\{f,g\}$ leads directly to a terminal node, i.e. an optimal sequence, with completion time 22. The value 22 is obtained for lower bound at two other nodes. The first one, near the root, is removed by bound $\sup\{f,k\}$ which gives the value 23. The second node, almost terminal, leads to another optimal sequence.

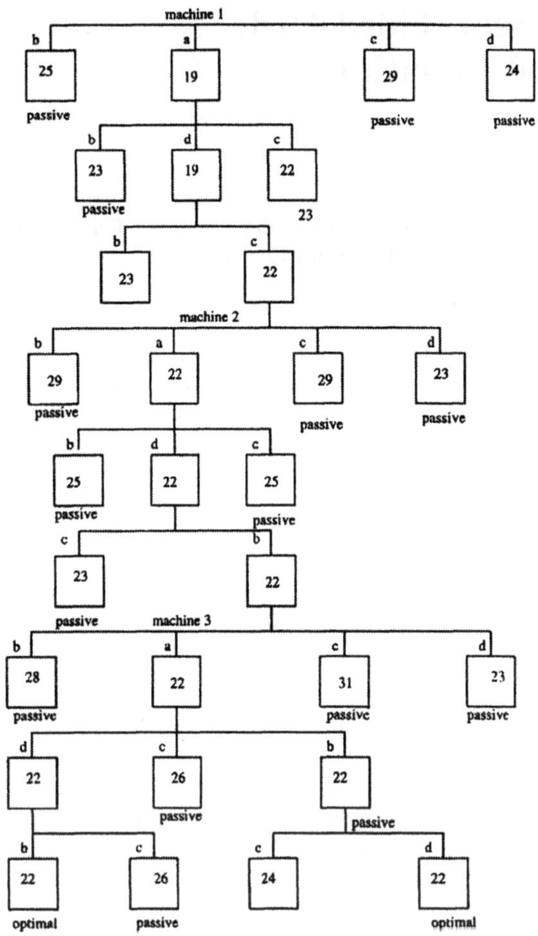

Fig. 1.

References

Baker, K.R. (1975): A comparative Survey of Flow-Shop Algorithms. Ops. Res. **23**, 62–73

Carlier, J. (1982): The one-machine sequencing problem. European J. Oper. Res. **11**, 42–47

Carlier, J., Pinson, E. (1989): An algorithm for solving the job-shop problem. Mgment Sc. **35**, 164–176

On the Linear Third-Order Differential Equation

Olivier Vallée

Centre Universitaire de Bourges, Faculté des Sciences - Université d'Orléans, Rue Gaston Berger, BP 4043, 18028 Bourges Cedex, France

Abstract. The generalised second order RICCATI equation is a nonlinear differential equation which can be transformed into a linear third order ODE. We are interested here in the search for asymptotic solutions of this equation in the case that there is a transition point (turning point). Then we consider the analytical solutions of $y''' - xy' - \mu y = 0$, which involve and generalise AIRY functions.

1 Introduction

Contrary to the linear second order ODE and the associated RICCATI equation, the literature concerning the linear third order ODE is rather poor. The explanation for this is quite simple: equations of physics and mechanics are very often of the second order and give analytic solutions. However, for example, fluid mechanics leads to the resolution of differential equations of higher order, such as the biharmonic equation, the ORR-SOMMERFELD equation .... Moreover the solution of nonlinear equations is mostly an open problem for which very few analytical results are known.

A system of three first order nonlinear differential equations may produce extremely complex behaviour, e.g. strange attractors, that cannot be produced by a system of two equations. Therefore the potential for linearisation of those equations is an important step in the understanding of their solutions. Also we are naturally inclined to the study of the linear third order ODE.

Amongst nonlinear differential equations one can observe a wide class of second order equations (see for instance Davis 1962):

$$z'' + (P_0 + P_1 z)z' + Q_0 + Q_1 z + Q_2 z^2 + Q_3 z^3 = 0 \tag{1}$$

in which one finds the first two PAINLEVÉ equations. The generalisation of the RICCATI equation to the second order is obtained by eliminating the constants k_i in the fraction:

$$Z = \frac{k_1 v_1 + k_2 v_2 + k_3 v_3}{k_1 w_1 + k_2 w_2 + k_3 w_3},$$

where the functions v_i and w_i are assumed to be linearly independent. A change of function then leads to equation (1). Actually, if the coefficients Q_2

and Q_3 satisfy the following conditions: $Q_2 = (P_0P_1 + P_1')/3$ and $Q_3 = Q_1^2/9$, then, if P_1 is not zero, one obtains (putting $z = (3y')/(P_1y)$) the linear third order ODE:

$$y''' + P(x)y'' + Q(x)y' + R(x)y = 0, \tag{2}$$

where the coefficients are given by the following expressions:

$$P = P_0 - 2\frac{P_1'}{P_1}; Q = Q_1 - P_0\frac{P_1'}{P_1} + 2(\frac{P_1'}{P_1})^2 - \frac{P_1''}{P_1}; R = \frac{1}{3}Q_0P_1.$$

The aim of this paper is to give asymptotic solutions to the third order linear equation. This will enable us to propose a particular equation the solutions of which we examine.

2 The Third-Order Differential Equation

We give in this section general results on the linear third order ODE (2) that will be useful in the following sections. Firstly it can be put into the canonical form (see Ince 1956):

$$z''' + f(x)z' + g(x)z = 0 \tag{3}$$

by making the change of function:

$$y(x) = z(x)\exp\left(-\frac{1}{3}\int^x P(x')\mathrm{d}x'\right).$$

Now, if L is the operator associated to the equation:

$$L[z] = z''' + f(x)z' + g(x)z, \tag{4}$$

the adjoint operator $\overline{L}$ may be written as:

$$\overline{L}[y] = -y''' - \frac{\mathrm{d}}{\mathrm{d}x}(f(x)y) + g(x)y. \tag{5}$$

Then we have the following theorem:

Theorem 1 *If $z_i, i = 1, 2, 3,$ are three linearly independent solutions of the equation $L[z] = 0$, then the adjoint equation $\overline{L}[y] = 0$ admits as solutions the minors of the Wronskian of the equation $L[z] = 0$: $y_i = \epsilon_{ijk}(z_jz_k' - z_j'z_k)$, where ϵ_{ijk} is the completely antisymmetric tensor.*

One defines the Wronskian of the equation $L[z] = 0$ by the determinant:

$$W(z_1, z_2, z_3) = \begin{vmatrix} z_1 & z_2 & z_3 \\ z_1' & z_2' & z_3' \\ z_1'' & z_2'' & z_3'' \end{vmatrix} = C,$$

where C is a constant.

The proof is given by a simple substitution. It is more likely that the present result can be generalised to canonical linear ODEs of any order; in particular, it is evident in the case of the self-adjoint second order equation: $z'' + h(x)z = 0$, where minors are $y_i = \delta_{ij}z_j, (i, j = 1, 2)$.
We turn now to the asymptotic solutions of (3).

3 Asymptotic Solutions

Let us consider the linear third order differential equation (cf Langer 1955a and Langer 1955b):

$$z''' + \lambda^2(f(x)z' + g(x)z) = 0, \tag{6}$$

where λ is a large parameter. We are looking for solutions of this equation when $\lambda \longrightarrow \infty$ and when it has a transition point (turning point) x_0, i.e. when $f(x_0) = 0$. We also suppose that: $f'(x_0) \neq 0$ and $g(x_0) \neq 0$. For this purpose we start by changing the dependent and the independent variables:

$$\begin{cases} u = u(x) \\ z = \dfrac{1}{u'}\phi(u). \end{cases} \tag{7}$$

Thus the equation (6) becomes:

$$\phi''' + \left[\lambda^2 \frac{f(x(u))}{u'^2} + S(u)\right]\phi' + \left[\lambda^2 \frac{g(x(u))}{u'^3} + \frac{1}{2}\frac{\mathrm{d}S(u)}{\mathrm{d}u}\right]\phi = 0, \tag{8}$$

where the derivatives of ϕ are relative to the variable u and the derivatives of u relative to the variable x. The quantity $S(u)$ is given in terms of the Schwartzian derivative $\{u, x\}$:

$$S(u) = -\frac{1}{u'^2}\{u, x\} = -\frac{1}{u'^2}\left(\frac{u'''}{u'} - \frac{3}{2}\left(\frac{u''}{u'}\right)^2\right).$$

Now we can fix the change of variable by putting: $u = f(x)/u'^2$, that is to say:

$$u = \left[\frac{3}{2}\int_{x_0}^{x} f(x')^{\frac{1}{2}}\mathrm{d}x'\right]^{\frac{2}{3}}. \tag{9}$$

The equation (8) then becomes:

$$\phi''' + \left[\lambda^2 u + S(u)\right]\phi' + \left[\lambda^2 h(u) + \frac{1}{2}\frac{\mathrm{d}S(u)}{\mathrm{d}u}\right]\phi = 0, \tag{10}$$

where $h(u) = g(x(u))/u'^3$. Thus far we have not used the limit: $\lambda \to \infty$ that will imply the consideration of the reference equation (where $\mu = C^t$):

$$\Phi''' + \lambda^2 u\Phi' + \lambda^2\mu\Phi = 0. \tag{11}$$

To this end we make the expansion: $\phi(u) = \varpi(u)\Phi(u) + O(\lambda^{-1})$. Inclusion of it in equation (11) leads for the first term (up to order $O(\lambda{-}1)$) to the identification:

$$u\varpi'(u) = (\mu - h(u))\varpi(u). \tag{12}$$

An integration gives us:

$$\varpi(u) = \exp\left[-\int_0^u \frac{h(v)-\mu}{v}\mathrm{d}v\right]. \tag{13}$$

The integral convergence of relation (13) is ensured by putting $h(0) = \mu$, allowing us to write the asymptotic solution of the equation (6) as:

$$\lim_{\lambda\to\infty} z(x,\lambda) = \frac{1}{u'}\exp\left[-\int_0^u \frac{h(v)-h(0)}{v}\mathrm{d}v\right]\Phi(u), \tag{14}$$

where $\Phi(u)$ is a solution of (11) and u is given by the relation (9). Moreover the solution obtained is uniform at the transition point x_0. Further, because we have in a neighbourhood of x_0: $f(x) \approx f'(x_0)(x - x_0)$, we derive from relation (9):

$$u \approx \left[\frac{3}{2}f'(x_0)^{\frac{1}{2}}\int_{x_0}^{x}\sqrt{(x'-x_0)}\mathrm{d}x'\right]^{\frac{2}{3}} = f'(x_0)^{\frac{1}{3}}(x-x_0).$$

Then $u'(x_0) = f'(x_0)^{\frac{1}{3}}$ and, consequently from $h(u) = g(x(u))/u'^3$, one finds $h(0) = g(x_0)/f'(x_0) = \mu$.

These results lead naturally to the search for solutions of equation (11) and this is the goal of the next section.

4 The Reference Equation

This section is devoted to the analytic solutions of (Langer 1955b)

$$y''' - xy' - \mu y = 0. \tag{15}$$

A first remark about this equation concerns a scaling change of the variable x that leads to the equation already considered (equation (11)) by putting $x = -\lambda^{\frac{2}{3}}\overline{x}$.

A general solution of this equation can be obtained by a method of LAPLACE. Putting:

$$y = \int_C \mathrm{e}^{zx} f(z)\mathrm{d}z,$$

with the condition:

$$\mathrm{e}^{-z^3/3}\Big|_C = 0, \tag{16}$$

we obtain the following integral representations:

$$y = \int_{C_i} \mathrm{e}^{-z^3/3+zx}\, z^{\mu-1}\,\mathrm{d}z, \tag{17}$$

where the integration paths C_i are given on the figure below.

Some particular cases of the reference equation (15) may also be quoted (cf Abramowitz 1965):

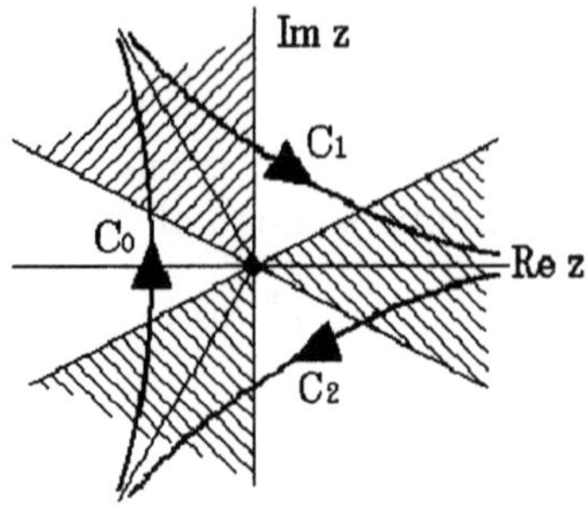

– The case where $\mu = 0$ comes from the AIRY equation: $(y')'' - xy' = 0$, leading to the general solution:

$$y_0 = a \int Ai(x)\,\mathrm{d}x + b \int Bi(x)\,\mathrm{d}x + c, \tag{18}$$

where a, b and c are constants of integration, and Ai and Bi are the AIRY functions.

– The case where $\mu = 1$ corresponds to the inhomogeneous AIRY equation: $y'' - xy = c/\pi$, leading to the solution:

$$y_1 = a\,Ai(x) + b\,Bi(x) + cHi(x). \tag{19}$$

$Hi(x)$ is the inhomogeneous AIRY function.

– The case where $\mu = 1/2$ involves the square of AIRY functions:

$$y_{1/2} = a\,Ai^2(2^{-2/3}x) + b\,Bi^2(2^{-2/3}x) + c\,Ai(2^{-2/3}x)Bi(2^{-2/3}x).$$

We turn now to the more general case where μ is any relative integer. We put $A(x) = a\,Ai(x) + b\,Bi(x)$, $A'(x)$ its derivative and $A_1(x)$ its primitive. Based on what we have just said, we can see that: $y_0(x) = A_1(x)$. Putting: $z_n = y_{-n}$ for the parameter $\mu = -n$, $n \in \mathbb{N}$, we have the recurrence relation: $z_n' = n\,z_{n-1}$. One can then check that the first solutions to equation (15) are:

$$\begin{aligned} z_1 &= xA_1(x) - A'(x), \\ z_2 &= x^2A_1(x) - A(x) - xA'(x), \\ z_3 &= (x^3 - 2)A_1(x) - xA(x) - x^2A'(x). \end{aligned}$$

Actually the solutions can be generated thanks to the recurrence relation:

$$z_n = xz_{n-1} - (n-1)(n-2)z_{n-3}, \tag{20}$$

which enables to write explicitly these solutions as:

$$z_n(x) = Pi_n(x)A_1(x) - Qi_{n-2}(x)A(x) - Ri_{n-1}(x)A'(x), \tag{21}$$

where the polynomials: Pi_n, Qi_n and Ri_n obey the recurrence relationship:

$$\begin{aligned}
Pi_n &= xPi_{n-1} - (n-1)(n-2)Pi_{n-3},\\
Qi_n &= xQi_{n-1} - n(n+1)Qi_{n-3},\\
Ri_n &= xRi_{n-1} - n(n-1)Ri_{n-3}.
\end{aligned}$$

The first few of these polynomials ($n = 0, 1, 2$) are respectively $1, x, x^2$. It is worth noting that, if the integration constants a and b are zero, the solution is given by the single polynomial $z_n = Pi_n(x)$. These solutions allow us also to express the solutions of the adjoint equation (see Theorem 1), i.e. for: $\mu = n+1$, $n \in \mathbb{N}$:

$$y''' - xy' - (n+1)y = 0. \tag{22}$$

This last equation has the following remarkable recurrence properties:

$$\begin{aligned}
y_n &= xy_{n-1} - (n-2)y_{n-3},\\
y_n &= y'_{n-1}.
\end{aligned}$$

As far as we have the solutions for $\mu = -n$, Theorem 1 allows us to find the solutions of (22). As an example, in the case where $\mu = 0$, the solution: $y_0(x) = a\int Ai(x)\,\mathrm{d}x + b\int Bi(x)\,\mathrm{d}x + c$, gives the case $\mu = 1$: $y_1(x) = \alpha\,Ai(x) + \beta\,Bi(x) + \gamma Hi(x)$ with: $\alpha = ca' - c'a, \beta = c'b - cb', \gamma = ab' - a'b$, thanks to the relation:

$$Hi(x) = Bi(x)\int_{-\infty}^{x} Ai(t)\,\mathrm{d}t - Ai(x)\int_{-\infty}^{x} Bi(t)\,\mathrm{d}t.$$

As another example the recurrence relations above permits the calculation of the nth derivative of an AIRY function. We have successively for the derivative: $Ai', Ai'' = xAi, Ai''' = Ai + xAi'$ and, for $n \geq 3$, we have:

$$\begin{aligned}
Ai^{(n)}(x) = \frac{1}{(n-1)!}&[(Pi_{n-1}Qi_{n-2} - Pi_nQi_{n-3})Ai(x)\\
&+(Pi_{n-1}Ri_{n-1} - Pi_nRi_{n-2})Ai'(x)].
\end{aligned}$$

By the same method, we can obtain the derivatives of the inhomogeneous AIRY function $Hi(x)$ with the following integral representation (for $n \geq 3$):

$$\begin{aligned}
Hi^{(n)}(x) = \frac{1}{\pi}\int_0^{\infty} \mathrm{e}^{-t^3/3+xt}\,t^n\,\mathrm{d}t = \frac{1}{(n-1)!}&[(Pi_{n-1}Qi_{n-2} - Pi_nQi_{n-3})Hi(x)\\
&+(Pi_{n-1}Ri_{n-1} - Pi_nRi_{n-2})Hi'(x)\\
&+\frac{1}{\pi}\,(Qi_{n-2}Ri_{n-2} - Qi_{n-3}Ri_{n-1})].
\end{aligned}$$

Some properties of the polynomials $Pi_n(x)$ are given in the appendix.

5 Conclusion

The results we have presented here were formerly devoted to quantum mechanics as a starting point of an inversion method. In fact, if ψ_1 et ψ_2 are two solutions of the stationary SCHRÖDINGER equation:

$$\psi'' + \frac{K(r)}{\hbar^2}\,\psi = 0,$$

then the product and the squares of these solutions obey the following third order differential equation:

$$\varphi''' + \frac{1}{\hbar^2}\,[4\,K\,\varphi' + 2\,K'\,\varphi] = 0,$$

where $1/\hbar^2$ corresponds to the parameter λ^2 we have introduced above, in the semiclassical limit $\hbar \to 0$. In the same fashion, it can easily be proven that the function $\chi = (\psi_1\psi_2)'$ is a solution of the third order ODE:

$$\chi''' - \frac{K''}{K'}\chi'' + 4\frac{K}{\hbar^2}\chi' + \frac{1}{\hbar^2}\left[6K' - 4\frac{K''K}{K'}\right]\chi = 0.$$

Moreover LANGER has used the reference equation (15) as the starting equation in the study of asymptotic solutions of the ORR-SOMMERFELD equation (fourth order), which describes the hydrodynamic instability of a POISEUILLE flow. In this case the parameter λ^2 corresponds to the REYNOLDS number of the fluid. A more sophisticated method (see for instance the book by Drazin and Reid 1995) involves not the reference equation (15) but the following one of the fourth order with two parameters:

$$y^{iv} - xy'' - \alpha y' - \beta y = 0. \tag{23}$$

In the case where $\beta = 0$, the (15) is recovered.

The general case again involves AIRY functions or their generalisations, but that is another story ....

Appendix: Properties of Airy Polynomials

We define an AIRY polynomial of degree n, $Pi_n(x)$, by the following generalised integral (convolution product of two distributions):

$$Pi_n(x) = \int_{-\infty}^{+\infty} y^n\,Ai(x-y)\,\mathrm{d}y.$$

By using the AIRY differential equation, $Ai''(z) = z\,Ai(z)$, we readily see that these polynomials obey the following recurrence relation:

$$Pi_{n+1}(x) = xPi_n(x) - Pi''_n(x).$$

From this definition we also have:

$$Pi'_n(x) = nPi_{n-1}(x).$$

Thus AIRY polynomials satisfy the third order differential equation:

$$Pi'''_n(x) - xPi'_n(x) + nPi_n(x) = 0.$$

From these last two equations we deduce the recurrence relation:

$$Pi_n(x) = xPi_{n-1}(x) - (n-1)(n-2)Pi_{n-3}(x).$$

An addition relation for the AIRY polynomials can be also established:

$$\sum_{k=0}^{n} \binom{n}{k} Pi_{n-k}(x)Pi_k(y) = 2^{n/3} Pi_n\left(\frac{x+y}{2^{1/3}}\right).$$

Amongst other properties we can quote the value at zero:

$$\begin{cases} Pi_{3n}(0) = \left(-\frac{1}{3}\right)^n \frac{(3n)!}{n!} \\ Pi_{3n+1}(0) = Pi_{3n+2}(0) = 0 \end{cases}$$

and a generating function:

$$e^{-t^3/3+xt} = \sum_{n=0}^{\infty} \frac{t^n}{n!} Pi_n(x).$$

Note finally that AIRY polynomials have properties very similar to those of HERMITE, $He_n(x)$ (Abramowitz 1965).

References

Abramowitz, M., Stegun, I. (1965): *Handbook of Mathematical Functions* (Dover, New York), 446–454

Davis, H.T. (1962): *Introduction to Nonlinear Differential and Integral Equations* (Dover, New York), 57–78

Drazin, P.G., Reid, W.H. (1995): *Hydrodynamic Stability* (Cambridge University Press, Cambridge)

Ince, E. (1956): *Ordinary Differential Equations* (Dover, New York)

Langer, R.E. (1955a): On the asymptotic forms of the solutions of ordinary differential equations of the third order in a region containing a turning point. Trans. Am. Math. Soc. **80**, 93–123

Langer, R.E. (1955b): The solutions of the differential equation $v''' + \lambda^2 w' + 3\mu\lambda^2 v = 0$. Duke Math. J. **22**, 525–541

Telecommunication Network Optimisation and Statistical Physics

Jean-Luc Lutton and Nicolas Preux

Département DAC/ORI, Centre National d'Etudes des Télécommunications, F-92794 Issy-Les-Moulineaux, France

Abstract. The subject of this paper concerns the design of telecommunication networks. Through a simple model of competition in the local access network, we illustrate the interest of some statistical physics concepts. In particular, simulated annealing as a computing method and dimensional analysis for optimal solution behaviour are physical contributions to such network studies. Moreover one can demonstrate a phase transition phenomenon while optimising two technology networks.

1 Introduction

Statistical physics seems to be far away from performance evaluation or telecommunication network optimisation. In fact several of these physical concepts can appear when considering telecommunication problems.

First of all the simulated annealing algorithm that is deduced from the Metropolis simulation algorithm (Metropolis et al. 1953) gives an efficient tool to optimise combinatorial problems (Bonomi and Lutton 1984; Bonomi and Lutton 1986; Bonomi and Lutton 1988; Chardaire and Lutton 1993; Lutton and Bonomi 1986). In the following we present an application of simulated annealing to a specific concentrator location problem. Related concepts such as the thermodynamic limit and mean field theory can be considered while studying performance evaluation in complex queuing systems.

The technique of stochastic geometry (Baccelli et al. 1996) is also a helpful tool to study network asymptotic behaviour. Such a method allows the determination of the mean value of some characteristic parameters in large networks (costs, distances, traffic) as a function of a small set of structural parameters. This determination is analytic and does not need any "microscopic" computing optimisation in particular instances.

Finally we shall see that a simple model of competition in the local loop network (part of the network between subscribers and the local exchange) can bring a phase transition phenomenon into light.

All these facts illustrate that some tools and methods devoted to statistical physics can also be profitable in nonphysical fields such as telecommunication network study and design. We can highlight the innovative and forerunner works of Marc Feix who initiated such work in his team twenty years ago (Bonomi et al. 1980; Bonomi et al. 1981; Bonomi et al. 1982; Bonomi et al. 1983; Lutton and Bonomi 1983; Lutton et al. 1986).

2 Problem Context

Due to important and numerous technological changes, the telecommunications sector is a restless, moving area. Between transmission and switching equipment with greater capacity, optic fibres allow gigabit/s, and soon terabit/s, connections. Even copper wire contributes to this evolution. The xDSL technology is now able to carry some megabit/s along a few kilometres. With the deregulation of the frequency band and with improvements in its utilisation (efficient antennae, numerical signal coding, cellular radio telephone systems, CDMA) the capacity of hertzian transmission systems has been considerably increased. The exponential development of mobile radio telephone systems is an achievement of this evolution.

The telecommunications market is also subject to regulation with the introduction of competition between operators. This competition implies the end of incoherences that exist between telecommunication service costs and tariffs. Future tariffs must be cost oriented. As a consequence current cross-subsidies cannot be maintained in a competitive environment. For example international and long distance telephone call subsidisation of the service access and local calls must disappear. Tariff changes recently introduced in France illustrate this trend: by decreasing the time unit for local calls, decreasing the tariffs of international and long distance calls and increasing the subscription price.

At first competition will take place on the international and long distance call markets because their tariffs still present inequalities. Cross subsidies profitable to the local call market have not yet completely vanished and balance will only be attained in the future. At that time competition will rise in the local access network, a strategic place for new competitors. The advantages of controlling the local access network are threefold. Firstly it is the most costly part of the network, secondly it brings the "marketing" contact with the clients and finally it will strongly determine the new service evolutions: internet access, video multicasting and so on.

Introduction of competition in the local access network is yet a difficult problem. In the United States the local network was considered as a natural monopoly by economists. As a consequence the regulation board decided to split the ATT local access network into "baby-bells". More intuitively it seems realistic that building two (or more) access networks in the same area is not a good idea. This would imply many economic wastages and civil engineering duplications. This financial inefficiency would undoubtedly increase service prices for the customers. Now the evolution of technology has rather modified this intuitive idea.

Alternative solutions to classical copper wired networks have appeared. They allow an access to subscribers without expensive civil engineering requirements. CATV networks are one of these alternatives, but their geographic capabilities are weak. They can be used only in dense urban zones where they have been opened. Two other solutions are promising as they allow a good covering of the territory.

The first one consists of a set of low orbit satellites (from 50 to 840 according to the projects: Globalstar, Iridium or Teledesic). The second one uses hertzian spreading technology. Then mobile operators can take advantage of their installations to offer fixed telephone services to professional and residential people. Likewise new operators can built new installations for covering significant and interesting zones without a too important paid-in capital. In fact the critical issue concerns the limited band resource available. ATT recently unveiled a construction project of such a hertzian access network in the technical press. In this context one can easily imagine that it is of great importance to study market sharing problems in the local access network.

In the following sections of this paper we consider a simple model of competition between two technologies (hertzian radio spreading versus copper wire) in the local access network.

3 The Model

Local access networks, either hertzian or copper wired ones, are almost always tree networks that connect a set of terminal points to a central point. In our model we assume that tree networks for both technologies are one-concentrator level trees.

The major difference between the two technologies lies in the difference between connecting costs and concentrator costs. Copper wire technology has a connecting cost greater than the hertzian connecting cost whereas copper wire concentrators are cheaper than hertzian concentrators. Copper wire connecting costs are more expensive than hertzian connecting costs due to civil engineering. To simplify our model we assume that both concentrator equipments do not have any capacity limit.

The first objective for our model is to find the minimal cost network and consequently determine if this optimal network is homogeneous (composed of only one technology) or not. Note that this minimal cost network is the network that optimises the social welfare. Indeed it provides the service (it connects all the terminals to the central exchange) at the lower cost.

Finding this optimal network can be formalised as follows

$$\mathrm{Min}_{X,Y} C = \sum_{k\in K}\sum_{i\in I}\sum_{j\in J} \alpha_k d_{ij} X_{ij}^k + \sum_{k\in K}\sum_{j\in J} F_k Y_j^k \tag{1}$$

under the constraints

$$\begin{array}{ll} \forall i \in I & \sum_{k\in K}\sum_{j\in J} X_{ij}^k = 1 \\ \forall k \in K, \forall i \in I, \forall j \in J & X_{ij}^k \le Y_j^k \\ \forall k \in K, \forall i \in I, \forall j \in J & X_{ij}^k \in \{0,1\} \\ \forall k \in K, \forall j \in J & Y_j^k \in \{0,1\} \end{array} \tag{2}$$

where I is the set of terminals and J the set of available concentrator sites. K stands for the set of possible technologies. X and Y are boolean decision variables: $X_{ij}^k = 1$ indicates that terminal i is connected to a k-technology concentrator located in j. If a k-technology concentrator is planted in a concentrator site j, then $Y_j^k = 1$. Connecting costs are proportional to the euclidean distance d_{ij} and α_k is the proportionality factor depending on the technology k. F_k is the fixed cost for a k-technology concentrator.

The first set of constraints force all terminals to be connected to one and only one concentrator. The second set of constraints imply that a terminal i can be connected to a concentrator site j using the technology k only if a concentrator of technology k is planted in site j. The last two constraints are integrity constraints upon decision variables X and Y.

One can observe the following property of the above described problem. This property will be used in the development of our simulated annealing algorithm below. Only the decision variables Y_j^k are relevant. Indeed, if the Y_j^k are known, one can immediately deduce connecting variables X_{ij}^k: a terminal i is connected to the cheaper $\alpha_k d_{ij}$ such that $Y_j^k = 1$. Note that this property does not hold when one has capacity limits on the concentrator equipments.

This problem is an extension of the Simple Plant Location Problem (SPLP) with more than one technology. Consequently some algorithms developed for the monotechnology case (Cornuéjols et al. 1977; Handler and Mirchandani 1979) can be modified without difficulty to treat the multitechnology case. Thus we have put in hand a lagrangean relaxation and a Dantzig Wolfe decomposition method for the problem. Both of these methods do not give always the exact optimal solution as they are relaxation methods. So one can couple them with a branch and bound to obtain the exact optimal solution. The lagrangean relaxation method gives very good results. In the following we compare simulated annealing results with the best lower bound of the optimal solution given by the lagrangean relaxation.

One can also use an heuristic method like the simulated annealing algorithm to solve this problem.

4 Simulated Annealing

Simulated annealing (SA) is based on a Monte Carlo simulation technique used in statistical physics: the Metropolis algorithm (Metropolis et al. 1953). This adaptation is due to Kirkpatrick et al. (1983) in the beginning of eighties. The basic idea stems from the similitude between finding the ground state in a spin glass system and partitioning an electronic circuit into two chips in order to minimise the number of connections between these two chips. Since this period simulated annealing has been intensively used to solve with success many problems. With the taboo search method and some others it belongs to the metaheuristic family. Variants of the basic method have been developed in order to improve its efficiency (Boissin and Lutton 1993; Chardaire et al. 1995).

In the domain of operations research simulated annealing is considered as a metaheuristic. Indeed SA is an excellent way to drive more elementary heuristics such as iterative improvement or greedy heuristics. Individually these methods often have poor performances. For the location problem one can use three elementary heuristics (Chardaire and Lutton 1993). One can add a concentrator in an empty site, drop one from a planted site or move one from its current position to a neighbouring empty site. For this multitechnology location problem we could try a fourth one: change the technology of an immobile planted concentrator.

One can summarise the simulated annealing for our problem as follows.

Initialisation:

- load the configuration parameters (terminal points, available concentrator sites, technology list, link cost matrices and concentrator costs)
- read annealing schedule parameters (initial temperature t_0, temperature decreasing factor ft, temperature step number *step_nb*, iteration number for every temperature steps *iter_nb* (*istep*))
- build an initial configuration U and compute its cost C

Algorithm :

for_all the *step_nb* temperature steps
  for_all the *iter_nb*(*istep*) iterations of the temperature step *istep*
    for_all the transformations (**add**, **drop** and **move**)
      Apply the transformation on the current configuration
      Compute the cost difference ΔC between the modified configuration and the current one.
      The modified configuration becomes the new current one
        - if $\Delta C < 0$
        - if $\varepsilon \leq e^{-\frac{\Delta C}{t}}$, where ε is randomly drawn in $[0, 1[$
    end_for_all {transformations}
  end_for_all {iterations}
  $t = t * ft$
end_for_all {temperature steps}

The **add** transformation proceeds on a randomly drawn empty site with an equally randomly chosen technology. The **drop** transformation is applied on a randomly chosen planted concentrator site. The **move** transformation makes a move at random from one built site towards an empty one.

The annealing schedule computed is driven by the following parameters. The decrease in temperature is geometric. The initial temperature is such that about 30% to 50% of the transformation attempts are accepted at the first temperature step. We take a temperature decreasing ratio ft equal to 0.9. The number of temperature steps is about thirty. Note that the iteration number per temperature step is also geometrically increased from step to step by a geometric ratio equal to 1.03. Under this condition the simulated annealing CPU-time is directly determined by the number of iterations in the first temperature step.

The next two figures illustrate the simulated annealing process. In Fig. 1 we report the total cost measured at each temperature step as a function of the temperature. Mean cost values for both technologies are also reported. The specific cases are constructed of 400 terminals randomly drawn in a square $[0, 10000]^2$.

Fixed costs F_1 and F_2 for concentrators are such that: $F_1 = 6000$ and $F_2 = 8000$. The connecting costs for the two technologies are such that $\alpha_1 = 1$ and $\alpha_2 = 0.84$. All terminal sites can be concentrator sites. In this figure one can observe the convergence of the simulated annealing visited configuration to the optimal solution.

Running the SA over 60 different instances (same parameters as above) we can compute a maximal bound of the SA relative error. In fact this maximal error is the difference between the value of the best solution *opt* achieved

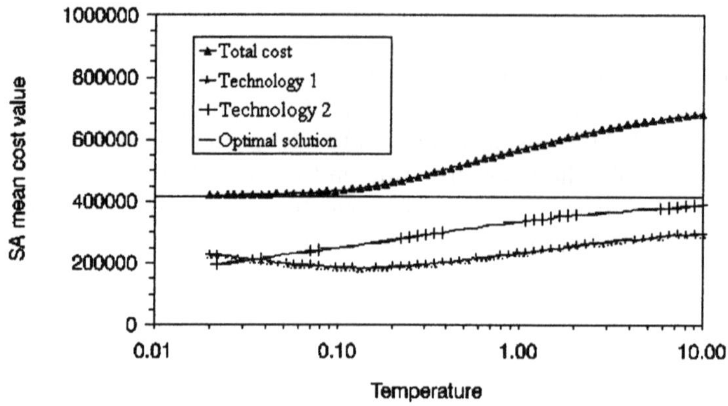

Fig. 1. Mean cost values versus the temperature measured during a simulated annealing algorithm run.

by the SA and the best lower bound L^* of the optimal solution obtained by the lagrangean relaxation algorithm. This difference is normalised by the value of the best bound to get the relative error value. In Fig. 2 we present a histogram that gives the probability distribution of this relative error for different values of CPU-times. The CPU-time of a SA algorithm directly depends on the total number of iterations. We consider four CPU-time values that have been obtained by taking 1250, 2500, 5000 and 10000 iterations in the first temperature step.

We note that the simulated annealing almost always gives good results. The error is always less than 2%. On average it is equal to 1.1% for the lower CPU-time and it is equal to 0.7% for the greatest one. The example with 10000 iterations seems to give the best solutions. The results with 20000 iterations were quite equivalent.

One must observe that the lagrangean relaxation method gives very good results on this simple location problem. For 75% of the cases it finds solutions less than 1% from the optimal one and the CPU-time is about 450 seconds (depending on the case). Faced with these results one can ask about the attraction of simulated annealing. It is true that simulated annealing is not necessarily the correct answer for this simple model. Our objective in this paper is essentially the presentation of a simple but nevertheless realistic network optimisation problem that can be solved. In fact simulated annealing is a powerful tool for practical problems. For example, if we add traffic requirements on terminals and if one has several types of concentrator equipment for each technology (they differ by their costs and capacities), then utilisation of classical methods such the lagrangean relaxation becomes more tedious

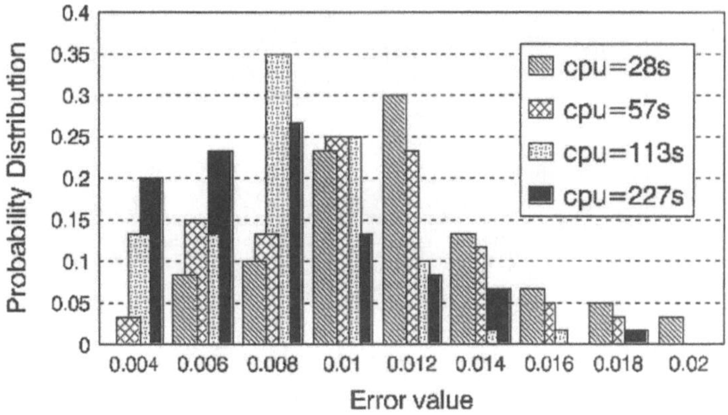

Fig. 2. Performance evaluation of the algorithm: probability distribution of the relative error: $(opt - L^*)/L^*$

(maybe impossible). The adaptation of a simulated annealing algorithm is generally easy and in consequence can be quickly developed. In an industrial context this flexibility is essential. It constitutes the main advantage of simulated annealing algorithms. The ability of the method to obtain good solutions comes afterwards.

5 Asymptotic Behaviour

In Fig. 3 we present the optimal configuration for a 2000-terminal example. The cost parameters are the same as in the previous section.

In Fig. 4 the example has 1000 terminals and the cost parameters are such that: $F_1 = 2000$, $\alpha_1 = 1$, $F_2 = 8000$, $\alpha_2 = 0.45$.

In these two figures the light grey stars correspond to technology 1 (copper wire) and the black ones to technology 2 (hertzian). For both of these problems we observe that the optimal solution is heterogeneous. The network is shared between the two technologies. In the first example the technology sharing factor $C_2/(C_1+C_2)$ is equal to 0.65. In the second example it is equal to 0.41. It would be interesting to characterise more precisely the network sharing between the two technologies especially when the size of the network becomes large.

Let us consider the monotechnology case. In the asymptotic limit of a large number n of terminals ($n \longrightarrow \infty$), a large number m of planted concentrators ($m \longrightarrow \infty$) and a large number of terminals connected to a concentrator

Fig. 3. Optimal configuration for an instance with 2000 terminals; $F_1 = 6000$, $\alpha_1 = 1$; $F_2 = 8000$, $\alpha_2 = 0.84$.

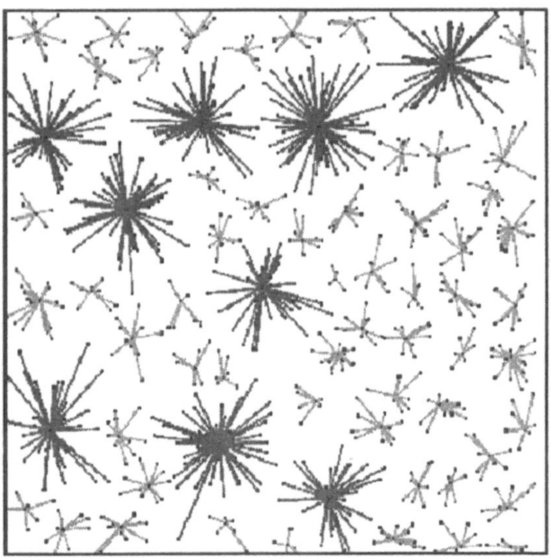

Fig. 4. Optimal configuration for an instance with 1000 terminals; $F_1 = 2000$, $\alpha_1 = 1$; $F_2 = 8000$, $\alpha_2 = 0.45$.

$(n/m \longrightarrow \infty)$, the optimal solution of the monotechnology problem can be written as

$$C_1^\infty = m_1 F_1 + n\alpha_1 \langle l \rangle \ . \tag{3}$$

The first part of the RHS gives the total concentrator cost. The second part gives the total connecting cost. $\langle l \rangle$ is the average length of the connection between the terminals and the concentrators. In the asymptotic limit described above the optimal solution looks like a honeycomb (Papadimitriou 1981) where each concentrator is located at the middle of a honeycomb cell (hexagonal cell). Let us consider an hexagonal cell with an unit surface area. If we denote by $\langle L \rangle$ the average length between the centre of this cell and any other point in it, (3) becomes

$$C_1 = m_1 F_1 + n\alpha_1 \langle L \rangle \sqrt{\frac{S}{m_1}}, \tag{4}$$

where S is the surface of the zone covered by the network. From (4) one can define the optimal solution. It is such that

$$\frac{\mathrm{d}C_1}{\mathrm{d}m_1} = F_1 - \frac{n\alpha_1 \langle L \rangle \sqrt{S}}{2} m_1^{-3/2} = 0 \ . \tag{5}$$

Equation (5) gives

$$m_1^* = \left(\frac{\alpha_1}{F_1} \frac{n \langle L \rangle \sqrt{S}}{2} \right)^{2/3} \tag{6}$$

$$C_1^* = 3 m_1^* F_1 \ . \tag{7}$$

One can remark on an interesting result. At the optimal solution the connecting cost is twice the concentrator cost. Now one can write the same results for a network with only technology 2 and one can seek the condition that equalises costs of the two monotechnology networks. This condition is

$$C_1^* = C_2^* \Rightarrow \alpha_1^2 F_1 = \alpha_2^2 F_2 \ . \tag{8}$$

Let us define the critical parameter $\alpha_1^2 F_1 / \alpha_2^2 F_2$. In Fig. 5 we report the market sharing factor $C_2/(C_1 + C_2)$ as a function of this critical parameter. The market sharing factor is computed from the optimal solutions. Different network sizes are considered. For each size we average the results over several cases. We consider 250 cases for the 10-terminal network, 50 cases for the 50-terminal network, 12 cases for the 200-terminal network and 2 cases for the 1000-terminal network.

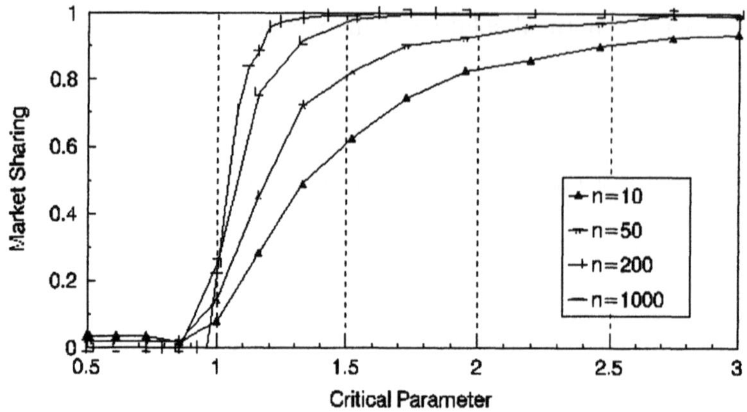

Fig. 5. Market sharing factor $C_2/(C_1 + C_2)$ as a function of the critical parameter $\alpha_1^2 F_1/\alpha_2^2 F_2$. Four network sizes are considered: $n = 10$, $n = 50$, $n = 200$, $n = 1000$.

The critical parameter value is computed by ranging the connecting cost coefficient α_2 over different values, leaving unchanged the other cost parameters ($F_1 = 6000$, $F_2 = 8000$, $\alpha_1 = 1$).

The results depicted in Fig. 5 are extremely interesting. For values of the critical parameter smaller than one the network only uses technology 1. On the other hand, for values of the critical parameter greater than one, the network is built with the technology 2. The transition between these two opposite situations happens around a critical parameter value equal to one. The greater the size of the network, the sharper is the transition.

One can presume that this transition would be sudden for a network of infinite size. Such behaviour characterises a phase transition phenomena. In the context of competitive networks this result should bring some anxiety to the competitors. Indeed their ability to gain a part of the market can vary a lot with a small change in the concentrator and connecting costs. This kind of result is very important when considering bandwidth purchase by auction. A potential purchaser must give particular attention to the maximal price he can pay in order to remain competitive against the copper wire network.

6 Conclusion

In this paper we have shown that some methods and tools which issued from the physics research domain can be very useful to study and optimise industrial systems such as telecommunication networks. Now systems become very complex. So microscopic analysis, done by classical operational research tools, becomes intractable. Faced with this complexity statistical physics can bring some answers to the behaviour of the macroscopic system.

For the multitechnology concentrator location problem studied above we have seen the attraction of such an approach. Simulated annealing and asymptotic limit calculation give some answers to this simple model of competition in the local access network. In particular they allow characterisation of the optimal solution (social welfare criteria). This solution implies a cooperative context between the two technology operators. In a noncooperative framework each operator tries to maximise his benefit regardless of the social welfare. The two technologies (which can now be viewed as firms) propose network strategies independently. As a consequence the attained solution is not necessarily the cooperative one.

The noncooperative game approach for studying such competition problem is extremely complex and almost surely not solvable. The 2^n strategies for each player become rapidly intractable. However, one can perhaps gain information using statistical physics.

References

Aarts, E., Korst, J. (1989): *Simulated annealing and Boltzmann machines* (Wiley, New York)

Baccelli, F., Klein, M., Zuyev, S. (1996): Géométrie aléatoire et architecture de réseaux de communications. Annales des télécommunications **51**, 158-179

Boissin, N., Lutton, J.-L. (1993): A parallel simulated annealing algorithm. Parallel Computing **19**, 859–872

Bonomi, E., Feix, M.R., Hébuterne, G. (1980): Théorie de l'information: structure et modélisation d'un réseau de connexion. Annales des Télécommunications **35**, 39–59

Bonomi, E., Lutton, J.-L., Feix, M.R. (1981): Telephone networks, statistical mechanics and non-random connecting procedures in *Numerical Methods in the Studies of Critical Phenomena*, Della Dora, J., Demongeot, J., Lacolle, B. eds. (Springer-Verlag, New York), 225–232

Bonomi, E., Lutton, J.-L., Feix, M.R. (1982): Concept d'extensivité pour un réseau de Clos et probabilité de blocage pour un système à trois étages. Annales des Télécommunications **37**, 239–256

Bonomi, E., Lutton, J.-L., Feix, M.R. (1983): Three-stage rearrangeable connecting networks in the thermodynamic limit: number of needed reswitchings. IEEE Transactions on Communications **COM-31**, 110–113

Bonomi, E., Lutton, J.-L. (1984): The N-city travelling salesman problem: statistical mechanics and the Metropolis algorithm. SIAM Review **26**, 551–568

Bonomi, E., Lutton, J.-L. (1986): The asymptotic behaviour of quadratic sum assignment problems: a statistical mechanics approach. EJOR **26**, 295–300

Bonomi, E., Lutton, J.-L. (1988): Le recuit simulé. Pour la Science **129**, 68–77

Chardaire, P., Lutton, J.-L., Sutter, A. (1995): Thermostatistical persistency: a powerful improving concept for simulated annealing algorithm. EJOR **86**, 565–579

Chardaire, P., Lutton, J.-L. (1993): Using Simulated annealing to solve concentrator location problems in telecommunication networks in *Applied Simulated Annealing, Lecture Notes in Economics and Mathematical Systems*, Vidal, R.V.V. ed. (Springer-Verlag, Heidelberg, Berlin), 175–199

Cornuéjols, G., Fisher, M. L., Nemhauser, G. L. (1977): Location of banks' accounts to optimize float: an analytic study of exact and approximated algorithm. Management Science **23**, 789–810

Handler, G.Y., Mirchandani, P.B. (1979): *Location on network: theory and algorithm* (MIT Press, Boston)

Kirkpatrick, S., Gelatt, C. D., Vecchi, M. M. (1983): Optimization by simulated annealing. Science **220**, 671–680

Lutton, J.-L., Bonomi, E. (1983): Blocking probability for a multistage Clos connecting network. Journal of Statistical Physics **31**, 155–167

Lutton, J.-L., Bonomi, E., Feix, M.R. (1986): Résolution d'un problème d'affectation à l'aide d'une procédure de réarrangement. Annales des Télécommunications **41**, 121–132

Lutton, J.-L., Bonomi, E. (1986): Simulated annealing algorithm for the minimum weighted perfect Euclidian matching problem. RAIRO Operation Research **20**, 177–197

Metropolis, N., Rosenbluth, A., Rosenbluth, M., Teller, A., Teller, E. (1953): Equation of state calculations by fast computing machines. Journal of Chemical Physics **21**, 1087–1092

Papadimitriou, C. H. (1981): Worst-case and probabilistic analysis of a geometric location problem. SIAM Journal of Computing **10**, 542–557

A Study of the Analytical and Local Semiclassical Wigner Distribution

Manuel Soares, Olivier Vallée, and Charles de Izarra

GREMI, Université d'Orléans, Faculté des Sciences, Rue Gaston Berger, BP 4043, 18028 Bourges Cedex, France

Abstract. An analytical and local expression of the semiclassical Wigner distribution is performed using the Airy uniform approximation in the neighbourhood of a turning point. At the classical limit the one-dimensional Wigner distribution leads to the microcanonical distribution. For the two-dimensional systems the variables separable problems (nonchaotic dynamics) and the variables nonseparable problems (possibility of chaos) are not locally distinguishable by using the uniform approximation. Meanwhile the examination of a section of the phase space allows us to see that the apparent complexity of the Wigner distribution is not a proof of chaos.

1 Introduction

For many years research has been conducted to establish a link between classical chaotic systems and their corresponding quantum equivalents, as well as the criteria which allow us to decide if a quantum system shows chaotic behaviour or not: irregular spectra (Percival 1973; Berry 1977), sensitivity to small perturbations (Pomphrey 1974; Grémaud et al. 1993), distribution of energy levels (Tabor 1977; McDonald and Kaufmann 1979), structures of stationary states (Heller 1984; McDonald and Kaufmann 1979).... Other authors, however, doubt the very existence of quantum chaos (Ford 1989; Ford and Ilg 1992; Ford and Mantica 1992) and they point out that, even if quantum systems manifest unusual behaviour when their classical equivalents are chaotic, there are good reasons to suppose that quantum chaos does not exist.

Indeed it has already been established that closed and finite quantum systems exhibit no chaotic behaviour (Ford and Ilg 1992). Nevertheless the main objection to the existence of quantum chaos remains the linear nature of the Schrödinger equation. Consequently it is not subject to exponential instability of the initial conditions (Ford and Ilg 1992; Dando and Monteiro 1994), which is the *sine qua non* condition for chaos. Moreover all the criteria for the recognition of classical chaos are founded upon the notion of trajectories, but any possibility for defining a trajectory is excluded by the Heisenberg uncertainty principle (see for instance Ford and Mantica 1992).

Otherwise analytical methods for the study of quantum chaos are lacking (Berry 1977) and, more generally, the correspondence principle between the chaotic classical systems and their quantum equivalent systems is unknown as well as the quantization of them, i.e. the resolution of the eigenvalues and eigenfunctions. Nonetheless Berry has shown that the Wigner distribution can be an attractive way to link the quantum and classical descriptions for an integrable system (see also Ozorio de Almeida and Hannay 1982; Meredith 1992).

Therefore our purpose is the study of a quantum probability distribution, the Wigner distribution, both in the classical phase space and at the semiclassical limit $\hbar \to 0$.

2 The Wigner Distribution

The Wigner distribution $W(r,p)$ allows us to represent a quantum state $|\psi\rangle$. This distribution is explained as the quantum equivalent of a density of the classical phase space (Wigner 1932).

In quantum mechanics the uncertainty principle does not allow us to know the conjugate variables (spatial and momentum coordinates) at the same time. Meanwhile Wigner has shown, by analogy with classical mechanics, that we have in terms of quantum operators:

$$\langle \hat{A} \rangle = \int_{-\infty}^{+\infty} \mathrm{d}^N r \int_{-\infty}^{+\infty} \mathrm{d}^N p \, A(r,p) \, W(r,p) \, ,$$

where $W(r,p)$ is the Wigner distribution defined by:

$$W(r,p) = \frac{1}{(\pi\hbar)^N} \int_{-\infty}^{+\infty} \psi(r+\eta) \, \psi^*(r-\eta) \exp\left[\mathrm{i}2p\frac{\eta}{\hbar}\right] \mathrm{d}^N \eta, \tag{1}$$

for a state described by the time-independent wave function $\psi(r)$.

This distribution (see for instance Berry 1977; O'Connel 1983) has many properties. In particular $W(r,p)$ is real for all r and p. Moreover $W(r,p)$ can be negative. This is the reason why this distribution is not strictly speaking a phase space density (Heller 1976; Berry 1977). Notwithstanding this the Wigner distribution $W(r,p)$ has the usual properties of a probability distribution and it allows us to obtain the usual densities by projection into the position space:

$$\int_{-\infty}^{+\infty} \mathrm{d}^N p \, W(r,p) = |\langle r \mid \psi \rangle|^2$$

and into the momentum space:

$$\int_{-\infty}^{+\infty} \mathrm{d}^N r \, W(r,p) = |\langle p \mid \psi \rangle|^2 .$$

Also note that, if $|\psi\rangle$ is normalised to unity, then:

$$\int_{-\infty}^{+\infty} \mathrm{d}^N r \int_{-\infty}^{+\infty} \mathrm{d}^N p \, W(r,p) = \langle \psi \mid \psi \rangle = 1.$$

At the semiclassical limit $\hbar \to 0$, if the system is classically integrable, $W(r,p)$ leads to the classical density and the equation of motion, leads to the linear Liouville equation which governs the time evolution of the classical phase space densities (Heller 1976). Indeed the time evolution of a quantum state by following the classical trajectories can be determined by the Heller semiclassical approach. Thus it is in this way that the connection between the classical and quantum domains is established: the wave function is represented by a quantum distribution moving on a classical path.

Finally it can be seen that, if the system in question is not integrable, then the dynamics of this system, the distribution of the energy levels and the nature of the stationary states are for the present unknown. Meanwhile a conjecture has been suggested by Voros (1976) and Berry (1977): the microcanonical hypothesis. Also for an ergodic system (i.e. irregular), at the semiclassical limit, the Wigner distribution leads to the microcanonical distribution:

$$W_{\mathrm{m}}(r,p) = \lim_{\hbar \to 0} \frac{\delta(E - H(r,p))}{\int\int \mathrm{d}p\mathrm{d}r \; \delta(E - H(r,p))}, \tag{2}$$

where δ is the Dirac delta function and H the hamiltonian of the system in the state of energy E.

3 The Airy Uniform Approximation of the Schrödinger Equation

The time-independent one-dimensional Schrödinger equation (or the radial equation in the three-dimensional case) is:

$$\psi'' + \frac{p^2(r)}{\hbar^2}\psi = 0, \tag{3}$$

where $p(r)$ is the momentum of the particle with mass $m = 1/2$ and energy E in the potential $V(r)$: $p(r) = \sqrt{E - V(r)}$ and $\hbar$ is the reduced Planck's constant. Using appropriate changes of functions and variables, at the semiclassical limit $\hbar \to 0$, we obtain the ordinary JWKB solutions (see Berry and Mount 1972). However, these solutions are not relevant in the neighbourhood of a turning point r_{t} (i.e. at a zero of the momentum: $p(r_{\mathrm{t}}) = 0$). Therefore, if the momentum has a turning point and if we wish to approach the solutions of the Schrödinger equation (3) even in the neighbourhood of r_{t}, we can use a semiclassical approximation of higher order: the uniform approximation. Under these conditions Airy functions are needed for the solutions of the differential equation (3).

The approximate solution of the Schrödinger equation, uniformly valid far or near a turning point, is written as (Berry and Mount 1972)

$$\psi(r) = \frac{\sqrt{\pi}}{\hbar^{1/6}} \left[\frac{\sigma(r)}{p^2(r)}\right]^{1/4} Ai\left[\frac{\sigma(r)}{\hbar^{2/3}}\right], \quad \sigma(r) = \left(\frac{3}{2}\int_{r_{\mathrm{t}}}^{r} p(u)\mathrm{d}u\right)^{2/3}. \tag{4}$$

Note that, if a number of turning points exist, the exact solution of the Schrödinger equation can be approached by connecting the wave functions of (4)-type with a good accuracy (Miller 1968). Note equally that we can obtain the usual JWKB wave functions from the asymptotic limit of the expression (4) with the help of the asymptotic formula of *Ai* (Abramowitz and Stegun 1965).

Moreover by expanding $\sigma'(r)$ in the neighbourhood of a turning point r_t we obtain a new expression for the wave function:

$$\psi(r) = \frac{\sqrt{\pi}}{\hbar^{1/6}\sqrt{\alpha}} Ai\left[\frac{\alpha}{\hbar^{2/3}}(r - r_t)\right], \tag{5}$$

where $r \approx r_t$ and α is a constant which depends upon the derivative of the potential at the turning point: $\alpha^3 = (\mathrm{d}V(r)/\mathrm{d}r)_{r=r_t}$. Below we use this expression, (5), of the wave function.

The turning point method is not suitable if the derivative of the potential is zero for r_t, i.e. $\alpha = 0$ (local minimum or maximum, saddle point of the potential...): then we must use the method of stationary phase which leads us to formally equivalent results.

4 The Semiclassical Wigner Distribution

The one-dimensional, time-independent Wigner, distribution $W_1(r,p)$, simply dependent of the position r and the momentum $p(r)$, is defined by

$$W_1(r,p) = \frac{1}{\pi\hbar}\int_{-\infty}^{+\infty} \psi(r+\eta)\,\psi^*(r-\eta)\exp\left[\mathrm{i}2p\frac{\eta}{\hbar}\right]\mathrm{d}\eta. \tag{6}$$

By substitution of the wave function ψ by its semiclassical and developed expression (5) we obtain

$$W_1(r,p) = \frac{1}{\hbar^{4/3}}\int_{-\infty}^{+\infty} \frac{1}{\alpha} Ai\left[\alpha(r - r_t - \eta)\right] Ai\left[\alpha(r - r_t + \eta)\right]\exp\left[\mathrm{i}2p\frac{\eta}{\hbar}\right]\mathrm{d}\eta, \tag{7}$$

or, after some analytical calculus (Balazs and Zipfel 1973):

$$W_1(r,p) = \frac{1}{2^{1/3}\hbar^{2/3}\alpha^2} Ai\left\{\frac{2^{2/3}}{\hbar^{2/3}\alpha^2}\left[\alpha^3(r - r_t) + p^2(r)\right]\right\}. \tag{8}$$

This expression of the local and semiclassical Wigner distribution must reduce, at the classical limit $\hbar \to 0$, to the microcanonical distribution (cf. formula (2)). Let us consider the argument of the Airy function in (8):

$$A = (r - r_t)V'(r_t) + p^2,$$

where $\alpha^3 = V'(r_t)$. By neglecting the terms of higher order in the development of the potential V (it is the JWKB principle) we find this argument to be

$$A = V(r) - V(r_t) + p^2(r) = (H - E)_{r=r_t},$$

where H is the classical Hamiltonian for the particle of energy E in the potential $V(r)$. Then the Wigner distribution becomes

$$W_1(r,p) = \frac{1}{2^{1/3}\hbar^{2/3}\alpha^2} Ai\left\{\frac{2^{2/3}}{\hbar^{2/3}\alpha^2}\left[(H-E)_{r=r_t}\right]\right\}.$$

Using $\beta = \frac{2^{2/3}}{\hbar^{2/3}\alpha^2}$ we obtain

$$W_1(r,p) = \frac{\beta}{2} Ai\left\{\beta\left[(H-E)_{r=r_t}\right]\right\}. \tag{9}$$

Formally we carry out the transition to the classical limit by letting Planck's constant $\hbar$ approach zero, i.e. by letting the constant β approach infinity. With the help of the relation:

$$\lim_{\beta\to\infty} \beta Ai\,(\beta x) = \lim_{\beta\to\infty} \frac{1}{2\pi}\int_{-\infty}^{+\infty} e^{i\left(\frac{t^3}{3\beta^3}+xt\right)}\,dt = \delta(x), \tag{10}$$

the Wigner distribution (8) becomes, at the classical limit:

$$W_1\,(r,p) = \frac{1}{2}\delta(H-E)_{r=r_t}. \tag{11}$$

Note that the factor 1/2 is due to the development at the neighbourhood of one turning point because only a half space is considered.

At the classical limit the semiclassical Wigner distribution, developed in the neighbourhood of a turning point, (8) is so reduced to the microcanonical distribution.

5 The Two-Dimensional Wigner Distribution

A rather long but relatively easy calculation allows us to obtain the analytical expression of the two-dimensional semiclassical Wigner distribution developed in the neighbourhood of a turning point $r_t = (x_t, y_t)$:

$$\begin{aligned} W_2\,(x,y,p_x,p_y) = {} & \frac{1}{2^{2/3}\hbar^{4/3}a^2d^2}\, Ai\left\{\frac{2^{2/3}}{\hbar^{2/3}}\left[(ax+by)+\left(\frac{p_xd-p_yc}{\Delta}\right)^2\right]\right\} \\ & \times\ Ai\left\{\frac{2^{2/3}}{\hbar^{2/3}}\left[(cx+dy)+\left(\frac{p_xb-p_ya}{\Delta}\right)^2\right]\right\}, \end{aligned} \tag{12}$$

where $\Delta = ad - bc \neq 0$ and the constants a, b, c and d are dependent on the potential surface.

Like the one-dimensional case this method cannot be used at a minimum, maximum or saddle point of the potential ($\Delta = 0$).

We perform the linear change of variables:

$$\begin{cases} X = ax + by \\ Y = cx + dy. \end{cases} \tag{13}$$

The determinant is not zero ($\Delta = ad - bc \neq 0$) and the inverse transformation is defined by

$$\begin{cases} x = \frac{1}{\Delta}\left(dX - bY\right) \\ \\ y = \frac{1}{\Delta}\left(-cX + aY\right). \end{cases} \tag{14}$$

However, the conjugate variables x and p_x satisfy the relation: $p_x = \partial L/\partial \dot{x}$, where $L(\dot{x}, x, \dot{y}, y)$ is the Lagrangian of the system. Thus after the transformation we obtain the new conjugate momentum:

$$P_X = \frac{\partial \mathcal{L}}{\partial \dot{X}} = \frac{\partial L}{\partial \dot{x}}\frac{\partial \dot{x}}{\partial \dot{X}} + \frac{\partial L}{\partial \dot{y}}\frac{\partial \dot{y}}{\partial \dot{Y}},$$

where $\mathcal{L}$ is the new Lagrangian. This relation leads to

$$P_X = p_x \frac{d}{\Delta} + p_y \frac{-c}{\Delta}. \tag{15}$$

In the same way we obtain

$$P_Y = p_x \frac{-b}{\Delta} + p_y \frac{a}{\Delta}. \tag{16}$$

The expression for the Wigner distribution (12) becomes:

$$W_2\left(X, P_X; Y, P_Y\right) = \frac{2^{-2/3}}{\hbar^{4/3} a^2 d^2}\, Ai\left\{\frac{2^{2/3}}{\hbar^{2/3}}\left(X + P_X^2\right)\right\} Ai\left\{\frac{2^{2/3}}{\hbar^{2/3}}\left(Y + P_Y^2\right)\right\}. \tag{17}$$

Thus the linear transformation, given in (13), which transforms the Hamiltonian $H\left(x, y, p_x, p_y\right)$ into the Hamiltonian $\mathcal{H}\left(X, P_X; Y, P_Y\right)$ is a canonical transformation. As in the case above we can perform the classical limit:

$$W_2^{\text{nonsep.}}\left(X, P_X; Y, P_Y\right) = \frac{1}{4}\delta(\mathcal{H} - E)_{X_t, Y_t}\,.$$

The linear inverse transform (14) allows us to return to the microcanonical distribution: $\delta(H\left(x, y, p_x, p_y\right) - E)$.

Even if the system is not a separable system, the development of the wave function in the neighbourhood of a turning point (or the linearisation of the potential in this neighbourhood) allows us to come down to the variables separable case with the help of a linear canonical transformation. Therefore, at the classical limit, the *local and semiclassical* Wigner distribution is always reduced to the microcanonical distribution.

6 Configuration of the Wigner Distribution in the Phase Space

Because we have an analytical expression for the semiclassical Wigner distribution, it seems only natural to display it in its domain, the phase space. The qualitative behaviour of this function is already well-known for an integrable system (Berry 1977), but the case of nonintegrable systems is unknown. However, Berry suggested the following hypothesis: that the Wigner function would, in this case, look like a series of minima and maxima randomly spread in the phase space. We therefore also illustrate this spread in a qualitative manner over a section (x, p_x) of the phase space.

The expression (17) for the Wigner distribution can be rewritten, except the constants, as follows:

$$W_2 = Ai\left(X + P_X^2\right). \tag{18}$$

We illustrate in the phase space (r, p), which is reduced here into (X, P_X), in Fig. 1. Note that this Wigner function is an expansion in the neighbourhood of a *single* turning point. Therefore this figure represents only the half of the distribution.

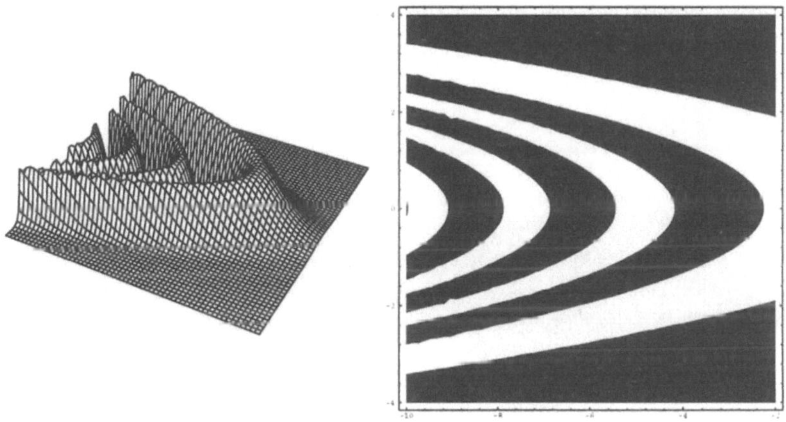

Fig. 1. *The two-dimensional Wigner distribution for the variables separable case and in the plane (X, P_X). On the right is shown a section of the distribution in which we can see the pattern of the placement of the minima and the maxima.*

The expression of the two-dimensional Wigner distribution is useful for our purpose because the components of unimodular transformation matrix will replace the constants a, b, c, d (cf. formula (12)).

We can choose a *priori* all matrices for which the determinant is 1 in order to insure the preservation of the space phase volume. It is well known that the volume taken up by a dynamical system is invariant during the evolution of this system. With the help of a transformation matrix the system will be disturbed by a local deformation of the space. Then we will observe the reaction of the distribution upon these constraints.

Among all the unimodular transformation matrices we consider the shearing matrix:

$$\begin{bmatrix} 1 & 0 \\ \gamma & 1 \end{bmatrix},$$

with which the deformations are more perceptible than with the matrices of rotation or hyperbolic rotation. For some different values of the parameter γ the Wigner distribution is shown in a section (x, p_x) of the phase space. The formula (12) allows us to obtain (except for the constants):

$$W_2 = Ai\left[x + p_x^2\right] Ai\left[\gamma x\right]. \tag{19}$$

For $\gamma = 0$ W_2 is reduced into the one-dimensional distribution in the plane (x, p_x). For other values of the parameter γ the distribution is subjected to very large deformations. In particular we are interested in the parameter value $\gamma = 1$ and we can observe the pattern of the minima and the maxima of W_2 on the Fig. 2. This pattern seems to be the same as that of the Berry conjecture (Berry 1977), where Berry assumed that the minima and the maxima of the semiclassical distribution are placed randomly if the system is ergodic. By qualitative means we can obtain the same distribution. However, now the variables are separable and the system is integrable.

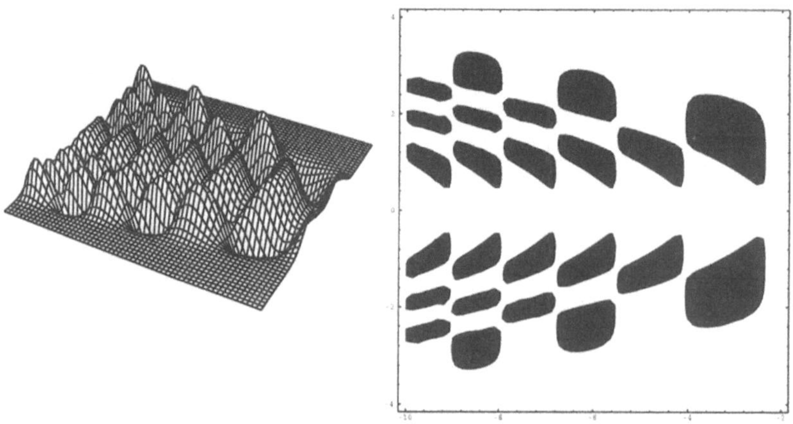

Fig. 2. *The two-dimensional Wigner distribution, locally deformed by a shearing constraint where $\gamma = 1$, in the plane (x, p_x). At the right of the figure the section of the distribution shows the peaks built by the minima and the maxima.*

Therefore it seems that it is not necessary to consider an ergodic system in order that the corresponding Wigner distribution display an irregular behaviour. Note that Fig. 2 is not completely irregular, but our model is the particular case of an expansion in the neighbourhood of a turning point. Nevertheless even in the context of this very simple model, where the variables are always separable (and therefore where the trajectories are always regular), the distribution need not have a simple structure, which seems to invalidate the reverse of the Berry conjecture, at least locally.

7 Conclusion

Contrary to classical chaos the concept of quantum chaos is hard to characterise. This is why the study of the transition classical/quantum (*i.e.* the semiclassical domain) is interesting. Recently Tomsovic and Heller (1993) built "chaotic" wave functions (the wave functions of a system where the dynamics of the corresponding classical system are chaotic) from purely semiclassical methods. Their results were applied to the stadium billard problem (classically chaotic) and stationary states were obtained which seem to be the same for the Wigner distribution as the Berry conjecture, i.e. the same we obtained by semiclassical methods, but where all possibility of chaotical behaviour is excluded.

Indeed, from a semiclasical method (the Airy uniform approximation), we obtained an analytical expression of the Wigner distribution, first for the one-dimensional case and then for the two-dimensional case. These analytical expressions allowed us to study locally the distribution in a section of the phase space. In this way we showed that, even if the variables of the system under consideration are separable (which is characteristic of a nonchaotic system), the Wigner distribution can have a complicated appearance in the phase space. It seems too that the chaos in a quantum system cannot be characterised either by its stationary states or by the apparent complexity of the Wigner distribution in the phase space.

References

Abramowitz, M., Stegun, I. (1965): *Handbook of Mathematical Functions* (Dover Publications, New York)

Balazs, N.L., Zipfel, G.G. (1973): Quantum oscillations in the semiclassical fermion μ-space density. Ann. Phys. **77**, 139–156

Berry, M.V. (1977): Semiclassical mechanics in phase space: a study of Wigner's function. Phil. Trans. R. Soc. London A **287**, 237–271

Berry, M.V., Mount, K.E. (1972): Semiclassical approximations in wave mechanics. Rep. Prog. Phys. **35**, 315–397

Dando, P.A., Monteiro, T.S. (1994): Quantum surfaces of section for the diamagnetic hydrogen atom: Husimi functions versus Wigner functions. J. Phys. B **27**, 2681–2692

Ford, J. (1989): Quantum chaos, is there any? in: *Directions in Chaos*, Hao Bai-Lin ed. (World Scientific, Singapore) Vol 2

Ford, J., Ilg, M. (1992): Eigenfunctions, eigenvalues, and time evolution of finite, bounded, undriven quantum systems are not chaotic. Phys. Rev. A **45**, 6165–6173

Ford, J., Mantica, G. (1992): Does quantum mechanics obey the correspondence principle ? Is it complete ? Am. J. Phys. **60**, 1086–1098

Grémaud, B., Delande, D., Gay, J.C. (1993): Origin of narrow resonances in the diamagnetic Rydberg spectrum. Phys. Rev. Lett. **70**, 1615–1618

Heller, E.J. (1976): Wigner phase space method: analysis for semiclassical applications. J. Chem. phys. **65**, 1289–1298

Heller, E.J. (1984): Bound-state eigenfunctions of classically chaotic hamiltonian systems: scars of periodic orbits. Phys. Rev. Lett. **53**, 1515–1518

McDonald, S.W., Kaufman, A.N. (1979): Spectrum and eigenfunctions for a hamiltonian with stochastic trajectories. Phys. Rev. Lett. **42**, 1189–1191

Meredith, D.C. (1992): Semiclassical wavefunctions of nonintegrable systems and localization on periodic orbits. J. Stat. Phys. **68**, 97–130

Ozorio de Almeida, A.M., Hannay, J.H. (1982): Geometry of two dimensional tori in phase space: projections, sections and the Wigner function. Ann. Phys. **138**, 115–154

Percival, I.C. (1973): Regular and irregular spectra. J. Phys. B **6**, L229–L232

Pomphrey, N. (1974): Numerical identification of regular and irregular spectra. J. Phys. B **7**, 1909–1915

Tabor, M. (1989): Chaos and integrability in semiclassical mechanics. in *Chaos and integrability in non-linear dynamics*, Tabor, M. ed. (Wiley, New York)

Tomsovic, S., Heller, E.J. (1993): Semiclassical construction of chaotic eigenstates. Phys. Rev. Lett. **70**, 1405–1408

Voros, A. (1976): Semi-classical approximations. Ann. Inst. Henri Poincaré A **XXIV**, 31–90

Wigner, E.P. (1932): On the quantum correction for thermodynamic equilibrium. Phys. Rev. **40**, 749–759

Quantum Transition and Irreversibility

Jean Salmon

Fondation Louis de Broglie; 23, Quai Conti, 75006 Paris, France
and
Conservatoire National des Arts et Métiers; Département de Physique
292, rue Saint Martin, 75003 Paris, France

Abstract. We consider the Schrödinger equation as the reversible approximation of a more general irreversible equation whose expression is derived. It becomes, therefore, possible to describe continously any quantum transition.

Keywords. Irreversibility, Quantum Mechanics, Schrödinger equation.

1 Introduction

The Schrödinger equation does not allow the time evolution of a quantum transition. Consider, for example, the electron in a hydrogen atom which executes a permitted transition from a state with energy E_2 to a state with energy E_1. The Schrödinger equation does not provide the continuous variation from the energy level E_2 to the energy level E_1. It gives only the probability transition between the two states and the associated life-time τ

This equation is reversible and therefore restricted in its results. Consequently we are going to propose a more general equation containing an irreversible contribution. The resolution of this new equation will lead to a continuous variation of the energy E from the value E_2 to the value E_1 as a fonction of time t and of the characteristic time τ

$$E(t) = E_1 + (E_2 - E_1)e^{-\frac{t}{\tau}} \tag{1}$$

2 Notations

The time is t and $\vec{r}$ the position vector. The permittivity of voïd space is ε_0, the light velocity c and the Planck constant h. We define

$$\hbar = \frac{h}{2\pi}. \tag{2}$$

The electron of mass m_e and charge $-e$ evolves into the Coulomb potential V created by the proton of mass m_p and charge $+e$. We have

$$V = -\left[\frac{e^2}{4\pi\varepsilon_0 r}\right] \tag{3}$$

and we define

$$m_0 = m_e\left(1+\frac{m_e}{m_p}\right)^{-1}. \tag{4}$$

3 Study of the Reversible Approximation

Let $\hat{H}_0$ the usual Hamiltonien operator and $\psi(t, \vec{r})$ the usual wave-function. One has

$$H_0 = -\frac{\hbar^2}{2m_0}\Delta - \frac{e^2}{4\pi\varepsilon_0 r} \tag{5}$$

and the Schrödinger 's equation becomes

$$i\hbar\frac{\partial\psi}{\partial t} = \hat{H}_0\psi. \tag{6}$$

This equation is reversible. To the energies E_1 and E_2 are associated, respectively, the functions ψ_1 and ψ_2 such that

$$\hat{H}_0\psi_1 = E_1\psi_1, \tag{7}$$

$$\hat{H}_0\psi_2 = E_2\psi_2. \tag{8}$$

The wave-functions are ψ_1 and ψ_2 with

$$E = E_2; \quad \psi = \psi_2 = e^{-\frac{iE_2 t}{\hbar}}\psi_2\,, \tag{9a}$$

$$\psi = \psi_1 = e^{-\frac{iE_1 t}{\hbar}} \psi_1 \quad . \tag{9b}$$

These wave-functions are reversible and normalised to unity. In addition, ψ_1 and ψ_2 are orthogonals, i.e.,

$$\int \psi_1^+ \psi_2 \underline{dr} = e^{-\frac{it}{\hbar}(E_2 - E_1)} \int \psi_1^+ \psi_2 \underline{dr} = 0 \tag{10}$$

The dipole associated to this transition is $\vec{d}$ given by

$$\vec{d} = -e\,\vec{\ell} \tag{12}$$

where

$$\vec{\ell} = \int \psi_1^+ \,\vec{r}\, \psi_2 \underline{dr} \tag{13}$$

$$\ell^2 = \vec{\ell} \cdot \vec{\ell} \,. \tag{14}$$

The corresponding pulsation is

$$\omega = \frac{E_2 - E_1}{\hbar} \tag{15}$$

and the life-time of the electron is

$$\tau = \frac{3\pi\varepsilon_0 \hbar c^3 \beta}{e^2 \omega^3 \ell^2}, \tag{16}$$

β being a coefficient close to unity (Durand 1970).

The Schrödinger equation cannot describe accurately the quantum transition because of its reversible character. It is therefore very interesting to find out an irreversible equation. An approach has been developped by Monsieur de BROGLIE and his collaborators (Silva et al. 1961; Silva et al. 1962; Diner 1984). They have proposed a non-linear irreversible equation.

Now, the wave-function will be called ψ_I and it will be governed by the equation (where γ is a coupling constant which can be choosen arbitrarily)

$$i\hbar\frac{\partial\psi_{\mathrm{I}}}{\partial t}=\hat{H}_0\psi_{\mathrm{I}}+\gamma\left(\psi_{\mathrm{I}}\psi_{\mathrm{I}}^{+}\right)\frac{\partial}{\partial t}\left(\psi_{\mathrm{I}}\psi_{\mathrm{I}}^{+}\right). \tag{17}$$

This equation possesses two remarkable properties. First of all, it provides the solutions of the reversible equation for the energies E_1 and E_2. Second, one is allowed to recover the usual relation

$$\frac{\partial\rho}{\partial t}+\vec{\nabla}\cdot\vec{j}=0 \tag{18}$$

with

$$\rho=\psi_{\mathrm{I}}\psi_{\mathrm{I}}^{+}, \tag{19}$$

$$\vec{j}=-i\frac{\hbar}{gm_0}\left[\psi_{\mathrm{I}}^{+}\vec{\nabla}\psi_{\mathrm{I}}-\psi_{\mathrm{I}}\vec{\nabla}\psi_{\mathrm{I}}^{+}\right] \tag{20}$$

4 Another Approach

This second approach has been suggested by the analogy of the problem with the statistical mechanics of a gas of particles. As a first aproximation, we consider that the behavior of a gas is reversible. Let t the time, $\vec{r}$ the position, $\vec{w}$ the velocity, m the mass of a particle and V_0 the interaction potential between them. The distribution function, $f\left(t,\vec{r},\vec{w}\right)$, is a solution of the reversible kinetic equation

$$\frac{\partial f}{\partial f}+\vec{w}\cdot\frac{\partial f}{\partial\vec{r}}-\frac{1}{m}\frac{\partial V_0}{\partial\vec{r}}\cdot\frac{\partial f}{\partial\vec{w}}=0\,;\quad f=f_{\mathrm{M}}. \tag{21}$$

Actually, the gas is not reversible (for example, because of the viscosity phenomena) and equation (21) must contain an additional irreversible term. Let ν the collision frequency and f_{M} the distribution function of the local pseudo-equilibrium (Maxwell distribution). The evolution of the gas is therefore described by the following non-reversible equation

$$\frac{\partial f_{\mathrm{I}}}{\partial t}+\vec{w}\cdot\frac{\partial f_{\mathrm{I}}}{\partial\vec{r}}-\frac{1}{m}\frac{\partial V_0}{\partial\vec{r}}\cdot\frac{\partial f_{\mathrm{I}}}{\partial\vec{w}}=\nu\left(f_{\mathrm{M}}-f_{\mathrm{I}}\right). \tag{22}$$

Equation (22) is not linear since v is an integral of the distribution function but it constrains the evolution to converge back to the equilibrium for an isolated gas. If initially the gas is not at equilibrium, we will see the distribution function to go to f_M. In analogy with (22), we consider a wave-function ψ_I made non-reversible by the introduction of the new term $(t/\tau)(E_1 - E)\psi_I$. As a matter of fact, the quantum transition is not reversible and this new irreversible term must appear in the equation governing ψ_I as a relaxation term, from going from the energy level E_2 to the energy level E_1. Moreover, it has to take into account the caracteristic time τ coming from the reversible approximation and its value must be zero at the initial time ($E = E_2$).

As a consequence, the equation for ψ_I becomes

$$i\hbar\frac{\partial\psi_I}{\partial t} = \hat{H}_0\psi_I + \frac{t}{\tau}(E_1 - E)\psi_I. \tag{23}$$

In order to obtain coherent and simple results, it is necessary to modify the interpretation of the integral

$$J = \int \psi_I^+ i\hbar\frac{\partial\psi_I}{\partial t}\underline{dr}\ , \tag{24}$$

and not to write

$$J=E, \tag{25}$$

but

$$J = E + t\frac{dE}{dt}. \tag{26}$$

For a stationary state, one recover $J=E$.

Let us look for a solution of (23) in the form

$$\psi_I = e^{-i\frac{tE(t)}{\hbar}}\left[C_1(t)\psi_1 + C_2(t)\psi_2\right]. \tag{27}$$

The wave-function ψ_I must be normalized and we assume that ψ_1 and ψ_2 are real. It turns out that

$$\int \psi_1^+ \psi_1 \underline{dr} = \int \left[C_1 C_1^+ \psi_1^2 + C_2 C_2^+ \psi_2^2 + \left(C_1 C_2^+ + C_2 C_1^+ \right) \psi_1 \psi_2 \right] \underline{dr}$$
$$= C_1 C_1^+ + C_2 C_2^+$$
$$= 1, \tag{28}$$

which is a constrain for the coefficients C_1 and C_2.

Multiplying (23) by ψ_1^+ and integrating over space, one has

$$\int \psi_1^+ i\hbar \frac{\partial \psi_1}{\partial t} \underline{dr} = E + t\frac{dE}{dt} = \int \psi_1^+ \hat{H}_0 \psi_1 \underline{dr} + \frac{t}{\tau}(E_1 - E) \int \psi_1^+ \psi_1 \underline{dr}, \tag{29}$$

which becomes

$$E + t\frac{dE}{dt} = \int \left[C_1^+ \psi_1 + C_2^+ \psi_2 \right] \hat{H}_0 [C_1 \psi_1 + C_2 \psi_2] \underline{dr} + \frac{t}{\tau}(E_1 - E). \tag{30}$$

However, since

$$\hat{H}_0 \psi_1 = E_1 \psi_1, \tag{31}$$

$$\hat{H}_0 \psi_2 = E_2 \psi_2, \tag{32}$$

we obtain from (30)

$$C_1 C_1^+ E_1 + C_2 C_2^+ E_2 + \frac{t}{\tau}(E_1 - E) = E + t\frac{dE}{dt}. \tag{33}$$

Making equals the terms in t, we get

$$\frac{dE}{dt} = \frac{E_1 - E}{\tau}, \tag{34}$$

and

$$C_1 C_1^+ E_1 + C_2 C_2^+ E_2 = E. \tag{35}$$

The solution of (34) is very simple

$$E = E_1 + (E_2 - E_1) e^{-\frac{t}{\tau}} \tag{36}$$

with the conditions

$$t = 0, \quad E = E_2 \quad ; \quad t = \infty, \quad E = E_1. \tag{37}$$

Equation (35) provides

$$C_1 C_1^+ E_1 + C_2 C_2^+ E_2 = E_1 + (E_2 - E_1) e^{-\frac{t}{\tau}} \tag{38}$$

and since we have $C_1 C_1^+ + C_2 C_2^+ = 1$ (normalisation condition), one gets

$$C_1 C_1^+ = 1 - e^{-\frac{t}{\tau}}, \tag{39}$$

and

$$C_2 C_2^+ = e^{-\frac{t}{\tau}}. \tag{40}$$

With C_1 and C_2 real, the irreversible wave-function is

$$\psi_I = e^{-i\frac{t}{\hbar}\left[E_1 + (E_2 - E_1) e^{-\frac{t}{\tau}}\right]} \left[\left(1 - e^{-\frac{t}{\tau}}\right)^{1/2} \psi_1 + \left(e^{-\frac{t}{\tau}}\right)^{1/2} \psi_2\right]. \tag{41}$$

Let $\vec{r_1}$ and $\vec{r_2}$ the mean values of $\vec{r}$ for, respectively, the two states 1 and 2, given by

$$\overline{r_1} = \int \psi_1^2 r \underline{dr}, \tag{42}$$

$$\overline{r_2} = \int \psi_2^2 r \underline{dr}. \tag{43}$$

We multiply (41) by $r\psi_I^+$ and after integration over space, one gets

$$\bar{r} = \bar{r}_1 \left(1 - e^{-\frac{t}{\tau}} \right)^{1/2} + \bar{r}_2 \left(e^{-\frac{t}{\tau}} \right)^{1/2} . \tag{44}$$

This equation provides the mean value of the position as a function of time during the transition.

5 Application to the Atom of Hydrogen

Let us consider the transition between the two states $n = 2 \; \ell = 1$ and $n = 1 \quad \ell = 0$.

The Bohr radius a_0 is given by

$$a_0 = \frac{4\pi\varepsilon_0 \hbar}{m_0 e^2} = 5,292 \cdot 10^{-11} m \tag{45}$$

and we obtain

$$\psi_1 = \pi^{-1/2} a_0^{-3/2} e^{-\frac{r}{a_0}} , \tag{46}$$

$$\psi_2 = \frac{1}{4} (2\pi)^{1/2} a_0^{-3/2} \cos\theta e^{-\frac{r}{2a_0}} . \tag{47}$$

As a consequence one gets

$$r_1 = 1,5 a_0 \tag{48}$$

and

$$r_2 = 5 a_0 . \tag{49}$$

The expression for τ is obtained by Durand (1970), i.e.,

$$\tau = 1,6 \cdot 10^{-9} s . \tag{50}$$

Expressing $\bar{r}$ in meters and t in seconds, it turns out from (44) that

$$\bar{r} = 7,938 \cdot 10^{-11}\left[1 - e^{6,25 \cdot 10^{8} t}\right]^{\frac{1}{2}} + 2,796 \cdot 10^{-10}\left[e^{-6,95 \cdot 10^{8} t}\right]^{\frac{1}{2}}. \qquad (51)$$

Conclusion

The Schrödinger equation does not provide a precise description of a quantum transition but it gives the life-time τ of the electron experiencing the transition. Introducing an irreversible term (which is a function of τ) in this equation, we obtain a new one from which the time-dependence of the transition is derived.

References

Diner, S. (1984) : *The Wave-Particle Dualism* (Reidel Publishing Company, Dordrecht), 149-172

Durand, E. (1970) : *Mécanique Quantique*, tome 1 (Masson, Paris)

Silva, J. A., Fer, F., Leruste, P. , Lochak, G. (1961): Quantification, Stationarité et Non-Linéarité. Cahiers de physique **129**, 1-15

Silva, J. A., Fer, F., Leruste, P. , Lochak, G. (1962): Problèmes de Stabilité pour des Systèmes à un Grand Nombre de Degrés de Liberté. Cahiers de physique **137**, 1-15

Part IV

Closure

Pioneering in Plasma Simulation

Charles K. Birdsall

EECS Dept., Univ. of California, Berkeley CA 94720-1770, USA

Abstract. Pioneering leads the way in science and is essential to those who follow, who reap the benefits. Plasma simulation, begun in the 1950s, has had numerous pioneering efforts over the past five decades. This short paper recalls some of these, apropos to recognizing the early contributions by Marc Feix. The next decade will probably see science simulations done in high schools with simulations of whole devices, small and large, accelerating immensely.

1 Introduction

Pioneering trials and errors are essential to all scientific progress. Someone must stick its neck out in proposing and trying new ideas, in sorting out good ideas from bad ones. So it has been with plasma and other physics and chemistry simulations, with progress made hand-in-hand with computers becoming faster, bigger, cheaper and easier to use. Add to this progress a commitment by many of the early pioneers to first-principles, self-consistent simulations, now continued, leading to application to more and more complex models.

Let "plasma" include neutral plasmas with at least two species (positively and negatively charged particles), ranging from low temperature discharges to high temperature fusion reactors as well as single species nonneutral plasmas (e.g., essentially pure electron or pure ion beams or rotating clouds).

What have been some of the steps along the way? We all have our own examples that struck us as milestones, that we deemed worth reporting and now deem worth recalling. I am no exception and am happy to recall some of the steps, including some by our honoree, Marc Feix.

My story begins with my work in industry, starting in 1951, and comes to the present, with some added thoughts about the next steps, going into the twenty-first century. The text is essentially a talk, not an archival journal article, not all-inclusive; such would be at least one hundred pages long!

What was new, done earlier by a few, not yet clear, not easy and slow is now very clear, easy, done by many and routine.

2 1951: Analog Computation of $zJ_1(z)/J_0(z)$

In dealing with microwave amplifiers using electron beams there was interest in finding the response of the beam to various kinds of bounding walls. A

traveling-wave tube uses a propagating wall, with phase velocity at about the speed of the beam. The impedance seen by an electron beam near the propagating wall is **resistive-inductive** or **resitive-capacitive**, depending on whether the beam velocity is larger or smaller than the wall phase velocity (approximately). A **resistive wall amplifier**, with a non-propagating wall, may be made with a thin resistive coating on the inside of a glass wall; the impedance "seen by the electron beam" in the vacuum region inside the coated wall is **resistive-capacitive**. Obviously, one may think up all kinds of walls, propagating and non-propagating, and then seek the beam-wall interactions.

In 1951, John Whinnery and I (Birdsall and Whinnery 1953) did that, limiting ourselves to the non-propagating variety. The impedance which we used was a **field impedance**, given by a ratio of axial electric field to circumferential magnetic field, E_z/H_ϕ, that "looking at the wall". In order to obtain the complex value of the wavenumber for the gain (growth) of the fields with distance along the beam we found that we needed values of the ratio of Bessel functions $zJ_1(z)/J_0(z)$ for complex values of z. We looked for help from our computational people at Hughes Aircraft. Their answer was that their computer did not have Bessel functions of complex argument. We found tables constructed during WWII, but thought them tedious to use. So we looked for another way to obtain our ratio.

We had access to an electrolytic (water) tank which was set up to follow equipotentials in two dimensions, which were solutions of Laplace's equation. Our function was analytic, with poles and zeros along the real z axis. We merely needed to put in electrodes to set these up, apply voltages and follow equipotentials. We set the value of the potential desired by a probe, cranked in the x coordinate by hand and let the servo produce the y location, all the time plotting the location of the probe. We took two days to make one plot.

Figure 1 is the resulting plot, made by this analog computer, because the local digital computer could not do it. Quite an experience and certainly not one to be repeated today!

Indeed, in 1982, Dr Han Uhm (Uhm 1982) published a similar plot for similar applications, done on a supercomputer. He says that a fast PC can do it now readily, probably in less than a minute.

3 1953: Cold Test of Slow-Wave Circuits by Resonances and Small Perturbations

Just a brief statement here. In designing microwave tubes using electron beams one needs to know quite a bit about the circuits to be used to interact with the beam, such as cavities for a klystron, slow-wave structures for traveling-wave tubes and on and on. In 1953-54 Tom Everhart and I (Birdsall and Everhart 1956) worked on a special circuit, like contra-wound helices, made in a ring-bar form (Fig. 2). We had our shop make three such

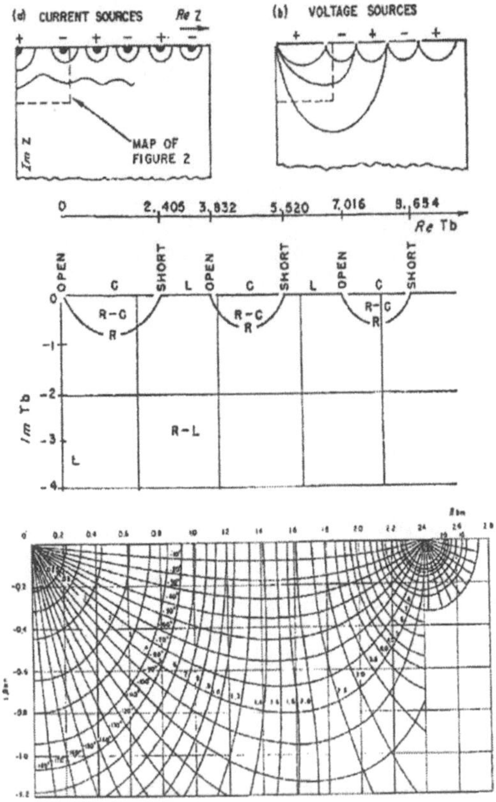

Fig. 1. Evolution of the scheme calculation of an analytic function in a water tank in 1951. The plot at the bottom took two days to make. It gives the level surfaces for the region marked in the uppermost figure.

circuits, with various thicknesses etc and found their dispersion (ω versus k) by resonating them and then found their fields by perturbing them. This was tedious work, not terribly accurate, time consuming (days), both in the laboratory and in the shop. We are now doing such numerically by solving the finite differenced wave equation **in a few minutes.**

4 1960: Plasma Simulations in One Dimension

In 1956 P.K. Tien and J. Moshman (Tien and Moshman 1956) published particle simulations of a thermionically emitted electron beam (half-Maxwellian initial velocity distribution) in a diode. They sought the spectrum of fluctuations in electron velocity and current from an electron gun for a microwave

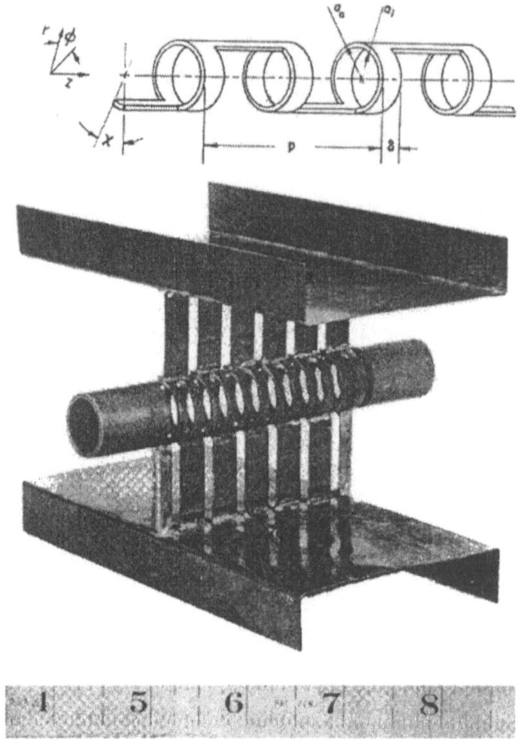

Fig. 2. Ring-bar helices, slow-wave circuits for traveling-wave tubes, made in 1953. ω vs k obtained by resonating these; E_z^2/W is obtained by perturbing the fields. Days of skilled effort.

tube. They found that the fluctuating current was much less than full shot noise ($1/\sqrt{N}$ fluctuations) for low frequencies, but was near full shot noise when the frequency was above the electron plasma frequency of the potential minimum (near the emitter). (This noise smoothing behaviour allowed triodes to have low noise figures at low frequencies.) The model used sheet charges in a shorted diode (for the space charge fields) with the applied battery voltage superposed. To many these simulations were the precursor of warm plasma simulations, having many of the techniques of plasma simulations which began several years later. To be sure these were primitive, with a very slow, limited memory, computer, allowing the following of only about 300 or 400 particles (really too few to obtain good statistics) with runs of hours.

In 1959 Oscar Buneman (Buneman 1959) published a beam-plasma instability particle simulation article, with cold particles, showing the evolution from small amplitude growth (linear theory) through to fully nonlinear saturation, which was unknown analytically. **This article indeed was a milestone** in showing the power of particle simulations to obtain the nonlinear end of exponential growth, which is very hard to obtain analytically. Similar studies of instability growths and saturations have followed for the past four decades, a tribute to Buneman leading the way.

In 1962 John Dawson (Dawson 1962) provided some physical justifications for doing the kinetics of a warm (Maxwellian velocity distribution) one-dimensional plasma by using only a relatively small number of particles (electron and ion sheet charges) in a Debye length. **This was a revolution.** All of us are taught that natural and laboratory plasmas "require" many particles in a Debye sphere to be a "real" plasma, that is $N\lambda_D^3 \gg 1$, in full three dimensions. Dawson showed that in one dimension (which is quite acceptable for studying many plasma phenomena) one only needs an $N_D = n\lambda_D$ of perhaps 5 or 10, immediately allowing one to do many simulations with far fewer than millions of particles (which would have been impossible with 1960 computers - and for a few more years to come!).

A few months later in 1962 Owen Eldridge and Marc Feix (Eldridge and Feix 1962a) provided a similar model, with more information, with $N_D = n\lambda_D$ varying from 0.25 to 10, complementing the almost simultaneous work by Dawson. They soon again (Eldridge and Feix 1962b) published a research note and another full article (Eldridge and Feix 1963), with especial attention to obtaining the correct Fokker-Planck coefficients (drag and diffusion) for a 1d plasma.

The article by Dawson and the three by Eldridge and Feix together constitute a very firm basis for using one dimensional charged particles (sheet charges) for obtaining plasma kinetics.

These are very fine examples of pioneering studies using simulations.

5 Electron Diodes, Virtual Cathode Oscillations

In 1959 William B Bridges, my student in Berkeley, and I were studying noise from thermionic electron emitters (cathodes). The prior analytic studies by many authors, and the simulations by Tien and Moshman mentioned above, were fine, but did not completely satisfy us as to the **dynamics** of the potential minimum formed about a Debye length, λ_D, in front of the emitter. [The **static** behavior had been given long before by Irving Langmuir in 1923 (Langmuir 1923) (which incidentally was duplicated in detail in 1d simulations by another of my students, about ten years ago)]. Earlier in 1959 I had made a simple analysis of the dynamics of the minimum, piecewise in time – very approximate. Bill and I decided to make a simple computational model in order to obtain the dynamics of a potential minimum caused by the space charge of a beam. We settled on injecting a cold beam across a short-circuited diode in 1d. Bill began simulations using a few particles, a large sheet of paper, trial and error trajectories, but self-consistent, using a mechanical calculator, pencil and lots of erasers. This trial lasted just a short while. Bill then put it all on an IBM 704 computer (appreciably faster and bigger than Tien's machine of three or four years earlier but still with only a few hundreds of particles). He observed the initial dynamic slowing down of the electrons due to their own space charge and then, low and beware, he saw the dynamic beginning of the electrons forming a **virtual cathode** and then being reflected back to the injection plane, with the process, called **virtual cathode oscillations**, then repeating. The frequency was near the plasma frequency of the (now dynamic) potential minimum. It blew our minds as the oscillations were the step missing from the **theory of limiting currents,** only indirectly connected to noise - a serendipitous discovery, made wholly by simulations (published in 1961 (Birdsall and Bridges 1961) and 1963 (Bridges and Birdsall 1963) with more in our book (Birdsall and Bridges 1966)). We dropped our noise studies and shifted to studying the dynamics of limiting currents. There are commercial devices now using VC oscillations, called **vircators**, some with pulsed power outputs up to terawatts (albeit not very efficient as we had also found). Our computer runs took hours.

Starting a few years ago we added Bill's simulation model to our input files for our PC simulations. I showed this to him recently at a seminar at Cal Tech, where he now teaches. His reponse was: **"There goes my Ph D thesis in one minute"**. I tell my current students of Bill's remark, 35 years after his thesis, reminding them that the same will happen with their simulations, so painstakingly generated (now running millions of particles, in 2d3v, some in 3d3v): in one minute!

6 1967: First Numerical Simulation of Plasmas Conference, Organised by Marc Feix

In 1967 Marc Feix organized the **Symposium on Computer Simulation of Plasma and Many-Body Problems**, held at the College of William and Mary, Williamsburg, Virginia, with NASA sponsorship, April 19-21. There were thirty-six papers from five areas. The attendance was excellent as were the discussions. So much so that there now have been fifteen Numerical Simulation of Plasmas Conferences, the last held in 1994 and the next planned for early 1998.

We all owe Dr Feix a very special tribute for having the foresight to set up the first such conference when plasma simulations were only about five years old, dating from the Dawson and Eldridge and Feix papers in 1962. Thank you Marc!

7 1960-1970-1980-1990s Fusion Progress

Simulation of fusion plasmas, dating from the early 1960s, covers an immense amount of ground, in sheer numbers of papers, in a vast variety of techniques and in a wide range of applications. Yet, still, we do not have a fusion reactor producing nett electrical power. Nobel laureate Glenn Seaborg some time ago said that solving the fusion problem was the greatest scientific challenge of the second half of the twentieth. He may have to add another half century to his prediction.

Fusion simulations have been well supported (in some sense) in the USA and abroad. LLNL long had its own fusion computer centre, recently called NERSC (and broadened to include more energy research) and recently moved to LBNL. A number to keep in mind is what percentage of the overall fusion budget has gone into laboratory experiments (which are invaluable) versus computer experiments; less than 3% for the latter is a guess.

My University research group and many others were tied into NERSC and its predecessors and were served well by numerous Cray machines for about thirty years. Several years ago we found that overnight runs on our workstations were competing well with a hundred users running on NERSC. We are now fully on (much faster yet) workstations. This trend, away from mainframes, is widespread for many simulations.

We are now close to simulating whole plasma reactors employing RF glow discharges (low temperature collisional plasmas) as used in 40% of semiconductor processing (a huge industry, in the tens of billions of dollars per year). We do fairly good plasma physics and quite a bit of volume and surface chemistry, but we have not yet convinced industry that these simulations are better than their current empiricism, which has been immensely successful. Some progress is being made.

Hence one might imagine simulating the plasma physics of an entire fusion reactor before building such a reactor. This approach is especially attractive when considering the price of ITER, at some ten billion dollars. Is 1% to 3% (as noted above) of this amount, some \$100 to \$300 million, enough to simulate ITER? One answer I find is: more than enough, but hurting for more people rather than for faster machines.The present ITER simulation effort worldwide is perhaps \$20 million per year. Should other alternative reactors also be simulated?

There is a current USA program, called the Numerical Tokamak Turbulence Project (NTTP) with ten to fifteen simulators, working on the tokamak core physics. Their computer runs, using 3d3v gyrokinetics, involve ten to one hundred million particles, lasting a millisecond of tokamak real time, on parallel computers with 64 to 256 processors. Let us wish them well, as fusion generated electric energy may well be one of our best hopes for future energy sources.

8 Industry Growing up on Simulation

Industry already uses simulations in many ways, from approximate models to fully first principles, self-consistent approaches. There are many scientists and engineers involved, over a wide range of industries, with many papers published and conferences held.

Already mentioned is the competition between empiricism and simulations for plasma processing of semiconductors. With chip dimensions becoming smaller and smaller, requiring more care, the race to be first and best may be won by reactor designers using faster, bigger and cheaper computers, capable of providing computer experiment results on a time scale comparable with doing laboratory experiments. An example of plasma science for plasma processing is Fig. 3, V Vahedi's snapshot (unpublished, early 1990s) of log of the electron distribution function, $f_e(x, v, t = 7)$, for a 70mTorr capacitively coupled RF discharge. The bounding walls are at $x = 0$ and $x = L$. Note the hot tail of the electrons which rotates at ω_{RF}.

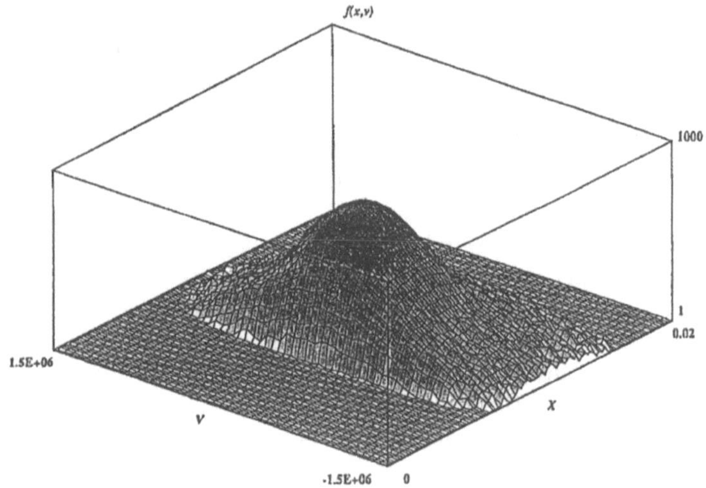

Fig. 3. Log of electron distribution function in x, v space for an RF capacitively coupled discharge, showing the cool bulk and the hot tail rotating at ω_{RF}, early 1990s.

Richard True, at Litton Electron Devices, says that they rarely cut metal for laboratory experiments **before** they have done the simulations and that about 20 to 25% of his laboratory budget goes to simulations (much more than 3%). Other similar high tech industries must follow, or lose out.

9 The Near Future: Science and Mathematics Programming in High School

About 1990, in my freshman orientation seminar, I asked how many students had their own computers; everyone did. I asked how many had used them on science or engineering problems; no one had. This was sad.

In 1963 I was assigned to teach our first course in computer programming (Fortran 4), alongside two much more experienced professors. I was recruited to add some physical problems, which I did. At that time I suggested to my colleagues that **we needed a beginning calculus book with finite-difference-equations (FDEs) taught on the left hand pages and infinitesimals (ODEs, PDEs) taught on the right hand pages.** The FDEs were then to be available for programming physics and mathematics problems on computers, all of which were mainframes then. Now with much faster PCs, much easier to program, we still need such a text. This text could be used in high schools to give each student its own **computer experiments.**

Let your imagination run ahead a little in time with FDEs taught in high schools. Then FDE computing skills would be applied in succeeding science and mathematics courses (high school and college). Evolutionary or revolutionary? No matter. Science and mathematics courses will never be the same. All students will have available the tools to verify the analytic models put forth by their teachers and have a tool to parallel their learning in laboratory experiments.

In 1995 and 1996 we ran a **plasma computer experiments laboratory** in parallel with a plasma lecture course, which had very good analytic modeling. Once a week we assigned computer experiments like the lecturer's analytic models and checked on his approximations and assumptions, doing calculations right on the graphical output as if these were real experiments. Each computer experiment was run from an input file of initial and boundary conditions, only 5% of which were numerical; that is, almost all the numbers looked like experimental inputs (voltages, currents, frequencies, lengths, pressures etc). In the fifth week, after completing four computer experiments, we went back to fundamentals for one week to demonstrate the accuracy and stability (Δt and Δx) limits. This scheme worked well for seniors and graduate students.

This is a glimpse of the future of simulations as I see it and it is already here!

10 The Farther Future (of the University)

Peter Drucker (Drucker 1997), a well known management consultant and author, was recently quoted in as saying: "Thirty years from now, the big university campuses will be relics... [The] totally uncontrollable expenditure, without any visible improvement in....education, mean that the system is rapidly becoming untenable.....We [can] deliver.... classes off campus via satellite or two-way video at a fraction of the cost. The college won't survive as a residential institution."

His statement can be given more detail, be expanded, be made more realistic by any of you.

In the relatively wee world of plasma physics we simply **know** that computer experiments will precede laboratory experiments, possibly obviating the need for many laboratory experiments. May we extraploate to the rest of the scientific world?

Food for thought!

References

Birdsall, C.K. (1991): Particle-in-Cell Charged-Particle Simulations, Plus Monte Carlo Collisions with Neutral Atoms, PIC-MCC. IEEE Transactions on Plasma Science **19**, 65–85

Birdsall, C.K., Bridges, W.B. (1961): Space-charge instabilities in electron diodes and plasma converters. J. Appl. Phys. **32**, 2611–2618

Birdsall, C.K., Bridges, W.K. (1996): *Electron Dynamics of Diode Regions* (Academic Press, New York)

Birdsall, C.K., Everhart, T.E. (1956): Modified contra-wound helix circuits for high-power traveling-wave tubes. IRE Transactions on Electron Devices ED-3:190

Birdsall, C.K., Whinnery, John R. (1953): Waves in an electron stream with general admittance walls. Journal of Applied Physics **24**, 314–323

Bridges, W.B., Birdsall, C.K. (1963): Space-charge instabilities in electron diodes ii. J. Appl. Phys. **34**, 2946–2955

Buneman, O. (1959): Dissipation of currents in ionized media. Phys. Rev. **115**, 503–517

Dawson, J.M. (1962): One-dimensional plasma model. Phys. Fluids **5**, 445–459

Drucker, Ṗ. (1997): citation from *Forbes Magazine*. Science **275**, 1745

Eldridge, O.C., Feix, M. (1962): One-dimensional plasma model at thermodynamic equilibrium. Phys. Fluids **5**, 1076–1080

Eldridge, O.C., Feix, M. (1962): Fokker-planck coefficients for a one-dimensional plasma. Phys. Fluids **5**, 1307–1308

Eldridge, O.C., Feix, M. (1963): Numerical experiments with a plasma model. Phys. Fluids **6**, 398–406

Langmuir, I. (1923): The effect of space charge and initial velocities on the potential distribution and thermionic current between parallel plane electrodes. Phys. Rev. **21**, 419–435

Tien, P.K., Moshman, J. (1956): Monte Carlo Calculation of noise near the potential minimum of a high frequncy diode. J. Appl. Phys. **27**, 1067–1078

Uhm, Han S. (1982): Space charge waves in a cylindrical waveguide with arbitary wall impedance. Phys. Fluids **25**, 690–696

List of Participants

Andrieux, Jean-Claude
Rue du Morvan, 21200 Beaune, France

Bégué, Marie-Line begue@lpmi.u-nancy.fr
LPMI, Université Henri Poincaré, BP 239, 54506 Vandoeuvre-les-Nancy Cedex, France

Ben Hadid, Yacine yacine.benhadid@labomath.univ-orleans.fr
MAPMO UMR 6628, UFR Sciences, Université d'Orléans, BP 6759, 45067 Orléans Cedex 2, France

Bertrand, Pierre bertrand@lpmi.u-nancy.fr
LPMI, Université Henri Poincaré, BP 239, 54506 Vandoeuvre-les-Nancy Cedex, France

Birdsall, Charles K. birdsall@eecs.Berkeley.edu
Plasma Theory and Simulation Group, Dept of Electrical Engineering and Computer Sciences, Cory Hall # 1770, University of California, Berkeley, CA 94720-1770, USA

Bonnaud, Guy bonnaud@limeil.cea.fr
CEA Limeil, DPTA, 94195 Villeneuve-St-Georges, France

Bonami, Aline aline.bonami@labomath.univ-orleans.fr
MAPMO UMR 6628, UFR Sciences, Université d'Orléans, BP 6759, 45067 Orléans Cedex 2, France

Bonomi, Ernesto ernesto@crs4.it
Center for Advanced Studies, Research and Development, Cagliari, Italy

Bouchut, François francois.bouchut@labomath.univ-orleans.fr
MAPMO UMR 6628, UFR Sciences, Université d'Orléans, BP 6759, 45067 Orléans Cedex 2, France

Bouquet, Serge bouquet@bruyeres.cea.fr
CEA/B3, BP 12, 91680 Bruyères-le-Châtel, France

Burgan, Jean-Robert burgan@bruyeres.cea.fr
CEA/BII, DRIF/Direction, BP 12, 91680 Bruyères-le-Châtel, France

Cairo, Laurent laurent.cairo@labomath.univ-orleans.fr
MAPMO UMR 6628, UFR Sciences, Université d'Orléans, BP 6759, 45067 Orléans Cedex 2, France

Chapelle, Joseph
GREMI, UFR Sciences, Université d'Orléans, BP 6759, 45067 Orléans Cedex 2, France

Chavarriga, Javier chava@eup.udl.es
Departmento de Matematicas, Universidad de Lleida, Plaza Victor Siurana 1, 25003 Lleida, Espagne

Chave, Jérôme chave@spec.saclay.cea.fr
CEA, SPEC, CEN Saclay, Orme des Merisiers, F-91191 Gif-sur-Yvette Cedex, France

Clérouin, Jean clerouin@limeil.cea.fr
CEA/LV, 94195 Villeneuve-St-Georges Cedex, France

Dubois, Marc mad@amoco.saclay.cea.fr
CEA, SPEC, CEN Saclay, Orme des Merisiers, F-91191 Gif-sur-Yvette Cedex, France

De Izarra, Charles Charles.DeIzarra@univ-orleans.fr
GREMI, Université d'Orléans, Faculté des Sciences, Rue Gaston Berger, BP 4043, 18028 Bourges Cedex, France

Feix, Marc R. marc.feix@labomath.univ-orleans.fr
MAPMO UMR 6628, UFR Sciences, Université d'Orléans, BP 6759, 45067 Orléans Cedex 2, France

Figua, Habiba hfigua@labomath.univ-orleans.fr
MAPMO UMR 6628, UFR Sciences, Université d'Orléans, BP 6759, 45067 Orléans Cedex 2, France

Fijalkow, Eric eric.fijalkow@labomath.univ-orleans.fr
MAPMO UMR 6628, UFR Sciences, Université d'Orléans, BP 6759, 45067 Orléans Cedex 2 , France

Frey, Jean-Marie
5, Place de la Liberté, 29200 Brest, France

Garcia, Isaac garcia@eup.udl.es
Departmento de Matematicas, Universidad de Lleida, Plaza Victor Siurana 1, 25003 Lleida, Espagne

Géronimi, Claude claude.geronimi@labomath.univ-orleans.fr
MAPMO UMR 6628, UFR Sciences, Université d'Orléans, BP 6759, 45067 Orléans Cedex 2 , France

Ghizzo, Alain ghizzo@lpmi.u-nancy.fr
LPMI, Université Henri Poincaré, BP 239, 54506 Vandoeuvre-les-Nancy Cedex, France

Giacomini, Hector giacomini@celfi.phys.univ-tours.fr
UFR Sciences, Université de Tours, Parc de Grandmont, 37200 Tours, France

Giné, Jaume gine@eup.udl.es
Departmento de Matematicas, Universidad de Lleida, Plaza Victor Siurana 1, 25003 Lleida, Espagne

Goedert, Joao goedert@if.ufrgs.br
Instituto de Fisica da UFRGS, Cx. Postal 15051, 91501-970 Porto Alegre RS, Brésil

Gutierrez, Edgardo dts@philonline.com.ph
Center for Fluid Dynamics, University of the Philippines Los Banos, College, Laguna, Philippines

Hiéblot, James
LPCE/CNRS, 3A Avenue de la Recherche Scientifique, 45071 Orléans Cedex, France

Izrar, Boujema izrar@cnrs-orleans.fr
Laboratoire d'Aérothermique, 1C Avenue de la Recherche Scientifique, 45071 Orléans Cedex, France

Krasnosselskikh, Volodia vkrasnos@cnrs-orleans.fr
LPCE/CNRS, 3A Avenue de la Recherche Scientifique, 45071 Orléans Cedex, France

James, François francois.james@labomath.univ-orleans.fr
MAPMO UMR 6628, UFR Sciences, Université d'Orléans, BP 6759, 45067 Orléans Cedex 2, France

Labrunie, Simon labrunie@ccr.jussieu.fr
LPTM, Université Paris 7 - Denis Diderot, Tour Centrale, 3eme Etage, 75251 PARIS Cedex 05, France

Lavaud, Michel michel.lavaud@univ-orleans.fr
GREMI,UFR Sciences, Université d'Orléans, BP 6759, 45067 Orléans Cedex 2 , France

Leach, Peter leach@scifs1.und.ac.za
Department of Mathematics and applied Mathematics, University of Natal, Durban 4041, South Africa

Lefebvre, Bertrand lefebvre@cnrs-orleans.fr
LPCE/CNRS, 3A Avenue de la Recherche Scientifique, 45071 Orléans Cedex, France

Lefebvre, Erik lefebver@limeil.cea.fr
CEA Limeil, DPTA, 94195 Villeneuve-St-Georges, France

Lewis, H.Ralph H.Ralph.Lewis@Dartmouth.edu
Dartmouth College, Department of Physics and Astronomy, Hanover, New Hampshire, USA

Llibre, Jaume jllibre@mat.uab.es
Departament de Matemàtiques, Universitat Autònoma de Barcelona, 08193 Bellaterra, Barcelona, Spain

Lutton, Jean-Luc jeanluc.lutton@francetelecom.fr
CNET ISSY, DAC/ORI, 38-40 rue du General-Leclerc, F-92131 Issy-les-Moulineaux, France

Manfredi, Giovanni manfredi@lpmi.u-nancy.fr
School of Cosmic Physics, Dublin Institute for Advanced Studies, 5 Merrion Square, Dublin 2, Ireland

Mas-Gallic, Sylvie masgalli@cmapx.polytechnique.fr
CMAP, Ecole Polytechnique, 91120 Palaiseau, France

Miller, Bruce N. bmiller@gamma.is.tcu.edu
Physics Department, Box 298840, Texas Christian University, Forth Worth, Texas 76129, USA

Minardi, Ettore minardi@ifp.mi.cnr.it
Instituto di Fisica del Plasma, Associazine EUR-ENEA-CNR, Via R. Cozzi, 20125 MILANO, Italy

Mineau, Pascal pascal_mineau@dassault-systemes.fr
Dassault Systemes, Département Analyse et Simulation, 9 Quai Marcel Dassault, BP 310, 92156 Suresnes Cedex, France

Munier, Alain munier@limeil.cea.fr
CEA Limeil, 94195 Villeneuve St Georges Cedex, France

Muriel, Amador dts@philonline.com.ph
Centre for Fluid Dynamics (World Laboratory), University of the Philippines at Los Banos, College, Laguna, Philippines

Nier, Francis nier@math.polytechnique.fr
Centre de Mathématique, URA-CNRS 169, Ecole Polytechnique, 91128 Palaiseau Cedex, France

Perdang, Jean perdang@astro.ulg.ac.be
Institut d'Astrophysique, 5, Avenue de Cointe, B-4000 Liège, Belgium

Perthame, Benoit perthame@ann.jussieu.fr
UPMC, BC187, 4 Place Jussieu, 75252 Paris Cedex, France

Preux, Nicolas nicolas.preux@issy.cnet.fr
CNET ISSY, DAC/ORI, 38-40 rue du General-Leclerc, F-92131 Issy-les-Moulineaux, France

Rouet, Jean-Louis jean-louis.rouet@labomath.univ-orleans.fr
MAPMO UMR 6628, UFR Sciences, Université d'Orléans, BP 6759, 45067 Orléans Cedex 2, France

Sabatier, Michel sabatier@bruyeres.cea.fr
CEA, DRIF/DCSA, BP 12, 91680 Bruyères-le-Châtel, France

Salmon, Jean Fax: (33) 01 40 51 08 65
Fondation Louis De Broglie, 23 quai de Conti, 75006 Paris, France

Skarka, Vladimir skarka@univ-angers.fr
Laboratoire POMA, UFR Sciences, Université d'Angers, 2 Bd. Lavoisier, 49045 Angers Cedex 1, France

Soares, Manuel Manuel.Soares@univ-orleans.fr
Centre Universitaire de Bourges, Rue Gaston Berger, BP 4043, 18028 Bourges Cedex, France

Strelcyn, Jean-Marie Jean-Marie.Strelcyn@univ-rouen.fr
Département de Mathématique, Université de Rouen, 768421 Mont Saint Aignan, France

Toupin, Catherine toupin@limeil.cea.fr
CEA Limeil, DPTA, 94195 Villeneuve-St-Georges, France

Vallée, Olivier Oliver.Vallee@univ-orleans.fr
Centre Universitaire de Bourges, Rue Gaston Berger, BP 4043, 18028 Bourges Cedex, France
Vo Khac, Khoan Fax: (33) 02 38 41 72 05
UMR 6628, UFR Sciences, Université d'Orléans, BP 6759, 45067 Orléans Cedex 2, France
Wilhelmsson, Hans Fax: (33) 05 56 01 13 40
45 rue Condillac, 33000 Bordeaux

Springer and the environment

At Springer we firmly believe that an international science publisher has a special obligation to the environment, and our corporate policies consistently reflect this conviction.

We also expect our business partners – paper mills, printers, packaging manufacturers, etc. – to commit themselves to using materials and production processes that do not harm the environment. The paper in this book is made from low- or no-chlorine pulp and is acid free, in conformance with international standards for paper permanency.

Springer

Lecture Notes in Physics

For information about Vols. 1–481 please contact your bookseller or Springer-Verlag

Vol. 482: X.-W. Pan, D. H. Feng, M. Vallières (Eds.), Contemporary Nuclear Shell Models. Proceedings, 1996. XII, 309 pages. 1997.

Vol. 483: G. Trottet (Ed.), Coronal Physics from Radio and Space Observations. Proceedings, 1996. XVII, 226 pages. 1997.

Vol. 484: L. Schimansky-Geier, T. Pöschel (Eds.), Stochastic Dynamics. XVIII, 386 pages. 1997.

Vol. 485: H. Friedrich, B. Eckhardt (Eds.), Classical, Semi-classical and Quantum Dynamics in Atoms. VIII, 341 pages. 1997.

Vol. 486: G. Chavent, P. C. Sabatier (Eds.), Inverse Problems of Wave Propagation and Diffraction. Proceedings, 1996. XV, 379 pages. 1997.

Vol. 487: E. Meyer-Hofmeister, H. Spruik (Eds.), Accretion Disks – New Aspects. Proceedings, 1996. XIII, 356 pages. 1997.

Vol. 488: B. Apagyi, G. Endrédi, P. Lévay (Eds.), Inverse and Algebraic Quantum Scattering Theory. Proceedings, 1996. XV, 385 pages. 1997.

Vol. 489: G. M. Simnett, C. E. Alissandrakis, L. Vlahos (Eds.), Solar and Heliospheric Plasma Physics. Proceedings, 1996. VIII, 278 pages. 1997.

Vol. 490: P. Kutler, J. Flores, J.-J. Chattot (Eds.), Fifteenth International Conference on Numerical Methods in Fluid Dynamics. Proceedings, 1996. XIV, 655 pages. 1997.

Vol. 491: O. Boratav, A. Eden, A. Erzan (Eds.), Turbulence Modeling and Vortex Dynamics. Proceedings, 1996. XII, 245 pages. 1997.

Vol. 492: M. Rubí, C. Pérez-Vicente (Eds.), Complex Behaviour of Glassy Systems. Proceedings, 1996. IX, 467 pages. 1997.

Vol. 493: P. L. Garrido, J. Marro (Eds.), Fourth Granada Lectures in Computational Physics. XIV, 316 pages. 1997.

Vol. 494: J. W. Clark, M. L. Ristig (Eds.), Theory of Spin Lattices and Lattice Gauge Models. Proceedings, 1996. XI, 194 pages. 1997.

Vol. 495: Y. Kosmann-Schwarzbach, B. Grammaticos, K.M. Tamizhmani (Eds.), Integrability of Nonlinear Systems. VII, 380 pages. 1997.

Vol. 496: F. Lenz, H. Grießhammer, D. Stoll (Eds.), Lectures on QCD. VII, 483 pages. 1997.

Vol. 497: J. P. Greve, R. Blomme, H. Hensberge (Eds.), Stellar Atmospheres: Theory and Observations. VIII, 352 pages. 1997

Vol. 498: Z. Horváth, L. Palla (Eds.), Conformal Field Theories and Integrable Models. Proceedings, 1996. X, 251 pages. 1997.

Vol. 499: K. Jungmann, J. Kowalski, I. Reinhard, F. Träger (Eds.), Atomic Physics Methods in Modern Research, IX, 448 pages. 1997.

Vol. 500: D. Joubert (Ed.), Density Functionals: Theory and Applications, XVI, 194 pages. 1998.

Vol. 501: J. Kertész, I. Kondor (Eds.), Advances in Computer Simulation. VIII, 166 pages. 1998.

Vol. 502: H. Aratyn, T.D. Imbo, W.-Y. Keung, U. Sukhatme (Eds.), Supersymmetry and Integrable Models. Proceedings, 1997. XI, 379 pages. 1998.

Vol. 503: J. Parisi, S. C. Müller, W. Zimmermann (Eds.), A Perspective Look at Nonlinear Media. From Physics to Biology and Social Sciences. VIII, 372 pages. 1998.

Vol. 504: A. Bohm, H.-D. Doebner, P. Kielanowski (Eds.), Irreversibility and Causality. Semigroups and Rigged Hilbert Spaces. XIX, 385 pages. 1998.

Vol. 505: D. Benest, C. Froeschlé (Eds.), Impacts on Earth. XVII, 223 pages. 1998.

Vol. 506: D. Breitschwerdt, M. J. Freyberg, J. Trümper (Eds.), The Local Bubble and Beyond. Proceedings, 1997. XXVIII, 603 pages. 1998.

Vol. 507: J. C. Vial, K. Bocchialini, P. Boumier (Eds.), Space Solar Physics. Proceedings, 1997. XIII, 296 pages. 1998.

Vol. 508: H. Meyer-Ortmanns, A. Klümper (Eds.), Field Theoretical Tools for Polymer and Particle Physics. XVI, 258 pages. 1998.

Vol. 509: J. Wess, V. P. Akulov (Eds.), Supersymmetry and Quantum Field Theory. Proceedings, 1997. XV, 405 pages. 1998.

Vol. 510: J. Navarro, A. Polls (Eds.), Microscopic Quantum Many-Body Theories and Their Applications. Proceedings, 1997. XIII, 379 pages. 1998.

Vol. 511: S. Benkadda, G. M. Zaslavsky (Eds.), Chaos, Kinetics and Nonlinear Dynamics in Fluids and Plasmas. Proceedings, 1997. VIII, 438 pages. 1998.

Vol. 512: H. Gausterer, C. Lang (Eds.), Computing Particle Properties. Proceedings, 1997. VII, 335 pages. 1998.

Vol. 513: A. Bernstein, D. Drechsel, T. Walcher (Eds.), Chiral Dynamics: Theory and Experiment. Proceedings, 1997. IX, 394 pages. 1998.

Vol. 514: F. W. Hehl, C. Kiefer, R.J.K. Metzler, Black Holes: Theory and Observation. Proceedings, 1997. XV, 519 pages. 1998.

Vol. 515: C.-H. Bruneau (Ed.), Sixteenth International Conference on Numerical Methods in Fluid Dynamics. Proceedings. XV, 568 pages. 1998.

Vol. 516: J. Cleymans, H. B. Geyer, F. G. Scholtz (Eds.), Hadrons in Dense Matter and Hadrosynthesis. Proceedings, 1998. XII, 253 pages. 1999.

Vol. 517: Ph. Blanchard, A. Jadczyk (Eds.), Quantum Future. Proceedings, 1997. X, 244 pages. 1999.

Vol. 518: P. G. L. Leach, S. E. Bouquet, J.-L. Rouet, E. Fijalkow (Eds.), Dynamical Systems, Plasmas and Gravitation. Proceedings, 1997. XII, 397 pages. 1999.

Vol. 519: A. Pękalski, K. Sznajd-Weron (Eds.), Anomalous Diffusion. From Basics to Applications. Proceedings, 1998. XVIII, 378 pages. 1999.

Vol. 520: J. A. van Paradijs, J. A. M. Bleeker (Eds.), X-Ray Spectroscopy in Astrophysics. EADN School X. Proceedings, 1997. XV, 530 pages. 1999.

Monographs

For information about Vols. 1–10 please contact your bookseller or Springer-Verlag

Vol. m 11: A. D. Yaghjian, Relativistic Dynamics of a Charged Sphere. XII, 115 pages. 1992.

Vol. m 12: G. Esposito, Quantum Gravity, Quantum Cosmology and Lorentzian Geometries. Second Corrected and Enlarged Edition. XVIII, 349 pages. 1994.

Vol. m 13: M. Klein, A. Knauf, Classical Planar Scattering by Coulombic Potentials. V, 142 pages. 1992.

Vol. m 14: A. Lerda, Anyons. XI, 138 pages. 1992.

Vol. m 15: N. Peters, B. Rogg (Eds.), Reduced Kinetic Mechanisms for Applications in Combustion Systems. X, 360 pages. 1993.

Vol. m 16: P. Christe, M. Henkel, Introduction to Conformal Invariance and Its Applications to Critical Phenomena. XV, 260 pages. 1993.

Vol. m 17: M. Schoen, Computer Simulation of Condensed Phases in Complex Geometries. X, 136 pages. 1993.

Vol. m 18: H. Carmichael, An Open Systems Approach to Quantum Optics. X, 179 pages. 1993.

Vol. m 19: S. D. Bogan, M. K. Hinders, Interface Effects in Elastic Wave Scattering. XII, 182 pages. 1994.

Vol. m 20: E. Abdalla, M. C. B. Abdalla, D. Dalmazi, A. Zadra, 2D-Gravity in Non-Critical Strings. IX, 319 pages. 1994.

Vol. m 21: G. P. Berman, E. N. Bulgakov, D. D. Holm, Crossover-Time in Quantum Boson and Spin Systems. XI, 268 pages. 1994.

Vol. m 22: M.-O. Hongler, Chaotic and Stochastic Behaviour in Automatic Production Lines. V, 85 pages. 1994.

Vol. m 23: V. S. Viswanath, G. Müller, The Recursion Method. X, 259 pages. 1994.

Vol. m 24: A. Ern, V. Giovangigli, Multicomponent Transport Algorithms. XIV, 427 pages. 1994.

Vol. m 25: A. V. Bogdanov, G. V. Dubrovskiy, M. P. Krutikov, D. V. Kulginov, V. M. Strelchenya, Interaction of Gases with Surfaces. XIV, 132 pages. 1995.

Vol. m 26: M. Dineykhan, G. V. Efimov, G. Ganbold, S. N. Nedelko, Oscillator Representation in Quantum Physics. IX, 279 pages. 1995.

Vol. m 27: J. T. Ottesen, Infinite Dimensional Groups and Algebras in Quantum Physics. IX, 218 pages. 1995.

Vol. m 28: O. Piguet, S. P. Sorella, Algebraic Renormalization. IX, 134 pages. 1995.

Vol. m 29: C. Bendjaballah, Introduction to Photon Communication. VII, 193 pages. 1995.

Vol. m 30: A. J. Greer, W. J. Kossler, Low Magnetic Fields in Anisotropic Superconductors. VII, 161 pages. 1995.

Vol. m 31 (Corr. Second Printing): P. Busch, M. Grabowski, P.J. Lahti, Operational Quantum Physics. XII, 230 pages. 1997.

Vol. m 32: L. de Broglie, Diverses questions de mécanique et de thermodynamique classiques et relativistes. XII, 198 pages. 1995.

Vol. m 33: R. Alkofer, H. Reinhardt, Chiral Quark Dynamics. VIII, 115 pages. 1995.

Vol. m 34: R. Jost, Das Märchen vom Elfenbeinernen Turm. VIII, 286 pages. 1995.

Vol. m 35: E. Elizalde, Ten Physical Applications of Spectral Zeta Functions. XIV, 224 pages. 1995.

Vol. m 36: G. Dunne, Self-Dual Chern-Simons Theories. X, 217 pages. 1995.

Vol. m 37: S. Childress, A.D. Gilbert, Stretch, Twist, Fold: The Fast Dynamo. XI, 406 pages. 1995.

Vol. m 38: J. González, M. A. Martín-Delgado, G. Sierra, A. H. Vozmediano, Quantum Electron Liquids and High-Tc Superconductivity. X, 299 pages. 1995.

Vol. m 39: L. Pittner, Algebraic Foundations of Non-Com-mutative Differential Geometry and Quantum Groups. XII, 469 pages. 1996.

Vol. m 40: H.-J. Borchers, Translation Group and Particle Representations in Quantum Field Theory. VII, 131 pages. 1996.

Vol. m 41: B. K. Chakrabarti, A. Dutta, P. Sen, Quantum Ising Phases and Transitions in Transverse Ising Models. X, 204 pages. 1996.

Vol. m 42: P. Bouwknegt, J. McCarthy, K. Pilch, The W3 Algebra. Modules, Semi-infinite Cohomology and BV Algebras. XI, 204 pages. 1996.

Vol. m 43: M. Schottenloher, A Mathematical Introduction to Conformal Field Theory. VIII, 142 pages. 1997.

Vol. m 44: A. Bach, Indistinguishable Classical Particles. VIII, 157 pages. 1997.

Vol. m 45: M. Ferrari, V. T. Granik, A. Imam, J. C. Nadeau (Eds.), Advances in Doublet Mechanics. XVI, 214 pages. 1997.

Vol. m 46: M. Camenzind, Les noyaux actifs de galaxies. XVIII, 218 pages. 1997.

Vol. m 47: L. M. Zubov, Nonlinear Theory of Dislocations and Disclinations in Elastic Body. VI, 205 pages. 1997.

Vol. m 48: P. Kopietz, Bosonization of Interacting Fermions in Arbitrary Dimensions. XII, 259 pages. 1997.

Vol. m 49: M. Zak, J. B. Zbilut, R. E. Meyers, From Instability to Intelligence. Complexity and Predictability in Nonlinear Dynamics. XIV, 552 pages. 1997.

Vol. m 50: J. Ambjørn, M. Carfora, A. Marzuoli, The Geometry of Dynamical Triangulations. VI, 197 pages. 1997.

Vol. m 51: G. Landi, An Introduction to Noncommutative Spaces and Their Geometries. XI, 200 pages. 1997.

Vol. m 52: M. Hénon, Generating Families in the Restricted Three-Body Problem. XI, 278 pages. 1997.

Vol. m 53: M. Gad-el-Hak, A. Pollard, J.-P. Bonnet (Eds.), Flow Control. Fundamentals and Practices. XII, 527 pages. 1998.

Vol. m 54: Y. Suzuki, K. Varga, Stochastic Variational Approach to Quantum-Mechanical Few-Body Problems. XIV, 324 pages. 1998.

Vol. m 55: F. Busse, S. C. Müller, Evolution of Spontaneous Structures in Dissipative Continuous Systems. X, 559 pages. 1998.

MIX
Papier aus verantwortungsvollen Quellen
Paper from responsible sources
FSC® C105338

If you have any concerns about our products,
you can contact us on
ProductSafety@springernature.com

In case Publisher is established outside the EU,
the EU authorized representative is:
Springer Nature Customer Service Center GmbH
Europaplatz 3, 69115 Heidelberg, Germany

Printed by Libri Plureos GmbH
in Hamburg, Germany